FOURIER ANALYSIS AND HAUSDORFF DIMENSION

During the past two decades there has been active interplay between geometric measure theory and Fourier analysis. This book describes part of that development, concentrating on the relationship between the Fourier transform and Hausdorff dimension.

The main topics concern applications of the Fourier transform to geometric problems involving Hausdorff dimension, such as Marstrand type projection theorems and Falconer's distance set problem, and the role of Hausdorff dimension in modern Fourier analysis, especially in Kakeya methods and Fourier restriction phenomena. The discussion includes both classical results and recent developments in the area. The author emphasizes partial results of important open problems, for example, Falconer's distance set conjecture, the Kakeya conjecture and the Fourier restriction conjecture. Essentially self-contained, this book is suitable for graduate students and researchers in mathematics.

Pertti Mattila is Professor of mathematics at the University of Helsinki and an expert in geometric measure theory. He has authored the book *Geometry of Sets and Measures in Euclidean Spaces* as well as more than 80 other scientific publications.

Fourier Analysis and Hausdorff Dimension

PERTTI MATTILA
University of Helsinki

CAMBRIDGE
UNIVERSITY PRESS

University Printing House, Cambridge CB2 8BS, United Kingdom

One Liberty Plaza, 20th Floor, New York, NY 10006, USA

477 Williamstown Road, Port Melbourne, VIC 3207, Australia

314-321, 3rd Floor, Plot 3, Splendor Forum, Jasola District Centre, New Delhi - 110025, India

79 Anson Road, #06-04/06, Singapore 079906

Cambridge University Press is part of the University of Cambridge.

It furthers the University's mission by disseminating knowledge in the pursuit of education, learning and research at the highest international levels of excellence.

www.cambridge.org
Information on this title: www.cambridge.org/9781107107359

© Pertti Mattila 2015

First published 2015

A catalogue record for this publication is available from the British Library

ISBN 978-1-107-10735-9 Hardback

To John Marstrand

Contents

Preface

This is a book on geometric measure theory and Fourier analysis. The main purpose is to present several topics where these areas meet including some of the very active recent interplay between them. We shall essentially restrict ourselves to questions involving the Fourier transform and Hausdorff dimension leaving many other aspects aside.

The book is intended for graduate students and researchers in mathematics. The prerequisites for reading it are basic real analysis and measure theory. Familiarity with Hausdorff measures and dimension and with Fourier analysis is certainly useful, but all that is needed will be presented in Chapters 2 and 3. Although most of the material has not appeared in book form, there is overlap with several earlier books. In particular, Mattila [1995] covers part of Chapters 4–7, Wolff [2003] of Chapters 14, 19, 20 and 22, and Stein [1993] of 14 and 19–21. Several other overlaps are mentioned in the text. The surveys Iosevich [2001], Łaba [2008], [2014], Mattila [2004], Mitsis [2003a] and Tao [2001], [2004] are closely related to the themes of the book.

Acknowledgements

This book grew out of several graduate courses I have taught at the Department of Mathematics and Statistics of the University of Helsinki. I am grateful to the department for excellent facilities and for the students and post docs attending the courses for inspiration and for their comments and questions. For financial support I am also indebted to the Academy of Finland. Many mathematicians have been of great help in the preparation of this book. In particular, my special thanks are due to Vasilis Chousionis, who helped me in many aspects, and to Kenneth Falconer, who read large parts of the manuscript and made a great number of valuable comments on mathematics, style and language. Terence Tao's blogs and lecture notes have been very useful as well as Ana Vargas's master's thesis, which she kindly sent to me. For many comments and corrections I would like to thank Anthony Carbery, Marton Elekes, Burak Erdoğan, Risto Hovila, Tuomas Hytönen, Alex Iosevich, Tamás Keleti, Sangyuk Lee, José María Martell, Tuomas Orponen, Keith Rogers, Tuomas Sahlsten, Andreas Seeger, Pablo Shmerkin, Ville Suomala, Terence Tao, Ana Vargas and Laura Venieri. Finally I am much obliged to Jon Billam, Clare Dennison, Samuel Harrison and others at the Cambridge University Press for their help and for accepting the book for publication.

1

Introduction

The main object of this book is the interplay between geometric measure theory and Fourier analysis on $\mathbb{R}^n$. The emphasis will be more on the first in the sense that on several occasions we look for the best known results in geometric measure theory while our goals in Fourier analysis will usually be much more modest. We shall concentrate on those parts of Fourier analysis where Hausdorff dimension plays a role. Much more between geometric measure theory and Fourier analysis has been and is going on. Relations between singular integrals and rectifiability have been intensively studied for more than two decades; see the books David and Semmes [1993], Mattila [1995] and Tolsa [2014], the survey Volberg and Eiderman [2013], and Nazarov, Tolsa and Volberg [2014] for recent break-through results. Relations between harmonic measure, partial differential equations (involving a considerable amount of Fourier analysis) and rectifiability have recently been very actively investigated by many researchers; see, for example, Kenig and Toro [2003], Hofmann, Mitrea and Taylor [2010], Hofmann, Martell and Uriarte-Tuero [2014], and the references given therein.

In this book there are two main themes. Firstly, the Fourier transform is a powerful tool on geometric problems concerning Hausdorff dimension, and we shall give many applications. Secondly, some basic problems of modern Fourier analysis, in particular those concerning restriction, are related to geometric measure theoretic Kakeya (or Besicovitch) type problems. We shall discuss these in the last part of the book. We shall also consider various particular constructions of measures and the behaviour of their Fourier transforms.

The contents of this book can be divided into four parts.

PART I Preliminaries and some simpler applications of the Fourier transform.

PART II Specific constructions.

PART III Deeper applications of the Fourier transform.

PART IV Fourier restriction and Kakeya type problems.

Parts I and III are closely linked together. They are separated by Part II only because much of the material in Part III is rather demanding and Part II might be more easily digestible. In any case, the reader may jump over Part II without any problems. On the other hand, the sections of Part II are essentially independent of each other and only rely on Chapters 2 and 3. Part IV is nearly independent of the others. In addition to the basics of the Fourier transform, given in Chapter 3, the reader is advised to consult Chapter 11 on Besicovitch sets and Chapter 14 on oscillatory integrals before reading Part IV.

The applicability of the Fourier transform on Hausdorff dimension stems from the following three facts. First, the Hausdorff dimension of a Borel set $A \subset \mathbb{R}^n$, dim A, can be determined by looking at the behaviour of Borel measures μ with compact support spt $\mu \subset A$. We denote by $\mathcal{M}(A)$ the family of such measures μ with $0 < \mu(A) < \infty$. More precisely, by Frostman's lemma dim A is the supremum of the numbers s such that there exists $\mu \in \mathcal{M}(A)$ for which

$$\mu(B(x, r)) \leq r^s \quad \text{for } x \in \mathbb{R}^n, \quad r > 0. \tag{1.1}$$

This is easily transformed into an integral condition. Let

$$I_s(\mu) = \iint |x - y|^{-s} \, d\mu x \, d\mu y$$

be the s-energy of μ. Then dim A is the supremum of the numbers s such that there exists $\mu \in \mathcal{M}(A)$ for which

$$I_s(\mu) < \infty. \tag{1.2}$$

For a given μ the conditions (1.1) and (1.2) may not be equivalent, but they are closely related: (1.2) implies that the restriction of μ to a suitable set with positive μ measure satisfies (1.1), and (1.1) implies that μ satisfies (1.2) for any $s' < s$. Defining the Riesz kernel k_s, $k_s(x) = |x|^{-s}$, the s-energy of μ can be written as

$$I_s(\mu) = \int k_s * \mu \, d\mu.$$

For $0 < s < n$ the Fourier transform of k_s (in the sense of distributions) is $\widehat{k_s} = \gamma(n, s)k_{n-s}$ where $\gamma(n, s)$ is a positive constant. Thus we have by Parseval's theorem

$$I_s(\mu) = \int \widehat{k_s} |\widehat{\mu}|^2 = \gamma(n, s) \int |x|^{s-n} |\widehat{\mu}(x)|^2 \, dx.$$

Consequently, dim A is the supremum of the numbers s such that there exists $\mu \in \mathcal{M}(A)$ for which

$$\int |x|^{s-n} |\widehat{\mu}(x)|^2 \, dx < \infty. \tag{1.3}$$

Thus, in a sense, a large part of this book is a study of measures satisfying one, or all, of the conditions (1.1), (1.2) or (1.3). As we shall see, in many applications using (1.1) or (1.2) is enough but often (1.3) is useful and sometimes indispensable. In the most demanding applications one has to go back and forth with these conditions.

The first application of Fourier transforms to Hausdorff dimension was Kaufman's [1968] proof for one part of Marstrand's projection theorem. This result, proved by Marstrand [1954], states the following.

Suppose $A \subset \mathbb{R}^2$ is a Borel set and denote by $P_e, e \in S^1$, the orthogonal projection onto the line $\{te : t \in \mathbb{R}\}$: $P_e(x) = e \cdot x$.

(1) If dim $A \le 1$, then dim $P_e(A) = \dim A$ for almost all $e \in S^1$.

(2) If dim $A > 1$, then $\mathcal{L}^1(P_e(A)) > 0$ for almost all $e \in S^1$.

Here $\mathcal{L}^1$ is the one-dimensional Lebesgue measure.

Marstrand's original proof was based on the definition and basic properties of Hausdorff measures. Kaufman used the characterization (1.2) for the first part and (1.3) for the second part. We give here Kaufman's proof to illustrate the spirit of the techniques used especially in Part I; many of the later arguments are variations of the following.

To prove (1) let $0 < s < \dim A$ and choose by (1.2) a measure $\mu \in \mathcal{M}(A)$ such that $I_s(\mu) < \infty$. Let $\mu_e \in \mathcal{M}(P_e(A))$ be the push-forward of μ under P_e: $\mu_e(B) = \mu(P_e^{-1}(B))$. Then

$$\int_{S^1} I_s(\mu_e) \, de = \int_{S^1} \iint |e \cdot (x - y)|^{-s} \, d\mu x \, d\mu y \, de$$
$$= \iint \int_{S^1} |e \cdot (\tfrac{x-y}{|x-y|})|^{-s} \, de |x - y|^{-s} \, d\mu x \, d\mu y = c(s) I_s(\mu) < \infty,$$

where for $v \in S^1$, $c(s) = \int_{S^1} |e \cdot v|^{-s} \, de < \infty$ as $s < 1$. Referring again to (1.2) we see that dim $P_e(A) \ge s$ for almost all $e \in S^1$. By the arbitrariness of $s, 0 < s < \dim A$, we obtain dim $P_e(A) \ge \dim A$ for almost all $e \in S^1$. The opposite inequality follows from the fact that the projections are Lipschitz mappings.

To prove (2) choose by (1.3) a measure $\mu \in \mathcal{M}(A)$ such that $\int |x|^{-1} |\widehat{\mu}(x)|^2 \, dx < \infty$. Directly from the definition of the Fourier transform we see that $\widehat{\mu_e}(t) = \widehat{\mu}(te)$ for $t \in \mathbb{R}, e \in S^1$. Integrating in polar coordinates we obtain

$$\int_{S^1} \int_{-\infty}^{\infty} |\widehat{\mu_e}(t)|^2 \, dt \, de = 2 \int_{S^1} \int_0^{\infty} |\widehat{\mu}(te)|^2 \, dt \, de = 2 \int |x|^{-1} |\widehat{\mu}(x)|^2 \, dx < \infty.$$

Thus for almost all $e \in S^1$, $\widehat{\mu_e} \in L^2(\mathbb{R})$ which means that μ_e is absolutely continuous with L^2 density and hence $\mathcal{L}^1(P_e(A)) > 0$.

The interplay between geometric measure theory and Fourier restriction, that we shall discuss in Part IV, has its origins in the following observations:

Let g be a function on the unit sphere S^{n-1}, for example the restriction of the Fourier transform of a smooth function f defined on $\mathbb{R}^n$. Let us fatten the sphere to a narrow annulus $S(\delta) = \{x : 1 - \delta < |x| < 1 + \delta\}$. We can write this annulus as a disjoint union of $\approx \delta^{(1-n)/2}$ spherical caps R_j, each of which is almost (for a small δ) a rectangular box with $n - 1$ side-lengths about $\sqrt{\delta}$ and one about δ. Suppose we could write $g = \sum_j g_j$ where each g_j is a smooth function with compact support in R_j (which of course we usually cannot do). Then $f = \sum_j f_j$ where f_j is the inverse transform of g_j, which is almost the same as the Fourier transform of g_j. A simple calculation reveals that f_j is like a smoothened version of the characteristic function of a dual rectangular box $\widetilde{R_j}$ of R_j; it decays very fast outside $\widetilde{R_j}$. This dual rectangular box is a rectangular box with $n - 1$ side-lengths about $1/\sqrt{\delta}$ and one about $1/\delta$, so it is like a long narrow tube. Thus studying f based on the information we have about the restriction of its Fourier transform on S^{n-1}, we are lead to study huge collections of narrow tubes and the behaviour of sums of functions essentially supported on them. These are typical Kakeya problems.

A concrete result along these lines is:

If the restriction conjecture is true, then all Besicovitch sets in $\mathbb{R}^n$ have Hausdorff dimension n.

The restriction conjecture, or one form of it, says that the Fourier transform of any function in $L^p(\mathbb{R}^n)$ can be meaningfully restricted to S^{n-1} when $1 \leq p < \frac{2n}{n+1}$. In the dual form this amounts to saying that the Fourier transform defines a bounded operator $L^\infty(S^{n-1}) \to L^q(\mathbb{R}^n)$ for $q > \frac{2n}{n-1}$ in the sense that the inequality

$$\|\widehat{f}\|_{L^q(\mathbb{R}^n)} \leq C(n, q) \|f\|_{L^\infty(S^{n-1})}$$

holds. For $n = 2$ this is known to be true and for $n > 2$ it is open. The restriction conjecture is related to many other questions of modern Fourier analysis and partial differential equations. We shall discuss some of these in this book.

Besicovitch sets are sets of Lebesgue measure zero containing a unit line segment in every direction. They exist in $\mathbb{R}^n$ for all $n \geq 2$. It is known, and we shall prove it, that all Besicovitch sets in the plane have Hausdorff dimension 2, but in higher dimensions it is an open problem whether they have full dimension n. Fattening Besicovitch sets slightly we end up with collections of narrow tubes as discussed above.

Now I give a brief overview of each chapter. Chapter 2 gives preliminaries on Borel measures in $\mathbb{R}^n$ and Chapter 3 on the Fourier transform, including the proofs for the characterization of Hausdorff dimension in terms of (1.1), (1.2) and (1.3). In Chapter 4 we repeat the above proof for Marstrand's theorem with more details and study Falconer's distance set problem: what can we say about the size of the distance set

$$D(A) = \{|x - y| : x, y \in A\}$$

if we know the Hausdorff dimension of a Borel set $A \subset \mathbb{R}^n$? For instance, we show that if $\dim A > (n + 1)/2$ then $D(A)$ contains an open interval. In Chapter 5 we sharpen Marstrand's projection theorem by showing that the Hausdorff dimension of the exceptional directions in (1) is at most $\dim A$ and in (2) at most $2 - \dim A$. We also give the higher dimensional versions and introduce the concept of Sobolev dimension of a measure, the use of which unifies and extends the results. In Chapter 6 we slice, or disintegrate, Borel measures in $\mathbb{R}^n$ by m-planes and apply this process to prove that typically if an m-plane V intersects a Borel set $A \subset \mathbb{R}^n$ with $\dim A > n - m$, it intersects it in dimension $\dim A + m - n$. We also prove here an exceptional set estimate and give an application to the Fourier transforms of measures on graphs. Chapter 7 studies the more general question of generic intersections of two arbitrary Borel sets. We prove that if $A, B \subset \mathbb{R}^n$ are Borel sets and $\dim B > (n + 1)/2$, then for almost all rotations $g \in O(n)$ the set of translations by $z \in \mathbb{R}^n$ such that $\dim A \cap (g(B) + z) \geq \dim A + \dim B - n - \varepsilon$ has positive Lebesgue measure for every $\varepsilon > 0$.

We start Part II by studying in Chapter 8 classical symmetric Cantor sets with dissection ratio d and the natural measures on them. We compute the Fourier transform and show that it goes to zero at infinity if and only if $1/d$ is not a Pisot number. Bernoulli convolutions are studied in Chapter 9. They are probability distributions of random sums $\sum_j \pm \lambda^j, 0 < \lambda < 1$. We prove part of Solomyak's theorem which says that they are absolutely continuous for almost all $\lambda \in (1/2, 1)$. In Chapter 10 we investigate projections of the one-dimensional Cantor set in the plane which is the product of two standard symmetric linear half-dimensional Cantor sets. We show in two ways that it projects into a set of Lebesgue measure zero on almost all lines and we also derive more detailed information about its projections. Using the aforementioned result we construct Besicovitch sets in Chapter 11. We shall also prove there that they have Hausdorff dimension at least 2. We shall consider Nikodym sets, too. They are sets of measure zero containing a line segment on some line through every point of the space. In Chapter 12 we find sharp information about the almost sure decay of Fourier transforms of some measures on trajectories of Brownian motion. The decay is as fast as the Hausdorff dimension allows, so the trajectories give examples of Salem sets. In Chapter 13 we study absolute continuity properties, both with respect to Lebesgue measure and Hausdorff dimension, of classical Riesz products. In Chapter 14 we derive basic decay properties for oscillatory integrals $\int e^{i\lambda\varphi(x)}\psi(x)\,dx$ and apply them to the Fourier transform of some surface measures.

Beginning Part III in Chapter 15 we return to the applications of Fourier transforms to geometric problems on Hausdorff dimension; we apply decay estimates of the spherical averages $\int_{S^{n-1}} |\widehat{\mu}(rv)|^2\,dv$ to distance sets. We continue this in Chapter 16 and prove deep estimates of Wolff and Erdoğan using Tao's bilinear restriction theorem (which is proved later) and Kakeya type methods. This will give us the best known dimension results for the distance set problem. In Chapter 17 we define fractional Sobolev spaces in terms of Fourier transforms. We study convergence questions for Sobolev functions and for solutions of the Schrödinger equation and estimate the Hausdorff dimension of the related exceptional sets. The Fourier analytic techniques of Peres and Schlag are introduced in Chapter 18 and they are applied to get considerable extensions of projection type theorems, both in terms of mappings and in terms of exceptional set estimates.

In Part IV we first introduce in Chapter 19 the restriction problems and prove the basic Stein–Tomas theorem. It says that

$$\|\widehat{f}\|_{L^q(\mathbb{R}^n)} \leq C(n,q)\|f\|_{L^2(S^{n-1})} \quad \text{for } q \geq 2(n+1)/(n-1).$$

In fact, we do not prove the end-point estimate for $q = 2(n+1)/(n-1)$, but we shall give a sketch for it in Chapter 20 using a stationary phase method. We shall also prove the restriction conjecture

$$\|\widehat{f}\|_{L^q(\mathbb{R}^2)} \leq C(q)\|f\|_{L^\infty(S^1)} \quad \text{for } q > 4$$

in the plane using this method.

In Chapter 21 we first prove Fefferman's multiplier theorem saying that for a ball B in $\mathbb{R}^n$, $n \geq 2$, the multiplier operator T_B, $\widehat{T_B f} = \chi_B \widehat{f}$, is not bounded in L^p if $p \neq 2$. This uses Kakeya methods and really is the origin for the applications of such methods in Fourier analysis. We shall also briefly discuss Bochner–Riesz multipliers. In Chapter 22 we introduce the Kakeya maximal function

$$\mathcal{K}_\delta f : S^{n-1} \to [0, \infty],$$

$$\mathcal{K}_\delta f(e) = \sup_{a \in \mathbb{R}^n} \frac{1}{\mathcal{L}^n(T_e^\delta(a))} \int_{T_e^\delta(a)} |f| \, d\mathcal{L}^n$$

and study its mapping properties. Here $T_e^\delta(a)$ is a tube of width δ and length 1, with direction e and centre a. The Kakeya maximal conjecture is

$$\|\mathcal{K}_\delta f\|_{L^n(S^{n-1})} \le C_\varepsilon \delta^{-\varepsilon} \|f\|_{L^n(\mathbb{R}^n)} \text{ for all } \varepsilon > 0, \ f \in L^n(\mathbb{R}^n).$$

We shall prove that it follows from the restriction conjecture and implies the Kakeya conjecture that all Besicovitch sets in $\mathbb{R}^n$ have Hausdorff dimension n. We shall also show that the analogue of the Kakeya conjecture is true in the discrete setting of finite fields.

In Chapter 23 we prove various estimates for the Hausdorff dimension of Besicovitch sets. In particular, we prove Wolff's lower bound $(n + 2)/2$ with geometric methods and the Bourgain–Katz–Tao lower bound $6n/11 + 5/11$ with arithmetic methods. In Chapter 24 we study (n, k) Besicovitch sets; sets of measure zero containing a positive measure piece of a k-plane in every direction. Following Marstrand and Falconer we first give rather simple proofs that they do not exist if $k > n/2$. Then we shall present Bourgain's proof which relies on Kakeya maximal function inequalities and extends this to $k > (n + 1)/3$, and even further with more complicated arguments which we shall only mention.

The last chapter, Chapter 25, gives a proof for Tao's sharp bilinear restriction theorem:

$$\|\widehat{f_1}\widehat{f_2}\|_{L^q(\mathbb{R}^n)} \lesssim \|f_1\|_{L^2(S^{n-1})}\|f_2\|_{L^2(S^{n-1})} \quad \text{for } q > (n + 2)/n,$$

when $f_j \in L^2(S^{n-1})$ with dist(spt f_1, spt f_2) ≈ 1. In fact, we shall prove a weighted version of this due to Erdoğan which is needed for the aforementioned distance set theorem. We shall also deduce a partial result for the restriction conjecture from this bilinear estimate.

PART I

Preliminaries and some simpler
applications of the Fourier transform

2

Measure theoretic preliminaries

Here we give some basic information about measure theory on $\mathbb{R}^n$. Many proofs for the statements of this section can be found in Mattila [1995], but also in several other standard books on measure theory and real analysis. We shall also derive the Hausdorff dimension characterizations (1.1) and (1.2) from the Introduction, that is, we shall prove Frostman's lemma.

2.1 Some basic notation

In any metric space X, $B(x, r)$ will stand for the closed ball with centre x and radius r. The diameter of a set A will be denoted by $d(A)$ and the minimal distance between two non-empty sets A and B by $d(A, B)$ and between a point x and a set A by $d(x, A)$. The open δ-neighbourhood of A is $A(\delta) = \{x : d(x, A) < \delta\}$. The closure of A is $\overline{A}$ and its interior is $\mathrm{Int}(A)$. The characteristic function of A is denoted by χ_A.

The space of continuous complex valued functions on X will be denoted by $C(X)$ and its subspace consisting of functions with compact support by $C_0(X)$. As usual, the support of f, spt f, is the closure of $\{x : f(x) \neq 0\}$. The sets $C^+(X)$ and $C_0^+(X)$ consist of non-negative functions in $C(X)$ and $C_0(X)$, respectively. For an open set U in a Euclidean space, $C^k(U)$ consists of k times continuously differentiable functions on U and $C^\infty(U)$ of infinitely differentiable functions on U; $C_0^k(U)$ and $C_0^\infty(U)$ are their subspaces of functions with compact support.

In the n-dimensional Euclidean space $\mathbb{R}^n$ *Lebesgue measure* is denoted by $\mathcal{L}^n$ and the volume of the unit ball will be

$$\alpha(n) = \mathcal{L}^n(B(0, 1)).$$

We denote by σ^{n-1} the surface measure on the unit sphere $S^{n-1} = \{x \in \mathbb{R}^n : |x| = 1\}$, and sometimes also by σ^m the surface measure on m-dimensional unit spheres in $\mathbb{R}^n$. For $r > 0$, σ_r^{n-1} will stand for the surface measure on the sphere $S(r) = S^{n-1}(r) = \{x \in \mathbb{R}^n : |x| = r\}$ of radius r.

The *Dirac measure* δ_a at a point a is defined by $\delta_a(A) = 1$, if $a \in A$, and $\delta_a(A) = 0$, if $a \notin A$.

The L^p space with respect to a measure μ is denoted by $L^p(\mu)$ and its norm by $\| \cdot \|_{L^p(\mu)}$. Sometimes we also write $\| f \|_{L^p(\mu,A)} = (\int_A |f|^p \, d\mu)^{1/p}$. When μ is a Lebesgue measure we usually write more simply L^p and $\| \cdot \|_p$, or $L^p(A)$ and $\| \cdot \|_{L^p(A)}$ when we consider L^p functions in a Lebesgue measurable set A. Often we also use the notation $L^p(S^{n-1})$ and $\| \cdot \|_{L^p(S^{n-1})}$ instead of $L^p(\sigma^{n-1})$ and $\| \cdot \|_{L^p(\sigma^{n-1})}$. These, as well as other function spaces considered in this book, are spaces of complex valued functions.

We shall mean by $a \lesssim_\alpha b$ that $a \le Cb$ where C is a constant depending on α. If it is clear from the context what C should depend on, we may write only $a \lesssim b$. In the notation $a \lesssim_\alpha b$ the parameters included in α do not always contain all that is needed. For example, we often do not write explicitly the dependence on the dimension n of $\mathbb{R}^n$. If $a \lesssim b$ and $b \lesssim a$ we write $a \approx b$. By $C(\alpha)$ and $c(\alpha)$ we shall always mean positive and finite constants depending only on α.

By $\mathbb{N}$ we denote the set of positive integers and by $\mathbb{N}_0$ the set of non-negative integers.

2.2 Borel and Hausdorff measures

We mean by a *measure* on a set X what is usually meant by outer measure, that is, a non-negative, monotone, countably subadditive function on $\{A : A \subset X\}$ that gives the value 0 for the empty set. As usual, the *Borel sets* in a metric space X form the smallest σ-algebra of subsets of X containing all closed subsets of X. By a *Borel measure* in X we mean a measure μ for which Borel sets are measurable and which is *Borel regular* in the sense that for any $A \subset X$ there is a Borel set B such that $A \subset B$ and $\mu(A) = \mu(B)$. The additional requirement of Borel regularity is not really restrictive for our purposes since if for a measure μ the Borel sets are μ measurable, then ν defined by $\nu(A) = \inf\{\mu(B) : A \subset B, B \text{ is a Borel set}\}$ is Borel regular and agrees with μ on Borel sets, as one easily checks. But requiring Borel regularity has the advantage that Borel measures are uniquely determined by their values on Borel sets. From this it follows that in $\mathbb{R}^n$ they are uniquely determined by integrals of continuous functions with compact support. A Borel measure is *locally finite* if compact sets have finite measure. Locally finite Borel measures are often called *Radon measures*.

The *support* of a measure μ on X is the smallest closed set F such that $\mu(X \setminus F) = 0$. It is denoted by spt μ. For $A \subset X$ the set of all Borel measures μ on X with $0 < \mu(A) < \infty$ and with compact spt $\mu \subset A$ will be denoted by $\mathcal{M}(A)$.

New measures can be created by restricting measures to subsets: if μ is a measure on X and $A \subset X$, the *restriction* of μ to A, $\mu \llcorner A$, is defined by

$$\mu \llcorner A(B) = \mu(A \cap B) \quad \text{for } B \subset X.$$

It is a Borel measure if μ is a Borel measure and A is a μ measurable set with $\mu(A) < \infty$.

The *image* or push-forward of a measure μ under a map $f : X \to Y$ is defined by

$$f_\sharp \mu(B) = \mu(f^{-1}(B)) \quad \text{for } B \subset Y.$$

It is a Borel measure if μ is a Borel measure and f is a Borel function. The definition is equivalent to saying that

$$\int g \, df_\sharp \mu = \int g \circ f \, d\mu$$

for all non-negative Borel functions g on X. This formula will be used repeatedly.

A measure μ is *absolutely continuous* with respect to a measure ν if $\nu(A) = 0$ implies $\mu(A) = 0$. We denote this by $\mu \ll \nu$. Borel measures μ and ν are *mutually singular* if there is a Borel set $B \subset X$ such that $\mu(X \setminus B) = \nu(B) = 0$.

The *integral* $\int f d\mu$ or $\int f(x) d\mu x$ means always the integral $\int_X f \, d\mu$ over the whole space X. In case μ is Lebesgue measure on $\mathbb{R}^n$ we often omit the measure and write simply, for example, $\int f = \int f \, d\mathcal{L}^n$ and $\int f(x) dx = \int f(x) d\mathcal{L}^n x$.

If g is a non-negative μ measurable function we denote by $g\mu$ the measure such that $g\mu(B) = \int_B g \, d\mu$ for Borel sets B. Thus $\mu \llcorner A = \chi_A \mu$. If g is complex valued, $g\mu$ means the obvious complex measure. Non-negative Lebesgue measurable functions g on $\mathbb{R}^n$ will be identified with the measures $g\mathcal{L}^n$.

We shall often use *Hausdorff measures* $\mathcal{H}^s$, $s \geq 0$. By definition,

$$\mathcal{H}^s(A) = \lim_{\delta \to 0} \mathcal{H}^s_\delta(A),$$

where, for $0 < \delta \leq \infty$,

$$\mathcal{H}^s_\delta(A) = \inf \left\{ \sum_j \alpha(s) 2^{-s} d(E_j)^s : A \subset \bigcup_j E_j, d(E_j) < \delta \right\}.$$

Here $\alpha(s)$ is a positive number. For integers n we have already fixed that $\alpha(n)$ is the volume of the n-dimensional unit ball (with $\alpha(0) = 1$). Then in $\mathbb{R}^n$, $\mathcal{H}^n = \mathcal{L}^n$. When s is not an integer, the value of $\alpha(s)$ is insignificant. To avoid unnecessary constants at some later estimates, let us choose $\alpha(s)2^{-s} = 1$, when s is not an integer.

The *Hausdorff dimension* of $A \subset \mathbb{R}^n$ is

$$\dim A = \inf\{s : \mathcal{H}^s(A) = 0\} = \sup\{s : \mathcal{H}^s(A) = \infty\}.$$

Since (as an easy exercise), $\mathcal{H}^s(A) = 0$ if and only if $\mathcal{H}^s_\infty(A) = 0$, we can replace $\mathcal{H}^s$ in the definition of dim by the simpler $\mathcal{H}^s_\infty$. So, more simply,

$$\dim A = \inf \left\{ s : \forall \varepsilon > 0 \, \exists E_1, E_2, \cdots \subset X \text{ such that} \right.$$

$$\left. A \subset \bigcup_j E_j \text{ and } \sum_j d(E_j)^s < \varepsilon \right\}.$$

For the definition of dimension, the sets E_j above can be restricted to be balls, because each E_j is contained in a ball B_j with $d(B_j) \leq 2d(E_j)$. The spherical measure obtained using balls is not the same as the Hausdorff measure but it is between $\mathcal{H}^s$ and $2^s \mathcal{H}^s$.

The m-dimensional Hausdorff measure restricted to a sufficiently nice, even just Lipschitz, m-dimensional surface is the standard surface measure, but we shall not really need this fact. We shall frequently use the surface measure σ^{n-1} on the unit sphere $S^{n-1} = \{x \in \mathbb{R}^n : |x| = 1\}$. A useful fact about it is that up to multiplication by a constant it is the unique Borel measure on S^{n-1} which is invariant under rotations. More precisely, the *orthogonal group* $O(n)$ of $\mathbb{R}^n$ consists of linear maps $g : \mathbb{R}^n \to \mathbb{R}^n$ which preserve the inner product: $g(x) \cdot g(y) = x \cdot y$ for all $x, y \in \mathbb{R}^n$. Then σ^{n-1} is determined, up to a multiplication by a constant, by the property

$$\sigma^{n-1}(g(A)) = \sigma^{n-1}(A) \quad \text{for all } A \subset \mathbb{R}^n, g \in O(n).$$

Since $O(n)$ is a compact group, it has a unique Haar probability measure θ_n. This means that θ_n is the unique Borel measure on $O(n)$ such that $\theta_n(O(n)) = 1$ and

$$\theta_n(\{g \circ h : h \in A\}) = \theta_n(\{h \circ g : h \in A\})$$
$$= \theta_n(A) \quad \text{for all } A \subset O(n), g \in O(n).$$

The measures σ^{n-1} and θ_n are related by the formula

$$\theta_n(\{g \in O(n) : g(x) \in A\}) = \sigma^{n-1}(A)/\sigma^{n-1}(S^{n-1}) \quad \text{for } A \subset S^{n-1}, x \in S^{n-1}.$$
$$(2.1)$$

This follows from the fact that both sides define a rotationally invariant Borel probability measure on S^{n-1} and such a measure is unique.

2.3 Minkowski and packing dimensions

We shall mainly concentrate on Hausdorff dimension, but in a few occasions we shall also discuss Minkowski and packing dimensions. The Minkowski dimension is often called the box counting dimension. Recall that

$$A(\delta) = \{x : (x, A) < \delta\}$$

is the open δ-neighbourhood of A.

Definition 2.1 The *lower Minkowski dimension* of a bounded set $A \subset \mathbb{R}^n$ is

$$\underline{\dim}_M A = \inf\{s > 0 : \liminf_{\delta \to 0} \delta^{s-n} \mathcal{L}^n(A(\delta)) = 0\},$$

and the *upper Minkowski dimension* of A is

$$\overline{\dim}_M A = \inf\{s > 0 : \limsup_{\delta \to 0} \delta^{s-n} \mathcal{L}^n(A(\delta)) = 0\}.$$

Let $N(A, \delta)$ be the smallest number of balls of radius δ needed to cover A. Then

$$\underline{\dim}_M A = \liminf_{\delta \to 0} \frac{\log N(A, \delta)}{\log(1/\delta)},$$

and

$$\overline{\dim}_M A = \limsup_{\delta \to 0} \frac{\log N(A, \delta)}{\log(1/\delta)}.$$

We have also that

$$\dim A \le \underline{\dim}_M A \le \overline{\dim}_M A.$$

These facts are easy to verify, or one can consult Mattila [1995], for example.

Definition 2.2 The *packing dimension* of $A \subset \mathbb{R}^n$ is

$$\dim_P A = \inf \left\{ \sup_j \overline{\dim}_M A_j : A = \bigcup_{j=1}^{\infty} A_j, \, A_j \text{ is bounded} \right\}.$$

Then

$$\dim A \le \dim_P A \le \overline{\dim}_M A.$$

2.4 Weak convergence

The proof of Frostman's lemma below and many other things are based on weak convergence:

Definition 2.3 The sequence (μ_j) of Borel measures on $\mathbb{R}^n$ converges weakly to a Borel measure μ if for all $\varphi \in C_0(\mathbb{R}^n)$,

$$\int \varphi \, d\mu_j \to \int \varphi \, d\mu.$$

The following weak compactness theorem is very important, though not very deep. It follows rather easily from the separability of the space $C_0(\mathbb{R}^n)$.

Theorem 2.4 *Any sequence (μ_j) of Borel measures on $\mathbb{R}^n$ such that $\sup_j \mu_j(\mathbb{R}^n) < \infty$ has a weakly converging subsequence.*

We shall mainly be interested in singular measures, but it will be very useful to approximate them with smooth functions. This can be done with approximate identities:

Definition 2.5 We say that the family $\{\psi_\varepsilon : \varepsilon > 0\}$ of non-negative continuous functions on $\mathbb{R}^n$ is an *approximate identity* if spt $\psi_\varepsilon \subset B(0, \varepsilon)$ and $\int \psi_\varepsilon = 1$ for all $\varepsilon > 0$.

Usually one generates an approximate identity by choosing a non-negative continuous function ψ with spt $\psi \subset B(0, 1)$, $\int \psi = 1$, and defining

$$\psi_\varepsilon(x) = \varepsilon^{-n} \psi(x/\varepsilon).$$

Such a C^∞-function ψ is $\psi(x) = c e^{1/(|x|^2-1)}$ for $|x| < 1$ and $\psi(x) = 0$ for $|x| \geq 1$, where c is chosen to make the integral 1.

The *convolution* $f * g$ of functions f and g is defined by

$$f * g(x) = \int f(x - y) g(y) \, dy,$$

and the convolution of a function f and a Borel measure μ by

$$f * \mu(x) = \int f(x - y) \, d\mu y,$$

whenever the integrals exist. The convolution of Borel measures μ and ν on $\mathbb{R}^n$ is defined by

$$\int \varphi \, d(\mu * \nu) = \iint \varphi(x + y) \, d\mu x \, d\nu y \quad \text{for } \varphi \in C_0^+(\mathbb{R}^n).$$

Theorem 2.6 *Let $\{\psi_\varepsilon : \varepsilon > 0\}$ be an approximate identity and μ a locally finite Borel measure on $\mathbb{R}^n$. Then $\psi_\varepsilon * \mu$ converges weakly to μ as $\varepsilon \to 0$, that is,*

$$\int \varphi(\psi_\varepsilon * \mu) \, d\mathcal{L}^n \to \int \varphi \, d\mu$$

for all $\varphi \in C_0(\mathbb{R}^n)$.

The proof is rather straightforward and can be found for example in Mattila [1995].

Note that the functions $\psi_\varepsilon * \mu$ are C^∞ if ψ_ε are and they have compact support if μ has. If μ has compact support, the convergence in Theorem 2.6 takes place for all $\varphi \in C(\mathbb{R}^n)$.

2.5 Energy-integrals and Frostman's lemma

Although bounding Hausdorff measures and dimension from above often is easy, one just needs to estimate some convenient coverings, it usually is much more difficult to find lower bounds; then one should estimate arbitrary coverings. Frostman's lemma transforms the problem to finding measures with good upper bounds for measures of balls. It has turned out to be extremely efficient. Its proof can be found in many sources, for example in Bishop and Peres [2016], Carleson [1967], Kahane [1985], Mattila [1995], Mörters and Peres [2010], Tolsa [2014] and Wolff [2003]. The proofs in Bishop and Peres [2016], Mörters and Peres [2000] and Tolsa [2014] are somewhat non-standard and the second proof given in Mattila [1995], Theorem 8.8, due to Howroyd, is quite different from others since it is based on the Hahn–Banach theorem and it applies in very general metric spaces. However, as this result is very central for this book, we prove it also here, although leaving some details to the reader. We give the proof only for compact sets, which is all that is really needed in this book. For Borel, and more general Suslin (or analytic) sets, see, e.g., Bishop and Peres [2016] or Carleson [1967].

Considering only compact sets in fact is not a restriction of generality for our purposes. By a result of Davies [1952b], see also Federer [1969], Theorem 2.10.48, or Mattila [1995], Theorem 8.13, any Borel (or even Suslin) set $A \subset \mathbb{R}^n$ with $\mathcal{H}^s(A) > 0$ contains a compact subset C with $0 < \mathcal{H}^s(C) < \infty$. The proof of this is rather complicated, but when one studies $\mathcal{H}^s$ measurable sets A with $\mathcal{H}^s(A) < \infty$ one gets this much easier by standard approximation theorems, see for example Mattila [1995], Theorem 1.10. Finally, the essence of our results is usually already present for compact sets.

Since Davies's result and Frostman's lemma hold for Suslin sets, essentially all our results formulated for Borel sets are valid for this more general class. The only reason for stating them for Borel sets is that these are better known.

Theorem 2.7 *[Frostman's lemma]*
Let $0 \leq s \leq n$. For a Borel set $A \subset \mathbb{R}^n$, $\mathcal{H}^s(A) > 0$ if and only there is $\mu \in \mathcal{M}(A)$ such that

$$\mu(B(x,r)) \leq r^s \quad \text{for all } x \in \mathbb{R}^n, \quad r > 0. \tag{2.2}$$

In particular,

$$\dim A = \sup\{s : \text{there is } \mu \in \mathcal{M}(A) \text{ such that (2.2) holds}\}.$$

A measure satisfying (2.2) is often called a *Frostman measure* or an *s* Frostman measure.

Proof One direction is very easy: if $\mu \in \mathcal{M}(A)$ satisfies (2.2) and $B_j, j = 1, 2, \ldots,$ are balls covering A, we have

$$\sum_j d(B_j)^s \geq \sum_j \mu(B_j) \geq \mu(A) > 0,$$

which implies $\mathcal{H}^s(A) > 0$.

For the other direction, suppose A is compact. Assume $\mathcal{H}^s(A) > 0$. Then there is $c > 0$ such that

$$\sum_j d(E_j)^s \geq c \tag{2.3}$$

for all coverings $E_j, j = 1, 2, \ldots,$ of A. We construct the measure μ as a weak limit of measures μ_k. To define μ_k look at the dyadic cubes of side-length 2^{-k} in a standard cubical partitioning of $\mathbb{R}^n$. First we define a measure $\mu_{k,1}$ which is a constant multiple of Lebesgue measure on each such cube Q. For Q such that $A \cap Q \neq \varnothing$, we normalize Lebesgue measure on Q so that $\mu_{k,1}(Q) = d(Q)^s$ and for the cubes Q such that $A \cap Q = \varnothing$ we let $\mu_{k,1}$ be the zero measure on Q. This measure would be fine for balls with diameter $< 2^{-k}$ but not necessarily for the bigger balls. Thus we modify it to a measure $\mu_{k,2}$ by investigating the dyadic cubes of side-length 2^{1-k}. On each such cube Q we let $\mu_{k,2}$ be $\mu_{k,1}$ if $\mu_{k,1}(Q) \leq d(Q)^s$, otherwise we make it smaller by normalizing $\mu_{k,1}$ on Q so that $\mu_{k,2}(Q) = d(Q)^s$. We continue this until we come to a single cube Q_0 which contains our compact set A (we may assume to begin with that the dyadic partitioning is chosen so that A is inside some cube belonging to it). Let μ_k be the final measure obtained in this way. Then, since we never increased the measure along the process, $\mu_k(Q) \leq d(Q)^s$ for all dyadic cubes with side-length at least 2^{-k}. In fact, this holds for all dyadic cubes by the first step of the construction. This implies easily that $\mu_k(B) \lesssim_n d(B)^s$ for all balls B. The construction yields that every $x \in A$ is contained in some dyadic subcube Q of Q_0 with side-length at least 2^{-k} such that

$$\mu_k(Q) = d(Q)^s.$$

Choosing maximal, and hence disjoint, such cubes Q_j, they cover A and thus by (2.3),

$$\mu_k(\mathbb{R}^n) = \sum_j \mu_k(Q_j) = \sum_j d(Q_j)^s \geq c. \tag{2.4}$$

We can now take some weakly converging subsequence of (μ_k) and consider the limit measure μ. Then it is immediate from the construction that spt $\mu \subset A$ (here we use that A is compact). It is also clear that $\mu(B) \lesssim_n d(B)^s$ for all balls B. The only danger is that μ might be the zero measure, but (2.4) shows that this cannot happen. $\qquad\square$

One of the most fundamental concepts in this book will be the *s-energy*, $s > 0$, of a Borel measure μ:

$$I_s(\mu) = \iint |x - y|^{-s}\, d\mu x \, d\mu y = \int k_s * \mu \, d\mu,$$

where k_s is the *Riesz kernel*:

$$k_s(x) = |x|^{-s}, \quad x \in \mathbb{R}^n.$$

If μ has compact support we have trivially,

$$I_s(\mu) < \infty \text{ implies } I_t(\mu) < \infty \quad \text{for } 0 < t < s.$$

We can quite easily relate the energies to the Frostman condition (2.2) using the standard formula

$$\int |x - y|^{-s} \, d\mu y = s \int_0^\infty \frac{\mu(B(x, r))}{r^{s+1}} \, dr.$$

This immediately gives that if $\mu \in \mathcal{M}(\mathbb{R}^n)$ satisfies (2.2), then for $0 < t < s$,

$$I_t(\mu) \leq t \iint_0^{d(\text{spt}\,\mu)} \frac{\mu(B(x, r))}{r^{t+1}} \, dr \, d\mu x \leq t\mu(\mathbb{R}^n) \int_0^{d(\text{spt}\,\mu)} r^{s-t-1} \, dr < \infty.$$

On the other hand, if $I_s(\mu) < \infty$, then $\int |x - y|^{-s} \, d\mu x < \infty$ for μ almost all $x \in \mathbb{R}^n$ and we can find $0 < M < \infty$ such that the set $A = \{x :$

$\int |x - y|^{-s} \, d\mu x < M\}$ has positive μ measure. Then one checks easily that $(\mu \llcorner A)(B(x, r)) \leq 2^s M r^s$ for all $x \in \mathbb{R}^n, r > 0$. This gives:

Theorem 2.8 *For a Borel set $A \subset \mathbb{R}^n$,*

$$\dim A = \sup\{s : there\ is\ \mu \in \mathcal{M}(A)\ such\ that\ I_s(\mu) < \infty\}.$$

Let us look at a few very easy examples:

Example 2.9

(i) Let $\mu = \mathcal{L}^1 \llcorner [0, 1]$. Then $\dim[0, 1] = 1$, $\mu \in \mathcal{M}([0, 1])$ and $I_s(\mu) < \infty$ if and only $s < 1$. Similarly, if $A \subset \mathbb{R}^n$ is Lebesgue measurable and bounded with $\mathcal{L}^n(A) > 0$ and $\mu = \mathcal{L}^n \llcorner A$, then $I_s(\mu) < \infty$ if and only $s < n$.
(ii) Let $\mu = \mathcal{H}^1 \llcorner \Gamma$ where Γ is a rectifiable curve. Again $I_s(\mu) < \infty$ if and only $s < 1$.
(iii) Let μ be the natural measure on the standard $1/3$-Cantor set C, that is, $\mu = \mathcal{H}^{s_0} \llcorner C$ where $s_0 = \log 2 / \log 3$ is the Hausdorff dimension of C. Then $\mu \in \mathcal{M}(C)$ and $I_s(\mu) < \infty$ if and only $s < s_0$.

As an easy application of Frostman's lemma we obtain the inequality for dimensions or product sets:

Theorem 2.10 *Let A and B be non-empty Borel sets in $\mathbb{R}^n$. Then*

$$\dim A \times B \geq \dim A + \dim B.$$

Proof Choose $0 \leq s < \dim A$, $0 \leq t < \dim B$, $\mu \in \mathcal{M}(A)$ with $\mu(B(x, r)) \leq r^s$ and $\nu \in \mathcal{M}(B)$ with $\nu(B(x, r)) \leq r^t$. Then the product measure $\mu \times \nu$ belongs to $\mathcal{M}(A \times B)$ with $\mu \times \nu(B((x, y), r)) \leq r^{s+t}$ from which the theorem follows. $\qquad\square$

2.6 Differentiation of measures

For $\mu \in \mathcal{M}(\mathbb{R}^n)$ define the *lower derivative* and *derivative* of μ at $x \in \mathbb{R}^n$ by

$$\underline{D}(\mu, x) = \liminf_{r \to 0} \frac{\mu(B(x, r))}{\alpha(n) r^n}$$

and

$$D(\mu, x) = \lim_{r \to 0} \frac{\mu(B(x, r))}{\alpha(n) r^n},$$

the latter if the limit exists. We shall make use of the following basic differentiation theorem of measures, for a proof, see, e.g., Mattila [1995], Theorem 2.12:

Theorem 2.11 *Let $\mu \in \mathcal{M}(\mathbb{R}^n)$. Then*

(a) the derivative $D(\mu, x)$ exists and is finite for $\mathcal{L}^n$ almost all $x \in \mathbb{R}^n$,
(b) $\int_B D(\mu, x)\,dx \le \mu(B)$ for all Borel sets $B \subset \mathbb{R}^n$ with equality if $\mu \ll \mathcal{L}^n$,
(c) $\mu \ll \mathcal{L}^n$ if and only if $\underline{D}(\mu, x) < \infty$ for μ almost all $x \in \mathbb{R}^n$.

Perhaps a lesser known fact in this theorem is part (c), its proofs given in Mattila [1995] and Federer [1969] both use Besicovitch's covering theorem; Bishop and Peres [2016] give a very simple proof without it in Section 3.5.

2.7 Interpolation

We shall review the basic interpolation theorems that will be used in the book. The proofs can be found in many sources and we skip them here.

Let (X, μ) and (Y, ν) be two measure spaces. The first interpolation theorem is the Riesz–Thorin theorem. For a proof, see for example Grafakos [2008] or Katznelson [1968].

Theorem 2.12 *Let $1 \le p_0, p_1, q_0, q_1 \le \infty$ and let T be a linear operator on $L^{p_0}(\mu) + L^{p_1}(\mu)$ taking values in the space of ν measurable functions on Y such that*

$$\|T(f)\|_{L^{q_0}(\nu)} \le C_0\|f\|_{L^{p_0}(\mu)} \quad \text{for all } f \in L^{p_0}(\mu)$$

and

$$\|T(f)\|_{L^{q_1}(\nu)} \le C_1\|f\|_{L^{p_1}(\mu)} \quad \text{for all } f \in L^{p_1}(\mu).$$

Then for all $0 < \theta < 1$,

$$\|T(f)\|_{L^q(\nu)} \le C_0^{1-\theta} C_1^\theta \|f\|_{L^p(\mu)} \quad \text{for all } f \in L^p(\mu),$$

where

$$\frac{1}{p} = \frac{1-\theta}{p_0} + \frac{\theta}{p_1} \quad \text{and} \quad \frac{1}{q} = \frac{1-\theta}{q_0} + \frac{\theta}{q_1}.$$

Various maximal operators are not linear but only sublinear: $|T(f + g)| \le |Tf| + |Tg|$. However, one can usually apply the Riesz–Thorin theorem to linearized operators and get essentially the same result. More precisely, suppose $Tf(y) = \sup_a T_a(|f|)(y)$, where each T_a is a linear operator with $Tg \ge 0$ when $g \ge 0$. Given p and q as above and a non-negative function $f \in L^p(\mu)$, choose for each $y \in Y$ a parameter $a(y)$ such that $Tf(y) \approx T_{a(y)}f(y)$ and the function $y \mapsto T_{a(y)}f(y)$ is ν measurable (which usually is possible). Defining $Lg(y) = T_{a(y)}g(y)$, L is linear, $\|Tf\|_{L^q(\nu)} \approx \|Lf\|_{L^q(\nu)}$ and $|Lg| \le Tg$. Then apply the theorem to L. Since we required f to be non-negative, this does not give precisely the constant $C_0^{1-\theta}C_1^\theta$, but we just need to multiply it by 3. That is enough, for example, for the applications to Kakeya maximal functions in the last part of the book.

Often one only has weak type inequalities to start with and the operator is just sublinear. The Marcinkiewicz interpolation theorem generalizes the Riesz–Thorin theorem to this setting with the expense of having weaker information on the constants. We say that T is of *weak type* (p, q) if there is a finite constant C such that

$$\nu(\{y \in Y : |T(f)(y)| > \lambda\}) \le (C\lambda^{-1}\|f\|_{L^p(\mu)})^q \quad \text{for all } f \in L^p(\mu), \quad \lambda > 0. \tag{2.5}$$

Theorem 2.13 *Let $1 \le p_0, p_1, q_0, q_1 \le \infty$, $p_0 \ne p_1, q_0 \ne q_1$ and let T be a sublinear operator on $L^{p_0}(\mu) + L^{p_1}(\mu)$ taking values in the space of ν measurable functions on Y such that T is of weak type (p_0, q_0) and of weak type (p_1, q_1). Then for all $0 < \theta < 1$,*

$$\|T(f)\|_{L^q(\nu)} \le C_\theta \|f\|_{L^p(\mu)} \quad \text{for all } f \in L^p(\mu),$$

where

$$\frac{1}{p} = \frac{1-\theta}{p_0} + \frac{\theta}{p_1} \quad \text{and} \quad \frac{1}{q} = \frac{1-\theta}{q_0} + \frac{\theta}{q_1}$$

and C_θ in addition to θ depends on p_0, p_1, q_0, q_1 and the constants in the (p_0, q_0) and (p_1, q_1) weak type inequalities of T.

For a proof, see for example Grafakos [2008], Theorem 1.4.19. Often one says that T is of *strong type* (p, q) if $\|T(f)\|_{L^q(\nu)} \lesssim \|f\|_{L^p(\mu)}$ for all $f \in L^p(\mu)$. Clearly strong type implies weak type.

One can generalize further: it is enough to assume the *restricted weak type*. This means that (2.5) is required to hold only for all characteristic functions $f = \chi_A$ of μ measurable sets A. Theorem 1.4.19 in Grafakos [2008] is proven in this generality; see also Stein and Weiss [1971], Section V.3.

2.8 Khintchine's inequality

We shall have a couple of applications for a probabilistic result called Khintchine's inequality. Let ω_j, $j = 1, 2, \ldots$, be independent random variables on a probability space (Ω, P) taking values ± 1 with equal probability $1/2$. One can take for example $\Omega = \{-1, 1\}^{\mathbb{N}}$, $\omega_j((x_k)) = x_j$, and P the natural measure on Ω, the infinite product of the measures $\frac{1}{2}(\delta_{-1} + \delta_1)$. Denote by $\mathbf{E}(f)$ the expectation (P-integral) of the random variable f. The independence of the ω_j implies that

$$\mathbf{E}(\omega_j \omega_k) = \mathbf{E}(\omega_j)\mathbf{E}(\omega_k) = 0 \quad \text{for } j \neq k,$$

and that for any finite subset J of $\mathbb{N}$ and any bounded Borel functions $g_j : \mathbb{R} \to \mathbb{C}$, $j \in J$, the random variables $g_j \circ \omega_j$, $j \in J$, are independent, in particular,

$$\mathbf{E}(\Pi_{j \in J} g_j \circ \omega_j) = \Pi_{j \in J} \mathbf{E}(g_j \circ \omega_j).$$

Theorem 2.14 *For any $a_1, \ldots, a_N \in \mathbb{C}$ and $0 < p < \infty$,*

$$\mathbf{E}\left(\left|\sum_{j=1}^{N} \omega_j a_j\right|^p\right) \approx_p \left(\sum_{j=1}^{N} |a_j|^2\right)^{p/2}.$$

Proof We shall prove this for $1 < p < \infty$, which is the only case we shall need. If $p = 2$, the claim follows from independence as equality. Next we prove the inequality '$\lesssim$'. We may obviously assume that the a_j are real. Let $t > 0$. For a fixed j, $\mathbf{E}(e^{ta_j\omega_j}) = \frac{1}{2}(e^{ta_j} + e^{-ta_j})$. Thus by the independence,

$$\mathbf{E}(e^{t \sum_j a_j \omega_j}) = \Pi_j \mathbf{E}(e^{ta_j\omega_j}) = \Pi_j \frac{1}{2}(e^{ta_j} + e^{-ta_j}).$$

The elementary inequality $\frac{1}{2}(e^x + e^{-x}) \leq e^{x^2/2}$ implies that

$$\mathbf{E}(e^{t \sum_j a_j \omega_j}) \leq e^{(t^2/2) \sum_j a_j^2}.$$

This gives for all $t > 0, \lambda > 0$, by Chebychev's inequality

$$P\left(\left\{\omega : \sum_j a_j \omega_j \geq \lambda\right\}\right) = P(\{\omega : e^{t \sum_j a_j \omega_j} \geq e^{\lambda t}\})$$

$$\leq e^{-\lambda t}\mathbf{E}(e^{t \sum_j a_j \omega_j}) \leq e^{-\lambda t + (t^2/2) \sum_j a_j^2}.$$

Take $t = \frac{\lambda}{\sum_j a_j^2}$. Then

$$P\left(\left\{\omega : \sum_j a_j \omega_j \geq \lambda\right\}\right) \leq e^{-\frac{\lambda^2}{2\sum_j a_j^2}}$$

and so

$$P\left(\left\{\omega : \left|\sum_j a_j \omega_j\right| \geq \lambda\right\}\right) \leq 2e^{-\frac{\lambda^2}{2\sum_j a_j^2}}.$$

Applying this and the formula (which follows from Fubini's theorem)

$$\mathbf{E}(|f|^p) = p \int_0^\infty \lambda^{p-1} P(\{\omega : |f(\omega)| \geq \lambda\}) \, d\lambda,$$

we get by a change of variable

$$\mathbf{E}\left(\left|\sum_j a_j \omega_j\right|^p\right) \leq 2p \int_0^\infty \lambda^{p-1} e^{-\frac{\lambda^2}{2\sum_j a_j^2}} \, d\lambda = c(p) \left(\sum_j a_j^2\right)^{p/2},$$

which is the desired inequality.

To prove the opposite inequality we use duality. Suppose $p > 1$ and let $q = \frac{p}{p-1}$. Then by the two previous cases, $p = 2$ and '$\lesssim$', and by Hölder's inequality,

$$\sum_j |a_j|^2 = \mathbf{E}\left(\left|\sum_j a_j \omega_j\right|^2\right) \leq \mathbf{E}\left(\left|\sum_j a_j \omega_j\right|^p\right)^{1/p} \mathbf{E}\left(\left|\sum_j a_j \omega_j\right|^q\right)^{1/q}$$

$$\lesssim \mathbf{E}\left(\left|\sum_j a_j \omega_j\right|^p\right)^{1/p} \left(\sum_j |a_j|^2\right)^{1/2},$$

which yields

$$\mathbf{E}\left(\left|\sum_j a_j \omega_j\right|^p\right)^{1/p} \gtrsim \left(\sum_j |a_j|^2\right)^{1/2}$$

and proves the theorem. $\qquad\square$

2.9 Further comments

An excellent source for basic measure theory is the book A. Bruckner, J. Bruckner and Thomson [1997]. Hausdorff measures and dimensions, Frostman's lemma and energy-integrals, and other dimensions are widely discussed, for example, in Bishop and Peres [2016], Falconer [1985a], [1990] and Mattila [1995].

Frostman proved his lemma, Theorem 2.7, in his thesis in 1935 with applications to potential theory in mind, see for example Carleson [1967] and Landkof [1972] for these. For the applications to harmonic functions the clue is that the fundamental solution of the Laplace equation in $\mathbb{R}^n$, $n \geq 3$, is $c(n)|x|^{2-n}$. This on many occasions leads to representations of harmonic functions as potentials $\int |x - y|^{2-n} d\mu x$ with suitable measures μ (in the plane one has to use logarithmic potentials) and further to connections with Hausdorff dimension via Theorem 2.8. This is not just restricted to harmonic functions, but similar features are present for other function classes and for the solutions of many other partial differential equations, in particular for complex analytic functions where the fundamental solution is the Cauchy kernel; see Tolsa's book [2014] for that.

The proof of Khintchine's inequality was taken from Wolff [2003]. It can also be stated in terms of Rademacher functions; see Grafakos [2008] for this and more.

3

Fourier transforms

This chapter is a quick introduction to Fourier transforms. We shall pay particular attention to Fourier transforms of measures. Apart from the standard theory, we will develop the formula for the Fourier transform of the surface measure on the unit sphere and we will prove the representation of the energy-integrals in terms of the Fourier transform, the crucial relation for this book which was already discussed in the Introduction.

3.1 Fourier transforms in L^1 and L^2

The Fourier transform of a Lebesgue integrable function $f \in L^1(\mathbb{R}^n)$ is defined by

$$\mathcal{F}(f)(\xi) = \widehat{f}(\xi) = \int f(x) e^{-2\pi i \xi \cdot x} \, dx, \quad \xi \in \mathbb{R}^n. \tag{3.1}$$

Then $\widehat{f}$ is a bounded continuous function. The following formulas easily follow by Fubini's theorem:

$$\int \widehat{f} g = \int f \widehat{g}, \ f, g \in L^1(\mathbb{R}^n), \quad \text{(product formula)}, \tag{3.2}$$

$$\widehat{(f * g)} = \widehat{f} \widehat{g}, \ f, g \in L^1(\mathbb{R}^n), \quad \text{(convolution formula)}. \tag{3.3}$$

Trivial changes of variables show how the Fourier transform behaves under simple transformations. For $a \in \mathbb{R}^n$ and $r > 0$ define the translation τ_a and dilation δ_r by

$$\tau_a(x) = x + a, \ \delta_r(x) = rx, \quad x \in \mathbb{R}^n.$$

Then for $f \in L^1(\mathbb{R}^n), \xi \in \mathbb{R}^n$,

$$\widehat{f \circ \tau_a}(\xi) = e^{2\pi i a \cdot \xi} \widehat{f}(\xi), \quad \mathcal{F}(e^{2\pi i a \cdot x} f)(\xi) = \widehat{f}(\xi - a), \tag{3.4}$$

$$\widehat{f \circ \delta_r}(\xi) = r^{-n} \widehat{f}(r^{-1}\xi). \tag{3.5}$$

Recall that the orthogonal group $O(n)$ of $\mathbb{R}^n$ consists of linear maps $g : \mathbb{R}^n \to \mathbb{R}^n$ which preserve inner product: $g(x) \cdot g(y) = x \cdot y$ for all $x, y \in \mathbb{R}^n$. Then

$$\widehat{f \circ g} = \widehat{f} \circ g \quad \text{for } g \in O(n). \tag{3.6}$$

The proof of the following *Riemann–Lebesgue lemma* is also easy:

$$\widehat{f}(\xi) \to 0 \quad \text{when } |\xi| \to \infty \text{ and } f \in L^1(\mathbb{R}^n). \tag{3.7}$$

The *inversion formula* is a bit trickier to prove:

$$f(x) = \int \widehat{f}(\xi) e^{2\pi i \xi \cdot x} \, d\xi \quad \text{if } f, \widehat{f} \in L^1(\mathbb{R}^n), \quad \text{(inversion formula)}. \tag{3.8}$$

Of course, we must interpret this being true after possibly redefining f in a set of measure zero.

Proof Define

$$\Psi(x) = e^{-\pi |x|^2}, \quad \Psi_\varepsilon(x) = e^{-\pi \varepsilon^2 |x|^2}.$$

Then $\widehat{\Psi} = \Psi$. This follows from the definitions by complex integration, or by observing that $\Psi(x) = e^{-\pi |x_1|^2} \cdots e^{-\pi |x_n|^2}$ and when $n = 1$, Ψ and $\widehat{\Psi}$ satisfy the same differential equation $f'(x) = -2\pi x f(x)$ with the initial condition $f(0) = 1$. We have by (3.5),

$$\widehat{\Psi_\varepsilon}(\xi) = \varepsilon^{-n} e^{-\pi |\xi|^2/\varepsilon^2}.$$

Write

$$I_\varepsilon(x) = \int \widehat{f}(\xi) e^{-\pi \varepsilon^2 |\xi|^2} e^{2\pi i \xi \cdot x} \, d\xi.$$

Then by Lebesgue's dominated convergence theorem,

$$I_\varepsilon(x) \to \int \widehat{f}(\xi) e^{2\pi i \xi \cdot x} \, d\xi \quad \text{as } \varepsilon \to 0.$$

On the other hand, setting $g_x(y) = e^{-\pi \varepsilon^2 |y|^2} e^{2\pi i x \cdot y}$, we have by (3.4) $\widehat{g_x}(y) = \widehat{\Psi_\varepsilon}(y - x) = \Psi^\varepsilon(x - y)$, where $\Psi^\varepsilon(y) = \varepsilon^{-n} \Psi(y/\varepsilon)$. By the product formula (3.2),

$$I_\varepsilon(x) = \int \widehat{f} g_x = \int f \widehat{g_x} = \Psi^\varepsilon * f(x).$$

As $\int \Psi = \widehat{\Psi}(0) = 1$, the functions Ψ^ε, $\varepsilon > 0$, provide an approximate identity for which $\Psi^\varepsilon * f \to f$ as $\varepsilon \to 0$ almost everywhere; they do not have compact support, but the rapid decay at infinity is enough. The combination of these two limits gives the inversion formula. $\qquad\qquad\qquad\qquad\qquad\qquad\qquad\qquad\quad$ $\square$

Corollary 3.1 *If f and $\widehat{f}$ are integrable, then f is continuous.*

We denote the *inverse Fourier transform* of $g \in L^1(\mathbb{R}^n)$ by

$$\mathcal{F}^{-1}(g)(x) = \check{g}(x) = \int g(\xi) e^{2\pi i \xi \cdot x} \, d\xi.$$

Then the inversion formula means that $\mathcal{F}^{-1}(\widehat{f}) = f$ if $f, \widehat{f} \in L^1(\mathbb{R}^n)$. Defining $\tilde{f}(x) = f(-x)$ each of the following three formulas is a restatement of the inversion formula:

$$\check{f} = \widehat{\tilde{f}} = \tilde{\widehat{f}}, \quad \widehat{\widehat{f}} = \tilde{f}, \quad \overline{\widehat{f}} = \widehat{\overline{\tilde{f}}}. \qquad (3.9)$$

Applying the inversion formula to the convolution formula (3.3) we get

$$\widehat{fg} = \widehat{f} * \widehat{g}, \quad \text{if } f, g, fg, \widehat{f}, \widehat{g} \in L^1(\mathbb{R}^n). \qquad (3.10)$$

The *Schwartz class* $\mathcal{S} = \mathcal{S}(\mathbb{R}^n)$ of rapidly decreasing functions is very convenient in Fourier analysis. It consists of infinitely differentiable complex valued functions on $\mathbb{R}^n$ which together with their partial derivatives of all orders tend to zero at infinity more quickly than $|x|^{-k}$ for all integers k. Observe that $C_0^\infty(\mathbb{R}^n) \subset \mathcal{S}(\mathbb{R}^n)$.

The first basic fact is that

$$f \in \mathcal{S}(\mathbb{R}^n) \quad \text{if and only if} \quad \widehat{f} \in \mathcal{S}(\mathbb{R}^n). \qquad (3.11)$$

This follows from the formulas for partial derivatives, which in turn follow easily by partial integration: if $f \in \mathcal{S}$ (or more generally under some obvious conditions):

$$\widehat{\partial^\alpha f}(\xi) = (2\pi i \xi)^\alpha \widehat{f}(\xi), \qquad (3.12)$$

$$\partial^\alpha \widehat{f}(\xi) = \mathcal{F}((-2\pi i x)^\alpha f)(\xi). \qquad (3.13)$$

Here $\alpha = (\alpha_1, \ldots, \alpha_n)$, $\alpha_j \in \{0, 1, \ldots\}$, $x^\alpha = x_1^{\alpha_1} \cdots x_n^{\alpha_n}$ and ∂^α means α_j partial derivatives with respect to x_j.

Secondly, we have

$$\int f \overline{g} = \int \widehat{f} \, \overline{\widehat{g}}, \quad f, g \in \mathcal{S}(\mathbb{R}^n), \quad \text{(Parseval)}, \qquad (3.14)$$

$$\|f\|_2 = \|\widehat{f}\|_2, \quad f, g \in \mathcal{S}(\mathbb{R}^n), \quad \text{(Plancherel)}. \qquad (3.15)$$

Parseval's formula (which of course gives Plancherel's formula) is an easy consequence of the inversion formula and the product formula:

$$\int f\overline{g} = \int \widehat{\widehat{f}}(-x)\overline{g}(x)\,dx = \int \widehat{\widehat{f}}(x)\overline{g}(-x)\,dx = \int \widehat{f}(x)\overline{h}(x)\,dx,$$

where $h(x) = \overline{g}(-x)$. We see immediately from the definition of the Fourier transform that $\widehat{h}(x) = \overline{\widehat{g}(x)}$, which proves Parseval's formula.

So the Fourier transform is a linear L^2 isometry of $S(\mathbb{R}^n)$ onto itself. The formula (3.1) cannot be used to define the Fourier transform for L^2 functions; the integral need not exist if f is not integrable. But $S(\mathbb{R}^n)$ is dense in $L^2(\mathbb{R}^n)$, so (3.11) and (3.15) give immediately a unique isometric linear extension of the Fourier transform to L^2. Thus we have $\widehat{f}$ defined for all $f \in L^1 \cup L^2$. Parseval's and Plancherel's formulas now extend at once to L^2:

$$\int f\overline{g} = \int \widehat{f}\,\overline{\widehat{g}}, \quad f, g \in L^2(\mathbb{R}^n), \tag{3.16}$$

$$\|f\|_2 = \|\widehat{f}\|_2, \quad f, g \in L^2(\mathbb{R}^n). \tag{3.17}$$

Hence the Fourier transform is a linear isometry of L^2 onto itself. Similarly, the translation and dilation formulas (3.4) and (3.5) continue to hold for L^2 functions almost everywhere.

If $f \in C_0^\infty(\mathbb{R}^n)$, then by (3.11) $\widehat{f} \in S(\mathbb{R}^n)$, but it cannot have compact support unless f is identically zero. In fact, we can say much more. For simplicity assume $n = 1$. The function g,

$$g(z) = \int e^{-2\pi i x z} f(x)\,dx, \quad z \in \mathbb{C},$$

agrees with $\widehat{f}$ on $\mathbb{R}$ and it is a non-constant complex analytic function in the whole complex plane provided $f \in C_0^\infty(\mathbb{R})$ is not the zero function. Hence its zero set is discrete and so also $\{x \in \mathbb{R} : \widehat{f}(x) = 0\}$ is discrete. The same argument and statement obviously hold also for measures $\mu \in \mathcal{M}(\mathbb{R})$ in place of f. These facts are a reflection of the *Heisenberg uncertainty principle*: a function and its Fourier transform cannot both be small. For more on this, see Havin and Jöricke [1995] and Wolff [2003].

Example 3.2 The fact that every Schwartz function is a Fourier transform of another Schwartz function is very useful for construction of various examples with desired properties. For example, we can find a non-negative function $\varphi \in S(\mathbb{R}^n)$ such that $\varphi \geq 1$ on $B(0, 1)$, $\widehat{\varphi} \geq 0$ and spt $\widehat{\varphi} \subset B(0, 1)$ (or vice versa, $\widehat{\varphi} \geq 1$ on $B(0, 1)$ and spt $\varphi \subset B(0, 1)$). To see this choose first a non-negative function $\psi \in S(\mathbb{R}^n)$ for which spt $\psi \subset B(0, 1/2)$ and $\int \psi = 2$ and set $\eta = \mathcal{F}^{-1}(\psi * \widetilde{\psi}) = |\widehat{\psi}|^2$ where $\widetilde{\psi}(x) = \psi(-x)$. Then $\widehat{\eta} = \psi * \widetilde{\psi}$, both η and $\widehat{\eta}$ are non-negative, and $\eta(0) = \widehat{\psi}(0)^2 = (\int \psi)^2 = 4$. It follows that spt $\widehat{\eta} \subset B(0, 1)$ and for some $0 < r < 1$, $\eta(x) > 1$ when $|x| \le r$. Define $\varphi(x) = \eta(rx)$. Then $\varphi \ge 1$ on $B(0, 1)$ and $\widehat{\varphi}(x) = r^{-n}\widehat{\eta}(x/r)$, whence spt $\widehat{\varphi} \subset B(0, r) \subset B(0, 1)$.

3.2 Fourier transforms of measures and distributions

The Fourier transform of a finite Borel measure μ on $\mathbb{R}^n$ is defined by

$$\widehat{\mu}(\xi) = \int e^{-2\pi i \xi \cdot x} \, d\mu x, \quad \xi \in \mathbb{R}^n. \tag{3.18}$$

When $\mu \in \mathcal{M}(\mathbb{R}^n)$, that is, μ has compact support, $\widehat{\mu}$ is a bounded Lipschitz continuous function:

$$\|\widehat{\mu}\|_\infty \le \mu(\mathbb{R}^n) \quad \text{and} \quad |\widehat{\mu}(x) - \widehat{\mu}(y)| \le R\mu(\mathbb{R}^n)|x - y| \quad \text{for } x, y \in \mathbb{R}^n, \tag{3.19}$$

if spt $\mu \subset B(0, R)$. This is an easy exercise. But $\widehat{\mu}$ need not be in L^p for any $p < \infty$; for example $\widehat{\delta_a}(\xi) = e^{-2\pi i \xi \cdot a}$.

The product and convolution formulas have by Fubini's theorem easy extensions for measures: for $f \in L^1(\mathbb{R}^n)$, $\mu, \nu \in \mathcal{M}(\mathbb{R}^n)$,

$$\int \widehat{\mu} f = \int \widehat{f} \, d\mu, \tag{3.20}$$

$$\int \widehat{\mu} \, d\nu = \int \widehat{\nu} \, d\mu, \tag{3.21}$$

$$\widehat{f * \mu} = \widehat{f}\widehat{\mu}, \tag{3.22}$$

$$\widehat{f\mu} = \widehat{f} * \mu, \tag{3.23}$$

$$\widehat{\mu * \nu} = \widehat{\mu}\widehat{\nu}. \tag{3.24}$$

As discussed in the previous chapter, we can approximate measures with smooth compactly supported functions using convolution. Let $\{\psi_\varepsilon : \varepsilon > 0\}$ be a C^∞ approximate identity such that

$$\psi_\varepsilon(x) = \varepsilon^{-n}\psi(x/\varepsilon), \ \varepsilon > 0, \ \psi \ge 0, \ \text{spt}\,\psi \subset B(0, 1), \ \int \psi = 1.$$

Then

$$\widehat{\psi_\varepsilon}(\xi) = \widehat{\psi}(\varepsilon\xi) \to \widehat{\psi}(0) = \int \psi = 1 \quad \text{as } \varepsilon \to 0.$$

Setting $\mu_\varepsilon = \psi_\varepsilon * \mu$ for a finite Borel measure μ, we have that μ_ε converges weakly to μ as $\varepsilon \to 0$ and

$$\widehat{\mu_\varepsilon} = \widehat{\psi_\varepsilon}\widehat{\mu} \to \widehat{\mu} \quad \text{uniformly.}$$

This immediately gives for $\mu, \nu \in \mathcal{M}(\mathbb{R}^n)$,

$$\widehat{\mu} = \widehat{\nu} \quad \text{implies} \quad \mu = \nu. \tag{3.25}$$

We further have for $\mu \in \mathcal{M}(\mathbb{R}^n)$,

$$\widehat{f\mu} = \widehat{f} * \widehat{\mu}, \quad f \in \mathcal{S}(\mathbb{R}^n), \tag{3.26}$$

$$\int \widehat{f} \, d\mu = \int \overline{\widehat{f}} \, \widehat{\mu}, \quad f \in \mathcal{S}(\mathbb{R}^n), \tag{3.27}$$

$$\int \widehat{f}\,\overline{\widehat{g}}\, d\mu = \int f(\widehat{\mu} * \overline{g}), \quad f, g \in \mathcal{S}(\mathbb{R}^n). \tag{3.28}$$

These follow by approximating μ as above by $\psi_\varepsilon * \mu$ and using the basic formulas for Schwartz class functions. We leave the easy details to the reader.

As usual, we shall identify absolutely continuous measures with functions: if μ is absolutely continuous (with respect to $\mathcal{L}^n$), it is by the Radon–Nikodym theorem of the form $\mu = f\mathcal{L}^n$ for some $f \in L^1(\mathbb{R}^n)$ and we shall identify μ and f.

Theorem 3.3 *Let $\mu \in \mathcal{M}(\mathbb{R}^n)$. If $\widehat{\mu} \in L^2(\mathbb{R}^n)$, then $\mu \in L^2(\mathbb{R}^n)$.*

Proof Since the Fourier transform maps $L^2(\mathbb{R}^n)$ onto $L^2(\mathbb{R}^n)$, there is $f \in L^2(\mathbb{R}^n)$ such that $\widehat{\mu} = \widehat{f}$. Write

$$\mu_\varepsilon = \psi_\varepsilon * \mu, \, f_\varepsilon = \psi_\varepsilon * f.$$

Then by the convolution formula,

$$\widehat{\mu_\varepsilon} = \widehat{\psi_\varepsilon}\widehat{\mu} = \widehat{\psi_\varepsilon}\widehat{f} = \widehat{f_\varepsilon},$$

and so $\mu_\varepsilon = f_\varepsilon$. As $\mu_\varepsilon \to \mu$ and $f_\varepsilon \to f$, we have $\mu = f$. $\square$

Theorem 3.4 *Let $\mu \in \mathcal{M}(\mathbb{R}^n)$. If $\widehat{\mu} \in L^1(\mathbb{R}^n)$, then μ is a continuous function.*

Proof Let μ_ε be as in the previous proof. Then $\mu_\varepsilon \in \mathcal{S}(\mathbb{R}^n)$ and by the inversion formula and the dominated convergence theorem,

$$\mu_\varepsilon(x) = \int \widehat{\mu_\varepsilon}(\xi)e^{2\pi i\xi \cdot x} \, d\xi = \int \widehat{\psi}(\varepsilon\xi)\widehat{\mu}(\xi)e^{2\pi i\xi \cdot x} \, d\xi$$

$$\to \int \widehat{\mu}(\xi)e^{2\pi i\xi \cdot x} \, d\xi =: g(x)$$

as $\varepsilon \to 0$. Since $\widehat{\mu} \in L^1$, the function g is continuous. On the other hand μ_ε converges weakly to μ, so $\mu = g$. $\qquad\qquad\qquad\square$

Definition 3.5 A *tempered distribution* is a continuous linear functional T : $\mathcal{S}(\mathbb{R}^n) \to \mathbb{C}$. Its Fourier transform is the tempered distribution $\widehat{T}$ defined by

$$\widehat{T}(\varphi) = T(\widehat{\varphi}) \quad \text{for } \varphi \in \mathcal{S}(\mathbb{R}^n).$$

We shall not make any real use of the theory of tempered distributions so we do not specify what continuity means here. This can be found in Duoandikoetxea [2001] and in many other Fourier analysis books.

All L^p functions, $1 \leq p \leq \infty$, and more generally all locally integrable functions f such that $|f(x)| \lesssim |x|^m$ when $|x| > 1$ for some fixed m, can be considered as tempered distributions T_f:

$$T_f(\varphi) = \int f\varphi, \quad \varphi \in \mathcal{S}(\mathbb{R}^n),$$

and so they have the Fourier transform as a tempered distribution. In particular this is true for the Riesz kernel k_s.

To define the Fourier transform of an L^p function when $1 < p < 2$, we can also make use of L^1 and L^2: any $f \in L^p$, $1 < p < 2$, can be written as $f = f_1 + f_2$, $f_1 \in L^1$, $f_2 \in L^2$. Then we can define $\widehat{f} = \widehat{f_1} + \widehat{f_2}$, and this agrees with the distributional definition. For $p = 2$ we have the Plancherel identity, $\|\widehat{f}\|_2 = \|f\|_2$, and for $p = 1$ we have the trivial estimate:

$$\|\widehat{f}\|_\infty \leq \|f\|_1.$$

From these the Riesz–Thorin interpolation theorem 2.12 gives the following *Hausdorff–Young inequality*:

$$\|\widehat{f}\|_q \leq \|f\|_p \quad \text{for } f \in L^p, 1 < p < 2, q = \frac{p}{p-1}. \qquad (3.29)$$

No such inequality holds when $p > 2$. This can be shown with the help of Khintchine's inequality; see Wolff [2003].

3.3 The Fourier transform of radial functions, Bessel functions

One of the goals of this section is to find the Fourier transform of the surface measure on the sphere $S^{n-1} = \{x \in \mathbb{R}^n : |x| = 1\}$ and of the Riesz kernels $k_s, 0 < s < n$. Let us first compute the Fourier transform of radial functions. We shall skip here some lengthy but elementary calculations which are well presented for example in the books of Grafakos [2008] and of Stein and Weiss [1971]. We assume here that $n \geq 2$.

Suppose $f \in L^1(\mathbb{R}^n)$, $f(x) = \psi(|x|)$, $x \in \mathbb{R}^n$, for some $\psi : [0, \infty) \to \mathbb{C}$. We shall use the following two Fubini-type formulas which can either be proven by basic calculus or deduced from a general coarea formula.

The first is the standard integration in polar coordinates formula: if $f \in L^1(\mathbb{R}^n)$, then

$$\int_{\mathbb{R}^n} f \, d\mathcal{L}^n = \int_{S^{n-1}} \left(\int_0^\infty f(rx) r^{n-1} \, dr \right) d\sigma^{n-1} x. \tag{3.30}$$

For the second, fix $e \in S^{n-1}$ and let $S_\theta = \{x \in S^{n-1} : e \cdot x = \cos \theta\}$ for $0 \leq \theta \leq \pi$. The set S_θ is an $(n-2)$-dimensional sphere of radius $\sin \theta$ (which is a 2-point set when $n = 2$), so

$$\sigma^{n-2}_{\sin \theta}(S_\theta) = b(n)(\sin \theta)^{n-2},$$

where $b(n) = \sigma^{n-2}(S^{n-2})$. Then for $g \in L^1(S^{n-1})$,

$$\int_{S^{n-1}} g \, d\sigma^{n-1} = \int_0^\pi \left(\int_{S_\theta} g(x) \, d\sigma^{n-2}_{\sin \theta} x \right) d\theta. \tag{3.31}$$

Applying (3.30) and Fubini's theorem,

$$\widehat{f}(re) = \int f(y) e^{-2\pi i r e \cdot y} \, dy = \int_0^\infty \psi(s) s^{n-1} \left(\int_{S^{n-1}} e^{-2\pi i r s e \cdot x} \, d\sigma^{n-1} x \right) ds.$$

The inside integral can be computed with the help of (3.31), since $e^{-2\pi i r s e \cdot x}$ is constant in S_θ:

$$\int_{S^{n-1}} e^{-2\pi i r s e \cdot x} \, d\sigma^{n-1} x$$
$$= \int_0^\pi e^{-2\pi i r s \cos \theta} \sigma^{n-2}_{\sin \theta}(S_\theta) \, d\theta = b(n) \int_0^\pi e^{-2\pi i r s \cos \theta} (\sin \theta)^{n-2} \, d\theta.$$

Changing variable $\cos \theta \mapsto -t$ and introducing for $m > -1/2$ the *Bessel functions* $J_m : [0, \infty) \to \mathbb{R}$:

$$J_m(u) := \frac{(u/2)^m}{\Gamma(m + 1/2)\Gamma(1/2)} \int_{-1}^1 e^{iut} (1 - t^2)^{m-1/2} \, dt, \tag{3.32}$$

with $\Gamma(x) = \int_0^\infty t^{x-1} e^{-t} \, dt$, we obtain

$$\int_{S^{n-1}} e^{-2\pi i r s e \cdot x} \, d\sigma^{n-1}(x) = b(n) \int_{-1}^1 e^{2\pi i r s t} (1 - t^2)^{(n-3)/2} \, dt$$
$$= c(n)(rs)^{-(n-2)/2} J_{(n-2)/2}(2\pi r s).$$

This leads to the formula for the Fourier transform of the radial function f:

$$\widehat{f}(x) = c(n)|x|^{-(n-2)/2} \int_0^\infty \psi(s) J_{(n-2)/2}(2\pi |x| s) s^{n/2} \, ds. \tag{3.33}$$

The following estimate is obvious:

$$|J_m(t)| \le C(m) t^m \quad \text{for } t > 0. \tag{3.34}$$

A basic property of Bessel functions is the following decay estimate:

$$|J_m(t)| \le C(m) t^{-1/2} \quad \text{for } t > 0. \tag{3.35}$$

We shall see in Chapter 14 that this follows from general results on oscillatory integrals. Here we derive it from explicit asymptotic formulas, which we shall later need anyway.

When $m = k - 1/2, k \in \{1, 2, \dots\}$, repeated partial integrations show that the Bessel function J_m can be written in terms of elementary functions in the form from which (3.35) easily follows. In particular,

$$J_{1/2}(t) = \frac{\sqrt{2}}{\sqrt{\pi t}} \sin t. \tag{3.36}$$

All Bessel functions behave roughly like this at infinity, that is,

$$J_m(t) = \frac{\sqrt{2}}{\sqrt{\pi t}} \cos(t - \pi m/2 - \pi/4) + O(t^{-3/2}), \quad t \to \infty. \tag{3.37}$$

This can be verified with a fairly simple integration, see Stein and Weiss [1971], pp. 158–159, or Grafakos [2008], Appendix B8. Both of these books, as well as Watson's classic [1944], contain much more information on Bessel functions. The above asymptotics is a special case of general asymptotic expansions of oscillatory integrals as derived in Chapter 6 of Wolff [2003] and Chapter VIII of Stein [1993].

We shall also need the following recursion formulas:

$$\frac{d}{dt}(t^{-m} J_m(t)) = -t^{-m} J_{m+1}(t), \tag{3.38}$$

$$\frac{d}{dt}(t^m J_m(t)) = t^m J_{m-1}(t). \tag{3.39}$$

Their proofs are rather straightforward differentiation, see Grafakos [2008], Appendix B2, for example.

A simple consequence of the formulas (3.33), (3.39) and (3.35) is the decay estimate for the characteristic function of the unit ball in $\mathbb{R}^n$:

$$|\widehat{\chi_{B(0,1)}}(x)| \le C(n)|x|^{-(n+1)/2} \quad \text{for } x \in \mathbb{R}^n. \tag{3.40}$$

We now return to the surface measure σ^{n-1} on the sphere S^{n-1}. One checks easily that σ^{n-1} is the weak limit of the measures $\delta^{-1}\mathcal{L}^n \llcorner (B(0, 1 + \delta) \setminus B(0, 1))$ as $\delta \to 0$.

Applying the formula (3.33) to the characteristic function of the annulus $B(0, 1 + \delta) \setminus B(0, 1)$ and letting $\delta \to 0$, we get

$$\widehat{\sigma^{n-1}}(x) = c(n)|x|^{(2-n)/2} J_{(n-2)/2}(2\pi|x|). \tag{3.41}$$

Consequently,

$$|\widehat{\sigma^{n-1}}(x)| \leq C(n)|x|^{(1-n)/2} \quad \text{for } x \in \mathbb{R}^n. \tag{3.42}$$

This is the best possible decay for any measure on a smooth hypersurface, in fact, on any set of Hausdorff dimension $n - 1$, cf. Section 3.6. The reason for getting such a good decay for $\widehat{\sigma^{n-1}}$ is curvature; for example segments are not curving at all but circles are curving uniformly. Also for more general surfaces curvature properties play a central role in the behaviour of Fourier transforms. We shall discuss this more in Chapter 14.

To illustrate the effect of the lack of curvature, let us compute the Fourier transform of the length measure λ on the line segment $I = [-(1, 0), (1, 0)]$ in $\mathbb{R}^2$:

$$\widehat{\lambda}(\eta, \xi) = \int_{-1}^{1} e^{-2\pi i(\eta x + \xi 0)} dx = \int_{-1}^{1} \cos(2\pi\eta x) dx = \frac{\sin(2\pi\eta)}{\pi\eta}.$$

We see that $\widehat{\lambda}(\eta, \xi)$ tends to 0 for a fixed ξ when η tends to ∞, but it remains constant for a fixed η when ξ tends to ∞, and hence does not tend to 0 when $|(\eta, \xi)| \to \infty$.

3.4 The Fourier transform of Riesz kernels

Now we compute the Fourier transform of the Riesz kernels $k_s, k_s(x) = |x|^{-s}, 0 < s < n$. This computation is valid in $\mathbb{R}$, too.

Theorem 3.6 *For $0 < s < n$ there is a positive and finite constant $\gamma(n, s)$ such that $\widehat{k_s} = \gamma(n, s)k_{n-s}$ as a tempered distribution, that is,*

$$\int k_s\widehat{\varphi} = \gamma(n, s) \int k_{n-s}\varphi \quad \text{for } \varphi \in \mathcal{S}(\mathbb{R}^n). \tag{3.43}$$

The constant $\gamma(n, s)$ will be fixed throughout the book.

Proof Suppose first that $n/2 < s < n$. Then

$$k_s \in L^1 + L^2 := \{f_1 + f_2 : f_1 \in L^1, f_2 \in L^2\},$$

because

$$\int_{B(0,1)} k_s < \infty \quad \text{and} \quad \int_{\mathbb{R}^n \setminus B(0,1)} k_s^2 < \infty.$$

As observed before, for any $f = f_1 + f_2 \in L^1 + L^2$ we can define

$$\widehat{f} = \widehat{f_1} + \widehat{f_2} \in L^\infty + L^2.$$

Thus for $n/2 < s < n$ we have defined $\widehat{k_s}$ as a function in $L^\infty + L^2$. Since k_s is radial and satisfies $k_s(rx) = r^{-s}k_s(x)$ for $r > 0$, it follows from (3.33) and (3.5) that $\widehat{k_s}$ is also radial and satisfies $\widehat{k_s}(rx) = r^{s-n}\widehat{k_s}(x)$. Thus it is of the above form $\gamma(n,s)k_{n-s}$. Using the product formula (verified by Fubini's theorem also in this case), we obtain for any $\varphi \in \mathcal{S}(\mathbb{R}^n)$,

$$\int k_s\widehat{\varphi} = \int \widehat{k_s}\varphi.$$

This means that $\widehat{k_s} = \gamma(n,s)k_{n-s}$ is also the Fourier transform of k_s as a tempered distribution.

Now we should show that $\gamma(n,s)k_{n-s}$ is the Fourier transform of k_s as a tempered distribution also when $0 < s \le n/2$. From the inversion formula, recall (3.9), we see that for a radial function $f \in \mathcal{S}$, the Fourier transform of $\widehat{f}$ is f. The analogous relation is valid also for tempered distributions by the following lemma:

Lemma 3.7 *Suppose that g is a locally integrable even function on $\mathbb{R}^n$ such that its distributional Fourier transform f is a locally integrable function. Then $\widehat{f} = g$.*

Proof Using the product formula and (3.9) we have for $\varphi \in \mathcal{S}$,

$$\widehat{f}(\varphi) = f(\widehat{\varphi}) = \int f\widehat{\varphi} = \int \widehat{g}\varphi = \int g\widehat{\varphi}$$

$$= \int g(x)\varphi(-x)\,dx = \int g(-x)\varphi(x)\,dx = \int g\varphi,$$

from which the lemma follows. $\square$

So for $0 < s < n/2$, the Fourier transform of $k_s = \gamma(n, n-s)^{-1}\widehat{k_{n-s}}$ (as $n/2 < n-s < n$) is $\gamma(n, n-s)^{-1}k_{n-s}$.

The case $s = n/2$ follows by a limiting argument: if $\widehat{k_{n/2}}$ is the Fourier transform of $k_{n/2}$ (as a tempered distribution), and $\varphi \in \mathcal{S}(\mathbb{R}^n)$, then

$$\widehat{k_{n/2}}(\varphi) = \int k_{n/2}\widehat{\varphi} = \lim_{s \to n/2} \int k_s\widehat{\varphi}$$

$$= \lim_{s \to n/2} \gamma(n, s) \int k_{n-s}\varphi = \int k_{n/2}\varphi.$$

The interchange of limit and integration can be verified by the dominated convergence theorem. There is a small problem: why is $\lim_{s \to n/2} \gamma(n, s) = 1$? To see that apply (3.43) for $s \neq n/2$ and $\varphi(x) = e^{-\pi|x|^2}$. Then $\widehat{\varphi} = \varphi$ and we obtain

$$\int k_s\varphi = \int k_s\widehat{\varphi} = \gamma(n, s) \int k_{n-s}\varphi,$$

that is,

$$\int |x|^{-s}e^{-\pi|x|^2}dx = \gamma(n, s) \int |x|^{s-n}e^{-\pi|x|^2}\,dx.$$

This gives immediately that $\lim_{s \to n/2} \gamma(n, s) = 1$ and completes the proof of the theorem.　　　□

Remark 3.8 Computing the integrals in the last formula of the above proof one finds that

$$\gamma(n, s) = \pi^{s-n/2}\frac{\Gamma\left(\frac{n-s}{2}\right)}{\Gamma\left(\frac{s}{2}\right)}. \tag{3.44}$$

Theorem 3.6 gives easily the following lemma:

Lemma 3.9 *If $0 < s < n$ and $\varphi \in \mathcal{S}(\mathbb{R}^n)$, then*

$$\widehat{\varphi k_s} = \gamma(n, s)\widehat{\varphi} * k_{n-s} \quad and \quad \widehat{\varphi * k_s} = \gamma(n, s)\widehat{\varphi}k_{n-s}.$$

Proof Clearly, $\varphi k_s \in L^1(\mathbb{R}^n)$. For any $\psi \in \mathcal{S}(\mathbb{R}^n)$ we have by Fubini's theorem, the formula (3.9); $\check{f} = \widehat{f}$, Theorem 3.6 and the convolution and product formulas,

$$\int \gamma(n, s)(\widehat{\varphi} * k_{n-s})\psi = \int \gamma(n, s)k_{n-s}(\widehat{\widetilde{\varphi}} * \psi) = \int \gamma(n, s)k_{n-s}(\widecheck{\varphi} * \psi)$$

$$= \int \gamma(n, s)k_{n-s}\mathcal{F}^{-1}(\varphi\psi) = \int k_s\varphi\widehat{\psi} = \int \widehat{k_s\varphi}\psi.$$

It follows that $\widehat{\varphi k_s} = \gamma(n, s)\widehat{\varphi} * k_{n-s}$. The second formula can be proven in the same way, or it can also be reduced to the first.　　　□

3.5 Fourier transforms and energy-integrals of measures

The following formula is the key to relating Hausdorff dimension to the Fourier transform:

Theorem 3.10 *Let $\mu \in \mathcal{M}(\mathbb{R}^n)$ and $0 < s < n$. Then*

$$I_s(\mu) = \gamma(n, s) \int |\widehat{\mu}(x)|^2 |x|^{s-n} \, dx. \tag{3.45}$$

Proof Let us try to prove this formally using the basic formulas. By the Parseval and convolution formulas and by Theorem 3.6,

$$I_s(\mu)$$
$$= \int k_s * \mu \, d\mu = \int \widehat{k_s * \mu} \overline{\widehat{\mu}} = \int \widehat{k_s} |\widehat{\mu}|^2 = \gamma(n, s) \int |\widehat{\mu}(x)|^2 |x|^{s-n} \, dx.$$

Since the Fourier transform of k_s exists only in distributional sense we have to be more careful and justify that we can use the Parseval and convolution formulas in this situation.

Let $\varphi \in S(\mathbb{R}^n)$ be real valued. Then changing $z = y - x$ below and denoting again $\widetilde{\varphi}(x) = \varphi(-x)$, we obtain

$$I_s(\varphi) = \iint k_s(y - x)\varphi(x)\varphi(y) \, dx \, dy$$
$$= \iint k_s(z)\varphi(y - z)\varphi(y) \, dz \, dy = \int k_s(\widetilde{\varphi} * \varphi).$$

By (3.10) and (3.9) $\widetilde{\varphi} * \varphi$ is the Fourier transform of $\widehat{\varphi}\,\overline{\widehat{\varphi}} = |\widehat{\varphi}|^2$, whence by Theorem 3.6,

$$I_s(\varphi) = \gamma(n, s) \int k_{n-s} |\widehat{\varphi}|^2 = \gamma(n, s) \int |x|^{s-n} |\widehat{\varphi}(x)|^2 \, dx.$$

Thus we have proved the theorem for such smooth measures φ.

To finish the proof we approximate μ with $\mu_\varepsilon = \psi_\varepsilon * \mu$ as before using a non-negative function $\psi \in C_0^\infty(\mathbb{R}^n)$ with $\int \psi = 1$. Applying the above to $\varphi = \mu_\varepsilon$ we get

$$\iint \left(\iint |x - y|^{-s} \psi_\varepsilon(x - z)\psi_\varepsilon(y - w) \, dx \, dy \right) d\mu z \, d\mu w$$
$$= \iint \left(|x - y|^{-s} \int \psi_\varepsilon(x - z) \, d\mu z \int \psi_\varepsilon(y - w) \, d\mu w \right) dx \, dy$$
$$= I_s(\mu_\varepsilon) = \gamma(n, s) \int |\widehat{\mu}(x)|^2 |\widehat{\psi}(\varepsilon x)|^2 |x|^{s-n} \, dx.$$

The last term approaches $\gamma(n, s) \int |\widehat{\mu}(x)|^2 |x|^{s-n} \, dx$ as $\varepsilon \to 0$. By the change of variables $u = (x - z)/\varepsilon$ and $v = (y - w)/\varepsilon$ we get for the inner integral in the first term,

$$\iint |x - y|^{-s} \psi_\varepsilon(x - z)\psi_\varepsilon(y - w) \, dx \, dy$$
$$= \iint |\varepsilon(u - v) + z - w|^{-s} \psi(u)\psi(v) \, du \, dv.$$

This tends to $|z - w|^{-s}$ when $\varepsilon \to 0$ and $z \neq w$. So it is enough to show that we can interchange the limit and integration above in the first term. With the help of the above identity, the following estimate is easy to check:

$$\iint |x - y|^{-s} \psi_\varepsilon(x - z)\psi_\varepsilon(y - w) \, dx \, dy \lesssim |z - w|^{-s}.$$

Using this we complete the proof applying the dominated convergence theorem provided $I_s(\mu) < \infty$. If $I_s(\mu) = \infty$, we get by Fatou's lemma

$$\infty = I_s(\mu) \leq \liminf_{\varepsilon \to 0} \iint \left(\iint |x - y|^{-s} \psi_\varepsilon(x - z)\psi_\varepsilon(y - w) \, dx \, dy \right) d\mu z \, d\mu w$$
$$= \gamma(n, s) \liminf_{\varepsilon \to 0} \int |\widehat{\mu}(x)|^2 |\widehat{\psi}(\varepsilon x)|^2 |x|^{s-n} \, dx = \gamma(n, s) \int |\widehat{\mu}(x)|^2 |x|^{s-n} \, dx.$$

This completes the proof of the theorem. $\qquad\square$

We can also obtain such a formula for signed measures. But since we shall only need it for bounded functions we give it for them. For $f, g \in L^\infty(\mathbb{R}^n)$ the *mutual energy* $I_s(f, g), 0 < s < n$, is

$$I_s(f, g) = \iint |x - y|^{-s} f(x)g(y) \, dx \, dy.$$

This is defined if f and g are non-negative. For general functions $I_s(f, g)$ is defined if $I_s(|f|, |g|) < \infty$. If in addition, $f, g \in L^1(\mathbb{R}^n)$, the proof of (3.45) gives

$$I_s(f, g) = \gamma(n, s) \int \widehat{f}(x)\overline{\widehat{g}(x)} |x|^{s-n} dx. \tag{3.46}$$

By approximation this remains valid for $f, g \in L^\infty(\mathbb{R}^n) \cap L^2(\mathbb{R}^n)$ with $I_s(|f|, |g|) < \infty$. Notice that when $f = g$ we have

$$I_s(f) := I_s(f, f) = \gamma(n, s) \int |\widehat{f}(x)|^2 |x|^{s-n} \, dx \geq 0, \tag{3.47}$$

even if f were not non-negative.

A natural setting for the mutual energy is the space of signed Borel measures μ for which $I_s(\mu)$ is finite. Then $I_s(\mu, \nu)$ defines an inner product in this space, see Landkof [1972].

3.6 Salem sets and Fourier dimension

Suppose $\mu \in \mathcal{M}(\mathbb{R}^n)$ and $I_s(\mu) = \gamma(n, s) \int |\widehat{\mu}(x)|^2 |x|^{s-n} \, dx < \infty$. Then

$$|\widehat{\mu}(x)| \leq |x|^{-s/2}$$

for 'most' x with large norm. Here 'most' simply means what is needed in order that the above integral would be finite. For example we must have

$$\lim_{R \to \infty} R^{-n} \mathcal{L}^n(\{x \in B(0, R) : |\widehat{\mu}(x)| > |x|^{-s/2}\}) = 0.$$

On the other hand, if

$$|\widehat{\mu}(x)| \leq |x|^{-s/2} \quad \text{for all } x \in \mathbb{R}^n, \tag{3.48}$$

then $I_t(\mu) < \infty$ for all $t < s$. This implies by Theorem 2.8 that $\dim(\operatorname{spt} \mu) \geq s$. Thus if $\mu \in \mathcal{M}(A)$ and $\dim A = s$, the best decay at infinity we can hope for the Fourier transform of μ is that given by (3.48). This motivates the following definition.

Definition 3.11 A set $A \subset \mathbb{R}^n$ is a *Salem set* if for every $s < \dim A$ there is $\mu \in \mathcal{M}(A)$ such that (3.48) holds.

Another way to say this is to define first the *Fourier dimension*:

Definition 3.12 The Fourier dimension of a set $A \subset \mathbb{R}^n$ is

$$\dim_F A = \sup\{s \leq n : \exists \mu \in \mathcal{M}(A) \text{ such that } |\widehat{\mu}(x)| \leq |x|^{-s/2} \ \forall x \in \mathbb{R}^n\}.$$

Then for Borel sets A, $\dim_F A \leq \dim A$, and A is a Salem set if and only if $\dim_F A = \dim A$.

Often Fourier dimension is defined slightly differently: instead of measures $\mu \in \mathcal{M}(A)$ one uses Borel probability measures μ such that $\mu(A) = 1$. These definitions agree for closed sets, but they do not agree for all Borel sets, not even for all F_σ-sets, as follows from Ekström, Persson and Schmeling [2015].

Notice that if $A \subset \mathbb{R}^n$ is a Salem set and $0 < s < \dim A$, we can always find $\mu \in \mathcal{M}(A)$ such that both (3.48) and $I_s(\mu) < \infty$ hold.

If $\mu \in \mathcal{M}([0, 1]^n)$ with $|\widehat{\mu}(z)| \lesssim |z|^{-s/2}$ for $z \in \mathbb{Z}^n$ and $\varphi \in \mathcal{S}(\mathbb{R}^n)$, then $|\widehat{\varphi\mu}(x)| \lesssim |x|^{-s/2}$ for $x \in \mathbb{R}^n$, see Lemma 9.A.4 in Wolff [2003]. It follows that the Fourier dimension of subsets of the unit cube $[0, 1]^n$ can be determined by just looking at the Fourier coefficients $\widehat{\mu}(z)$, $z \in \mathbb{Z}^n$, of the measures μ in $\mathcal{M}(A)$.

By (3.42) spheres are Salem sets, but subsets of m-dimensional planes in $\mathbb{R}^n$, $m < n$, are not. The Fourier dimension, and thus the property of being a Salem set, depends on the space where the set is embedded in: if $A \subset \mathbb{R}^m \subset \mathbb{R}^n$ and $m < n$, then for any $\mu \in \mathcal{M}(A)$, the Fourier transform $\widehat{\mu}(x)$ does not tend to zero as $x \in \mathbb{R}^n$, $|x| \to \infty$, because $\widehat{\mu}(x)$ depends only on the $\mathbb{R}^m$ coordinates of x. Hence all subsets of hyperplanes have zero Fourier dimension. We shall encounter more interesting examples of sets with positive Hausdorff dimension and zero Fourier dimension in Chapter 8.

Körner [2011] showed that for any $0 \leq t \leq s \leq 1$ there exists a compact set of the real line which has Hausdorff dimension s and Fourier dimension t.

There are many random Salem sets; we shall come to this in Chapter 12. Non-trivial deterministic fractal Salem sets are however hard to construct. The following result was proved by Kaufman [1981]:

Theorem 3.13 *Let $\alpha > 0$ and let E_α be the set of $x \in \mathbb{R}$ such that for infinitely many rationals p/q,*

$$|x - p/q| \leq q^{-(2+\alpha)}.$$

Then E is a Salem set with $\dim E = 2/(2 + \alpha)$.

We shall not prove this result, a proof can be found in Kaufman [1981] and also in Wolff [2003], Chapter 9. Let us quickly see what kind of set this is. By a classical theorem of Dirichlet on Diophantine approximation, for every irrational x there are infinitely many rationals p/q such that $|x - p/q| \leq q^{-2}$, and this is essentially the best one can say for all x. The set E_α consists of real numbers which are much better approximable by rationals. The upper bound $2/(2 + \alpha)$ for the Hausdorff dimension of E_α is easily derived using coverings of $E \cap [-N, N]$, $N = 1, 2 \ldots$, with intervals of the type $[p/q - q^{-(2+\alpha)}, p/q + q^{-(2+\alpha)}]$ where p and q are suitable integers. The lower bound is harder. It can be derived without Fourier transforms; see Section 10.3 of Falconer [1990]. But in order to verify that E_α is a Salem set, one needs to construct $\mu \in \mathcal{M}(E_\alpha)$ with sufficient decay for the Fourier transform, and this will automatically also give the lower bound. Kaufman constructed such a μ with

$$|\widehat{\mu}(x)| \lesssim \log |x| |x|^{-1/(2+\alpha)}, \quad |x| > 2.$$

Could one construct non-integral dimensional Salem sets E with more structure than just the knowledge of the dimension? For example, could they be Ahlfors–David regular? This means that E would be the support of a measure ν such that $\nu(B(x, r)) \approx r^s$ for $x \in E, 0 < r < d(E)$, and for every $t < s$ there would exist a measure $\mu \in \mathcal{M}(E)$ for which $|\widehat{\mu}(x)| \lesssim |x|^{-t/2}$ for $x \in \mathbb{R}^n$. One could also hope to find a single measure satisfying both conditions: Mitsis [2002b] asked for which values of s do there exist measures $\mu \in \mathcal{M}(\mathbb{R}^n)$ such that $\mu(B(x, r)) \approx r^s$ for $x \in \mathrm{spt}\,\mu, 0 < r < 1$, and $|\widehat{\mu}(x)| \lesssim |x|^{-s/2}$ for $x \in \mathbb{R}^n$? Presently any examples of this type are only known for integers s and they are measures on smooth s-dimensional surfaces. Partial results have been obtained by Łaba and Pramanik [2009] and by Chen [2014a]. In particular, Chen constructs measures as in Mitsis's question, except that he needs a logarithmic factor in one of the conditions. Related results can also be found in Körner [2011] and Shmerkin and Suomala [2014].

From the above we know that if a set has zero s-dimensional Hausdorff measure, then it cannot support a non-trivial measure whose Fourier transform would tend to zero at infinity faster than $|x|^{-s/2}$. But how quickly can they tend to zero in terms of $\varphi(|x|)$ for various functions φ? And what if $\mathcal{H}^s$ is replaced by Hausdorff measures defined by general gauge functions in place of r^s? Recent results on this delicate question were obtained by Körner [2014]. This paper also contains an excellent brief survey on the topic.

The existence of measures with a certain speed of decay of Fourier transforms has various consequences for the Hausdorff dimension. We shall return to this for instance in the case of distance sets, but now we give one simple application as an example. Denote here by $A^k = A + \cdots + A$ and $\mu^k = \mu * \cdots * \mu$ (k times) the k-fold sum-set and convolution product.

Proposition 3.14 *Let $A \subset \mathbb{R}$ be a Borel set and k be a positive integer.*

(a) If $\dim_F A > 1/k$, then $\mathcal{L}^1(A^k) > 0$.
(b) If $\dim_F A > 2/k$, then A^k contains an open interval.

Proof Let $0 < s < \dim_F A$ and $\mu \in \mathcal{M}(A)$ such that

$$|\widehat{\mu}(x)| \leq |x|^{-s/2} \quad \text{for } x \in \mathbb{R}.$$

We have $\mu^k \in \mathcal{M}(A^k)$ and $|\widehat{\mu^k}(x)| \leq |x|^{-ks/2}$ for $x \in \mathbb{R}$. If $ks > 1$, this implies that $\widehat{\mu^k} \in L^2$. Hence by Theorem 3.3 μ^k is absolutely continuous, whence $\mathcal{L}^1(A^k) > 0$. If $ks > 2$, $\widehat{\mu^k} \in L^1$, so we have by Theorem 3.4 that μ^k is a continuous function which implies that the interior of A^k is non-empty. $\qquad \square$

3.7 Spherical averages

For $\mu \in \mathcal{M}(\mathbb{R}^n)$, $n \geq 2$, we define the L^2 spherical averages of μ by

$$\sigma(\mu)(r) = \int_{S^{n-1}} |\widehat{\mu}(rv)|^2 \, d\sigma^{n-1} v = r^{1-n} \int_{S(r)} |\widehat{\mu}(v)|^2 d\sigma_r^{n-1} v \qquad (3.49)$$

for $r > 0$. Using integration in polar coordinates and the formula (3.45), the energy-integrals of μ can be written in terms of these:

$$I_s(\mu) = \gamma(n, s) \int_0^\infty \sigma(\mu)(r) r^{s-1} \, dr, \quad 0 < s < n. \qquad (3.50)$$

Although the Fourier transform need not tend to zero at infinity for measures with finite energy, the spherical averages behave better: they do tend to zero and we have quantitative estimates. We return to these estimates and their applications to distance sets and intersections in Chapter 15. Here we only give the following simple estimate:

Lemma 3.15 *If $0 < s \leq (n-1)/2$ and $\mu \in \mathcal{M}(\mathbb{R}^n)$ with $I_s(\mu) < \infty$, then for $r > 0$,*

$$\sigma(\mu)(r) \leq C(n, s) I_s(\mu) r^{-s}.$$

Proof We can assume, by approximation with $\psi_\varepsilon * \mu$ as before, that μ is a smooth non-negative function f with compact support. By the formula (3.28)

$$\sigma(f)(r) = r^{1-n} \int_{S^{n-1}(r)} |\widehat{f}(v)|^2 \, d\sigma_r^{n-1} v = r^{1-n} \int (f * \widehat{\sigma_r^{n-1}}) f.$$

Since $r^{1-n}\widehat{\sigma_r^{n-1}}(x) = \widehat{\sigma^{n-1}}(rx)$, we have

$$\sigma(\mu)(r) = \iint \widehat{\sigma^{n-1}}(r(x-y)) f(y) f(x) \, dy \, dx. \qquad (3.51)$$

Evidently,

$$|\widehat{\sigma^{n-1}}(r(x-y))| \lesssim 1 \leq (r|x-y|)^{-s},$$

if $r|x - y| \leq 1$, and by (3.42)

$$|\widehat{\sigma^{n-1}}(r(x-y))| \lesssim (r|x-y|)^{-(n-1)/2} \leq (r|x-y|)^{-s},$$

if $r|x - y| \geq 1$. Inserting the estimate $|\widehat{\sigma^{n-1}}(r(x-y))| \lesssim (r|x-y|)^{-s}$ into the formula (3.51), we obtain the desired inequality for f, and hence also for μ. $\square$

It is clear from (3.50) that the decay r^{-s} is the best we can hope for.

One can also show that without any energy assumptions the averages $\sigma(\mu)(r)$ tend to zero as $r \to \infty$ for every continuous measure $\mu \in \mathcal{M}(\mathbb{R}^n)$, $n \geq 2$; see Mattila [1987].

Instead of spheres one could also look at the convergence along lines through the origin. Kaufman [1973] proved that if $\mu \in \mathcal{M}(\mathbb{R}^2)$ with $I_1(\mu) < \infty$, then $\widehat{\mu}$ tends to zero along almost all lines through the origin. Moreover, if μ satisfies the Frostman condition $\mu(B(x,r)) \leq r^s$, $x \in \mathbb{R}^2$, $r > 0$, for some $1 < s < 2$, then the exceptional set of the lines has Hausdorff dimension at most $2 - s$. This is sharp as Kaufman showed using number theoretic examples similar to those in Section 3.6.

Simple as it is, Lemma 3.15 is not completely trivial: it is essential that we consider non-negative measures and functions. Stated in terms of Fourier transforms the inequality of Lemma 3.15 is

$$\int_{S^{n-1}} |\widehat{\mu}(rv)|^2 \, d\sigma^{n-1} v \leq C(n,s) r^{-s} \int_{\mathbb{R}^n} |\widehat{\mu}(x)|^2 |x|^{s-n} \, dx.$$

It is clear that such an estimate cannot hold even for all smooth compactly supported functions with varying sign.

3.8 Ball averages

It is much easier to control averages over solid balls than over spheres. First, if $\mu \in \mathcal{M}(\mathbb{R}^n)$ with $I_s(\mu) < \infty$ we have by (3.45),

$$\int_{B(0,R)} |\widehat{\mu}(x)|^2 \, dx \leq R^{n-s} \int_{B(0,R)} |x|^{s-n} |\widehat{\mu}(x)|^2 \, dx \lesssim R^{n-s} I_s(\mu).$$

But we can easily obtain such an estimate also from the Frostman condition

$$\mu(B(x,r)) \leq Cr^s \quad \text{for } x \in \operatorname{spt} \mu, \quad r > 0, \tag{3.52}$$

which does not imply that $I_s(\mu) < \infty$.

To see this, choose $\varphi \in \mathcal{S}(\mathbb{R}^n)$ such that $\widehat{\varphi} \geq 0$ and $\widehat{\varphi}(x) = 1$ when $|x| \leq 1$. Define $\varphi_R(x) = R^n \varphi(Rx)$ when $R > 0$. Then $\widehat{\varphi_R}(x) = \widehat{\varphi}(x/R)$ and we obtain by (3.22) and (3.27),

$$\int_{B(0,R)} |\widehat{\mu}|^2 \leq \int \widehat{\varphi_R} |\widehat{\mu}|^2 = \int \widehat{\varphi_R * \mu \overline{\mu}} = \int \varphi_R * \mu \, d\mu \lesssim R^{n-s},$$

where the last inequality follows from

$$\varphi_R * \mu(x) = R^n \int \varphi(R(x-y)) \, d\mu y \lesssim R^n \mu(B(x, 1/R))$$

$$+ R^n \sum_{j=1}^{\infty} 2^{-j(s+1)} \mu(B(x, 2^j/R) \setminus B(x, 2^{j-1}/R)) \lesssim R^{n-s},$$

using (3.52) and the fast decay of φ.

If μ satisfies with some positive constant c the lower regularity

$$\mu(Bx, r)) \geq cr^s \quad \text{for } x \in \operatorname{spt}\mu, \quad 0 < r < 1, \tag{3.53}$$

then for $R > 1$,

$$\int_{B(0,R)} |\widehat{\mu}|^2 \gtrsim R^{n-s}.$$

The proof is a slight modification of the above: recalling Example 3.2 choose φ so that $\varphi \geq 0$, $\varphi \geq 1$ on $B(0, 1)$ and $\operatorname{spt}\widehat{\varphi} \subset B(0, 1)$, and observe that then $\int_{B(0,R)} |\widehat{\mu}|^2 \gtrsim \int \widehat{\varphi_R} |\widehat{\mu}|^2$ and $\varphi_R * \mu(x) \gtrsim R^{n-s}$. In particular if μ is Ahlfors–David regular, that is, both (3.52) and (3.53) hold, we have $\int_{B(0,R)} |\widehat{\mu}|^2 \approx R^{n-s}$ for $R > 1$.

Strichartz [1989] and [1990a] made a much more detailed study of such ball averages and related matters. For instance, he showed that if μ satisfies (3.52) and the limit $\lim_{r\to 0} r^{-s} \mu(B(x, r))$ exists and is positive for μ almost all $x \in \mathbb{R}^n$, then for all $f \in L^2(\mu)$,

$$\lim_{R\to\infty} R^{s-n} \int_{B(0,R)} |\widehat{f\mu}|^2 = c(n, s) \int |f|^2 \, d\mu,$$

for some positive and finite constant $c(n, s)$. To get an idea when $f = 1$, notice that if $\widehat{\varphi}$ approximates well the characteristic function of $B(0, 1)$ and φ_R is as above, then $R^{s-n} \int_{B(0,R)} |\widehat{f\mu}|^2$ is close to $R^{s-n} \int \varphi_R * \mu \, d\mu$ by the above arguments, and the convergence of $r^{-s} \mu(B(x, r))$ as $r \to 0$ implies that $R^{s-n} \int \varphi_R * \mu \, d\mu$ converges as $R \to \infty$.

The existence of the positive and finite limit $\lim_{r\to 0} r^{-s} \mu(B(x, r))$ for μ almost all $x \in \mathbb{R}^n$ is a very restrictive condition. It forces s to be an integer by Marstrand's theorem, see Mattila [1995], Theorem 14.10, and μ to be a rectifiable measure by Preiss's theorem, see Mattila [1995], Theorem 17.8, or Preiss [1987]. On the other hand, rectifiable measures include all surface measures on smooth surfaces and much more.

3.9 Fourier transforms and rectangular boxes

At later stages of this book it will be essential to understand how the Fourier transform of a smooth function supported in a rectangular box behaves. The answer is: it lives essentially in a dual box, defined below. By the Heisenberg uncertainty principle it cannot have compact support so we mean by this that it decays quickly outside such a box.

Let us first quickly look at balls. If φ is infinitely differentiable on $\mathbb{R}^n$, $n \geq 2$, with $\operatorname{spt} \varphi \subset B(0, 1)$ and for $a \in \mathbb{R}^n$, $r > 0$, $\varphi_{a,r}(x) = \varphi((x - a)/r)$, then $\operatorname{spt} \varphi_{a,r} \subset B(a, r)$,

$$\widehat{\varphi_{a,r}}(x) = \int e^{-2\pi i x \cdot y} \varphi((y - a)/r)\, dy = r^n e^{-2\pi i x \cdot a} \widehat{\varphi}(rx),$$

and for all $N = 1, 2, \ldots$,

$$|\widehat{\varphi_{a,r}}(x)| \lesssim_N r^{n-N} |x|^{-N}.$$

So $\widehat{\varphi_{a,r}}$ decays fast outside $B(0, 1/r)$; this is our dual ball for $B(a, r)$.

Let R be a rectangular box in $\mathbb{R}^n$ (as in (3.54) below). We say that it is an $(r_1, \ldots, r_n)$-box if $r_1 \leq \cdots \leq r_n$ are its side-lengths. The $(\frac{1}{r_n}, \ldots, \frac{1}{r_1})$-box centred at the origin with the $\frac{1}{r_j}$ side parallel to the r_j side of R is called the dual of R and denoted by $\widetilde{R}$. More formally, let $Q_0 = [0, 1]^n$ and fix for the $(r_1, \ldots, r_n)$-box R an affine mapping A_R which maps Q_0 onto R written as

$$A_R(x) = g(Lx) + a, g \in O(n), a \in \mathbb{R}^n, Lx = (r_1 x_1, \ldots, r_n x_n) \quad \text{for } x \in \mathbb{R}^n.$$

Then

$$R = A_R(Q_0) \quad \text{and} \quad \widetilde{R} = g(L^{-1}(Q_0)). \tag{3.54}$$

Fix a non-negative function $\varphi \in \mathcal{S}(\mathbb{R}^n)$ such that $\varphi = 1$ on Q_0 and $\operatorname{spt} \varphi \subset 2Q_0$. For $t > 0$ we shall denote by tR the rectangular box which has the same centre as R and the side-lengths equal to those of R multiplied by t. We define

$$\varphi_R = \varphi \circ A_R^{-1} \text{ so that } \varphi_R = 1 \text{ on } R \text{ and } \operatorname{spt} \varphi_R \subset 2R. \tag{3.55}$$

Lemma 3.16 *With the above notation*

$$\widehat{\varphi_R}(x) = r_1 \cdots r_n e^{-2\pi i x \cdot a} \widehat{\varphi}(L(g^{-1}(x))). \tag{3.56}$$

For any $M \in \mathbb{N}$ and for any $(r_1, \ldots, r_n)$-box R,

$$|\widehat{\varphi_R}(x)| \leq C(\varphi, M) r_1 \cdots r_n \sum_{j=1}^{\infty} 2^{-Mj} \chi_{2^j \widetilde{R}}(x) \quad \text{for } x \in \mathbb{R}^n. \tag{3.57}$$

Moreover,

$$\|\widehat{\varphi_R}\|_1 = \|\widehat{\varphi}\|_1. \tag{3.58}$$

Proof Let g, L and a be as above. The Fourier transform of φ_R is

$$\widehat{\varphi_R}(x) = \int e^{-2\pi i x \cdot \xi} \varphi_R(\xi) \, d\xi = \int e^{-2\pi i x \cdot \xi} \varphi(A_R^{-1}(\xi)) d\xi.$$

Setting $\eta = A_R^{-1}(\xi)$ we get $d\xi = |\det(g \circ L)| d\eta = r_1 \cdots r_n d\eta$. Furthermore, since $g(x) \cdot g(y) = x \cdot y$ for all $x, y \in \mathbb{R}^n$ and L satisfies $L(x) \cdot y = x \cdot L(y)$ for all $x, y \in \mathbb{R}^n$,

$$\begin{aligned} \widehat{\varphi_R}(x) &= r_1 \cdots r_n \int e^{-2\pi i x \cdot A_R(\eta)} \varphi(\eta) \, d\eta \\ &= r_1 \cdots r_n e^{-2\pi i x \cdot a} \int e^{-2\pi i x \cdot g(L(\eta))} \varphi(\eta) \, d\eta \\ &= r_1 \cdots r_n e^{-2\pi i x \cdot a} \int e^{-2\pi i L(g^{-1}(x)) \cdot \eta} \varphi(\eta) \, d\eta \\ &= r_1 \cdots r_n e^{-2\pi i x \cdot a} \widehat{\varphi}(L(g^{-1}(x))), \end{aligned}$$

which proves (3.56).

For $\widehat{\varphi}$ we have by its fast decay,

$$|\widehat{\varphi}(x)| \lesssim_M \sum_{j=1}^{\infty} 2^{-Mj} \chi_{2^j Q_0}(x) \quad \text{for } x \in \mathbb{R}^n.$$

Hence

$$|\widehat{\varphi_R}(x)| \lesssim_M r_1 \cdots r_n \sum_{j=1}^{\infty} 2^{-Mj} \chi_{2^j Q_0}(Lg^{-1}(x)) = r_1 \cdots r_n \sum_{j=1}^{\infty} 2^{-Mj} \chi_{2^j \widetilde{R}}(x),$$

proving (3.57). Finally (3.58) is obvious by a change of variable. $\qquad\square$

In Chapter 16 we shall convolve Frostman measures with the above functions $\widehat{\varphi_R}$ and make use of the following lemma.

Lemma 3.17 *Let* $\mu \in \mathcal{M}(\mathbb{R}^n)$, $0 < s \leq n$, *and suppose that*

$$\mu(B(x, r)) \leq r^s \quad \text{for all } x \in \mathbb{R}^n, r > 0. \tag{3.59}$$

Let $R \subset \mathbb{R}^n$ *be an* $(r_1, r_2, \ldots, r_2)$-*box with* $r_1 \leq r_2$. *Define*

$$\mu_R = |\widehat{\varphi_R}| * \mu.$$

Then,

$$\|\mu_R\|_\infty \le C(\varphi) r_2^{n-s}, \tag{3.60}$$

$$\|\mu_R\|_1 = \|\widehat{\varphi}\|_1 \mu(\mathbb{R}^n), \tag{3.61}$$

and

$$\int_{K\widetilde{R}} \mu_R(x+y)\, dy \le C(\varphi) K^s r_1^{-1} r_2^{1-s} \quad \text{for all } K \ge 1, \quad x \in \mathbb{R}^n. \tag{3.62}$$

For any cube $Q \subset \mathbb{R}^n$,

$$\int_{B(x,r)} \mu_Q(y)\, dy \le C(\varphi) r^s \quad \text{for all } x \in \mathbb{R}^n, \quad r > 0. \tag{3.63}$$

Proof By (3.57) for any $M \ge 1$,

$$\mu_R(x) = \int |\widehat{\varphi}_R(x-y)|\, d\mu y \lesssim_M r_1 r_2^{n-1} \sum_{j=1}^\infty 2^{-Mj} \int \chi_{2^j \widetilde{R}}(x-y)\, d\mu y.$$

As $2^j \widetilde{R}$ is a $(\frac{2^j}{r_2}, \ldots, \frac{2^j}{r_2}, \frac{2^j}{r_1})$-box it can be covered with roughly $\frac{r_2}{r_1}$ balls of radius $\frac{2^j}{r_2}$. Taking $M = s+1$ and using (3.59) this gives

$$\mu_R(x) \lesssim r_1 r_2^{n-1} \sum_{j=1}^\infty 2^{-(s+1)j} \frac{r_2}{r_1} \left(\frac{2^j}{r_2}\right)^s = r_2^{n-s},$$

proving (3.60).

Furthermore by (3.58),

$$\|\mu_R\|_1 = \int |\widehat{\varphi}_R| * \mu = \int |\widehat{\varphi}_R(x-y)|\, dy\, d\mu x = \|\widehat{\varphi}_R\|_1 \mu(\mathbb{R}^n) = \|\widehat{\varphi}\|_1 \mu(\mathbb{R}^n),$$

which proves (3.61).

Next we prove (3.62). By (3.57) and the fact that if $y \in K\widetilde{R}$ and $x + y - z \in 2^j \widetilde{R}$, then $z = y - (x+y-z) + x \in K\widetilde{R} - 2^j \widetilde{R} + x = (K + 2^j)\widetilde{R} + x$, as $\widetilde{R}$ is centred at the origin, we obtain

$$\int_{K\widetilde{R}} \mu_R(x+y)\, dy = \int_{K\widetilde{R}} \int |\widehat{\varphi}_R(x+y-z)|\, d\mu z\, dy$$

$$\lesssim_M r_1 r_2^{n-1} \sum_{j=1}^\infty 2^{-Mj} \iint \chi_{K\widetilde{R}}(y) \chi_{2^j \widetilde{R}}(x+y-z)\, d\mu z\, dy$$

$$\le r_1 r_2^{n-1} \sum_{j=1}^\infty 2^{-Mj} \iint \chi_{(K+2^j)\widetilde{R}+x}(z) \chi_{2^j \widetilde{R}}(x+y-z)\, dy\, d\mu z$$

$$= r_1 r_2^{n-1} \sum_{j=1}^\infty 2^{-Mj} \mathcal{L}^n(2^j \widetilde{R}) \mu((K+2^j)\widetilde{R}+x)).$$

As in the proof of (3.60) $(K + 2^j)\widetilde{R} + x$ can be covered with roughly $\frac{r_2}{r_1}$ balls of radius $(K + 2^j)r_2^{-1}$, whence

$$\int_{K\widetilde{R}} \mu_R(x + y)\,dy$$

$$\lesssim_M r_1 r_2^{n-1} \sum_{j=1}^{\infty} 2^{-Mj} \frac{2^{nj}}{r_1 r_2^{n-1}} \frac{r_2}{r_1} \left(\frac{K + 2^j}{r_2}\right)^s$$

$$= \sum_{j=1}^{\infty} 2^{(n-M)j} r_1^{-1} r_2^{1-s} (K + 2^j)^s$$

$$\leq \sum_{j=1}^{\infty} 2^{(n+s-M)j} r_1^{-1} r_2^{1-s} (2K)^s = (2K)^s r_1^{-1} r_2^{1-s},$$

where we used also that $K + 2^j \leq 2^{j+1}K$ and we chose $M = n + s + 1$.

Finally (3.63) follows from (3.62) when $r \geq 1/r_1$, where r_1 is the side-length of Q: choose $K = r_1 r$, then $B(x, r) \subset K\widetilde{Q} + x$. When $r < 1/r_1$, it follows from (3.60). $\qquad\square$

The rectangular boxes will enter when we study restrictions of Fourier transforms on spheres. The reason is simple; a spherical cap on S^{n-1} of radius δ is contained in a $C\delta^2 \times C\delta \times \cdots \times C\delta$-box where C depends only on n.

In Lemma 3.16 the Fourier transform of our function φ_R becomes small when we go far away from the dual box of R, but we did not get any information how it behaves on the box itself. It will be useful to have functions which are large on those dual boxes. This is the content of the following lemma, often called the *Knapp example*. Here, as well as later, by the Fourier transform of a function $f \in L^1(S^{n-1})$ we mean the Fourier transform of the measure $f\sigma^{n-1}$:

$$\widehat{f}(\xi) = \int f(x)e^{-2\pi i \xi \cdot x}\,d\sigma^{n-1}x, \quad \xi \in \mathbb{R}^n.$$

Lemma 3.18 *Let $e_n = (0, \ldots, 0, 1) \in \mathbb{R}^n$, $n \geq 2$, and set for $0 < \delta < 1$,*

$$C_\delta = \{x \in S^{n-1} : 1 - x \cdot e_n \leq \delta^2\}, \quad D_\delta = \{x \in C_\delta : |x_{n-1}| \leq \delta^2\}.$$

Then with $c = 1/(12n)$ we have for $f = \chi_{C_\delta}$, $g = \chi_{D_\delta}$,

$$|\widehat{f}(\xi)| \geq \sigma^{n-1}(C_\delta)/2 \quad \text{for } \xi \in R_\delta$$

and

$$|\widehat{g}(\xi)| \geq \sigma^{n-1}(D_\delta)/2 \quad \text{for } \xi \in S_\delta,$$

where

$$R_\delta = \{\xi \in \mathbb{R}^n : |\xi_j| \le c/\delta \text{ for } j = 1, \dots, n - 1, |\xi_n| \le c/\delta^2\},$$

$$S_\delta = \{\xi \in \mathbb{R}^n : |\xi_j| \le c/\delta \text{ for } j = 1, \dots, n - 2, |\xi_{n-1}| \le c/\delta^2, |\xi_n| \le c/\delta^2\}.$$

Proof Notice that C_δ is a spherical cap of radius roughly δ, more precisely, $|x_j| \le \sqrt{2}\delta$ for $x \in C_\delta$, $j = 1, \dots, n - 1$. For $\xi \in \mathbb{R}^n$,

$$|\widehat{f}(\xi)| = \left| \int_{C_\delta} e^{-2\pi i \xi \cdot x} \, d\sigma^{n-1} x \right|$$

$$= \left| \int_{C_\delta} e^{-2\pi i \xi \cdot (x - e_n)} \, d\sigma^{n-1} x \right| \ge \int_{C_\delta} \cos(2\pi \xi \cdot (x - e_n)) \, d\sigma^{n-1} x.$$

We only used that $|e^{-2\pi \xi \cdot e_n}| = 1$ and that the absolute value of a complex number is at least its real part. One checks easily that

$$|2\pi \xi \cdot (x - e_n)| < \pi/3 \quad \text{for } x \in C_\delta, \, \xi \in R_\delta,$$

whence

$$\cos(2\pi \xi \cdot (x - e_n)) > 1/2 \quad \text{for } x \in C_\delta, \xi \in R_\delta,$$

and so

$$|\widehat{f}(\xi)| \ge \sigma^{n-1}(C_\delta)/2 \quad \text{for } \xi \in R_\delta.$$

The argument for g is exactly the same using that

$$|2\pi \xi \cdot (x - e_n)| < \pi/3 \quad \text{for } x \in D_\delta, \xi \in S_\delta. \qquad \square$$

We shall use the first part of this example to show the sharpness of the Stein–Tomas restriction theorem 19.4 and the second part to show the sharpness of Tao's bilinear restriction theorem 25.3.

3.10 Fourier series

Much of the theory of the Fourier transform has analogues for the Fourier series. We shall use one-dimensional Fourier series only twice, in connection with Cantor measures, Chapter 8, and Riesz products, Chapter 13. In the last chapter we shall also need higher dimensional Fourier series in the form of the Poisson summation formula. We now state a couple of fundamental results in general dimensions without proofs and then prove a few others. For the basics of the one-dimensional Fourier series, see, for example, Katznelson [1968], Chapter 1, and for the multi-dimensional theory Grafakos [2008], Chapter 3.

Let

$$Q_n = \{x \in \mathbb{R}^n : 0 \le x_j \le 1 \text{ for } j = 1, \dots, n\}$$

be the unique cube. For $\mu \in \mathcal{M}(Q_n)$ the *Fourier coefficients* of μ are

$$\widehat{\mu}(z) = \int_{Q_n} e^{-2\pi i z \cdot x} \, d\mu x, \quad z \in \mathbb{Z}^n.$$

Then we have again Parseval's formula for $f, g \in L^2(Q_n)$,

$$\sum_{z \in \mathbb{Z}^n} \widehat{f}(z)\overline{\widehat{g}(z)} = \int_{Q_n} f(x)\overline{g}(x) \, dx. \tag{3.64}$$

If f is continuous on Q_n and μ is a finite signed Borel measure on Q_n, this remains valid provided the series $\sum_{z \in \mathbb{Z}^n} \widehat{f}(z)\overline{\widehat{\mu}(z)}$ converges. Then

$$\sum_{z \in \mathbb{Z}^n} \widehat{f}(z)\overline{\widehat{\mu}(z)} = \int_{Q_n} f \, d\mu. \tag{3.65}$$

See Katznelson [1968], Section 1.7, for the one-dimensional case, which is all that we shall need for this fact. As a corollary we have that the Fourier coefficients determine uniquely measures on Q_n: if $\mu, \nu \in \mathcal{M}(Q_n)$, then

$$\widehat{\mu}(z) = \widehat{\nu}(z) \text{ for all } z \in \mathbb{Z}^n \text{ implies } \mu = \nu. \tag{3.66}$$

We have also the Fourier inversion formula: if $f \in L^1(Q_n)$ and $\sum_{z \in \mathbb{Z}^n} |\widehat{f}(z)| < \infty$, then f is continuous and

$$f(x) = \sum_{z \in \mathbb{Z}^n} \widehat{f}(z) e^{2\pi i z \cdot x} \quad \text{for } x \in Q_n. \tag{3.67}$$

More precisely, f can be redefined so that it becomes continuous and 1-periodic: $f(x) = f(y)$ for $x, y \in Q_n$ with $x_j - y_j = 0, 1$ or -1 for all $j = 1, \dots, n$. In fact, for the theory of Fourier series instead of Q_n it is more natural to use the torus $(S^1)^n$ as the underlying space, or consider 1-periodic functions on $\mathbb{R}^n$, but this will not be essential for us.

The Fourier inversion formula gives easily the *Poisson summation formula*:

Theorem 3.19 *If $f \in \mathcal{S}(\mathbb{R}^n)$, then*

$$\sum_{z \in \mathbb{Z}^n} f(x + z) = \sum_{z \in \mathbb{Z}^n} \widehat{f}(z) e^{2\pi i z \cdot x} \quad \text{for } x \in \mathbb{R}^n.$$

Proof Define the periodic function F by

$$F(x) = \sum_{z \in \mathbb{Z}^n} f(x + z).$$

Then for $z \in \mathbb{Z}^n$,

$$\widehat{F}(z) = \int_{Q_n} F(x) e^{-2\pi i z \cdot x} \, dx = \sum_{z \in \mathbb{Z}^n} \int_{Q_n - z} f(x) e^{-2\pi i z \cdot x} \, dx = \widehat{f}(z).$$

Hence $\sum_{z \in \mathbb{Z}^n} |\widehat{F}(z)| < \infty$ and the result follows from the inversion formula.
□

Corollary 3.20 *If $f \in \mathcal{S}(\mathbb{R}^n)$ and* spt $\widehat{f} \subset B(0, 1)$, *then*

$$\sum_{z \in \mathbb{Z}^n} f(x + z) = \int f \quad \text{for } x \in \mathbb{R}^n.$$

Proof We now have $\widehat{f}(0) = \int f$ and $\widehat{f}(z) = 0$ for $z \in \mathbb{Z}^n, z \neq 0$.
□

The s-energy of $\mu \in \mathcal{M}(Q_n)$ can essentially be written also in terms of the Fourier coefficients:

Theorem 3.21 *If $0 < s < n$ and $\mu \in \mathcal{M}(Q_n)$, then*

$$I_s(\mu)/C(n, s) \le \mu(\mathbb{R}^n)^2 + \sum_{z \in \mathbb{Z}^n \setminus \{0\}} |\widehat{\mu}(z)|^2 |z|^{s-n} \le C(n, s) I_s(\mu). \qquad (3.68)$$

Proof Since we are not going to use this formula, we shall only prove it in the special case where spt μ is contained in the interior of Q_n; for the general case, see Hare and Roginskaya [2002]. Then we may assume that μ is a smooth non-negative function with compact support in Q_n: let, as before, $\mu_\varepsilon = \psi_\varepsilon * \mu$ where $\psi_\varepsilon, \varepsilon > 0$, is a standard approximate identity. Then, as $\varepsilon \to 0$,

$$I_s(\mu_\varepsilon)$$
$$= \gamma(n, s) \int |\widehat{\psi}(\varepsilon x) \widehat{\mu}(x)|^2 |x|^{s-n} \, dx \to \gamma(n, s) \int |\widehat{\mu}(x)|^2 |x|^{s-n} \, dx = I_s(\mu)$$

by Theorem 3.10, and

$$\sum_{z \in \mathbb{Z}^n \setminus \{0\}} |\widehat{\mu_\varepsilon}(z)|^2 |z|^{s-n} = \sum_{z \in \mathbb{Z}^n \setminus \{0\}} |\widehat{\psi}(\varepsilon z) \widehat{\mu}(z)|^2 |z|^{s-n} \to \sum_{z \in \mathbb{Z}^n \setminus \{0\}} |\widehat{\mu}(z)|^2 |z|^{s-n}.$$

So let $f \in C^\infty(\mathbb{R}^n)$ with $f \ge 0$ and spt $f \subset \text{Int}(Q_n)$. Recalling Example 3.2 we choose $\varphi \in \mathcal{S}(\mathbb{R}^n)$ such that both φ and $\widehat{\varphi}$ are non-negative, $\varphi \ge 1$ on $Q_n - Q_n$ and spt $\widehat{\varphi} \subset B(0, 1/2)$. Then

$$I_s(f) \approx \int ((\varphi k_s) * f) f.$$

Set $g_\varepsilon = \psi_\varepsilon * (\varphi k_s)$. Then $g_\varepsilon \to \varphi k_s$ in $L^1(\mathbb{R}^n)$ as $\varepsilon \to 0$ which implies

$$\int (g_\varepsilon * f) f \to \int ((\varphi k_s) * f) f.$$

Using Lemma 3.9 we have

$$\widehat{g_\varepsilon}(z) = \gamma(n, s)\widehat{\psi}(\varepsilon z)\widehat{\varphi} * k_{n-s}(z) \to \gamma(n, s)\widehat{\varphi} * k_{n-s}(z) = \widehat{\varphi} * \widehat{k_s}(z).$$

By the properties of φ; $\widehat{\varphi} \geq 0$ and $\operatorname{spt}\widehat{\varphi} \subset B(0, 1/2)$, we have readily $\widehat{\varphi} * \widehat{k_s}(z) \approx |z|^{s-n}$, when $|z| \geq 1$, and $\widehat{\varphi} * \widehat{k_s}(0) \approx 1$. Therefore by Parseval's formula (3.64) for the Fourier series,

$$\int (g_\varepsilon * f)f = \sum_{z \in \mathbb{Z}^n} \widehat{g_\varepsilon}(z)|\widehat{f}(z)|^2 \to \sum_{z \in \mathbb{Z}^n} \widehat{\varphi} * \widehat{k_s}(z)|\widehat{f}(z)|^2 \approx |\widehat{f}(0)|^2$$
$$+ \sum_{z \in \mathbb{Z}^n \setminus \{0\}} |\widehat{f}(z)|^2 |z|^{s-n}.$$

Since $\widehat{f}(0) = \int f$, the combination of these formulas yields the theorem. □

3.11 Further comments

Duoandikoetxea's book [2001] and Strichartz's book [1994] are excellent first quick guides to Fourier analysis. Grafakos [2008] does the same and, combining with Grafakos [2009], gives a very wide view of Fourier analysis. The presentation of this chapter is largely based on Wolff's lecture notes [2003].

Bessel functions are extensively studied in Grafakos [2008], Stein and Weiss [1971] and Watson [1944].

In one dimension the expression of energy-integrals in terms of the Fourier transform and Fourier series and applications to Hausdorff dimension goes back at least to the works of Kahane and Salem, see Kahane and Salem [1963], and in higher dimensions to Carleson's [1967] book. For $n = 1$ the Fourier series formula (3.68) appears essentially already in the first volume of Zygmund's book [1959], page 70. In higher dimensions it was proved by Hare and Roginskaya [2002].

Hare and Roginskaya [2003] proved a formula analogous to (3.45) on Riemannian manifolds. In Hare and Roginskaya [2004] they studied energies of complex measures and their relations to Hausdorff dimension. Hare, Parasar and Roginskaya [2007] investigated energies with respect to more general kernels than the Riesz kernel k_s.

Salem sets in $\mathbb{R}^n$ can have any dimension $s \in [0, n]$. Salem [1951] was first to construct them in $\mathbb{R}$ in this generality as random Cantor sets. A related construction was given by Bluhm [1996]. We shall discuss other random Salem sets in Chapter 12. The first non-trivial deterministic fractal Salem set, but only with dimension 1 in $\mathbb{R}$, was found by Kahane [1970]. Kaufman's [1981] result,

Theorem 3.13, gave deterministic Salem sets in $\mathbb{R}$ with dimensions filling $(0, 1)$. A modification of Kaufman's construction was made by Bluhm [1998].

There is a rich literature on number theoretic sets, such as the set used by Kaufman, their Hausdorff dimensions and Fourier transforms of measures on them. This topic was pioneered by Jarnik and Besicovitch in the 1920s and 1930s. In particular, the Hausdorff dimension of the set E_α in Theorem 3.13 was found by Jarnik [1928] and [1931]. Dimension formulas for some other sets of this type can be found in Section 8.5 of Falconer [1985a], in Chapter 10 of Falconer [1990] and in Chapter 1 of Bishop and Peres [2016]. These books contain many references for the work done on this topic. Often these questions also have relationships to ergodic theory, see Jordan and Sahlsten [2013] for recent results and references.

Fourier dimension has not been much investigated systematically, but recently such a study was made by Ekström, Persson and Schmeling [2015]. They considered two definitions of the Fourier dimension: the one above and another one using Borel probability measures μ such that $\mu(A) = 1$ instead of $\mu \in \mathcal{M}(A)$. These two definitions do not always agree. Among other things they showed that for both definitions Fourier dimension is not finitely stable: $\max\{\dim_F A, \dim_F B\} \leq \dim_F(A \cup B)$ by the obvious monotonicity but the inequality may be strict; for the latter definition an example was given by Ekström [2014]. The above authors also defined the *modified Fourier dimension*

$$\dim_{MF} A = \sup\{s \leq n : \exists \mu \in \mathcal{M}(\mathbb{R}^n) \text{ such that }$$

$$\mu(A) > 0 \text{ and } |\widehat{\mu}(x)| \leq |x|^{-s/2} \; \forall x \in \mathbb{R}^n\},$$

and showed that it is countably stable.

Fourier transforms and series of measures and distributions on the real line and on the circle have deep connections to many other topics, such as number theory, complex analysis and operator theory. The books of Kahane and Salem [1963], Salem [1963], Travaglini [2014] and of Havin and Jöricke [1995] are good sources. Recent interesting papers are those of Poltoratski [2012] and Kozma and Olevskii [2013]. Measures whose Fourier transform tends to zero at infinity are called *Rajchman measures*. Lyons [1995] gives an excellent survey on them, concentrating on measures on the circle.

Lemma 3.17 is due to Erdoğan [2004].

4

Hausdorff dimension of projections
and distance sets

In this chapter we give the first applications of the Fourier transform to geometric problems on the Hausdorff dimension. We begin by considering orthogonal projections and prove Marstrand's projection theorem stating that almost all projections of a Borel set are as big as the dimension of the set allows. We shall prove this here only for the projections onto lines in order to bring forth the basic ideas in the simplest cases. In the next chapter we shall give various extensions of these results including projections onto m-dimensional planes in $\mathbb{R}^n$. Our second application will be on Falconer's problem on the size of the distance sets. We shall also prove that there are no Borel subrings of $\mathbb{R}$ with the Hausdorff dimension strictly between 0 and 1.

4.1 Projections

For $e \in S^{n-1}$, $n \geq 2$, define the projection

$$P_e : \mathbb{R}^n \to \mathbb{R}, \quad P_e(x) = e \cdot x.$$

This is essentially the orthogonal projection onto the line $\{te : t \in \mathbb{R}\}$. As P_e is Lipschitz,

$$\dim P_e(A) \leq \dim A \quad \text{for all } A \subset \mathbb{R}^n.$$

In the plane we shall often parametrize these projections with the angle the line makes with the positive x-axis and use the notation:

$$p_\theta : \mathbb{R}^2 \to \mathbb{R}, \quad p_\theta(x, y) = x \cos\theta + y \sin\theta, \theta \in [0, \pi).$$

Theorem 4.1 *Let $A \subset \mathbb{R}^n$ be a Borel set and $s = \dim A$. If $s \leq 1$, then*

$$\dim P_e(A) = s \quad \text{for } \sigma^{n-1} \text{ almost all } e \in S^{n-1}. \tag{4.1}$$

If $s > 1$, then

$$\mathcal{L}^1(P_e(A)) > 0 \quad \text{for } \sigma^{n-1} \text{ almost all } e \in S^{n-1}. \tag{4.2}$$

Proof If $\mu \in \mathcal{M}(A)$ and $e \in S^{n-1}$, the image $\mu_e = P_{e\sharp}\mu$ of μ under the projection P_e is defined by

$$\mu_e(B) = \mu(P_e^{-1}(B)), \quad B \subset \mathbb{R}.$$

Then $\mu_e \in \mathcal{M}(P_e(A))$ and

$$\widehat{\mu_e}(r) = \int_{-\infty}^{\infty} e^{-2\pi i r x} \, d\mu_e x = \int_{\mathbb{R}^n} e^{-2\pi i r(y \cdot e)} \, d\mu y = \widehat{\mu}(re) \tag{4.3}$$

for all $r \in \mathbb{R}$. To prove (4.1), suppose $0 < s = \dim A \leq 1$. Fix $0 < t < s$ and pick by Theorem 2.8 $\mu \in \mathcal{M}(A)$ such that $I_t(\mu) < \infty$. Using Theorem 3.10, (4.3) and (3.30) we obtain,

$$\int_{S^{n-1}} I_t(\mu_e) \, d\sigma^{n-1} e = \gamma(1, t) \int_{S^{n-1}} \left(\int_{-\infty}^{\infty} |\widehat{\mu_e}(r)|^2 r^{t-1} \, dr \right) d\sigma^{n-1} e$$

$$= 2\gamma(1, t) \int_{S^{n-1}} \left(\int_0^{\infty} |\widehat{\mu}(re)|^2 r^{t-1} \, dr \right) d\sigma^{n-1} e$$

$$= 2\gamma(1, t) \int_{\mathbb{R}^n} |\widehat{\mu}(x)|^2 |x|^{t-n} \, dx$$

$$= 2\gamma(1, t)\gamma(n, t)^{-1} I_t(\mu) < \infty.$$

In particular, $I_t(\mu_e) < \infty$ for σ^{n-1} almost all $e \in S^{n-1}$ and $\dim P_e(A) \geq t$ for such e. Considering a sequence (t_i), $t_i < s$, $t_i \to s$, we find that $\dim P_e(A) \geq s$ for almost all $e \in S^{n-1}$.

Suppose now that $s > 1$. Then there is $\mu \in \mathcal{M}(A)$ such that $I_1(\mu) < \infty$. Arguing as above with $t = 1$,

$$\int_{S^{n-1}} \left(\int_{-\infty}^{\infty} |\widehat{\mu_e}(r)|^2 \, dr \right) d\sigma^{n-1} e = 2\gamma(n, 1)^{-1} I_1(\mu) < \infty, \tag{4.4}$$

whence $\widehat{\mu_e} \in L^2(\mathbb{R})$ for σ^{n-1} almost all $e \in S^{n-1}$. Thus by Theorem 3.3, $\mu_e \in L^2(\mathbb{R})$ for σ^{n-1} almost all $e \in S^{n-1}$. In particular, μ_e is absolutely continuous with respect to $\mathcal{L}^1$ for σ^{n-1} almost all $e \in S^{n-1}$. As $\mu_e \in \mathcal{M}(P_e(A))$ we have $\mathcal{L}^1(P_e(A)) > 0$ for such e. $\qquad\square$

For a proof of the previous theorem without Fourier transforms, see Mattila [1995], Chapter 9.

Theorem 4.2 *Let $A \subset \mathbb{R}^n$ be a Borel set and $\dim A > 2$. Then $P_e(A)$ has non-empty interior for σ^{n-1} almost all $e \in S^{n-1}$.*

Proof Let $2 < s < \dim A$ and choose $\mu \in \mathcal{M}(A)$ such that $I_s(\mu) < \infty$. Defining μ_e as in the previous proof, we obtain by Schwartz's inequality

$$\int_{S^{n-1}} \int_{\mathbb{R}} |\widehat{\mu_e}(r)| \, dr \, d\sigma^{n-1} e$$

$$\leq 2 \int_{S^{n-1}} \int_1^{\infty} |\widehat{\mu_e}(r)| \, dr \, d\sigma^{n-1} e + 2\mu(\mathbb{R}^n)\sigma^{n-1}(S^{n-1})$$

$$\leq 2 \left(\int_{S^{n-1}} \int_1^{\infty} |\widehat{\mu}(re)|^2 r^{s-n+n-1} \, dr \, d\sigma^{n-1} e \right)^{1/2}$$

$$\times \left(\int_{S^{n-1}} \int_1^{\infty} r^{1-s} \, dr \, d\sigma^{n-1} \right)^{1/2} + C(\mu)$$

$$\leq 2 \left(\frac{\sigma^{n-1}(S^{n-1})}{s-2} \right)^{1/2} \left(\int_{\mathbb{R}^n} |\widehat{\mu}(x)|^2 |x|^{s-n} \, dx \right)^{1/2} + C(\mu)$$

$$\leq C(n, s) I_s(\mu)^{1/2} + C(\mu) < \infty.$$

Hence $\widehat{\mu_e} \in L^1(\mathbb{R})$ for σ^{n-1} almost all $e \in S^{n-1}$ and by Theorem 3.4 μ_e is a continuous function for such e. As $\mu_e \in \mathcal{M}(P_e(A))$, we conclude that the interior of $P_e(A)$ is non-empty for σ^{n-1} almost all $e \in S^{n-1}$. □

I do not know any proof without Fourier transforms for this theorem, although I am not sure if anyone has seriously tried to find one. The bound 2 is sharp: using Besicovitch sets we shall give in Chapter 11 an example of a Borel set in the plane whose complement has Lebesgue measure zero and all of whose projections have empty interior.

Let us derive as another consequence of the proof of Theorem 4.1 a quantitative estimate for the average length of projections:

Theorem 4.3 *Let $A \subset \mathbb{R}^n$ be Lebesgue measurable and let $\mu \in \mathcal{M}(A)$ with $\mu(A) = 1$ and $I_1(\mu) < \infty$. Then*

$$\int \mathcal{L}^1(P_e(A)) \, d\sigma^{n-1} e \geq \frac{\gamma(n, 1)\sigma^{n-1}(S^{n-1})^2}{2 I_1(\mu)}.$$

Proof The measurability of the function $e \mapsto \mathcal{L}^1(P_e(A))$ is easily checked for compact sets A and from that it follows for measurable sets by approximation. From the formula (4.4) we see that for σ^{n-1} almost all $e \in S^{n-1}$ the projection $\mu_e = P_{e\sharp}\mu$ is absolutely continuous and, using Parseval's theorem, it moreover belongs to $L^2(\mathbb{R})$ with

$$\int_{S^{n-1}} \left(\int_{-\infty}^{\infty} \mu_e(r)^2 \, dr \right) d\sigma^{n-1} e = 2\gamma(n, 1)^{-1} I_1(\mu).$$

By Schwartz's inequality,

$$1 = P_{e\sharp}\mu(\mathbb{R})^2 = \left(\int_{P_e(A)} \mu_e d\mathcal{L}^1\right)^2 \leq \mathcal{L}^1(P_e(A)) \int \mu_e^2 d\mathcal{L}^1.$$

A combination of these two inequalities gives

$$\int \mathcal{L}^1(P_e(A))^{-1} d\sigma^{n-1}e \leq \iint \mu_e^2 d\mathcal{L}^1 d\sigma^{n-1}e = 2\gamma(n,1)^{-1} I_1(\mu).$$

Thus by Schwartz's inequality,

$$\int \mathcal{L}^1(P_e(A)) d\sigma^{n-1}e \geq \left(\int \mathcal{L}^1(P_e(A))^{-1} d\sigma^{n-1}e\right)^{-1} \sigma^{n-1}(S^{n-1})^2$$

$$\geq \tfrac{1}{2}\gamma(n,1)\sigma^{n-1}(S^{n-1})^2 I_1(\mu)^{-1}. \qquad \square$$

4.2 Distance sets

Now we study another geometric problem on Hausdorff dimension, estimating the size of distance sets. The *distance set* of $A \subset \mathbb{R}^n$ is

$$D(A) = \{|x - y| : x, y \in A\} \subset [0, \infty).$$

The following Falconer's conjecture seems plausible:

Conjecture 4.4   If $n \geq 2$ and $A \subset \mathbb{R}^n$ is a Borel set with $\dim A > n/2$, then $\mathcal{L}^1(D(A)) > 0$, or even $\mathrm{Int}(D(A)) \neq \varnothing$.

This is open in all dimensions $n \geq 2$. In $\mathbb{R}$ it is false; it is easy to construct examples of compact sets $A \subset \mathbb{R}$ with $\dim A = 1$ and $\mathcal{L}^1(D(A)) = 0$. Below we shall give an example to show that $n/2$ could not be replaced by any smaller number.

A weaker conjecture on dimensional level is

Conjecture 4.5   If $n \geq 2$ and $A \subset \mathbb{R}^n$ is a Borel set with $\dim A > n/2$, then $\dim D(A) = 1$.

This too is open in all dimensions $n \geq 2$. But it is true for example for many self-similar sets. Theorem 4.6 below only gives $\dim D(A) \geq 1/2$ if $\dim A \geq n/2$. This has been improved by Bourgain to $\dim D(A) \geq 1/2 + c_n, c_n > 0$. We shall briefly discuss these partial results in Section 4.4.

Steinhaus's theorem, a simple application of Lebesgue's density theorem, says that if $A \subset \mathbb{R}^n$ is Lebesgue measurable with $\mathcal{L}^n(A) > 0$, then the difference set $\{x - y : x, y \in A\}$ contains a ball centred at the origin. It is easy to give examples which show that no such statement holds, neither for the difference set nor for the distance set, under the assumption that $\dim A$ is large. In Theorem 4.6 there may be no interval $(0, \varepsilon)$, $\varepsilon > 0$, inside $D(A)$ even if $\dim A = n$.

First we shall prove some weaker partial results. In Chapters 15 and 16 we shall prove the best known results.

Theorem 4.6 *Let $A \subset \mathbb{R}^n$, $n \geq 2$, be a Borel set.*

(a) If $\dim A > (n + 1)/2$, then $\mathrm{Int}(D(A)) \neq \varnothing$.
(b) If $(n - 1)/2 \leq \dim A \leq (n + 1)/2$, then $\dim D(A) \geq \dim A - (n - 1)/2$.

We use a similar technique as with the projections; we map a measure $\mu \in \mathcal{M}(A)$ to its *distance measure* $\delta(\mu) \in \mathcal{M}(D(A))$ defined for Borel sets $B \subset \mathbb{R}$ by

$$\delta(\mu)(B) = \int \mu(\{y : |x - y| \in B\}) \, d\mu x. \tag{4.5}$$

In other words, $\delta(\mu)$ is the image of $\mu \times \mu$ under the distance map $(x, y) \to |x - y|$, or equivalently, for any continuous function φ on $\mathbb{R}$,

$$\int \varphi d\delta(\mu) = \iint \varphi(|x - y|) \, d\mu x \, d\mu y.$$

Let us first see some simple properties of distance measures. Obviously,

$$\mathrm{spt}\, \delta(\mu) \subset D(\mathrm{spt}\, \mu). \tag{4.6}$$

Another simple observation is that

$$\delta(\mu_i) \to \delta(\mu) \quad \text{weakly if} \quad \mu_i \to \mu \text{ weakly.} \tag{4.7}$$

Recall that σ_r^{n-1} is the surface measure on the sphere $\{y \in \mathbb{R}^n : |y| = r\}$. Its Fourier transform is (recall (3.42))

$$\widehat{\sigma_r^{n-1}}(x) = r^{n-1}\widehat{\sigma^{n-1}}(rx) \quad \text{with} \quad |\widehat{\sigma_r^{n-1}}(x)| \lesssim r^{(n-1)/2}|x|^{(1-n)/2}.$$

For a smooth function f with compact support, $\delta(f)$ is also a function. It is given by

$$\delta(f)(r) = \int (\sigma_r^{n-1} * f)f. \tag{4.8}$$

To prove this one can check by Fubini's theorem and integration in polar coordinates that for any continuous function g with compact support in $\mathbb{R}$,

$$\int g(r) \int (\sigma_r^{n-1} * f)(x) f(x) \, dx \, dr = \iint g(|x - y|) f(x) f(y) \, dx \, dy,$$

which is also $\int g\delta(f)$ by the definition of $\delta(f)$.

Let ψ be a smooth function with compact support in $\mathbb{R}^n$, $\int \psi = 1$, $\psi_\varepsilon(x) = \varepsilon^{-n}\psi(x/\varepsilon)$ and $\mu_\varepsilon = \psi_\varepsilon * \mu$. Then $\mu_\varepsilon \to \mu$ weakly, as $\varepsilon \to 0$, whence $\delta(\mu_\varepsilon) \to \delta(\mu)$ weakly. Moreover $\widehat{\mu_\varepsilon}(x) = \widehat{\psi}(\varepsilon x)\widehat{\mu}(x) \to \widehat{\mu}(x)$ for all $x \in \mathbb{R}^n$.

We have now by (4.8) and Parseval's formula,

$$\delta(\mu_\varepsilon)(r) = \int (\sigma_r^{n-1} * \mu_\varepsilon)\mu_\varepsilon = \int \widehat{\sigma_r^{n-1}}|\widehat{\mu_\varepsilon}|^2 = \int \widehat{\sigma_r^{n-1}}(x)|\widehat{\psi}(\varepsilon x)|^2|\widehat{\mu}(x)|^2 dx.$$

(4.9)

Suppose then, recalling (3.45), that

$$I_{(n+1)/2}(\mu) = \gamma(n, (n+1)/2) \int |x|^{(1-n)/2}|\widehat{\mu}(x)|^2 \, dx < \infty.$$

Then, as $\varepsilon \to 0$, the right hand side of (4.9) converges to $\int \widehat{\sigma_r^{n-1}}|\widehat{\mu}|^2$ by Lebesgue's dominated convergence theorem, because

$$|\widehat{\sigma_r^{n-1}}(x)||\widehat{\psi}(\varepsilon x)|^2|\widehat{\mu}(x)|^2 \lesssim_r |x|^{(1-n)/2}|\widehat{\mu}(x)|^2.$$

On the other hand, the left hand side of (4.9) converges weakly to $\delta(\mu)$. So if $I_{(n+1)/2}(\mu) < \infty$, $\delta(\mu)$ is a function given by

$$\delta(\mu)(r) = \int \widehat{\sigma_r^{n-1}}|\widehat{\mu}|^2 = r^{n-1} \int \widehat{\sigma^{n-1}}(rx)|\widehat{\mu}(x)|^2 \, dx.$$

(4.10)

This is all that is needed to prove the first part of Theorem 4.6:

Proof of Theorem 4.6(a) If $\dim A > (n+1)/2$ we can find a measure $\mu \in \mathcal{M}(A)$ with $I_{(n+1)/2}(\mu) < \infty$ by Theorem 2.8. Then $\delta(\mu)$ is the function given by (4.10) which is easily seen to be continuous by Lebesgue's dominated convergence theorem. As $\operatorname{spt}\delta(\mu) \subset D(A)$ by (4.6), it follows that $D(A)$ has non-empty interior.   □

For the second part of Theorem 4.6 we need some estimate of the $\delta(\mu)$-measure of the intervals $[r, r + \eta]$. Let $R > 0$ be such that $\operatorname{spt} \mu \subset B(0, R)$. Then $\operatorname{spt}\delta(\mu) \subset [0, 2R]$. Let $0 < \eta < r < 2R$. By the definition of $\delta(\mu)$,

$$\delta(\mu)([r, r + \eta]) = \int \mu(\{y \in \mathbb{R}^n : r \le |x - y| \le r + \eta\})d\mu x = \int g_{r,\eta} * \mu d\mu,$$

where $g_{r,\eta}$ is the characteristic function of the annulus $\{x \in \mathbb{R}^n : r \le |x| \le r + \eta\}$.

Letting μ_ε be as above, we have by Parseval's formula

$$\delta(\mu_\varepsilon)([r, r + \eta]) = \int (g_{r,\eta} * \mu_\varepsilon)\mu_\varepsilon = \int \widehat{g_{r,\eta}}|\widehat{\mu_\varepsilon}|^2.$$

Letting $\varepsilon \to 0$, we find that

$$\delta(\mu)([r, r + \eta]) = \int \widehat{g_{r,\eta}}|\widehat{\mu}|^2.$$

(4.11)

In fact, we first get this by the weak convergence for all but at most countably many r and η, for those with $\delta(\mu)(\{r, r + \eta\}) = 0$, but since the right hand side is continuous in r and η, this holds for all r and η. Since $g_{r,\eta}$ is radial, we have by (3.33),

$$
\begin{aligned}
\widehat{g_{r,\eta}}(x) &= c(n)|x|^{-(n-2)/2} \int_r^{r+\eta} J_{(n-2)/2}(2\pi|x|s)s^{n/2}\,ds \\
&= c(n)|x|^{-n} \int_{r|x|}^{(r+\eta)|x|} J_{(n-2)/2}(2\pi u)u^{n/2}\,du.
\end{aligned}
\tag{4.12}
$$

This gives by (3.35)

$$
|\widehat{g_{r,\eta}}(x)| \lesssim |x|^{-n} \int_{r|x|}^{(r+\eta)|x|} u^{(n-1)/2}\,du \lesssim r^{(n-1)/2}|x|^{(1-n)/2}\eta.
\tag{4.13}
$$

To get another estimate we use the formula (3.39),

$$
\frac{d}{ds}(s^m J_m(s)) = s^m J_{m-1}(s)
$$

and again (3.35) getting

$$
\begin{aligned}
|\widehat{g_{r,\eta}}(x)| &= \left| c(n)(2\pi)^{-1}|x|^{-n} \int_{r|x|}^{(r+\eta)|x|} \frac{d}{du}(u^{n/2}J_{n/2}(2\pi u))\,du \right| \\
&= |c(n)(2\pi)^{-1}|x|^{-n}|((r+\eta)|x|)^{n/2}J_{n/2}(2\pi(r+\eta)|x|) \\
&\quad - (r|x|)^{n/2}J_{n/2}(2\pi r|x|)| \\
&\lesssim r^{(n-1)/2}|x|^{-(n+1)/2}.
\end{aligned}
$$

Using (4.11), these two estimates for $|\widehat{g_{r,\eta}}(x)|$ and (3.45) we obtain for $0 < t \le 1$,

$$
\begin{aligned}
\delta(\mu)([r, r+\eta]) &\lesssim r^{(n-1)/2}\eta \int_{\{x:|x|\le 1/\eta\}} |x|^{(1-n)/2}|\widehat{\mu}(x)|^2\,dx \\
&\quad + r^{(n-1)/2} \int_{\{x:|x|>1/\eta\}} |x|^{-(n+1)/2}|\widehat{\mu}(x)|^2\,dx \\
&\le r^{(n-1)/2}\eta^t \left(\int_{\{x:|x|\le 1/\eta\}} |x|^{(1-n)/2+t-1}|\widehat{\mu}(x)|^2\,dx \right. \\
&\qquad\qquad \left. + \int_{\{x:|x|>1/\eta\}} |x|^{-(n+1)/2+t}|\widehat{\mu}(x)|^2\,dx \right) \\
&= r^{(n-1)/2}\eta^t \int |x|^{(n-1)/2+t-n}|\widehat{\mu}(x)|^2\,dx \\
&= \gamma(n, (n-1)/2+t)^{-1} r^{(n-1)/2}\eta^t I_{(n-1)/2+t}(\mu).
\end{aligned}
$$

Proof of Theorem 4.6(b) We may assume that $\dim A > (n-1)/2$. Let $0 \le t \le 1$ be such that $(n-1)/2 + t < \dim A$. Then we can find a measure $\mu \in \mathcal{M}(A)$ with $I_{(n-1)/2+t}(\mu) < \infty$. The above estimate yields that $\delta(\mu)([r, r+\eta]) \lesssim_R \eta^t$

when $0 < \eta < r < 2R$. By a slight modification of the easy part of Frostman's lemma 2.7 this implies that $\mathcal{H}^t(D(A)) > 0$ and completes the proof.   □

For later use we derive the following consequence of the above arguments:

Lemma 4.7 *If $s \geq (n+1)/2$ and $\mu \in \mathcal{M}(\mathbb{R}^n)$ with $I_s(\mu) < \infty$, then for all $0 < \eta < r$,*

$$\mu \times \mu(\{(x, y) : r \leq |x - y| \leq r + \eta\}) \leq C(n, s)I_s(\mu)\eta r^{s-1}. \tag{4.14}$$

Moreover,

$$\|\delta(\mu)\|_\infty \leq C(n, s)d(\text{spt }\mu)^{s-1}I_s(\mu). \tag{4.15}$$

Proof Let $g_{r,\eta}$ be again the characteristic function of the annulus $\{x \in \mathbb{R}^n : r \leq |x| \leq r + \eta\}$. If $x \in \mathbb{R}^n$ and $r|x| \geq 1$, we have by (4.13),

$$|\widehat{g_{r,\eta}}(x)| \lesssim \eta r^{(n-1)/2}|x|^{(1-n)/2} = \eta|rx|^{(n+1)/2-s}r^{s-1}|x|^{s-n} \leq \eta r^{s-1}|x|^{s-n}.$$

If $r|x| \leq 1$, we have by (4.12) and (3.34),

$$|\widehat{g_{r,\eta}}(x)| = |c(n)|x|^{-n} \int_{r|x|}^{(r+\eta)|x|} J_{(n-2)/2}(2\pi u)u^{n/2}\,du|$$

$$\lesssim |x|^{-n}(r|x|)^{n-1}\eta|x| = \eta(r|x|)^{n-s}r^{s-1}|x|^{s-n} \leq \eta r^{s-1}|x|^{s-n}.$$

Using these inequalities and (4.11), we obtain

$$\delta(\mu)([r, r + \eta]) = \int \widehat{g_{r,\eta}}|\widehat{\mu}|^2 \lesssim \eta r^{s-1} \int |x|^{s-n}|\widehat{\mu}(x)|^2 = \gamma(n, s)^{-1}I_s(\mu)\eta r^{s-1},$$

which is (4.14). (4.15) follows immediately from (4.14).   □

No proof without Fourier transforms is known for Theorem 4.6. It is not known if the bound $(n + 1)/2$ is the best possible in order for $D(A)$ to have non-empty interior. It is not sharp in order for $D(A)$ to have positive Lebesgue measure; we shall discuss some improvements later. Recall Conjecture 4.4 saying that dim $A > n/2$ should be enough for $\mathcal{L}^1(D(A)) > 0$, and perhaps also for non-empty interior. This would be the best possible. We now show this by an example.

Example 4.8 For $n \geq 2$ and $0 < s < n/2$ there exists a compact set $C \subset \mathbb{R}^n$ with dim $C = s$ and dim $D(C) \leq 2s/n$.

Proof To find such a set we could start by trying to find large finite sets with few distances, that is, many distances realized by many pairs of points. Subsets of scaled copies of the integer lattice have this property. The example is obtained with a Cantor type construction using cubes centred at such sets.

More precisely, let (m_k) be a rapidly increasing sequence of positive integers, say $m_{k+1} > m_k^k$, and define

$$C_k = \{x \in \mathbb{R}^n : 0 \le x_j \le 1, |x_j - p_j/m_k| \le m_k^{-n/s}$$
$$\text{for some integers } p_j \text{ and for } j = 1, \ldots, n\},$$

$$C = \bigcap_{k=1}^{\infty} C_k.$$

Then dim $C = s$; this can be checked by for example modifying the method that is used for the Cantor sets C_d in Chapter 8, or one can consult Falconer [1985a], Theorem 8.15. Clearly, $D(C) \subset \bigcap_{k=1}^{\infty} D(C_k)$. Let $d \in D(C_k), d > 0$, say $d = |x - x'|$ with integers $p_j, p'_j, j = 1, \ldots, n$, satisfying $|x_j - p_j/m_k| \le m_k^{-n/s}$ and $|x'_j - p'_j/m_k| \le m_k^{-n/s}$. Then, with $p = (p_1, \ldots, p_n)$ and $p' = (p'_1, \ldots, p'_n)$,

$$|p/m_k - p'/m_k| - 2nm_k^{-n/s} \le d \le |p/m_k - p'/m_k| + 2nm_k^{-n/s}.$$

Here $|m_k^{-1} p| \le 2n$ and $|m_k^{-1} p'| \le 2n$, so $|p - p'|^2$ is an integer at most $16n^2 m_k^2$. It follows that $D(C_k)$ is covered with at most $16n^2 m_k^2$ intervals $I_{k,i}$ of length $4nm_k^{-n/s}$, whence

$$\mathcal{H}^{2s/n}(D(C)) \le \liminf_{k \to \infty} \sum_i d(I_{k,i})^{2s/n}$$
$$\le \liminf_{k \to \infty} 16n^2 m_k^2 (4nm_k^{-n/s})^{2s/n} = 16n^2 (4n)^{2s/n},$$

which gives dim $D(C) \le 2s/n$. $\qquad\square$

It is not difficult to modify the above construction to get dim $C = n/2$ and $\mathcal{L}^1(D(C)) = 0$.

Our second example shows that, at least in the plane, we need $s \ge (n + 1)/2$ in order that $I_s(\mu) < \infty$ would imply $\delta(\mu) \in L^\infty(\mathbb{R})$ as in Lemma 4.7:

Example 4.9 For any $0 < s < 3/2$ there exists $\mu \in \mathcal{M}(\mathbb{R}^2)$ such that $I_s(\mu) < \infty$ and $\delta(\mu) \notin L^\infty(\mathbb{R})$.

Proof We may assume $s > 1$. Let $s - 1 < t < 1/2$ and let $\nu \in \mathcal{M}(\mathbb{R}), C = \operatorname{spt} \nu$, be such that with some positive constants a and b,

$$ar^t \le \nu([x - r, x + r]) \le br^t \quad \text{for } x \in C, \quad 0 < r < 1.$$

For example μ_d in Chapter 8 with $t = \log 2/\log(1/d)$ is fine. Consider the measure λ which is obtained essentially summing ν and its translate by 1:

$$\lambda(A) = \nu(A \cup (A - 1)), \quad A \subset \mathbb{R}.$$

Let μ be the product measure of λ and Lebesgue measure on the unit interval:

$$\mu = \lambda \times (\mathcal{L}^1 \llcorner [0, 1]).$$

Then $\mu \in \mathcal{M}(\mathbb{R}^2)$ with $I_s(\mu) < \infty$ and spt $\mu = F := (C \cup (C + 1)) \times [0, 1]$.

Let $x = (x_1, x_2) \in F$. By simple geometry we see that for small $\delta > 0$ the annulus $\{y : 1 - \delta < |x - y| < 1 + \delta\}$ contains a rectangle $I \times J$ where the interval I has length δ and centre (either $x_1 - 1$ or $x_1 + 1$) in $C \cup (C + 1)$ and J has length $c\sqrt{\delta}$ for some absolute positive constant c. Hence

$$\mu(\{y : 1 - \delta < |x - y| < 1 + \delta\}) \geq ac\delta^{t+1/2},$$

and so

$$\delta(\mu)((1 - \delta, 1 + \delta)) \geq ac\mu(F)\delta^{t+1/2}.$$

Since $t < 1/2$ and this holds for arbitrarily small δ, $\delta(\mu)$ cannot have a bounded Radon–Nikodym derivative. $\qquad\square$

4.3 Dimension of Borel rings

As an application of the projection theorems we prove here that there are no Borel subrings of $\mathbb{R}$ having a Hausdorff dimension strictly between 0 and 1:

Theorem 4.10 *Let $E \subset \mathbb{R}$ be a Borel set which is also an algebraic subring of $\mathbb{R}$. Then either E has Hausdorff dimension zero or $E = \mathbb{R}$.*

Proof The proof is based on the study of the effect of linear functionals $\varphi : \mathbb{R}^k \to \mathbb{R}$ on the k-fold Cartesian product E^k. Suppose that $\dim E > 0$. Then by the basic inequality for product sets, Theorem 2.10, we have $\dim E^k \geq k \dim E$, so that we can choose k for which $\dim E^k > 2$. Then by Theorem 4.2 there is a linear functional $\varphi : \mathbb{R}^k \to \mathbb{R}$, say $\varphi = P_e$ for some $e \in \mathcal{S}^{k-1}$, such that $\varphi(E^k)$ has non-empty interior and so, as $\varphi(E^k)$ is a subgroup of $\mathbb{R}$, $\varphi(E^k) = \mathbb{R}$.

Lemma 4.11 *Let $E \subset \mathbb{R}$ be a subring. Assume that there is a positive integer k and a linear functional $\varphi : \mathbb{R}^k \to \mathbb{R}$ such that $\varphi(E^k) = \mathbb{R}$. Then such k and φ may be found so that φ maps E^k bijectively onto $\mathbb{R}$.*

Proof Let k be the least positive integer such that there is a linear functional $\varphi : \mathbb{R}^k \to \mathbb{R}$ with $\varphi(E^k) = \mathbb{R}$. We claim that φ is injective on E^k. Let $\{e_1, \ldots, e_k\}$ be the standard basis of $\mathbb{R}^k$ and write $r_j = \varphi(e_j)$. Now $\varphi(E^k) = \mathbb{R}$ implies that

$$\left\{ \sum_{j=1}^{k} a_j r_j : a_1, \ldots, a_k \in E \right\} = \mathbb{R}. \tag{4.16}$$

Assume that φ is not injective on E^k. Then there are $b_1, \ldots, b_k \in E$, not all zero, so that $\sum_{j=1}^{k} b_j r_j = 0$. We may assume that $b_k \neq 0$, so

$$r_k = \sum_{j=1}^{k-1} \frac{-b_j}{b_k} r_j.$$

Let $s \in \mathbb{R}$. Then $s/b_k \in \mathbb{R}$ and by (4.16) there exist $a_1, \ldots, a_k \in E$ such that $s/b_k = \sum_{j=1}^{k} a_j r_j$. Therefore

$$s = \sum_{j=1}^{k-1} b_k a_j r_j + b_k a_k \sum_{j=1}^{k-1} \frac{-b_j}{b_k} r_j = \sum_{j=1}^{k-1} (b_k a_j - a_k b_j) r_j.$$

This implies that

$$\left\{ \sum_{j=1}^{k-1} a_j r_j : a_1, \ldots, a_{k-1} \in E \right\} = \mathbb{R}.$$

So restricting φ to the first $k-1$ coordinates we have a linear functional $\mathbb{R}^{k-1} \to \mathbb{R}$ which maps E^{k-1} onto all of $\mathbb{R}$. This contradicts the minimality of k and proves that φ is injective on E^k. $\qquad\square$

Lemma 4.12 *Let $E \subset \mathbb{R}$ be a subring and a Borel set. Let k be a positive integer and $\varphi : \mathbb{R}^k \to \mathbb{R}$ a linear functional such that φ maps E^k bijectively onto $\mathbb{R}$. Then $k = 1$ and $E = \mathbb{R}$.*

Proof Let $\psi : \mathbb{R} \to E^k$ be the inverse of the restriction of φ to E^k. Since $\varphi : \mathbb{R}^k \to \mathbb{R}$ is continuous and one-to-one on E^k, it maps Borel subsets of E^k onto Borel sets by a standard result on Borel sets, see, e.g., Federer [1969], p. 67, or Bruckner, Bruckner and Thomson [1997], Theorem 11.12. Thus ψ is a Borel measurable group homomorphism. Let $\{e_1, \ldots, e_k\}$ be the standard basis of $\mathbb{R}^k$ and write $r_j = \varphi(e_j)$. Let $\pi_1 : \mathbb{R}^k \to \mathbb{R}$ be the projection onto the first coordinate. Then $\tau = \pi_1 \circ \psi$ maps $\mathbb{R} \to \mathbb{R}$, $\tau(x + y) = \tau(x) + \tau(y)$ for all $x, y \in \mathbb{R}$ and τ is Borel measurable. Therefore there is a constant c such that $\tau(x) = cx$ for all $x \in \mathbb{R}$. This can be seen as follows. The equality $\tau(x + y) = \tau(x) + \tau(y)$ for all $x, y \in \mathbb{R}$ immediately yields $\tau(q) = \tau(1)q$ for all $q \in \mathbb{Q}$. If we show that τ is continuous, it follows that $\tau(x) = cx$ for all $x \in \mathbb{R}$ with $c = \tau(1)$. It is enough to show that τ is continuous at 0. Let $\varepsilon > 0$. Since

$$\bigcup_{q \in \mathbb{Q}} \tau^{-1} B(q, \varepsilon/2) = \mathbb{R},$$

there is $q_0 \in \mathbb{Q}$ for which $\mathcal{L}^1(\tau^{-1} B(q_0, \varepsilon/2)) > 0$. Then by Steinhaus's theorem there exists $\delta > 0$ such that

$$
\begin{aligned}
B(0, \delta) &\subset \tau^{-1} B(q_0, \varepsilon/2) - \tau^{-1} B(q_0, \varepsilon/2) \\
&\subset \tau^{-1} (B(q_0, \varepsilon/2) - B(q_0, \varepsilon/2)) \\
&= \tau^{-1} B(0, \varepsilon),
\end{aligned}
$$

which shows that τ is continuous. Now $\tau(r_1) = 1$, so $c \neq 0$. But if $k > 1$, there would be $r_2 \neq 0$ with $\tau(r_2) = 0$, which is a contradiction. Therefore $k = 1$, so the linear functional $\varphi : \mathbb{R} \to \mathbb{R}$ has the form $\varphi(x) = ax$ for some constant a. Since φ maps E onto all of $\mathbb{R}$, we have $E = \mathbb{R}$. $\qquad\square$

The proof of Theorem 4.10 now follows combining Lemmas 4.11 and 4.12 and the observation preceding them. $\qquad\square$

Instead of using Theorem 4.2 we could have used Steinhaus's theorem and part (4.2) of Theorem 4.1, whose proof does not require Fourier transforms.

4.4 Further comments

Dimensions of projections have been studied actively from many perspectives. Recent surveys are given by Falconer, Fraser and Jin [2014] and Mattila [2014].

Theorem 4.1 is due to Marstrand [1954]. Kaufman [1968] gave a simple potential-theoretic proof, in particular the Fourier analytic argument for the second part is due to him. Theorem 4.2 was found independently by Falconer and O'Neil [1999] and by Peres and Schlag [2000]. Lima and Moreira [2011] gave a combinatorial proof of Marstrand's theorem and discussed its significance for dynamical systems. Much of this stems from the fact that the sum set $A + B$ is essentially the projection of the product set $A \times B$ and sum sets and their dimensions play an important role in dynamical systems. Thus results on the dimensions of the projections of product sets have a particular interest; see, for example, Peres and Shmerkin [2009], and Hochman and Shmerkin [2012].

The dimension preservation which holds for the Hausdorff dimension fails for the Minkowski and packing dimensions. However sharp inequalities and other related results have been proven by Falconer and Howroyd [1996], [1997], M. Järvenpää [1994] and Falconer and Mattila [1996]. For projection theorems in infinite dimensional Banach spaces, see Ott, Hunt and Kaloshin [2006] and the references given there.

Often it is difficult to determine the dimension of the projections in all directions, or in some specified directions. In Chapter 10 we shall discuss this problem in light of a particular example. However, for many self-similar and random constructions one can get such more precise information; see, for example, Falconer and Jin [2014a], Rams and Simon [2014a], [2014b], Simon and Vágó [2014], and Peres and Rams [2014].

Theorem 4.3 along with its higher dimensional versions was proved in Mattila [1990]. The constant in it is sharp at least when $n = 2$ and $n = 3$ with equality when A is a ball and μ is the normalized equilibrium measure for the capacity related to the Riesz kernel k_1. The formula for $\gamma(n, 1)$ was given in (3.44). In particular the sharp constant of Theorem 4.3 is π^2 when $n = 2$. For further discussions on the isoperimetric type questions related to average projections and capacities, see Mattila [1990], [1995], Remarks 9.11, and [2004], and with connections to stochastic processes, Betsakos [2004] and Banuelos and Méndez-Hernández [2010].

Theorem 4.6 was proved by Falconer [1985b] but with Int($D(A)$) $\neq \varnothing$ replaced by $\mathcal{L}^1(D(A)) > 0$. Falconer also gave the example 4.8, including the case dim $A = n/2$ and $\mathcal{L}^1(D(A)) = 0$. In Chapter 15 we shall see that $\mathcal{L}^1(D(A)) > 0$ already follows from dim $A > n/2 + 1/3$. The existence of interior points with Hölder continuity and smoothness estimates for the distance measures was obtained by Mattila and Sjölin [1999].

Mitsis [2002a] improved Theorem 4.6 in the range $n = 2$, $1/2 < \dim A < 1$, from the dimension statement to the following measure statement: if $1/2 < s < 1$ and $A \subset \mathbb{R}^2$ is a Borel set with $\mathcal{H}^s(A) > 0$, then $\mathcal{H}^{s-1/2}(D(A)) > 0$. The proof is rather simple and does not use Fourier transforms. To my knowledge this is the only general result of this type not relying on Fourier transforms.

The above proof for the first part of the Theorem 4.6 uses only the decay estimate $|\widehat{\sigma^{n-1}}(x)| \lesssim |x|^{(1-n)/2}$ for the Fourier transform of the surface measure on the unit sphere. So the proof and the result hold for any norm with such a decay property. For this it is enough that this surface has non-vanishing Gaussian curvature, as we shall discuss later. The problem in this generality was studied by Iosevich, Mourgoglou and Taylor [2012]. They also derived Hölder continuity and smoothness estimates for the corresponding distance measures. The proof of the second part uses more explicitly the Euclidean sphere in terms of Bessel functions. Perhaps this part could also be generalized by studying the derivatives of the respective Fourier transform.

Greenleaf, Iosevich, Liu and Palsson [2013] gave a proof for Falconer's theorem; dim $A > (n + 1)/2$ implies $\mathcal{L}^1(D(A)) > 0$, without using the decay properties of $\widehat{\sigma^{n-1}}$. Instead they used the rotational symmetry of the problem in the sense that for any $x_1, x_2, y_1, y_2 \in \mathbb{R}^n$ with $|x_1 - x_2| = |y_1 - y_2|$ there corresponds $g \in O(n)$ for which $x_1 - x_2 = g(y_1 - y_2)$. If $n > 2$, there are many such rotations but they form a lower dimensional submanifold of $O(n)$. The relation to distance sets comes from the following. If $\mu \in \mathcal{M}(\mathbb{R}^n)$, define for $g \in O(n)$ the measure ν_g by

$$\int f \, d\nu_g = \iint f(x - g(y)) \, d\mu x \, d\mu y.$$

That is, ν_g is the image of $\mu \times \mu$ under the map $(x, y) \mapsto x - g(y)$. Greenleaf, Iosevich, Liu and Palsson showed that

$$\int_{O(n)} \int_{\mathbb{R}^n} \nu_g^2 \, d\theta_n g \approx \int_{\mathbb{R}^n} \delta(\mu)^2,$$

assuming that the measures in question are absolutely continuous. The Fourier transform of ν_g is $\widehat{\nu_g}(x) = \widehat{\mu}(x)\overline{\widehat{\mu}(g^{-1}(x))}$. Using the easy indentity

$$\int |\widehat{\mu}(x)|^2 \int |\widehat{\mu}(g^{-1}(x))|^2 \, d\theta_n g d \, x = \int_0^\infty \sigma(\mu)(r)^2 r^{n-1} \, dr / \sigma^{n-1}(S^{n-1})$$

and the easy estimate of Proposition 15.8, this leads to the proof of Falconer's theorem (cf. also the proof of Lemma 7.1).

The example 4.9 in $\mathbb{R}^2$ showing that for $\delta(\mu) \in L^\infty$ one needs $I_s(\mu) < \infty$ with $s \geq (n + 1)/2$ was given in Mattila [1985]. Iosevich and Senger [2010] observed that it can be modified also to $\mathbb{R}^3$, but it is not clear if such an example can be constructed in higher dimensions. However, Iosevich and Senger proved in the same paper that in any dimension there are norms whose unit sphere is smooth and has non-vanishing Gaussian curvature such that for no $s < (n + 1)/2$ does $I_s(\mu) < \infty$ imply that the corresponding distance measure would be in L^∞.

Falconer [2005] investigated the distance set problem for polyhedral norms; the unit ball is a symmetric polytope with finitely many faces. Then it may happen that the distance set of A has measure 0 although $\dim A = n$. Falconer's method was not constructive. Konyagin and Łaba [2006] constructed explicit examples. The distance set problem for non-Euclidean norms was also studied by Iosevich and Łaba [2004], [2005] and Iosevich and Rudnev [2005], and for random norms by Hofmann and Iosevich [2005] and Arutyunyants and Iosevich [2004]. Eswarathasan, Iosevich and Taylor [2011] proved the statement $\dim A > (n + 1)/2$ implies $\mathcal{L}^1(D(A)) > 0$ for some metrics with curvature conditions not necessarily coming from a norm.

Orponen [2012a] proved for arbitrary self-similar planar sets K with $\mathcal{H}^1(K) > 0$ that $\dim D(K) = 1$. Generalizations and related results were obtained by Falconer and Jin [2014a] and by Ferguson, Fraser and Sahlsten [2013]. Rams and Simon [2014b] proved for a class of random sets arising from percolation that $\dim K > 1/2$ is sufficient to guarantee that $D(K)$ contains an interval. Some of these results were based on the powerful techniques developed by Hochman and Shmerkin [2012]. Other results on distance sets of special classes of sets can be found in Iosevich and Łaba [2004] and Iosevich and Rudnev [2005], [2007a].

There is an analogous difficult discrete Erdős distance problem: given N points in the plane (or in $\mathbb{R}^n$), how many different distances must there at least be between these points for large N? Denoting this minimal number by $g(N)$, Erdős [1946] proved that $g(N) \lesssim N/\sqrt{\log N}$. Guth and Katz [2011] obtained a nearly optimal bound by showing that $g(N) \gtrsim N/\log N$. Although this and the continuous distance set problem, which we have discussed in this chapter, are analogous, methods developed for one have not appeared to be useful for the other. There is an exception for this in finite fields: Iosevich and Rudnev [2007c] found a way of modifying the spherical averages method to prove estimates for distance sets in finite fields. We shall discuss this a bit more in Section 15.4.

Iosevich and Łaba [2005] and Iosevich, Rudnev and Uriarte-Tuero [2014] proved that results of the type 'dim $A > s$ implies $\mathcal{L}^1(D(A)) > 0$' imply results on some particular discrete sets.

The monograph Garibaldi, Iosevich and Senger [2011] discusses various aspects of the Erdős distance problem in an easily accessible manner.

D. M. Oberlin and R. Oberlin [2013a] studied the corresponding unit distance problem estimating the size of $\{(x, y) \in A \times A : |x - y| = 1\}$ both in the discrete case and continuous case. Bennett, Iosevich and Taylor [2014] investigated sets of finite chains

$$\{(|x_1 - x_2|, \dots, |x_k - x_{k+1}|) \in \mathbb{R}^k : x_j \in A\}$$

and showed that for any $k \geq 2$ they have positive k-dimensional Lebesgue measure provided $A \subset \mathbb{R}^n$ is a Borel set with dim $A > (n + 1)/2$. For $k = 1$ this is Falconer's distance set result. Greenleaf, Iosevich and Pramanik [2014] studied sets of necklaces of constant length $t > 0$, that is, sequences $(x_1, \dots, x_k), x_j \in A, x_i \neq x_j$ for $i \neq j$, such that $|x_j - x_{j+1}| = t$ for $j = 1, \dots k - 1$ and $|x_k - x_1| = t$. They showed that if $n \geq 4, k$ is even and $A \subset \mathbb{R}^n$ is a Borel set with dim $A > (n + 3)/2$, then there is an open interval $I \subset \mathbb{R}$ so that such a necklace exists for all $t \in I$. In $\mathbb{R}^3$ this is false for compact sets A with dim $A = 3$ due to an example of Maga [2010], but they proved a related result for all $n \geq 3$ involving also a hypothesis on the Fourier dimension of A.

Theorem 4.10 was proved with the above argument by Edgar and Miller [2003], and independently by Bourgain [2003] with a different argument which we shall discuss below. This answered a question of Erdős and Volkmann [1966]. In that paper Erdős and Volkmann proved that there exist Borel subgroups of $\mathbb{R}$ of any dimension between 0 and 1. Edgar and Miller also proved, with a rather similar method as was presented above, that any Borel subring of $\mathbb{C}$ of positive Hausdorff dimension is either $\mathbb{R}$ or $\mathbb{C}$. These results hold for Suslin subrings, too.

The result of Erdős and Volkmann immediately extends to $\mathbb{R}^n$: there are dense Borel subgroups of any dimension between 0 and n. In Lie groups this is sometimes true and sometimes false as shown by de Saxcé [2013], [2014] and by Lindenstrauss and de Saxcé [2014].

Falconer [1984] showed that assuming the continuum hypothesis there exist non-Borel subrings of $\mathbb{R}$ of any dimension between 0 and 1. He also gave a very simple proof in Falconer [1985b] using his distance set result showing that there exist no Borel subrings of $\mathbb{R}$ with dimension strictly between $1/2$ and 1.

Katz and Tao [2001] formulated discrete, discretizing at a level δ, versions of the distance set problem, the Furstenberg problem (see 11.5) and the above ring problem. They showed that these discretized problems are in a sense equivalent. Unfortunately this does not seem to help for the continuous problems: although we have now a relatively simple proof for the ring conjecture, it has not led to any progress on the other two questions. Tao [2000] gave a simpler presentation in finite fields of these connections.

When one discretizes at a level δ, one approximates sets with finite unions of balls with radius δ. Let us call such sets δ-discrete. A natural analogue of Frostman measure, recall (2.2), is a $(\delta, s)_n$-set. This is a δ-discrete set $A \subset \mathbb{R}^n$ satisfying

$$\mathcal{L}^n(A \cap B(x,r)) \lesssim \delta^n (r/\delta)^s \quad \text{for all } x \in \mathbb{R}^n, r \geq \delta.$$

Katz and Tao formulated discrete conjectures involving $(\delta, s)_n$-sets which corresponded (but are not necessarily equivalent) to the following questions, the first of them is a special case of Conjecture 4.5.

(1) Does $\dim A \geq 1$ imply $\dim D(A) \geq 1/2 + c_0$ for Borel sets $A \subset \mathbb{R}^2$ and for some constant $c_0 > 0$?

(2) Are there Borel subrings of $\mathbb{R}$ of Hausdorff dimension $1/2$?

We skip here the formulation of the discrete analogue of (1), as well as the discrete Furstenberg conjecture. The discrete ring conjecture corresponding to (2) is:

Let $0 < \delta < 1$ and let $A \subset \mathbb{R}$ be a $(\delta, 1/2)_1$-set of measure $\approx \sqrt{\delta}$. Then

$$\mathcal{L}^n(A + A) + \mathcal{L}^n(A \cdot A) \gtrsim \delta^{1/2 - c_1},$$

where $c_1 > 0$ is an absolute constant. Bourgain [2003] proved this and even more replacing $1/2$ by $\sigma, 0 < \sigma < 1$, see also Bourgain [2010]. As a consequence he got the positive answer to (1) and negative answer to (2), and more generally that there are no Borel subrings of $\mathbb{R}$ with dimension strictly between

0 and 1. Bourgain's proof is much more complicated than that of Edgar and Miller [2003], but the discrete result seems to be much deeper and more influential. Bourgain's paper led to further developments on several questions in Lie groups, see de Saxcé [2013] and Lindenstrauss and de Saxcé [2014], and the references given there.

In a way, the distance set question asks how the Hausdorff dimension of a set affects the distribution of pairs of points taken from that set. In addition to looking at distances, one can study many other configurations. For example, directions $\frac{x-y}{|x-y|} \in S^{n-1}$, $x, y \in A$, $x \neq y$. It follows immediately from the line intersection theorem as discussed in Chapter 6 and in Mattila [1995], Chapter 10, that the set Dir(A) of such directions has $\sigma^{n-1}(\text{Dir}(A)) > 0$ if $A \subset \mathbb{R}^n$ is a Borel set with dim $A > n - 1$. This is best possible, because $\sigma^{n-1}(\text{Dir}(A)) = 0$ if A lies in a hyperplane. Iosevich, Mourgoglou and Senger [2012] studied the induced direction measure, analogous to the distance measure. Considering triples of points one can ask about angles. This was done by Harangi, Keleti, Kiss, Maga, Máthé, Mattila and Strenner [2013] and by Iosevich, Mourgoglou and Palsson [2011].

A special, but very interesting and delicate, case of k point configurations is that of the existence of arithmetic progressions in various types of sets. Classical number theory problems deal with the existence of arithmetic progressions in subsets of the integers, but Hausdorff dimension versions also make perfect sense. Large Hausdorff dimension alone does not help, due to examples of Keleti [2008] and Maga [2010], but combined assumptions on Hausdorff and Fourier dimensions do help. Łaba and Pramanik [2009] proved deep results of this type for subsets of the reals. Chan, Łaba and Pramanik [2013] established very general extensions of these results to higher dimensions covering many interesting particular cases. Körner [2009] proved some sharp results on algebraic relations for points in the support of a measure with a given Fourier decay.

For other results on k-point sets and associated geometric configurations, such as k-simplices, see Erdoğan, Hart and Iosevich [2013], Eswarathasan, Iosevich and Taylor [2011], Grafakos, Greenleaf, Iosevich and Palsson [2012], Greenleaf and Iosevich [2012], Greenleaf, Iosevich and Mourgoglou [2014], Greenleaf, Iosevich, Liu and Palsson [2013], and Liu [2014].

5

Exceptional projections and Sobolev dimension

Here we shall extend the projection results of the previous chapter in several ways proving estimates for the dimension of the exceptional sets of projections, introducing the Sobolev dimension to unify such estimates, and proving corresponding results in general dimensions.

5.1 Exceptional sets for one-dimensional projections

We shall first give a different proof, without Fourier transforms, to the first part of Theorem 4.1 and we improve it by estimating the Hausdorff dimension of the exceptional set. Here again $P_e : \mathbb{R}^n \to \mathbb{R}$, $P_e(x) = e \cdot x$, is the orthogonal projection for $e \in S^{n-1}, n \geq 2$.

Theorem 5.1 *Let $A \subset \mathbb{R}^n$ be a Borel set with $s = \dim A \leq 1$. Then for all $t \in [0, s]$,*

$$\dim\{e \in S^{n-1} : \dim P_e(A) < t\} \leq n - 2 + t. \tag{5.1}$$

Proof Let $\sigma < t \leq s$. By Theorem 2.8 there exists $\mu \in \mathcal{M}(A)$ such that $I_\sigma(\mu) < \infty$. For $e \in S^{n-1}$ let $\mu_e \in \mathcal{M}(P_e(A))$ be as before:

$$\mu_e(B) = \mu(P_e^{-1}(B)), \quad B \subset \mathbb{R}. \tag{5.2}$$

By Theorem 2.8 it suffices to show that

$$\dim\{e \in S^{n-1} : I_\sigma(\mu_e) = \infty\} \leq n - 2 + t.$$

Suppose this is false. Then

$$\mathcal{H}^{n-2+t}(\{e \in S^{n-1} : I_\sigma(\mu_e) = \infty\}) > 0.$$

72

By Frostman's lemma 2.7 there is $\nu \in \mathcal{M}(S_\infty)$, where $S_\infty = \{e \in S^{n-1} : I_\sigma(\mu_e) = \infty\}$, such that $\nu(B(x, r)) \leq r^{n-2+t}$ for all $x \in \mathbb{R}^n$ and $r > 0$. In order to apply Frostman's lemma we should check that S_∞ is a Borel set, but we leave it as an exercise. We shall use the general formula for the integral $(f \geq 0)$,

$$\int f \, d\lambda = \int_0^\infty \lambda(\{x : f(x) \geq r\}) \, dr,$$

and the estimate,

$$\nu(\{e \in S^{n-1} : |P_e(x)| \leq \delta\}) \lesssim (\delta/|x|)^t, \tag{5.3}$$

which is trivial for $n = 2$ and follows for $n > 2$ by checking that the belt $\{e \in S^{n-1} : |P_e(x)| \leq \delta\}$ can be covered with roughly $(\delta/|x|)^{2-n}$ balls of radius $\delta/|x|$. We obtain for all $x \in \mathbb{R}^n \setminus \{0\}$,

$$\int_{S^{n-1}} |P_e(x)|^{-\sigma} \, d\nu e = \int_0^\infty \nu(\{e \in S^{n-1} : |P_e(x)|^{-\sigma} \geq r\}) \, dr$$

$$\lesssim \nu(S^{n-1})|x|^{-\sigma} + |x|^{-t} \int_{|x|^{-\sigma}}^\infty r^{-t/\sigma} \, dr$$

$$= \left(\nu(S^{n-1}) + \frac{\sigma}{t - \sigma}\right) |x|^{-\sigma}.$$

Hence by Fubini's theorem,

$$\int_{S^{n-1}} I_\sigma(\mu_e) \, d\nu e = \int_{S^{n-1}} \left(\iint |P_e(x - y)|^{-\sigma} \, d\mu x \, d\mu y\right) d\nu e$$

$$= \iint \left(\int_{S^{n-1}} |P_e(x - y)|^{-\sigma} \, d\nu e\right) d\mu x \, d\mu y \lesssim I_\sigma(\mu) < \infty.$$

In particular, $\nu(S_\infty) = 0$, which contradicts the assumption $\nu \in \mathcal{M}(S_\infty)$ and proves the theorem. $\qquad\square$

5.2 Sobolev dimension

We want to prove similar results for the exceptional sets of the second part of Theorem 4.1 and of Theorem 4.2. For this we need Fourier transforms. We can give a unified treatment and prove these two results simultaneously by introducing the Sobolev dimension of a measure:

Definition 5.2 The *Sobolev dimension* of a measure $\mu \in \mathcal{M}(\mathbb{R}^n)$, $n \geq 1$, is

$$\dim_S \mu = \sup\{s \in \mathbb{R} : \int_{\mathbb{R}^n} |\widehat{\mu}(x)|^2 (1 + |x|)^{s-n} \, dx < \infty\}.$$

Observe that $\dim_S \mu \geq 0$, because $\int_{\mathbb{R}^n} |\widehat{\mu}(x)|^2 (1 + |x|)^{s-n} \, dx < \infty$ if $s < 0$ due to the boundedness of $\widehat{\mu}$. Thus $0 \leq \dim_S \mu \leq \infty$. If μ is a function in $\mathcal{S}(\mathbb{R}^n)$, then $\dim_S \mu = \infty$.

Using $(1 + |x|)^{s-n}$ instead of $|x|^{s-n}$ is often just a technical convenience of having a locally bounded factor instead of a locally integrable one. For $s > 0$ the integrals $\int_{\mathbb{R}^n} |\widehat{\mu}(x)|^2 (1 + |x|)^{s-n} \, dx$ and $\int_{\mathbb{R}^n} |\widehat{\mu}(x)|^2 |x|^{s-n} \, dx$ are comparable, but for example for the Dirac measure δ_0, for which $\widehat{\delta_0} = 1$, the latter integral is infinite for all $s \in \mathbb{R}$, whereas $\int_{\mathbb{R}^n} |\widehat{\mu}(x)|^2 (1 + |x|)^{s-n} \, dx < \infty$ if $s < 0$. In particular, $\dim_S \delta_0 = 0$.

The term comes from Sobolev spaces. A function $f \in L^2(R^n)$ belongs to the Sobolev space $H^{k,2}(\mathbb{R}^n)$ if the kth order distributional partial derivatives of f belong to $L^2(\mathbb{R}^n)$. By the formula for the Fourier transform of the partial derivatives and by Parseval's formula

$$\int \sum_{|\alpha|=k} |\partial^\alpha f|^2 = c(n,k) \int |\widehat{f}(x)|^2 |x|^{2k} \, dx.$$

Replacing the exponent $2k$ on the right hand side with $2\sigma, \sigma \in \mathbb{R}$, leads to 'fractional order' Sobolev spaces. We shall study these in Chapter 17. We have used the exponent of the form $s - n$ instead of 2σ, because then s relates more naturally to the Hausdorff dimension.

The Sobolev dimension for us is motivated by its relation to energy-integrals coming from the formula

$$I_s(\mu) = \gamma(n,s) \int |\widehat{\mu}(x)|^2 |x|^{s-n} \, dx$$

of Theorem 3.10. Let us extend this notion to all $s \in \mathbb{R}$ using the right hand side:

Definition 5.3 The *Sobolev energy* of degree $s \in \mathbb{R}$ of a measure $\mu \in \mathcal{M}(\mathbb{R}^n)$ is

$$\mathcal{I}_s(\mu) = \int |\widehat{\mu}(x)|^2 |x|^{s-n} \, dx.$$

Then

$$\dim_S \mu = \sup\{s : \mathcal{I}_s(\mu) < \infty\}, \tag{5.4}$$

where we have interpreted $\sup \varnothing = 0$.

The formula for $\mathcal{I}_s(\mu)$ as a double integral $\iint |x - y|^{-s} \, d\mu x \, d\mu y$ does not extend beyond $0 < s < n$; for instance for $s = n$ this double integral is infinite for smooth non-negative functions f, not identically zero, whereas

$\mathcal{I}_n(f) = \|f\|_2^2$, and for $s = 0$ the double integral is $\|f\|_1^2$ and $\mathcal{I}_n(f) = \infty$ if $\widehat{f}(0) = \int f \neq 0$.

The greater the Sobolev dimension is, the smoother the measure is in some sense. The following result captures some parts of this principle:

Theorem 5.4 *Let $\mu \in \mathcal{M}(\mathbb{R}^n)$.*

(a) If $0 < \dim_S \mu < n$, then $\dim_S \mu = \sup\{s > 0 : I_s(\mu) < \infty\}$.
(b) If $\dim_S \mu > n$, then $\mu \in L^2(\mathbb{R}^n)$.
(c) If $\dim_S \mu > 2n$, then μ is a continuous function.

Proof Part (a) follows readily from Theorem 3.10 and the definition of the Sobolev dimension. In part (b) $\widehat{\mu} \in L^2(\mathbb{R}^n)$, and so also $\mu \in L^2(\mathbb{R}^n)$ by Theorem 3.3. Part (c) is proven as Theorem 4.2 with Schwartz's inequality: when $s \in (2n, \dim_S \mu)$,

$$\int_{\mathbb{R}^n} |\widehat{\mu}| \leq \left(\int_{\mathbb{R}^n} |\widehat{\mu}(x)|^2 (1 + |x|)^{s-n} \, dx \right)^{1/2} \left(\int_{\mathbb{R}^n} (1 + |x|)^{n-s} \, dx \right)^{1/2} < \infty,$$

and μ is a continuous function by Theorem 3.4. $\qquad\qquad\square$

Now we prove a result on the dimension of the exceptional sets involving Sobolev dimension of projected measures. As before we denote by μ_e the image of $\mu \in \mathcal{M}(\mathbb{R}^n)$ under the projection $P_e, e \in S^{n-1}$.

Theorem 5.5 *Let $\mu \in \mathcal{M}(\mathbb{R}^n), 0 < s < n$ and $I_s(\mu) < \infty$. Then for all $t, 0 < t \leq s$,*

$$\dim\{e \in S^{n-1} : \dim_S \mu_e < t\} \leq n - 2 + t \quad \text{if} \quad s \leq 1,$$
$$\dim\{e \in S^{n-1} : \dim_S \mu_e < t\} \leq n - 1 + t - s \quad \text{if} \quad 1 \leq s \leq n - 1 + t.$$

The first inequality is essentially Theorem 5.1 and follows by the same proof. The second is by part (a) of Theorem 5.4 a special case of the following more general statement:

Theorem 5.6 *Let $\mu \in \mathcal{M}(\mathbb{R}^n)$. Then for all $t > 0$,*

$$\dim\{e \in S^{n-1} : \dim_S \mu_e < t\} \leq \max\{0, n - 1 + t - \dim_S \mu\}. \qquad (5.5)$$

Proof Let $S_t = \{e \in S^{n-1} : \dim_S \mu_e < t\}$ and $s = \dim_S \mu$. Then S_t is a Borel set; we leave the poof of this as an exercise. Suppose that (5.5) is false for some $t > 0$ and choose $\tau > 0$ such that $n - 1 + t - s < \tau < \dim S_t$. Then Frostman's lemma gives us a measure $\nu \in \mathcal{M}(S_t)$ for which $\nu(B(x, r)) \leq r^\tau$

for all $x \in \mathbb{R}^n$ and $r > 0$. We shall show that

$$\int_{S^{n-1}} \int_{\mathbb{R}} |\widehat{\mu}_e(u)|^2 (1 + |u|)^{t-1} \, du \, dve < \infty. \tag{5.6}$$

This will give a contradiction with the definitions of S_t and v and proves the theorem.

In order to get to the integrals defining the Sobolev dimension of μ we choose an auxiliary function $\varphi \in \mathcal{S}(\mathbb{R}^n)$ such that $\varphi(x) = 1$ for all $x \in \operatorname{spt} \mu$. Then $\mu = \varphi\mu$ and $\widehat{\mu} = \widehat{\varphi\mu} = \widehat{\varphi} * \widehat{\mu}$. Hence by Schwartz's inequality

$$|\widehat{\mu}(x)|^2 \leq \left(\int |\widehat{\mu}(x - y)\widehat{\varphi}(y)| \, dy \right)^2$$

$$\leq \int |\widehat{\varphi}| \int |\widehat{\mu}(x - y)|^2 |\widehat{\varphi}(y)| \, dy \lesssim |\widehat{\varphi}| * |\widehat{\mu}|^2(x),$$

when $x \in \mathbb{R}^n$. As $\widehat{\varphi} \in \mathcal{S}(\mathbb{R}^n)$, we have for all $N \in \mathbb{N}$, $|\widehat{\varphi}(x)| \lesssim_{\varphi, N} (1 + |x|)^{-N}$, $x \in \mathbb{R}^n$. By (4.3), $\widehat{\mu}_e(u) = \widehat{\mu}(ue) = \widehat{\varphi\mu}(ue)$ for $u \in \mathbb{R}$. Using this, the above estimates and Fubini's theorem, we obtain

$$\int_{S^{n-1}} \int_{\mathbb{R}} |\widehat{\mu}_e(u)|^2 (1 + |u|)^{t-1} \, du \, dve$$

$$\lesssim \int_{S^{n-1}} \int_{\mathbb{R}} |\widehat{\varphi}| * |\widehat{\mu}|^2(ue)(1 + |u|)^{t-1} \, du \, dve$$

$$= \int_{S^{n-1}} \int_{\mathbb{R}} \left(\int_{\mathbb{R}^n} |\widehat{\varphi}(ue - x)| |\widehat{\mu}(x)|^2 \, dx \right) (1 + |u|)^{t-1} \, du \, dve$$

$$= \int_{\mathbb{R}^n} |\widehat{\mu}(x)|^2 \left(\int_{S^{n-1}} \int_{\mathbb{R}} |\widehat{\varphi}(ue - x)| (1 + |u|)^{t-1} \, du \, dve \right) dx$$

$$\lesssim \int_{\mathbb{R}^n} |\widehat{\mu}(x)|^2 \left(\int_{S^{n-1}} \int_{\mathbb{R}} (1 + |ue - x|)^{-N} (1 + |u|)^{t-1} \, du \, dve \right) dx.$$

In order to complete the proof we need to show that the last integral is finite. Set

$$L_e = \{ue : u \in \mathbb{R}\} \quad \text{for } e \in S^{n-1}.$$

Then for any $r > 0$,

$$v(\{e \in S^{n-1} : d(x, L_e) \leq r\}) \lesssim (r/|x|)^\tau. \tag{5.7}$$

This follows from the easy fact that the set in question can be covered with two balls of radius $\sqrt{2}(r/|x|)$.

We shall now show that for large enough N and for $x \in \mathbb{R}^n$, $x \neq 0$,

$$\int_{S^{n-1}} \int_{\mathbb{R}} (1 + |ue - x|)^{-N} (1 + |u|)^{t-1} \, du \, dve \lesssim (1 + |x|)^{t-1-\tau}. \tag{5.8}$$

This will complete the proof, because then

$$\int_{S^{n-1}} \int_{\mathbb{R}} |\widehat{\mu_e}(u)|^2 (1 + |u|)^{t-1} \, du \, dve$$
$$\lesssim \int_{\mathbb{R}^n} |\widehat{\mu}(x)|^2 (1 + |x|)^{t-1-\tau} \, dx \lesssim I_s(\mu) < \infty,$$

since $t - 1 - \tau < s - n$.

Fix $N > \max\{1 + \tau, t\}$. In addition to (5.7) we shall use the elementary inequality

$$\int_{\mathbb{R}} (1 + |ue - x|)^{-N} \, du \lesssim 1. \tag{5.9}$$

We split the integration into dyadic annuli centred at x and estimate

$$\int_{S^{n-1}} \int_{\mathbb{R}} (1 + |ue - x|)^{-N} (1 + |u|)^{t-1} \, du \, dve$$

$$= \iint_{\{u:|ue-x|\leq 1/2\}} (1 + |ue - x|)^{-N} (1 + |u|)^{t-1} \, du \, dve$$

$$+ \sum_{j=0}^{\infty} \iint_{\{u:2^{j-1}<|ue-x|\leq 2^j\}} (1 + |ue - x|)^{-N} (1 + |u|)^{t-1} \, du \, dve$$

$$\lesssim (1 + |x|)^{t-1} \int_{\{e:d(L_e,x)\leq 1/2\}} \int (1 + |ue - x|)^{-N} \, du \, dv(e)$$

$$+ \sum_{j\geq 0,|x|>2^{j+1}} \iint_{\{u:2^{j-1}<|ue-x|\leq 2^j\}} (1 + |ue - x|)^{-N} (1 + |u|)^{t-1} \, du \, dve$$

$$+ \sum_{j\geq 0,|x|\leq 2^{j+1}} \iint_{\{u:2^{j-1}<|ue-x|\leq 2^j\}} (1 + |ue - x|)^{-N} (1 + |u|)^{t-1} \, du \, dve$$

$$\lesssim (1 + |x|)^{t-1-\tau} + (1 + |x|)^{t-1} \sum_{j=0}^{\infty} 2^j 2^{-Nj} v(\{e : d(x, L_e) \leq 2^j\})$$

$$+ \sum_{j\geq 0,|x|\leq 2^{j+1}} 2^{-Nj} \int_{|s|\leq 2^{j+2}} (1 + |u|)^{t-1} \, du \lesssim (1 + |x|)^{t-1-\tau}$$

$$+ (1 + |x|)^{t-1-\tau} \sum_{j=0}^{\infty} 2^{(1-N+\tau)j} + \sum_{j\geq 0,|x|\leq 2^{j+1}} 2^{(t-N)j} \lesssim (1 + |x|)^{t-1-\tau},$$

because $N > 1 + \tau$ and $N > t$. Thus we have verified (5.8) and the proof is complete.

$\square$

We shall now combine Theorem 5.1 and the three previous theorems to get:

Corollary 5.7 *Let $A \subset \mathbb{R}^n, n \geq 2$, be a Borel set and $s = \dim A$.*

(a) If $s \leq 1$ and $t \in (0, s]$, then

$$\dim\{e \in S^{n-1} : \dim P_e(A) < t\} \leq n - 2 + t.$$

(b) If $s > 1$, then

$$\dim\{e \in S^{n-1} : \mathcal{L}^1(P_e(A)) = 0\} \leq n - s.$$

(c) If $s > 2$, then

$$\dim\{e \in S^{n-1} : \text{the interior of } P_e(A) \text{ is empty}\} \leq n + 1 - s.$$

5.3 Higher dimensional projections

The above results have rather straightforward generalizations to projections onto m-dimensional planes in $\mathbb{R}^n$ where $0 < m < n$. We just need some basic information about the *Grassmannian*:

$$G(n, m) = \{V : V \text{ is an } m - \text{dimensional linear subspace of } \mathbb{R}^n\}.$$

It is a smooth $m(n - m)$-dimensional compact submanifold of some Euclidean space. This can been seen using the following local coordinates. If $V_0 \in G(n, m)$, the planes $V \in G(n, m)$ in a neighbourhood of V_0 can be written as graphs over V_0:

$$V = \{x + Lx : x \in V_0, L : V_0 \rightarrow V_0^{\perp} \text{ linear}\},$$

and the correspondence between V and L is one-to-one.

There is a unique orthogonally invariant Borel probability measure $\gamma_{n,m}$ on $G(n, m)$. It can be obtained conveniently from the Haar measure θ_n on $O(n)$ by the formula

$$\gamma_{n,m}(A) = \theta_n(\{g \in O(n) : g(V_0) \in A\}),$$

where V_0 is any fixed plane in $G(n, m)$, see, e.g., Mattila [1995], Section 3.9.

We shall denote by $P_V : \mathbb{R} \rightarrow V$ the *orthogonal projection* from $\mathbb{R}^n$ onto V. Recall that $\mathcal{H}^m \llcorner V$ is the Lebesgue measure on $V \in G(n, m)$. Theorem 4.1 has the following higher dimensional generalization.

Theorem 5.8 *Let $A \subset \mathbb{R}^n$ be a Borel set and $s = \dim A$.*

(a) If $s \leq m$, then

$$\dim P_V(A) = s \quad \text{for } \gamma_{n,m} \text{ almost all } V \in G(n, m).$$

(b) If s > m, then

$$\mathcal{H}^m(P_V(A)) > 0 \quad \text{for } \gamma_{n,m} \text{ almost all } V \in G(n, m).$$

As before, this is an immediate consequence of the following measure version.

Theorem 5.9 *Let $\mu \in \mathcal{M}(\mathbb{R}^n)$ with $I_s(\mu) < \infty$.*

(a) If s ≤ m, then

$$I_s(P_{V\sharp}\mu) < \infty \quad \text{for } \gamma_{n,m} \text{ almost all } V \in G(n, m).$$

(b) If s > m, then

$$P_{V\sharp}\mu \ll \mathcal{H}^m \llcorner V \quad \text{for } \gamma_{n,m} \text{ almost all } V \in G(n, m).$$

This is proven in Mattila [1995] without Fourier transforms, a similar Fourier-analytic proof as that of Theorem 4.1 can also easily be given. But now we shall discuss more general results, the higher dimensional versions of the previous exceptional set estimates.

For any $x \in \mathbb{R}^n \setminus \{0\}$ the set $\{V \in G(n, m) : x \in V\}$ is a smooth submanifold of dimension $(m - 1)(n - 1 - (m - 1)) = (m - 1)(n - m)$: its elements are $W + L_x$, where $L_x \in G(n, 1)$ is the line through x, and W runs through the $(m - 1)$-planes in $L_x^\perp$. This implies that for $x \in \mathbb{R}^n \setminus \{0\}$ and for any $\delta > 0$, the set, essentially a $\delta/|x|$-neighbourhood of $\{V \in G(n, m) : x \in V\}$ when $|x| \geq \delta$,

$$\{V \in G(n, m) : d(x, V) \leq \delta\}$$

can be covered with roughly $(\delta/|x|)^{-(m-1)(n-m)}$ balls of radius $\delta/|x|$. Thus if ν is a Borel measure on $G(n.m)$ which satisfies

$$\nu(B(V, r)) \leq r^\tau \quad \text{for all } V \in G(n, m) \text{ and } r > 0, \tag{5.10}$$

we have

$$\nu(\{V \in G(n, m) : d(x, V) \leq \delta\}) \lesssim (\delta/|x|)^{\tau-(m-1)(n-m)}. \tag{5.11}$$

In the same way (5.10) implies

$$\nu(\{V \in G(n, m) : |P_V(x)| \leq \delta\}) \lesssim (\delta/|x|)^{\tau-m(n-m-1)}, \tag{5.12}$$

because $|P_V(x)| = d(x, V^\perp)$.

This is essentially all we need in order to generalize the proofs of Theorems 5.5 and 5.6 to get the following results:

Theorem 5.10 *Let $\mu \in \mathcal{M}(\mathbb{R}^n)$, $0 < s < n$ and $I_s(\mu) < \infty$. Then for all t, $0 < t \leq s$, with $\mu_V = P_{V\sharp}\mu$,*

$$\dim\{V \in G(n, m) : \dim_S \mu_V < t\} \leq m(n - m - 1) + t \quad \text{if } s \leq m,$$
$$\dim\{V \in G(n, m) : \dim_S \mu_V < t\} \leq m(n - m) + t - s$$
$$\text{if } m \leq s \leq m(n - m) + t.$$

Theorem 5.11 *Let $\mu \in \mathcal{M}(\mathbb{R}^n)$ with $n \geq 2$. Then*

$$\dim\{V \in G(n, m) : \dim_S \mu_V < t\} \leq \max\{0, m(n - m) + t - \dim_S \mu\} \tag{5.13}$$

for all $t > 0$.

The first part of Theorem 5.10 follows with essentially the same proof as the first part of Theorem 5.5: for a measure ν satisfying $\nu(B(V, r)) \leq r^{m(n-m-1)+t}$ we use (5.12) to replace (5.3). The second part is again a special case of Theorem 5.11. This in turn can be proven with small modifications of the proof of Theorem 5.6 using (5.11). We give now some details for that.

Let $s = \dim_S \mu$, $\tau > m(n - m) + t - s$ and let $\nu \in \mathcal{M}(G(n, m))$ be such that

$$\nu(B(V, r)) \leq r^\tau \quad \text{for } V \in G(n, m).$$

As in the proof of Theorem 5.6 it is enough to prove that

$$\int_{G(n,m)} \int_V |\widehat{P_{V\sharp}\mu}(u)|^2 (1 + |u|)^{t-m} \, d\mathcal{H}^m u \, d\nu V < \infty. \tag{5.14}$$

The proof for the one-dimensional projections relied on the formula $\widehat{\mu_e}(u) = \widehat{\mu}(ue)$. This is now replaced by

$$\widehat{P_{V\sharp}\mu}(u) = \widehat{\mu}(u) \quad \text{for } u \in V \in G(n, m), \tag{5.15}$$

which follows from

$$\widehat{P_{V\sharp}\mu}(u) = \int e^{-2\pi i u \cdot P_V(x)} d\mu x = \int e^{-2\pi i u \cdot x} d\mu x = \widehat{\mu}(u).$$

Let $\varphi \in \mathcal{S}(\mathbb{R}^n)$ be such that $\varphi(x) = 1$ for all $x \in \operatorname{spt} \mu$ so that $\widehat{\mu} = \widehat{\varphi\mu} = \widehat{\varphi} * \widehat{\mu}$. We have again

$$|\widehat{\mu}(x)|^2 \lesssim |\widehat{\varphi}| * |\widehat{\mu}|^2(x),$$

and for all $N \in \mathbb{N}$, $|\widehat{\varphi}(x)| \lesssim_{\varphi,N} (1 + |x|)^{-N}$, $x \in \mathbb{R}^n$. Hence

$$\int_{G(n,m)} \int_V |\widehat{P_{V\sharp}\mu}(u)|^2 (1 + |u|)^{t-m} \, d\mathcal{H}^m u \, d\nu V$$

$$= \int_{G(n,m)} \int_V |\widehat{\mu}(u)|^2 (1 + |u|)^{t-m} \, d\mathcal{H}^m u \, d\nu V$$

$$\lesssim \int_{G(n,m)} \int_V \left(\int_{\mathbb{R}^n} |\widehat{\varphi}(u - x)| |\widehat{\mu}(x)|^2 \, dx \right) (1 + |u|)^{t-m} \, d\mathcal{H}^m u \, d\nu V$$

$$= \int_{\mathbb{R}^n} |\widehat{\mu}(x)|^2 \left(\int_{G(n,m)} \int_V |\widehat{\varphi}(u - x)| (1 + |u|)^{t-m} \, d\mathcal{H}^m u \, d\nu V \right) dx$$

$$\lesssim \int_{\mathbb{R}^n} |\widehat{\mu}(x)|^2 \left(\int_{G(n,m)} \int_V (1 + |u - x|)^{-N} (1 + |u|)^{t-m} \, d\mathcal{H}^m u \, d\nu V \right) dx.$$

Now we need to show that for $x \in \mathbb{R}^n$, $x \neq 0$,

$$\int_{G(n,m)} \int_V (1 + |u - x|)^{-N} (1 + |u|)^{t-m} \, d\mathcal{H}^m u \, d\nu V \lesssim |x|^{t+m(n-m)-n-\tau}.$$

This will complete the proof as for Theorem 5.6. The proof of this estimate is a routine modification of the proof (5.8) using (5.11) and

$$\int_V (1 + |u - x|)^{-N} \, d\mathcal{H}^m u \lesssim 1$$

in place of (5.9). We leave the details to the reader.

Again we have the corollary:

Corollary 5.12 *Let $A \subset \mathbb{R}^n$ be a Borel set and $s = \dim A$.*

(a) If $s \leq m$ and $t \in (0, s]$, then

$$\dim\{V \in G(n, m) : \dim P_V(A) < t\} \leq m(n - m - 1) + t.$$

(b) If $s > m$, then

$$\dim\{V \in G(n, m) : \mathcal{H}^m(P_V(A)) = 0\} \leq m(n - m) + m - s.$$

(c) If $s > 2m$, then

$$\dim\{V \in G(n, m) : \text{the interior of } P_V(A) \text{ is empty}\} \leq m(n - m) + 2m - s.$$

(d) In particular if $s > 2m$, the interior of $P_V(A)$ is non-empty for $\gamma_{n,m}$ almost all $V \in G(n, m)$.

The upper bound in (a) is sharp when $t = s$, but not in general. We shall discuss this a bit more below. The upper bound in (b) is sharp, as we shall soon see. I do not know if the upper in (c) is sharp. For $m = 1$ the assumption

$s > 2$ in (d) is necessary: as remarked in the previous chapter, Besicovitch sets can be used to give examples of sets of dimension 2 whose projections on all lines have empty interior. Probably the condition $s > 2m$ in (d) is not sharp when $m > 1$, but no example is known. We shall now show the sharpness of the upper bound in (b):

Example 5.13 For any $m < s < n$ there exists a compact set $C \subset \mathbb{R}^n$ such that $\dim C = s$ and

$$\dim\{V \in G(n, m) : \mathcal{H}^m(P_V(C)) = 0\} = m(n - m) + m - s.$$

Proof We first assume that $m = 1$. We shall use sets defined by Diophantine approximation properties. Let $0 < \delta < 1$. Fix a rapidly increasing sequence (m_j) of positive integers, for instance $m_{j+1} > m_j^j$ for all $j \in \mathbb{N}$ suffices. Denote by $\|x\|$ the distance of the real number x to the nearest integer and define the sets

$$C = \{x \in [0, 1]^n : \|m_j x_i\| \le m_j^{1-n/s} \text{ for all } j \in \mathbb{N}, i = 1, \dots, n\},$$

$$E_\delta = \{(y_1, \dots, y_{n-1}) \in \mathbb{R}^{n-1} : \text{for infinitely many } j \in \mathbb{N} \text{ there is}$$

$$q_j \in \mathbb{N} \cap [1, m_j^{(1-\delta)(n-s)/s}] \text{ such that } \|q_j y_i\| \le q_j m_j^{-n/s}$$

$$\text{for all } i = 1, \dots, n - 1\}.$$

Then

$$\dim C = s \quad \text{and} \quad \dim E_\delta = (1 - \delta)(n - s).$$

Here C is the same set we used in Example 4.8. We shall not prove the second formula. When $n = 2$ the set E_δ is a slight modification of the set E_α in Kaufman's Theorem 3.13 and the proof given for it in Wolff [2003], Theorem 9.A.2, works also in this case. One can easily check that Falconer's argument for Jarnik's Theorem 10.3 in Falconer [1990] applies, too. Moreover, it extends readily to higher dimensions. Other references are given before Theorem 8.16 in Falconer [1985a].

For every $j \in \mathbb{N}$ we have

$$C \subset \bigcup_{z \in Z_j} B(z/m_j, nm_j^{-n/s}) \quad \text{where } Z_j = \mathbb{Z}^n \cap [0, m_j]^n. \tag{5.16}$$

We shall now show that

$$\mathcal{L}^1(\pi_Y(C)) = 0 \quad \text{for all } y \in E_\delta, \tag{5.17}$$

where $Y = (y, 1) \in \mathbb{R}^n$ and π_Y,

$$\pi_Y(x) = Y \cdot x, \quad x \in \mathbb{R}^n,$$

is essentially the orthogonal projection onto the line $\{tY : t \in \mathbb{R}\}$. From this it follows that

$$\dim\{L \in G(n, 1) : \mathcal{H}^1(P_L(C)) = 0\} \geq (1 - \delta)(n - s).$$

Letting $\delta \to 0$ will then complete the proof in the case $m = 1$.

Let $y \in E_\delta$ and let $q_j \in \mathbb{N}$, $1 \leq q_j \leq m_j^{(1-\delta)(n-s)/s}$, be related to y for infinitely many $j \in \mathbb{N}$ as in the definition of E_δ. Then, for these j, there are integers $p_{j,i}, i = 1, \ldots, n - 1$, such that

$$|y_i - p_{j,i}/q_j| \leq m_j^{-n/s}.$$

Then $|p_{j,i}| \lesssim m_j^{(1-\delta)(n-s)/s}$, the implicit constant is allowed to depend on y. Again let $Y = (y_1, \ldots, y_{n-1}, 1)$ and $P_j = (p_{j,1}/q_j, \ldots, p_{j,n-1}/q_j, 1)$. Then $|Y - P_j| \leq \sqrt{n}m_j^{-n/s}$, which implies that

$$|\pi_Y(x) - \pi_{P_j}(x)| \leq nm_j^{-n/s} \quad \text{for all } x \in C. \tag{5.18}$$

We shall now estimate the number of points in $\pi_{P_j}(Z_j)$ where Z_j is as in (5.16). Let $z = (z_i) \in \mathbb{Z}_j$. Then

$$\pi_{P_j}(z) = \tfrac{1}{q_j}(p_{j,1}z_1 + \cdots + p_{j,n-1}z_{n-1} + q_j z_n),$$

where $|p_{j,1}z_1 + \cdots + p_{j,n-1}z_{n-1} + q_j z_n| \lesssim m_j^{1+(1-\delta)(n-s)/s}$. Thus $\pi_{P_j}(z)$ can take $\lesssim m_j^{1+(1-\delta)(n-s)/s}$ values. Recalling (5.16) we get that $\pi_{P_j}(C)$ is covered with $\lesssim m_j^{1+(1-\delta)(n-s)/s}$ intervals of length $\lesssim m_j^{-n/s}$. Combining this with (5.18) we find that $\pi_Y(C)$ is covered with $\lesssim m_j^{1+(1-\delta)(n-s)/s}$ intervals of length $\lesssim m_j^{-n/s}$, which gives

$$\mathcal{L}^1(\pi_Y(C)) \lesssim \liminf_{j\to\infty} m_j^{1+(1-\delta)(n-s)/s-n/s} = 0,$$

because the exponent $1 + ((1 - \delta)(n - s) - n)/s$ is negative.

We have now finished the proof in the case $m = 1$. Suppose then that $m > 1$. Let $C_1 \subset \mathbb{R}^{n-m+1}$ with $\dim C_1 = s - m + 1$ be the set we found above for the case $m = 1$ with the exceptional set

$$E_1 = \{L \in G(n - m + 1, 1) : \mathcal{H}^1(P_L(C_1)) = 0\}, \quad \dim E_1 = n - s.$$

Then

$$C = C_1 \times [0, 1]^{m-1} \subset \mathbb{R}^n$$

serves the purpose in the general case.

It is obvious that $\dim C = s$. We should check that

$$E = \{V \in G(n,m) : \mathcal{H}^m(P_V(C)) = 0\}$$

has dimension at least $m(n-m) + m - s$. By simple linear algebra, identifying $\mathbb{R}^{n-m+1} = \mathbb{R}^{n-m+1} \times \{0\} \subset \mathbb{R}^n$, E contains all m-planes $L + W, L \in E_1$, $W \in G(n, m-1)$ with $W \subset L^\perp$. The set $\{W \in G(n, m-1) : W \subset L^\perp\}$ is essentially $G(n-1, m-1)$ and has dimension $(m-1)(n-1-(m-1)) = (m-1)(n-m)$. Thus (this requires a small argument which we leave to the reader)

$$\dim E \geq \dim E_1 + (m-1)(n-m) = n - s + (m-1)(n-m)$$
$$= m(n-m) + m - s$$

as required.   □

5.4 Further comments

Theorem 5.1 (which is Corollary 5.7(a)) was proved by Kaufman [1968] in the plane; the higher dimensional generalization Corollary 5.12(a) was done by Mattila [1975]. The example proving the sharpness of the upper bound was constructed by Kaufman and Mattila [1975], extending a previous example of Kaufman [1969]. It is given (for $m = 1, n = 2$) in Falconer [1985a], Theorem 8.17. Corollaries 5.7(b) and 5.12(b) were proven by Falconer [1982]. Example 5.13 above proving their sharpness is rather similar to the one of Kaufman and Mattila [1975]; its details were written down by Peltomäki [1987] in his licentiate thesis. Peres and Schlag [2000] introduced the Sobolev dimension and proved the results of this chapter related to it, together with Corollaries 5.7(c) and 5.12(c) as their consequences. They proved their results in a much more general setting which we shall discuss in Chapter 18. Orponen [2012b] proved various results on exceptional sets involving packing dimension and Baire category. See also Sections 9.2 and 10.5 for exceptional set results concerning self-similar measures and sets.

As far as I know the sharp bound in Theorem 5.1 for $t < \dim A$ is unknown. The upper bound t in Theorem 5.1 in the plane is not always sharp due to the following result of Bourgain [2003], [2010] and D. M. Oberlin [2012]:

Theorem 5.14 *Suppose $A \subset \mathbb{R}^2$ is a Borel set. Then*

$$\dim\{e \in S^1 : \dim P_e(A) < \dim A/2\} = 0.$$

The construction of Kaufman and Mattila [1975] can be used to get for any $0 < t \leq s < 2$ a compact set $A \subset \mathbb{R}^2$ with dim $A = s$ such that

$$\dim\{e \in S^1 : \dim P_e(A) \geq t\} \geq 2t - s.$$

Could $2t - s$ be the sharp upper bound in the range $s/2 \leq t \leq \min\{1, s\}$? In any case this shows that to get dimension 0 for the exceptional set, the bound dim $A/2$ is the best possible.

Bourgain's estimate is somewhat stronger than the above. He obtained his result as part of deep investigations in additive combinatorics, whereas Oberlin's proof is much simpler and more direct. Oberlin also had another exceptional set estimate in Oberlin [2014a].

Orponen [2014c] has a related discrete level result for product sets.

There have been some interesting recent developments on restricted families of projections and projection-type transformations. For example, one can take some smooth submanifold G of $G(n, m)$ and ask how projections P_V, $V \in G$, affect Hausdorff dimension. Such restricted families appear quite naturally in Heisenberg groups, see Balogh, Durand Cartagena, Fässler, Mattila and Tyson [2013], Balogh, Fässler, Mattila and Tyson [2012] and Fässler and Hovila [2014]. Another motivation for studying them comes from the work of E. Järvenpää, M. Järvenpää and Ledrappier and their co-workers on measures invariant under geodesic flows on manifolds; see E. Järvenpää, M. Järvenpää and Leikas [2005] and Hovila, E. Järvenpää, M. Järvenpää and Ledrappier [2012b]. They are also connected to Kakeya-type questions. A very simple example is the one where $G \subset G(3, 1)$ corresponds to projections π_θ onto the lines $\{t(\cos\theta, \sin\theta, 0) : t \in \mathbb{R}\}, \theta \in [0, 2\pi)$. Since $\pi_\theta(A) = \pi_\theta((\pi(A))$ where $\pi(x, y, z) = (x, y)$, and dim $A \leq \dim \pi(A) + 1$, it is easy to conclude using Marstrand's projection Theorem 4.1 that for any Borel set $A \subset \mathbb{R}^3$, for almost all $\theta \in [0, 2\pi)$,

$$\dim \pi_\theta(A) \geq \dim A - 1 \quad \text{if } \dim A \leq 2,$$
$$\dim \pi_\theta(A) = 1 \quad \text{if } \dim A \geq 2.$$

This is sharp by trivial examples; consider product sets $A = B \times C$, $B \subset \mathbb{R}^2$, $C \subset \mathbb{R}$. The reader can easily state and check the corresponding result for projections onto the orthogonal complements of the above lines. E. Järvenpää, M. Järvenpää, Ledrappier and Leikas [2008] showed that such sets of inequalities remain in force for any smooth, in a suitable sense non-degenerate, one-dimensional families of orthogonal projections onto lines and planes in $\mathbb{R}^3$. In fact, they proved such inequalities in more general dimensions and E. Järvenpää, M. Järvenpää and Keleti [2014] found the complete solution in all dimensions;

sharp inequalities for smooth non-degenerate families of orthogonal projections onto m-planes in $\mathbb{R}^n$.

However, this solution is not always sharp for a given family. In particular, the results remain true if one replaces the projections π_θ with the projections p_θ onto the lines $\{t(\cos\theta, \sin\theta, 1) : t \in \mathbb{R}\}$, but the trivial counter-examples do not work anymore. Actually one can now improve the above estimates relatively easily by showing that if $A \subset \mathbb{R}^3$ is a Borel set with $\dim A \leq 1/2$, then

$$\dim p_\theta(A) \geq \dim A \quad \text{for almost all } \theta \in [0, 2\pi).$$

The restriction $1/2$ comes because using Kaufman's method one is now led to estimate integrals of the type

$$\int_0^{2\pi} |a + \sin\theta|^{-s} \, d\theta$$

for $s < \dim A$, and they are bounded only if $s < 1/2$. So this is the best one can get without new ideas. Introducing some new geometric arguments Fässler and Orponen [2014] and Orponen [2013a] were able to improve these results. A little later D. M. Oberlin and R. Oberlin [2013b] obtained other improvements using the deep decay estimate theorem 15.5 of Erdoğan for spherical averages.

One reason for the possibility of such improvements is that the second family is more curved than the first one. That is, the set of the unit vectors generating the first family is the planar curve $\{(\cos\theta, \sin\theta, 0) : \theta \in [0, 2\pi]\}$ while for the second it spans the whole space $\mathbb{R}^3$.

There are also constancy results for projections: the dimension of the projections is the same for almost all planes. For the full Grassmannian and Hausdorff dimension this is obvious by Marstrand's projection theorem. For the packing and Minkowski dimension it is not obvious but true as shown by Falconer and Howroyd [1997]. Fässler and Orponen [2013] proved such results for certain restricted families of projections and Hausdorff, packing and Minkowski dimensions.

What more in addition to dimension estimates could be said about the exceptional sets? Are there interesting cases where there are no exceptions or where the exceptional set is countable? We shall discuss this in Chapter 10 in light of a particular example and with comments on more general self-similar sets. Can something be said about their structure, for example, could smooth sets or simple self-similar sets appear as exceptional sets or are they necessarily more complicated as in Example 5.13?

In addition to Sobolev dimension, there are many different dimensions for measures; see, for example, Falconer's book [1997], Chapter 10, and Bishop and Peres [2016], Chapter 1. In particular, in dynamical systems they are widely used. For instance, one can define the Hausdorff dimension of $\mu \in \mathcal{M}(\mathbb{R}^n)$ as

$$\dim \mu = \sup\{s : \liminf_{r \to 0} \log \mu(B(x, r))/\log r \geq s \text{ for } \mu \text{ almost all } x \in \mathbb{R}^n\}.$$

It follows that (see Falconer [1997])

$$\dim \mu = \inf\{\dim A : A \text{ is a Borel set with } \mu(A) > 0\}.$$

Then it is easy to show that $\dim_S \mu \leq \dim \mu$ and that strict inequality can occur, even with $\dim_S \mu = 0$, $\dim \mu = n$. However, if $0 < s < \dim \mu$, there is a Borel set with $\mu(A) > 0$ and $\dim_S(\mu \llcorner A) > s$.

6

Slices of measures and intersections with planes

Let $A \subset \mathbb{R}^n$ be a Borel set with dim $A > m$. We know from Theorem 5.8 that

$$\mathcal{H}^m(P_V(A)) > 0 \quad \text{for } \gamma_{n,m} \text{ almost all } V \in G(n,m).$$

This means that for $\gamma_{n,m}$ almost all $V \in G(n,m)$ the set of $a \in V$ for which the plane section $A \cap (V^\perp + a)$ is non-empty has positive $\mathcal{H}^m$ measure. But how large are these plane sections typically? The answer is that typically they have dimension dim $A - m$. A proof without the Fourier transform is given in Mattila [1995], Chapter 10. Here we give a Fourier analytic proof and estimate the dimension of the exceptional set of the planes V.

6.1 Sliced measures and estimates for energy-integrals

Let $\mu \in \mathcal{M}(\mathbb{R}^n)$. For any $V \in G(n,m)$ and $\mathcal{H}^m$ almost all $a \in V$, we can define sliced measures $\mu_{V,a}$ with the properties that

$$\text{spt}\mu_{V,a} \subset \text{spt}\mu \cap V_a^\perp \quad \text{where } V_a^\perp = V^\perp + a, \tag{6.1}$$

and for $\varphi \in C_0(\mathbb{R}^n)$,

$$\int_V \int \varphi \, d\mu_{V,a} \, d\mathcal{H}^m a = \int \varphi \, d\mu \quad \text{if } P_{V\sharp}\mu \ll \mathcal{H}^m \llcorner V. \tag{6.2}$$

We follow the construction of Mattila [1995], Section 10.1, where a few more details are given. Recall that if $I_m(\mu) < \infty$, then by Theorem 5.9, the push-forward measure $P_{V\sharp}\mu$ is absolutely continuous with respect to the Hausdorff m-measure $\mathcal{H}^m \llcorner V$ for $\gamma_{n,m}$ almost all $V \in G(n,m)$.

We start with a continuous non-negative compactly supported function φ on $\mathbb{R}^n$ and define a Radon measure ν_φ by setting

$$\nu_\varphi(A) = \int_A \varphi \, d\mu$$

for all Borel sets $A \subset \mathbb{R}^n$. Then $P_{V\sharp}\nu_\varphi$ is a Radon measure on V and by Theorem 2.11 the limit, the Radon–Nikodym derivative $D(P_{V\sharp}\nu_\varphi, a)$,

$$\mu_{V,a}(\varphi) := \lim_{\delta \downarrow 0} \alpha(m)^{-1} \delta^{-m} P_{V\sharp}\nu_\varphi(B(a, \delta)) = \lim_{\delta \downarrow 0} \alpha(m)^{-1} \delta^{-m} \int_{P_V^{-1}(B(a,\delta))} \varphi \, d\mu$$

exists for $\mathcal{H}^m$ almost all $a \in V$. In the above construction we first fixed φ and then defined $\mu_{V,a}(\varphi)$ for $\mathcal{H}^m$ almost all a. The exceptional set of the points a for which the limit does not exist will a priori depend on the choice of φ. However, by the separability of $C_0^+(\mathbb{R}^n)$, one can easily eliminate the dependence on φ. Thus we can define for $\mathcal{H}^m$ almost all $a \in V$ a non-negative functional on $C_0^+(\mathbb{R}^n)$ by

$$\varphi \mapsto \lim_{\delta \downarrow 0} \alpha(m)^{-1} \delta^{-m} \int_{P_V^{-1}(B(a,\delta))} \varphi \, d\mu.$$

This functional extends to a positive linear functional on $C_0(\mathbb{R}^n)$ and it follows by the Riesz representation theorem that for $\mathcal{H}^m$ almost all $a \in V$ there exists a Radon measure $\mu_{V,a}$ so that

$$\int \varphi \, d\mu_{V,a} = \lim_{\delta \downarrow 0} \alpha(m)^{-1} \delta^{-m} \int_{P_V^{-1}(B(a,\delta))} \varphi \, d\mu$$

for all $\varphi \in C_0(\mathbb{R}^n)$. This gives immediately (6.1). We call $\mu_{V,a}$ the *sliced measure* associated to the subspace V at the point a.

Theorem 2.11 implies that for any Borel set $B \subset V$ and any $\varphi \in C_0^+(\mathbb{R}^n)$,

$$\int_B D(P_{V\sharp}\nu_\varphi, a) \, d\mathcal{H}^m a \leq P_{V\sharp}\nu_\varphi(B) = \int_{P_V^{-1}(B)} \varphi \, d\mu \qquad (6.3)$$

with equality if $P_{V\sharp}\nu_\varphi \ll \mathcal{H}^m$. This means that

$$\int_B \int \varphi \, d\mu_{V,a} \, d\mathcal{H}^m a \leq \int_{P_V^{-1}(B)} \varphi \, d\mu \qquad (6.4)$$

with equality if $P_{V\sharp}\mu \ll \mathcal{H}^m$, since $P_{V\sharp}\mu \ll \mathcal{H}^m$ implies $P_{V\sharp}\nu_\varphi \ll \mathcal{H}^m$ for all $\varphi \in C_0^+(\mathbb{R}^n)$. Hence (6.2) holds, and in particular

$$\int_V \mu_{V,a}(\mathbb{R}^n) \, d\mathcal{H}^m a = \mu(\mathbb{R}^n) \quad \text{if } P_{V\sharp}\mu \ll \mathcal{H}^m. \qquad (6.5)$$

Using the fact that every non-negative lower semicontinuous function on $\mathbb{R}^n$ is a non-decreasing limit of functions in $C_0^+(\mathbb{R}^n)$ we conclude that (6.4) holds for functions which are merely lower semicontinuous: for each lower semicontinuous $g : \mathbb{R}^n \to [0, \infty]$ we have

$$\int_B \int g \, d\mu_{V,a}, \, d\mathcal{H}^m a \leq \int_{P_V^{-1}(B)} g \, d\mu \qquad (6.6)$$

for all Borel sets $B \subset V$, with equality if $P_{V\sharp}\mu \ll \mathcal{H}^m$.

Let ψ be a non-negative C^∞-function on $\mathbb{R}^n$ with compact support and such that $\psi(x) \geq (\alpha(m)\alpha(n-m))^{-1}$ for $x \in B(0,2)$. For $\varepsilon > 0$ define $\psi_\varepsilon(x) = \varepsilon^{-n}\psi(x/\varepsilon)$ and set for $\mu \in \mathcal{M}(\mathbb{R}^n)$, $\mu_\varepsilon = \psi_\varepsilon * \mu$.

Lemma 6.1 *Let $\mu \in \mathcal{M}(\mathbb{R}^n)$ with $I_m(\mu) < \infty$, $V \in G(n,m)$ and $a \in V$. Then for any lower semicontinuous function $g : \mathbb{R}^n \times \mathbb{R}^n \to [0,\infty]$,*

$$\iint g(x,y)\,d\mu_{V,a}x\,d\mu_{V,a}y$$

$$\leq \liminf_{\varepsilon \to 0} \int_{V^\perp} \int_{V^\perp} g(u+a, v+a)\mu_\varepsilon(u+a)\mu_\varepsilon(v+a)\,d\mathcal{H}^{n-m}u\,d\mathcal{H}^{n-m}v$$

provided the sliced measure $\mu_{V,a}$ exists.

This lemma follows immediately from the following lemma since $\tilde{\mu}_\varepsilon \leq \mu_\varepsilon$.

Lemma 6.2 *Let $\mu \in \mathcal{M}(\mathbb{R}^n)$ with $I_m(\mu) < \infty$, $V \in G(n,m)$, $a \in V$ and suppose that the sliced measure $\mu_{V,a}$ exists. Define*

$$C = \{x \in \mathbb{R}^n : |P_V(x)| \leq 1 \text{ and } |P_{V^\perp}(x)| \leq 1\},$$

$$\chi_\varepsilon(x) = (\alpha(m)\alpha(n-m))^{-1}\varepsilon^{-n}\chi_C(x/\varepsilon)$$

and

$$\tilde{\mu}_\varepsilon = \chi_\varepsilon * \mu.$$

(i) For any continuous function $\varphi : \mathbb{R}^n \to [0,\infty]$ with compact support,

$$\int \varphi\,d\mu_{V,a} = \lim_{\varepsilon \to 0} \int_{V^\perp} \varphi(u+a)\tilde{\mu}_\varepsilon(u+a)\,d\mathcal{H}^{n-m}u.$$

(ii) For any continuous function $\varphi : \mathbb{R}^n \times \mathbb{R}^n \to [0,\infty]$ with compact support,

$$\iint \varphi(x,y)\,d\mu_{V,a}x\,d\mu_{V,a}y$$

$$= \lim_{\varepsilon \to 0} \int_{V^\perp} \int_{V^\perp} \varphi(u+a, v+a)\tilde{\mu}_\varepsilon(u+a)\tilde{\mu}_\varepsilon(v+a)\,d\mathcal{H}^{n-m}u\,d\mathcal{H}^{n-m}v.$$

Proof The statement in (i) means that the measures $\tilde{\mu}_\varepsilon d\mathcal{H}^{n-m} \llcorner V_a^\perp$ converge weakly to $\mu_{V,a}$ as $\varepsilon \to 0$. Similarly, in the product space $\mathbb{R}^n \times \mathbb{R}^n$, the statement in (ii) means that the product measures $\tilde{\mu}_\varepsilon d\mathcal{H}^{n-m} \llcorner V_a^\perp \times \tilde{\mu}_\varepsilon d\mathcal{H}^{n-m} \llcorner V_a^\perp$ converge weakly to $\mu_{V,a} \times \mu_{V,a}$ as $\varepsilon \to 0$. Thus (ii) follows from (i) and the following general fact.

If $\sigma_\varepsilon \in \mathcal{M}(\mathbb{R}^p)$, $\tau_\varepsilon \in \mathcal{M}(\mathbb{R}^q)$, $\varepsilon > 0$, $\sigma_\varepsilon \to \sigma$ and $\tau_\varepsilon \to \tau$ weakly as $\varepsilon \to 0$, then $\sigma_\varepsilon \times \tau_\varepsilon \to \sigma \times \tau$.

This is easily verified. The convergence of $\int \varphi\,d\sigma_\varepsilon \times \tau_\varepsilon$ to $\int \varphi\,d\sigma \times \tau$ is immediate when $\varphi \in C_0(\mathbb{R}^{p+q})$ is of the form $\varphi(x,y) = \varphi_1(x)\varphi_2(y)$, and the general case follows since finite linear combinations of such products are dense in $C_0(\mathbb{R}^{p+q})$, either by the Stone–Weierstrass approximation theorem or by some simple direct argument.

So we have left to prove (i). To do this let $\varphi \in C_0(\mathbb{R}^n)$ and $\varepsilon > 0$. Then by Fubini's theorem, the definitions of $\widetilde{\mu}_\varepsilon$ and χ_ε, and the fact that $\int \chi_\varepsilon = 1$,

$$\alpha(m)^{-1}\varepsilon^{-m} \int_{P_V^{-1}(B(a,\varepsilon))} \varphi\, d\mu - \int_{V^\perp} \varphi(u+a)\widetilde{\mu}_\varepsilon(u+a)\, d\mathcal{H}^{n-m}u$$

$$= \alpha(m)^{-1}\varepsilon^{-m} \int_{P_V^{-1}(B(a,\varepsilon))} \varphi\, d\mu - \int_{V^\perp} \varphi(u+a)\int \chi_\varepsilon(x-u-a)\, d\mu x\, d\mathcal{H}^{n-m}u$$

$$= \alpha(m)^{-1}\varepsilon^{-m} \int_{P_V^{-1}(B(a,\varepsilon))} \varphi\, d\mu - \iint_{V^\perp} \varphi(u+a)\chi_\varepsilon(x-u-a)\, d\mathcal{H}^{n-m}u\, d\mu x$$

$$= \alpha(m)^{-1}\varepsilon^{-m} \int_{P_V^{-1}(B(a,\varepsilon))} \varphi\, d\mu - \alpha(m)^{-1}\varepsilon^{-m}$$

$$\times \int_{P_V^{-1}(B(a,\varepsilon))} \alpha(n-m)^{-1}\varepsilon^{m-n} \int_{V^\perp \cap B(P_{V^\perp}(x),\varepsilon)} \varphi(u+a)\, d\mathcal{H}^{n-m}u\, d\mu x$$

$$= \alpha(m)^{-1}\varepsilon^{-m} \int_{P_V^{-1}(B(a,\varepsilon))} \alpha(n-m)^{-1}\varepsilon^{m-n}$$

$$\times \int_{V^\perp \cap B(P_{V^\perp}(x),\varepsilon)} (\varphi(x)-\varphi(u+a))\, d\mathcal{H}^{n-m}u\, d\mu x.$$

In the last integrals $|u - P_{V^\perp}(x)| \le \varepsilon$ and $|a - P_V(x)| \le \varepsilon$, whence $|u + a - x| = |u - P_{V^\perp}(x) + a - P_V(x)| \le 2\varepsilon$. Thus given $\eta > 0$, we have $|\varphi(x) - \varphi(u+a)| < \eta$ when ε is sufficiently small. Then the last double integral is less than $\eta\alpha(m)^{-1}\varepsilon^{-m}\mu(P_V^{-1}(B(a)))$ in absolute value. As $\alpha(m)^{-1}\varepsilon^{-m}\int_{P_V^{-1}(B(a,\varepsilon))} \varphi\, d\mu$ converges to $\int \varphi\, d\mu_{V,a}$ and $\eta\alpha(m)^{-1}\varepsilon^{-m} \times \mu(P_V^{-1}(B(a)))$ converges to $\eta\mu_{V,a}(\mathbb{R}^n)$, which are finite, (i) follows. This completes the proof of the lemma. $\qquad\square$

Proposition 6.3 *Let* $m < s < n$, $\mu \in \mathcal{M}(\mathbb{R}^n)$ *and* $V \in G(n,m)$. *Then*

$$\int_V I_{s-m}(\mu_{V,a})\, d\mathcal{H}^m a \le C(n,m,s) \int_{\mathbb{R}^n} |P_{V^\perp}(x)|^{s-n} |\widehat{\mu}(x)|^2\, dx. \tag{6.7}$$

Proof Let ψ_ε and $\mu_\varepsilon = \psi_\varepsilon * \mu$ be as above. By Lemma 6.1 we have for $a \in V$,

$$I_{s-m}(\mu_{V,a}) = \iint |x-y|^{m-s}\, d\mu_{V,a}x\, d\mu_{V,a}y$$

$$\le \liminf_{\varepsilon\to 0} \int_{V^\perp} \int_{V^\perp} |u-v|^{m-s}\mu_\varepsilon(u+a)\mu_\varepsilon(v+a)\, d\mathcal{H}^{n-m}u\, d\mathcal{H}^{n-m}v. \tag{6.8}$$

Write $\mu_{\varepsilon,a}(u) = \mu_{\varepsilon}(u + a)$ and $\gamma = \gamma(n - m, s - m)$. Applying Theorem 3.10 in the $(n - m)$-space $V^{\perp}$, we have

$$\int_{V^{\perp}} \int_{V^{\perp}} |u - v|^{m-s} \mu_{\varepsilon}(u + a)\mu_{\varepsilon}(v + a) \, d\mathcal{H}^{n-m}u \, d\mathcal{H}^{n-m}v$$

$$= \int_{V^{\perp}} (k_{s-m} * \mu_{\varepsilon,a})\mu_{\varepsilon,a} \, d\mathcal{H}^{n-m} = \gamma \int_{V^{\perp}} |u|^{s-n} |\widehat{\mu_{\varepsilon,a}}(u)|^2 \, d\mathcal{H}^{n-m}u$$

$$= \gamma \int_{V^{\perp}} |u|^{s-n} \left| \int_{V^{\perp}} e^{-2\pi i u \cdot v} \mu_{\varepsilon}(v + a) \, d\mathcal{H}^{n-m}v \right|^2 d\mathcal{H}^{n-m}u.$$

Integrating over V and using Parseval's theorem on V and Fubini's theorem, we obtain

$$\int_V \int_{V^{\perp}} \int_{V^{\perp}} |u - v|^{m-s} \mu_{\varepsilon}(u + a)\mu_{\varepsilon}(v + a) \, d\mathcal{H}^{n-m}u \, d\mathcal{H}^{n-m}v \, d\mathcal{H}^m a$$

$$= \gamma \int_{V^{\perp}} |u|^{s-n} \int_V \left| \int_{V^{\perp}} e^{-2\pi i u \cdot v} \mu_{\varepsilon}(v + a) \, d\mathcal{H}^{n-m}v \right|^2 d\mathcal{H}^m a \, d\mathcal{H}^{n-m}u$$

$$= \gamma \int_{V^{\perp}} |u|^{s-n} \int_V \left| \int_V e^{-2\pi i a \cdot b} \int_{V^{\perp}} e^{-2\pi i u \cdot v} \mu_{\varepsilon}(v + b) \, d\mathcal{H}^{n-m}v \, d\mathcal{H}^m b \right|^2$$
$$\times \, d\mathcal{H}^m a \, d\mathcal{H}^{n-m}u$$

$$= \gamma \int_{V^{\perp}} |u|^{s-n} \int_V \left| \int_V \int_{V^{\perp}} e^{-2\pi i (u+a) \cdot (v+b)} \mu_{\varepsilon}(v + b) \, d\mathcal{H}^{n-m}v \, d\mathcal{H}^m b \right|^2$$
$$\times \, d\mathcal{H}^m a \, d\mathcal{H}^{n-m}u$$

$$= \gamma \int_{\mathbb{R}^n} |P_{V^{\perp}}(x)|^{s-n} \left| \int_{\mathbb{R}^n} e^{-2\pi i x \cdot y} \mu_{\varepsilon}(y) \, d\mathcal{L}^n y \right|^2 d\mathcal{L}^n x$$

$$= \gamma \int_{\mathbb{R}^n} |P_{V^{\perp}}(x)|^{s-n} |\widehat{\mu_{\varepsilon}}(x)|^2 \, dx$$

$$= \gamma \int_{\mathbb{R}^n} |P_{V^{\perp}}(x)|^{s-n} |\widehat{\psi_{\varepsilon}}(x)|^2 |\widehat{\mu}(x)|^2 \, dx$$

$$\lesssim \int_{\mathbb{R}^n} |P_{V^{\perp}}(x)|^{s-n} |\widehat{\mu}(x)|^2 \, dx.$$

Combining this with (6.8) we get

$$\int_V I_{s-m}(\mu_{V,a}) \, d\mathcal{H}^m a \lesssim \int_{\mathbb{R}^n} |P_{V^{\perp}}(x)|^{s-n} |\widehat{\mu}(x)|^2 \, dx,$$

as desired. $\square$

Proposition 6.4 *Let $m < s < n$, $\mu \in \mathcal{M}(\mathbb{R}^n)$ and let $\nu \in \mathcal{M}(G(n, m))$ be such that for some $t > m(n - m) + m - s$,*

$$\nu(B(V, r)) \le r^t \quad \text{for } V \in G(n, m), r > 0.$$

Then

$$\iint_V I_{s-m}(\mu_{V,a}) \, d\mathcal{H}^m a \, dvV \leq C(n, m, s) I_s(\mu).\tag{6.9}$$

Proof Integrating (6.7) with respect to v we obtain by Fubini's theorem

$$\iint_V I_{s-m}(\mu_{V,a}) \, d\mathcal{H}^m a \, dvV \lesssim \int_{\mathbb{R}^n} \int |P_{V^\perp}(x)|^{s-n} \, dvV |\widehat{\mu}(x)|^2 \, dx.$$

To estimate the inner integral we observe first that for $x \in \mathbb{R}^n \setminus \{0\}$,

$$\int |P_{V^\perp}(x)|^{s-n} \, dvV = |x|^{s-n} \int |P_{V^\perp}(x/|x|)|^{s-n} \, dvV,$$

then write $v = x/|x|$ and use (5.11) to get

$$\int |P_{V^\perp}(v)|^{s-n} \, dvV = \int_0^\infty v(\{V : |P_{V^\perp}(v)|^{s-n} > u\}) \, du$$
$$\leq v(G(n, m)) + \int_1^\infty v(\{V : d(v, V) < u^{1/(s-n)}\}) \, du$$
$$\lesssim 1 + \int_1^\infty u^{\frac{t-(m-1)(n-m)}{s-n}} \, du \lesssim 1,$$

since $t > m(n - m) + m - s$. Thus

$$\iint_V I_{s-m}(\mu_{V,a}) \, d\mathcal{H}^m a \, dvV \lesssim \int |x|^{s-n} |\widehat{\mu}(x)|^2 \, dx = \gamma(n, s)^{-1} I_s(\mu). \qquad \square$$

As before Proposition 6.4 immediately gives with the help of Frostman's lemma

Theorem 6.5 *Let $m < s < n$ and $\mu \in \mathcal{M}(\mathbb{R}^n)$ with $I_s(\mu) < \infty$. Then*

$$\dim\{V \in G(n, m) : \int I_{s-m}(\mu_{V,a}) \, d\mathcal{H}^m a = \infty\} \leq m(n - m) + m - s.\tag{6.10}$$

6.2 Dimension of plane sections

Now we are ready to get information about the dimension of plane sections of sets. First we derive easily an upper bound:

Proposition 6.6 *Let $s \geq m$ and $A \subset \mathbb{R}^n$ with $\mathcal{H}^s(A) < \infty$. Then for any $V \in G(n, m)$,*

$$\mathcal{H}^{s-m}(A \cap (V^\perp + a)) < \infty \quad \text{for } \mathcal{H}^m \text{ almost all } a \in V.$$

Proof Cover A for every $k = 1, 2, \ldots$ with compact sets $E_{k,i}, i = 1, 2, \ldots$, such that $d(E_{k,i}) < 1/k$ and

$$\sum_i \alpha(s)2^{-s}d(E_{k,i})^s < \mathcal{H}^s_{1/k}(A) + 1/k.$$

Let

$$F_{k,i} = \{a \in V : E_{k,i} \cap (V^\perp + a) \neq \varnothing\}.$$

Then $d(F_{k,i}) \leq d(E_{k,i})$, whence

$$\mathcal{H}^m(F_{k,i}) \leq \alpha(m)d(E_{k,i})^m.$$

Denoting by $\int^*$ the upper integral and using Fatou's lemma we obtain

$$\int^* \mathcal{H}^{s-m}(A \cap (V^\perp + a))\,d\mathcal{H}^m a = \int^* \lim_{k\to\infty} \mathcal{H}^{s-m}_{1/k}(A \cap (V^\perp + a))\,d\mathcal{H}^m a$$

$$\leq \int \liminf_{k\to\infty} \sum_i \alpha(s-m)2^{m-s}d(E_{k,i} \cap (V^\perp + a))^{s-m}\,d\mathcal{H}^m a$$

$$\leq \liminf_{k\to\infty} \sum_i \alpha(s-m)2^{m-s} \int_{F_{k,i}} d(E_{k,i} \cap (V^\perp + a))^{s-m}\,d\mathcal{H}^m a$$

$$\leq \liminf_{k\to\infty} \sum_i \alpha(s-m)2^{m-s}d(E_{k,i})^{s-m}\mathcal{H}^m(F_{k,i})$$

$$\leq \alpha(s-m)2^{m-s}\alpha(m) \liminf_{k\to\infty} \sum_i d(E_{k,i})^s$$

$$\leq C(m,s) \liminf_{k\to\infty}(\mathcal{H}^s_{1/k}(A) + 1/k) = C(m,s)\mathcal{H}^s(A) < \infty.$$

This gives the proposition. $\qquad\square$

The following theorem improves Corollary 5.12(b). Notice in particular that the exceptional set E has $\gamma_{n,m}$ measure zero.

Theorem 6.7 *Let $m < s \leq n$ and let $A \subset \mathbb{R}^n$ be a Borel set with $0 < \mathcal{H}^s(A) < \infty$. Then there is a Borel set $E \subset G(n,m)$ such that*

$$\dim E \leq m(n - m) + m - s$$

and

$$\mathcal{H}^m(\{a \in V : \dim A \cap (V^\perp + a) = s - m\}) > 0 \quad \text{for all } V \in G(n,m) \setminus E.$$

Proof By Frostman's lemma there is $\mu \in \mathcal{M}(A)$ such that $\mu(B(x,r)) \leq r^s$ for all $x \in \mathbb{R}^n$ and $r > 0$. Then $I_u(\mu) < \infty$ for all $m < u < s$. By Theorem 5.10 $P_{V\sharp}\mu \ll \mathcal{H}^m \llcorner V$ for all $V \in G(n,m)$ outside a set of Hausdorff dimension at most $m(n - m) + m - s$. By Theorem 6.5 $\int I_{u-m}(\mu_{V,a})\,d\mathcal{H}^m a < \infty$ outside a set of Hausdorff dimension at most $m(n - m) + m - u$. Thus setting for $i = 1, 2, \ldots, s - 1/i > m$,

$$E_i = \left\{ V \in G(n, m) : P_{V\sharp}\mu \not\ll \mathcal{H}^m \llcorner V \text{ or } \int I_{s-1/i-m}(\mu_{V,a}) \, d\mathcal{H}^m a = \infty \right\},$$

$$E = \bigcap_{j=1}^{\infty} \bigcup_{i=j}^{\infty} E_i,$$

we have $\dim E_i \leq m(n - m) + m - s + 1/i$ and $\dim E \leq m(n - m) + m - s$.

Let $V \in G(n, m) \setminus E$. Then there is j such that $V \notin E_i$ for all $i \geq j$, whence $P_{V\sharp}\mu \ll \mathcal{H}^m \llcorner V$ and $I_{s-1/i-m}(\mu_{V,a}) < \infty$ for $\mathcal{H}^m$ almost all $a \in V$. The first of these statements implies by (6.5) that $\mu_{V,a}(\mathbb{R}^n) > 0$ for $a \in V$ in a set of positive $\mathcal{H}^m$ measure, and the second that for $\mathcal{H}^m$ almost all such a, $\dim A \cap (V^\perp + a) \geq s - 1/i - m$. It follows that for $V \in G(n, m) \setminus E$,

$$\mathcal{H}^m(\{a \in V : \dim A \cap (V^\perp + a) \geq s - m\}) > 0.$$

The theorem follows now combining this with Proposition 6.6. $\qquad\square$

Theorem 6.8 *Let $m \leq s \leq n$ and let $A \subset \mathbb{R}^n$ be a Borel set with $\dim A > s$. Then there is a Borel set $E \subset G(n, m)$ such that*

$$\dim E \leq m(n - m) + m - s$$

and

$$\mathcal{H}^m(\{a \in V : \dim A \cap (V^\perp + a) > s - m\}) > 0 \quad \text{for all } V \in G(n, m) \setminus E.$$

The same proof as that of Theorem 6.7 gives this.

We state without proof an alternative version of Theorem 6.7. This follows from Theorem 6.7 by the argument for the proof of Theorem 10.10 in Mattila [1995]:

Theorem 6.9 *Let $m < s \leq n$ and let $A \subset \mathbb{R}^n$ be a Borel set with $\mathcal{H}^s(A) < \infty$. Then there is a Borel set $E \subset G(n, m)$ such that*

$$\dim E \leq m(n - m) + m - s$$

and for $\mathcal{H}^s$ almost all $x \in A$,

$$\dim A \cap (V^\perp + x) = s - m \quad \text{for all } V \in G(n, m) \setminus E.$$

6.3 Measures on graphs

We give here an application of the inequality (6.7) to measures on fractal graphs. The graph of a continuous function $f : [0, 1] \to \mathbb{R}$ can have large Hausdorff dimension, even equal to 2. Thus one could expect that at least some such fractal graphs in $\mathbb{R}^2$ would support measures whose Fourier transforms decay at infinity more quickly than $1/\sqrt{|x|}$. The surprising result of Fraser, Orponen and Sahlsten [2014] shows that the opposite is true:

Theorem 6.10 *For any function $f : A \to \mathbb{R}^{n-m}, A \subset \mathbb{R}^m$, we have for the graph $G_f = \{(x, f(x)) : x \in A\}$,*

$$\dim_F G_f \leq m.$$

Recall from Section 3.6 the definition of the Fourier dimension $\dim_F$. The theorem means that for any $\mu \in \mathcal{M}(G_f)$ the decay estimate $|\widehat{\mu}(x)| \lesssim |x|^{-s/2}$ can hold only if $s \leq m$. No measurability for f is required. This is because the only property of the graph that is used is that it intersects the $(n - m)$-planes $\{(a, y) : y \in \mathbb{R}^{n-m}\}, a \in \mathbb{R}^m$, in at most one point.

Proof Suppose that $s > 0$ and $\mu \in \mathcal{M}(G_f)$ are such that

$$|\widehat{\mu}(x)| \lesssim (1 + |x|)^{-s/2} \quad \text{for } x \in \mathbb{R}^n. \tag{6.11}$$

We have to show that $s \leq m$. Suppose on the contrary that $s > m$ and let $m < t < s$. We shall apply the inequality (6.7) with $V = \mathbb{R}^m \times \{0\}$ identified with $\mathbb{R}^m$. As before in (5.15), $\widehat{P_{V\sharp}\mu}(x) = \widehat{\mu}(x)$ for $x \in V$. Hence the decay estimate (6.11), with $s > m$, implies that $\widehat{P_{V\sharp}\mu} \in L^2(V)$ and so $P_{V\sharp}\mu \ll \mathcal{H}^m \llcorner V$. By (6.7)

$$\int_V I_{t-m}(\mu_{V,a}) \, d\mathcal{H}^m a \lesssim \int_{\mathbb{R}^n} |P_{V^\perp}(x)|^{t-n} |\widehat{\mu}(x)|^2 \, dx$$

$$\lesssim \int_{\mathbb{R}^n} |P_{V^\perp}(x)|^{t-n} (1 + |x|)^{-s} \, dx < \infty.$$

The finiteness of the last integral easily follows by Fubini's theorem since $t < s$. Thus $I_{t-m}(\mu_{V,a}) < \infty$ for $\mathcal{H}^m$ almost all $a \in V$. By (6.5) $\mu_{V,a}(\mathbb{R}^n) > 0$ for $a \in V$ in a set of positive $\mathcal{H}^m$ measure. Since also by (6.1) $\operatorname{spt}\mu_{V,a} \subset \operatorname{spt}\mu \cap V_a^\perp \subset G_f \cap V_a^\perp$, we get $\dim G_f \cap V_a^\perp \geq t - m > 0$ by Theorem 2.8. But this is impossible as $G_f \cap V_a^\perp$ contains at most one point for every $a \in V$. $\qquad\square$

6.4 Further comments

Theorem 6.8 without the exceptional set estimate, that is with $\gamma_{n,m}(E) = 0$, was proved by Marstrand [1954] in the plane, and in general dimensions by

Mattila [1975]. A proof with sliced measures and energy-integrals was given by Mattila [1981]. The exceptional set estimates and the Fourier analytic proof are due to Orponen [2014a].

Results on dimensions of sliced measures were proven by Falconer and Mattila [1996], M. Järvenpää and Mattila [1998] and E. Järvenpää, M. Järvenpää and Llorente [2004].

The upper bound $m(n - m) + m - s$ in Theorem 6.7 for the dimension of the exceptional set is sharp, since it was sharp already for Corollary 5.12(b). The exceptional set estimates in Theorems 6.7 – 6.9 concern only the Grassmannian part. Could it be possible to obtain some dimension estimates also for subsets of A, for example in Theorem 6.9 replacing $\mathcal{H}^s$ almost all by $\mathcal{H}^t$ almost all for some $t < s$? One could also ask for dimension estimates for sets where $\dim A \cap (V^\perp + x) < t$ or $\dim A \cap (V^\perp + x) > u$ when $t < s - m < u$. An easy estimate of this sort says that we can improve Proposition 6.6 to the the statement $\mathcal{H}^{s-t}(A \cap (V^\perp + a)) < \infty$ for $\mathcal{H}^t$ almost all $a \in V$ for $0 \le t \le s$. This follows by a straightforward modicifation of the proof of Proposition 6.6 or using the general inequality in Theorem 2.10.25 of Federer [1969].

There are more precise results for particular self-similar and related sets. Many of them say that in a fixed direction the dimension of the sections typically is a constant depending on the direction. Typically here refers to almost all planes in that direction meeting the set. A rather general result of this type was obtained by Wen and Xi [2010]. But often in some special directions, for example, in the direction of the coordinate planes or diagonal in $\mathbb{R}^n \times \mathbb{R}^n$ or in a countable dense set of 'rational' directions, this constant is smaller than the one for generic directions given by the results of this chapter. This is the case, for example, by the results of Hawkes [1975] for $C_{1/3} \times C_{1/3}$, where $C_{1/3}$ is the classical one third Cantor set, by Kenyon and Peres [1991] for products $C \times D$ of more general Cantor sets, by Liu, Xi and Zhao [2007] and Manning and Simon [2013] for the Sierpinski carpet, and by Barany, Ferguson and Simon [2012] for the Sierpinski gasket. Benjamini and Peres [1991] estimated the dimension of vertical sections in a planar fractal costruction with sharp dimension bounds for the corresponding exceptional set. Classes of self-similar sets were found by Wen, Wu and Xi [2013] for which some explicit directions could be determined such that the sections typically have exactly the generic value $\dim A - m$. Typically could also refer to almost all lines with respect to the projected measure instead of the Lebesgue measure. Then the results are often different and the dimension may be bigger than the generic value, see Manning and Simon [2013], Barany, Ferguson and Simon [2012] and Barany and Rams [2014].

The above mentioned results for products of Cantor sets C and D in $\mathbb{R}$ actually give dimensions of typical intersections $C \cap (D + z)$, $z \in \mathbb{R}$, but this is the same as intersecting $C \times D$ with lines parallel to the diagonal. Bishop and Peres [2016] give a detailed discussion on such intersections.

Let $A \subset \mathbb{R}^n$ be a Borel set with $0 < \mathcal{H}^s(A) < \infty$ for some $m < s < n$. Proposition 6.6 tells us that the intersections of A with $n - m$ planes typically have finite $(s - m)$-dimensional Hausdorff measure. Marstrand [1954] gave an example showing that almost all intersections may have zero measure. In general, determining whether the measure is positive or zero seems to be difficult even for simple self-similar sets. Kempton [2013] managed to prove the positivity for some self-similar sets. Orponen [2013b] proved that for many self-similar sets the generic intersection has infinite packing measure. For non-integral s it does not seem to be easy to find sets A with positive $\mathcal{H}^s$ measure for which $\mathcal{H}^{s-m}$ measure of $A \cap V$ would be finite for all $(n - m)$-planes V. Shmerkin and Suomala [2012] succeeded in this: they used random constructions to show that for any $n - 1 < s < n$ there exist compact sets $F \subset \mathbb{R}^n$ with $0 < \mathcal{H}^s(F) < \infty$ such that $\mathcal{H}^{s-n+1}(F \cap L) \leq C(F) < \infty$ for every line $L \subset \mathbb{R}^n$. In Shmerkin and Suomala [2014] they develop an interesting theory for a very general class of random measures with many results on projections, intersections and Fourier transforms. This paper also is a good source for references for related work.

In a way the slicing of a measure with plane sections is a special case of Rokhlin's [1962] general disintegration theorem, but it is essential for us in addition to have the concrete limit formulas for the sliced measures. Rokhlin's theorem gives for a map $f : X \to Y$ and a measure μ on X the disintegration formula

$$\int f \, d\mu = \int \left(\int f \, d\mu_y \right) df_\# \mu y$$

under very general conditions. Here the measures μ_y are called conditional measures and they are carried by the level sets $f^{-1}(y); \mu_y(X \setminus f^{-1}(y)) = 0$. They are defined for $f_\# \mu$ almost all $y \in Y$, whence setting $\mu_{f,x} = \mu_{f(x)}$, these measures are defined for μ almost all $x \in X$.

Furstenberg [2008] proved a general dimension conservation formula for homogeneous fractals, which include many self-similar fractals. Often this formula can be stated as

$$\dim f_\# \mu + \dim \mu_{f,x} = \dim \mu \quad \text{for } \mu \text{ almost all } x \in X.$$

In particular, for many self-similar measures and for typical measures in dynamical zooming processes which Furstenberg defined, this holds for every projection $f = P_V$, $V \in G(n, m)$, not only for almost every projection. Hochman [2014] developed this much further. See also Barany, Ferguson and Simon [2012] for a discussion about Furstenberg's formula in connection with the line sections of the Sierpinski gasket and Falconer and Jin [2014a], [2014b] in connection with a general setting including many deterministic and random self-similar sets.

Theorem 6.10 was proved by Fraser, Orponen and Sahlsten [2014]. In this paper it is also shown that for typical, in the Baire category sense, continuous functions $f : [0, 1] \to \mathbb{R}$, $\dim_F G_f = 0$, and more precisely

$$\limsup_{|x| \to \infty} |\widehat{\mu}(x)| \geq 1/5$$

for any probability measure $\mu \in \mathcal{M}(G_f)$. This paper answers the question of Kahane, see Shieh and Xiao [2006]: the graphs of the one-dimensional Brownian motion are almost surely not Salem sets. This is in contrast to trajectories which are almost surely Salem sets; see Chapter 12 for that. However, the interesting question about the almost sure Fourier dimension of the graphs of the Brownian motion is left open; Fraser, Orponen and Sahlsten [2014] only says that it is at most 1. It also seems to be open whether the level sets of Brownian motion are almost surely Salem sets.

We shall study smooth surfaces with non-zero Gaussian curvature in Section 14.3 and we show there that the Fourier transform of the surface measure has similar optimal decay as for the spheres. So the result of Fraser, Orponen and Sahlsten tells us that no better decay can take place on fractal surfaces than on smooth ones.

7

Intersections of general sets and measures

In this chapter we look at the general case where we have two arbitrary Borel sets A and B in $\mathbb{R}^n$, we keep A fixed, we move B by rotations $g \in O(n)$ and translations $\tau_z : x \mapsto x + z, z \in \mathbb{R}^n$, and we try to say something about the dimension of the intersections $A \cap \tau_z(g(B))$.

7.1 Intersection measures and energy estimates

Recall that θ_n is the unique Haar measure on the orthogonal group $O(n)$ with $\theta_n(O(n)) = 1$.

Let $\mu, \nu \in \mathcal{M}(\mathbb{R}^n)$. For $g \in O(n)$ and $x, y \in \mathbb{R}^n$ define

$$S_g(x, y) = x - g(y).$$

Observe that $x \in A \cap \tau_z(g(B))$ if and only $x \in A$ and $x = g(y) + z$, that is, $S_g(x, y) = z$, for some $y \in B$. Now we try to define intersection measures supported in $\operatorname{spt} \mu \cap \tau_z(g(\operatorname{spt} \nu)) = \pi(\{(x, y) \in \operatorname{spt} \mu \times \operatorname{spt} \nu, S_g(x, y) = z\})$, where $\pi(x, y) = x$. This is done by slicing as in the previous chapter. In fact, the process below is exactly slicing the product measure $\mu \times g_\sharp \nu$ with the n-planes parallel to the diagonal $\{(x, y) \in \mathbb{R}^{2n} : x = y\}$ and then projecting with π.

Let $\varphi \in C_0^+(\mathbb{R}^n)$ and define the measure $\lambda_\varphi \in \mathcal{M}(\mathbb{R}^n)$ setting for Borel sets $A \subset \mathbb{R}^n$,

$$\lambda_\varphi(A) = \iint_{S_g^{-1}(A)} \varphi(x) \, d\mu x \, d\nu y.$$

This means that λ_φ is the image of $(\varphi \mu) \times \nu$ under the map S_g. Then by the differentiation theorem 2.11 the limit

$$\lim_{\delta \downarrow 0} \alpha(n)^{-1} \delta^{-n} \lambda_\varphi(B(z, \delta)) = \lim_{\delta \downarrow 0} \alpha(n)^{-1} \delta^{-n} \iint_{\{(x,y):|x-g(y)-z|\leq\delta\}} \varphi(x) \, d\mu x \, d\nu y$$

100

exists and is finite for $\mathcal{L}^n$ almost all $z \in \mathbb{R}^n$. Thus as in the previous chapter we can define for $\mathcal{L}^n$ almost all $z \in \mathbb{R}^n$ the intersection measures $\mu \cap (\tau_z \circ g)_\sharp \nu$ with the properties

$$\int \varphi \, d\mu \cap (\tau_z \circ g)_\sharp \nu = \lim_{\delta \downarrow 0} \alpha(n)^{-1} \delta^{-n} \iint_{\{(x,y):|x-g(y)-z| \le \delta\}} \varphi(x) \, d\mu x \, d\nu y$$

for $\varphi \in C_0(\mathbb{R}^n)$, whence

$$\int h \, d\mu \cap (\tau_z \circ g)_\sharp \nu \le \lim_{\delta \downarrow 0} \alpha(n)^{-1} \delta^{-n} \iint_{\{(x,y):|x-g(y)-z| \le \delta\}} h(x) \, d\mu x \, d\nu y \tag{7.1}$$

for any lower semicontinuous $h : \mathbb{R}^n \to [0, \infty]$,

$$\operatorname{spt} \mu \cap (\tau_z \circ g)_\sharp \nu \subset \operatorname{spt} \mu \cap (g(\operatorname{spt} \nu) + z), \tag{7.2}$$

$$\int_B \int h \, d\mu \cap (\tau_z \circ g)_\sharp \nu \, d\mathcal{L}^n z \le \int_{S_g^{-1}(B)} h(x) \, d(\mu \times \nu)(x, y) \tag{7.3}$$

for any Borel set $B \subset \mathbb{R}^n$ and any lower semicontinuous $h : \mathbb{R}^n \to [0, \infty]$, with equality if $S_{g\sharp}(\mu \times \nu) \ll \mathcal{L}^n$, in particular,

$$\int \mu \cap (\tau_z \circ g)_\sharp \nu(\mathbb{R}^n) \, d\mathcal{L}^n z = \mu(\mathbb{R}^n) \nu(\mathbb{R}^n) \quad \text{if } S_{g\sharp}(\mu \times \nu) \ll \mathcal{L}^n. \tag{7.4}$$

Lemma 7.1 *Suppose* $0 < s < n, 0 < t < n, s + t \ge n$ *and* $t \ge (n+1)/2$. *If* $\mu, \nu \in \mathcal{M}(\mathbb{R}^n)$, $I_s(\mu) < \infty$ *and* $I_t(\nu) < \infty$, *then* $S_{g\sharp}(\mu \times \nu) \ll \mathcal{L}^n$ *for* θ_n *almost all* $g \in O(n)$.

Proof We prove this with S_g replaced by T_g, $T_g(x, y) = g^{-1}(x) - y$, which is equivalent. For $u \in \mathbb{R}^n$,

$$\widehat{T_{g\sharp}(\mu \times \nu)}(u) = \iint e^{-2\pi i u \cdot (g^{-1}(x) - y)} \, d\mu x \, d\nu y$$

$$= \int e^{-2\pi i u \cdot g^{-1}(x)} \, d\mu x \int e^{2\pi i u \cdot y} \, d\nu y = \widehat{\mu}(g(u)) \overline{\widehat{\nu}(u)}.$$

We shall use the identity

$$\int_{O(n)} f(g(x)) \, d\theta_n g = \int_{S^{n-1}} f(|x|v) \, d\sigma^{n-1} v / \sigma^{n-1}(S^{n-1}).$$

It is enough to prove this when $|x| = 1$ and then it is a consequence of (2.1). Since $n - t \leq (n-1)/2$ and $n - t \leq s$ we have by Lemma 3.15

$$\iint |T_{g\sharp}\widehat{(\mu \times \nu)}(u)|^2 \, du \, d\theta_n g$$

$$= \iint |\widehat{\mu}(g(u))\widehat{\nu}(u)|^2 \, du \, d\theta_n g$$

$$\approx \int \sigma(\mu)(|u|)|\widehat{\nu}(u)|^2 \, du \lesssim \int |u|^{t-n}|\widehat{\nu}(u)|^2 \, du \, I_{n-t}(\mu)$$

$$= \gamma(n, t)^{-1} I_t(\nu) I_{n-t}(\mu) \lesssim I_t(\nu) I_s(\mu) < \infty.$$

Hence $T_{g\sharp}(\mu \times \nu) \in L^2(\mathbb{R}^n)$ for θ_n almost all $g \in O(n)$, and the lemma follows.   □

Since the support of $S_{g\sharp}(\mu \times \nu)$ is contained in the algebraic difference set $\operatorname{spt}\mu - g(\operatorname{spt}\nu)$, this lemma gives immediately for sets (cf. the proof of Theorem 7.4 below).

Corollary 7.2 *If A and B are Borel sets in $\mathbb{R}^n$ with $\dim A + \dim B > n$ and $\dim B > (n+1)/2$, then for θ_n almost all $g \in O(n)$,*

$$\mathcal{L}^n(A - g(B)) > 0.$$

Lemma 7.3 *Suppose $0 < s < n, 0 < t < n, s + t > n$ and $t \geq (n+1)/2$. If $\mu, \nu \in \mathcal{M}(\mathbb{R}^n)$, $I_s(\mu) < \infty$ and $I_t(\nu) < \infty$, then*

$$\iint I_{s+t-n}(\mu \cap (\tau_z \circ g)_\sharp \nu) \, d\mathcal{L}^n z \, d\theta_n g \lesssim I_s(\mu) I_t(\nu).$$

Proof Set $r = s + t - n > 0$ and for $g \in O(n), z \in \mathbb{R}^n$,

$$W_{g,z}(\delta) = \{(x, y) : |S_g(x, y) - z| \leq \delta\}.$$

Using (7.1), Fatou's lemma, Fubini's theorem and (7.3) we have

$$\iint I_r(\mu \cap (\tau_z \circ g)_\sharp \nu) \, d\mathcal{L}^n z \, d\theta_n g$$

$$= \iiiint |x - u|^{-r} d(\mu \cap (\tau_z \circ g)_\sharp \nu)x \, d(\mu \cap (\tau_z \circ g)_\sharp \nu)u \, d\mathcal{L}^n z \, d\theta_n g$$

$$\leq \liminf_{\delta \downarrow 0} \alpha(n)^{-1} \delta^{-n} \iiiint_{W_{g,z}(\delta)} |x - u|^{-r}$$

$$\times d(\mu \times \nu)(x, y) \, d(\mu \cap (\tau_z \circ g)_\sharp \nu)u \, d\mathcal{L}^n z \, d\theta_n g$$

$$= \liminf_{\delta \downarrow 0} \alpha(n)^{-1} \delta^{-n} \iiint_{\{z : |x - g(y) - z| \le \delta\}} \int |x - u|^{-r}$$

$$\times \, d(\mu \cap (\tau_z \circ g)_\sharp v) u \, d\mathcal{L}^n z \, d(\mu \times v)(x, y) \, d\theta_n g$$

$$\le \liminf_{\delta \downarrow 0} \alpha(n)^{-1} \delta^{-n} \iiint_{\{(u,v) : |(x - g(y)) - (u - g(v))| \le \delta\}} |x - u|^{-r}$$

$$\times \, d(\mu \times v)(u, v) \, d(\mu \times v)(x, y) \, d\theta_n g$$

$$= \liminf_{\delta \downarrow 0} \alpha(n)^{-1} \delta^{-n} \int \theta_n(\{g : |(x - g(y)) - (u - g(v))| \le \delta\}) |x - u|^{-r}$$

$$\times \, d(\mu \times v)(u, v) \, d(\mu \times v)(x, y).$$

For the θ_n measure we have the estimate

$$\theta_n(\{g : |(x - g(y)) - (u - g(v))| \le \delta\}) \lesssim \delta^{n-1} |x - u|^{1-n}.$$

Moreover,

$$\theta_n(\{g : |(x - g(y)) - (u - g(v))| \le \delta\}) = 0$$

if $||x - u| - |y - v|| > \delta$.

The second of these is obvious. The first follows easily from (2.1): for $a, b \in \mathbb{R}^n, a \ne 0$,

$$\theta_n(\{g : |a - g(b)| \le \delta\}) = \theta_n(\{g : |a/|a| - g(b/|a|)| \le \delta/|a|\})$$

$$= \theta_n(\{g : |g^{-1}(a/|a|) - b/|a|| \le \delta/|a|\})$$

$$= \theta_n(\{g : |g(a/|a|) - b/|a|| \le \delta/|a|\})$$

$$= \sigma^{n-1}(S^{n-1} \cap B(b/|a|, \delta/|a|)) \lesssim \delta^{n-1} |a|^{1-n}.$$

Define

$$A_\delta = \{(u, v, x, y) \in (\mathbb{R}^n)^4 : ||x - u| - |y - v|| \le \delta \le |x - u|/2\},$$

$$B_\delta = \{(u, v, x, y) \in (\mathbb{R}^n)^4 : |x - u| \le 2\delta, |y - v| \le 3\delta\}.$$

Then

$$\iint I_r(\mu \cap (\tau_z \circ g)_\sharp v) \, d\mathcal{L}^n z \, d\theta_n g$$

$$\lesssim \liminf_{\delta \downarrow 0} \delta^{-1} \int_{A_\delta} |x - u|^{1-s-t} \, d(\mu \times v \times \mu \times v)(u, v, x, y)$$

$$+ \limsup_{\delta \downarrow 0} \delta^{-n} \int_{B_\delta} |x - u|^{-r} \, d(\mu \times v \times \mu \times v)(u, v, x, y)$$

$$=: S + T.$$

For S we use the estimate (4.14):

$$S \leq \liminf_{\delta \downarrow 0} \delta^{-1} \int_{\{(u,x):|x-u|\geq 2\delta\}} |x-u|^{1-s-t}$$
$$\times \nu \times \nu(\{(v,y) : ||x-u|-|y-v|| \leq \delta\}) \, d(\mu \times \mu)(u,x)$$
$$\lesssim I_t(\nu) \iint |x-u|^{-s} \, d\mu u \, d\mu x = \gamma(n,s)^{-1} I_s(\mu) I_t(\nu).$$

To estimate T, observe that by Fubini's theorem,

$$T \leq \limsup_{\delta \downarrow 0} \delta^{t-n} \int_{\{(u,x):|u-x|\leq 2\delta\}} |x-u|^{-r} \, d(\mu \times \mu)(u,x)$$
$$\times \delta^{-t} \nu \times \nu(\{(v,y) : |v-y| \leq 3\delta\}).$$

For the first factor we have for $0 < \delta < 1$,

$$\delta^{t-n} \int_{\{(u,x):|u-x|\leq 2\delta\}} |x-u|^{-r} \, d(\mu \times \mu)(u,x)$$
$$\leq 2^{n-t} \int_{\{(u,x):|u-x|\leq 2\delta\}} |x-u|^{-s} \, d(\mu \times \mu)(u,x),$$

which goes to zero as $\delta \to 0$ since $I_s(\mu) < \infty$. For the second factor,

$$\delta^{-t} \nu \times \nu(\{(v,y) : |v-y| \leq 3\delta\}) \leq 3^s \int_{\{(v,y):|v-y|\leq 3\delta\}} |v-y|^{-t} \, d(\nu \times \nu)(v,y),$$

which also goes to 0 as $\delta \to 0$, since $I_t(\nu) < \infty$. Hence $T = 0$. This completes the proof of the lemma. $\qquad\square$

7.2 Dimension of intersections of sets

Theorem 7.4 *Suppose $0 < s < n, 0 < t < n, s + t > n$ and $t > (n+1)/2$. If A and B are Borel sets in $\mathbb{R}^n$ with $\mathcal{H}^s(A) > 0$ and $\mathcal{H}^t(B) > 0$, then for θ_n almost all $g \in O(n)$,*

$$\mathcal{L}^n(\{z \in \mathbb{R}^n : \dim A \cap (\tau_z \circ g)(B) \geq s + t - n\}) > 0.$$

Proof By Frostman's lemma there are $\mu \in \mathcal{M}(A)$ and $\nu \in \mathcal{M}(B)$ such that $\mu(B(x,r)) \leq r^s$ and $\nu(B(x,r)) \leq r^t$ for $x \in \mathbb{R}^n$ and $r > 0$. Then $I_p(\mu) < \infty$ for $0 < p < s$ and $I_q(\nu) < \infty$ for $0 < q < t$. When in addition $p + q > n$ and $q > (n+1)/2$ we have by Lemma 7.3 for θ_n almost all $g \in O(n)$,

$$I_{p+q-n}(\mu \cap (\tau_z \circ g)_\sharp \nu) < \infty \quad \text{for } \mathcal{L}^n \text{ almost all } z \in \mathbb{R}^n.$$

Using Lemma 7.1 and (7.4) we have

$$\int \mu \cap (\tau_z \circ g)_\sharp \nu(\mathbb{R}^n) \, d\mathcal{L}^n z = \mu(\mathbb{R}^n)\nu(\mathbb{R}^n) > 0$$

for θ_n almost all $g \in O(n)$, whence

$$\mathcal{L}^n(E_g) > 0 \quad \text{where} \quad E_g = \{z \in \mathbb{R}^n : \mu \cap (\tau_z \circ g)_\sharp \nu(\mathbb{R}^n) > 0\}.$$

This gives $\dim A \cap (\tau_z \circ g)(B) \geq p + q - n$ for $\mathcal{L}^n$ almost all $z \in E_g$. Since E_g is independent of p and q, we can let $p \to s$ and $q \to t$ to complete the proof. $\qquad\square$

7.3 Further comments

The above results were also presented in Mattila [1995], where one can find more comments and examples. They were originally proven in Mattila [1985]. It is not known if the condition $t > (n + 1)/2$ is needed. Of course, we could replace it with $s > (n + 1)/2$. This restriction is not needed and the results are valid also in $\mathbb{R}$ if the orthogonal group $O(n)$ is replaced by a larger transformation group, for example with the maps $x \mapsto rg(x), x \in \mathbb{R}^n, g \in O(n), r > 0$, as proven by Kahane [1986] and Mattila [1984]. Such results also hold in $\mathbb{R}$, but they fail completely if only translations are used, as shown by the examples of Mattila [1984] and Keleti [1998]. More generally Kahane showed that any closed group of linear bijections of $\mathbb{R}^n$ acting transitively in $\mathbb{R}^n \setminus \{0\}$ is fine. Kahane also applied such intersection results to multiple points of stochastic processes.

Results on Hausdorff and packing dimensions of intersection measures were proven by M. Järvenpää [1997a], [1997b].

Bishop and Peres [2016] discuss in Chapter 1 the dimension of the intersections of some Cantor sets with their translates. Such results play an important role in dynamical systems, see Moreira and Yoccoz [2001]. Other results on intersections of Cantor sets are due to, among others, Peres and Solomyak [1998] and Elekes, Keleti and Máthé [2010].

Donoven and Falconer [2014] proved similar results as above for subsets of a fixed self-similar Cantor set; the group of transformations consists now of the intrinsic similarities of the Cantor set.

Minkowski dimensions of intersections with estimates on the exceptional sets were studied by Eswarathasan, Iosevich and Taylor [2013].

PART II

Specific constructions

8

Cantor measures

In this chapter we study Fourier transforms of measures on some Cantor sets.

8.1 Symmetric Cantor sets C_d and measures μ_d

We begin with standard symmetric Cantor sets. For $0 < d < 1/2$ we define the Cantor set with dissection ratio d by the usual process. Let $I = [0, 1]$. Delete from the middle of I an open interval of length $1 - 2d$ and denote by $I_{1,1}$ and $I_{1,2}$ the two remaining intervals of length d. Next delete from the middle of each $I_{1,j}$ an open interval of length $(1 - 2d)d$ and denote by $I_{2,i}, i = 1, 2, 3, 4$, all the four remaining intervals of length d^2. Continuing this we have after k steps 2^k closed intervals $I_{k,i}, i = 1, \ldots, 2^k$, of length d^k. Define

$$C_d = \bigcap_{k=1}^{\infty} \bigcup_{i=1}^{2^k} I_{k,i}.$$

Let μ_d be the 'natural' probability measure on C_d. This is the unique Borel measure $\mu_d \in \mathcal{M}(C_d)$ which is uniformly distributed in the sense that

$$\mu_d(I_{k,i}) = 2^{-k} \quad \text{for } i = 1, \ldots, 2^k, k = 1, 2 \ldots. \tag{8.1}$$

The uniqueness follows easily by, for example, checking that this condition fixes the values of integrals of continuous functions. The existence can be verified by showing (easily) that the probability measures

$$(2d)^{-k} \sum_{i=1}^{2^k} \mathcal{L}^1 \llcorner I_{k,i}$$

converge weakly as $k \to \infty$ to such a uniformly distributed measure μ_d.

Define

$$s_d = \log 2/\log(1/d), \quad \text{that is,} \quad 2d^{s_d} = 1.$$

Notice that then

$$\mu_d(I_{k,i}) = d(I_{k,i})^{s_d} \quad \text{for } i = 1, \ldots, 2^k, k = 1, 2 \ldots.$$

Using μ_d we can now check that

$$0 < \mathcal{H}^{s_d}(C_d) \le 1 \quad \text{and} \quad \dim C_d = s_d.$$

The upper bound $\mathcal{H}^{s_d}(C_d) \le 1$ is trivial since

$$\sum_{i=1}^{2^k} d(I_{k,i})^{s_d} = 2^k (d^k)^{s_d} = 1$$

for all k. To prove that $\mathcal{H}^{s_d}(C_d) > 0$ it is enough by Frostman's lemma to show that $\mu_d(J) \lesssim d(J)^{s_d}$ for every open interval $J \subset \mathbb{R}$. To prove this we may assume that $J \subset [0, 1]$ and $C_d \cap J \ne \varnothing$. Let $I_{l,j}$ be the largest (or one of them) of all the intervals $I_{k,i}$ contained in J. Then $J \cap C_d$ is contained in four intervals $I_{l,j_1} = I_{l,j_1}, \ldots, I_{l,j_4}$, whence

$$\mu_d(J) \le 4\mu_d(I_{l,j}) = 4d(I_{l,j})^{s_d} \le 4d(J)^{s_d},$$

and so $\mathcal{H}^{s_d}(C_d) > 0$.

By a modification of the above argument one can show that in fact

$$\mathcal{H}^{s_d} \lfloor C_d = \mu_d \quad \text{and} \quad \mathcal{H}^{s_d}(C_d) = 1.$$

This argument also easily yields with some positive constants a and b,

$$ar^{s_d} \le \mu_d([x - r, x + r]) \le br^{s_d} \quad \text{for } x \in C_d, \quad 0 < r < 1. \tag{8.2}$$

In order to compute the Fourier transform of μ_d it is helpful to express μ_d as a weak limit of finite linear combinations of Dirac measures indexed by binary sequences. To do this we observe that

$$C_d = \left\{ \sum_{j=1}^{\infty} \varepsilon_j (1-d) d^{j-1} : \varepsilon_j = 0 \text{ or } \varepsilon_j = 1 \right\}. \tag{8.3}$$

Let

$$\mathcal{E}_k = \{(\varepsilon_1, \ldots, \varepsilon_k) : \varepsilon_j = 0 \text{ or } \varepsilon_j = 1\},$$

$$a(\varepsilon) = \sum_{j=1}^{k} \varepsilon_j (1-d) d^{j-1} \quad \text{for } \varepsilon = (\varepsilon_j) \in \mathcal{E}_k,$$

and define

$$\nu_k = 2^{-k} \sum_{\varepsilon \in \mathcal{E}_k} \delta_{a(\varepsilon)}.$$

Then

$$\nu_k \to \mu_d \text{ weakly as } k \to \infty.$$

By the definition of the Fourier transform,

$$\widehat{\delta_a}(u) = e^{-2\pi i a u} \quad \text{for } a, u \in \mathbb{R},$$

so

$$\widehat{\nu_k}(u) = 2^{-k} \sum_{\varepsilon \in \mathcal{E}_k} \widehat{\delta_{a(\varepsilon)}}(u) = 2^{-k} \sum_{\varepsilon \in \mathcal{E}_k} e^{-2\pi i a(\varepsilon) u} = 2^{-k} \sum_{\varepsilon \in \mathcal{E}_k} e^{i \sum_{j=1}^k \varepsilon_j u_j}$$

where $u_j = -2\pi(1-d)d^{j-1}u$. Here

$$\sum_{\varepsilon \in \mathcal{E}_k} e^{i \sum_{j=1}^k \varepsilon_j u_j} = \Pi_{j=1}^k (1 + e^{i u_j})$$

as one can see by expanding the right hand side as a sum and checking that it agrees with the left hand side. Thus

$$\widehat{\nu_k}(u) = \Pi_{j=1}^k \frac{(1 + e^{i u_j})}{2} = \Pi_{j=1}^k e^{i u_j/2} \Pi_{j=1}^k \cos(u_j/2)$$
$$= e^{\sum_{j=1}^k i u_j/2} \Pi_{j=1}^k \cos(u_j/2),$$

where we have used the formula

$$\frac{1 + e^{ix}}{2} = e^{ix/2} \cos(x/2).$$

Recalling the definition of u_j we see that

$$\sum_{j=1}^k i u_j/2 = \sum_{j=1}^k -\pi i(1-d)d^{j-1}u = -\pi i(1 - d^k)u.$$

Therefore we obtain

$$\widehat{\nu_k}(u) = e^{-\pi i(1-d^k)u} \Pi_{j=1}^k \cos(\pi(1-d)d^{j-1}u).$$

Letting $k \to \infty$ we finally obtain

$$\widehat{\mu_d}(u) = e^{-\pi i u} \Pi_{j=1}^\infty \cos(\pi(1-d)d^{j-1}u). \tag{8.4}$$

When $d = 1/3$ we have for the classical ternary Cantor set

$$\widehat{\mu_{1/3}}(u) = e^{-\pi i u} \Pi_{j=1}^\infty \cos(2\pi 3^{-j}u).$$

It follows that $\widehat{\mu_{1/3}}(u)$ does not tend to 0 as u tends to ∞; look at $u = 3^k, k = 1, 2, \ldots$.

We shall now show that if $1/d \geq 3$ is an integer, then there is no measure in $\mathcal{M}(C_d)$ whose Fourier transform would tend to zero at infinity. The proof relies on the following fact: letting $I = (d, 1 - d)$ and $N = 1/d$,

$$[N^k x] \notin I \quad \text{for all } x \in C_d, k = 1, 2, \ldots, \tag{8.5}$$

where for $y \geq 0$, $[y]$ stands for the fractional part of y, that is, $[y] \in [0, 1)$ and $y - [y] \in \mathbb{N}$. To see this recall that by (8.3) C_d consists of points

$$x = \sum_{j=1}^{\infty} \varepsilon_j (1 - d) d^{j-1} = (N - 1) \sum_{j=1}^{\infty} \varepsilon_j N^{-j}$$

where $\varepsilon_j = 0$ or $\varepsilon_j = 1$. Then

$$N^k x = (N - 1) \sum_{j=1}^{\infty} \varepsilon_j N^{k-j} = (N - 1) \left(\sum_{j=0}^{k-1} \varepsilon_{k-j} N^j + \sum_{j=1}^{\infty} \varepsilon_{k+j} N^{-j} \right).$$

Thus

$$[N^k x] = (N - 1) \sum_{j=1}^{\infty} \varepsilon_{k+j} N^{-j} \in C_d \subset [0, 1] \setminus I.$$

Theorem 8.1 *If* $1/d \geq 3$ *is an integer, then for any* $\mu \in \mathcal{M}(C_d)$, $\limsup_{|x| \to \infty} |\widehat{\mu}(x)| > 0$.

Proof It is more convenient to use Fourier series than transform and we shall show a bit more: the Fourier coefficients $\widehat{\mu}(k)$ do not tend to zero as $k \in \mathbb{Z}, |k| \to \infty$. Suppose there exists $\mu \in \mathcal{M}(C_d)$ such that $\widehat{\mu}(k) \to 0$ as $k \in \mathbb{Z}, |k| \to \infty$. Choose a function $\varphi \in \mathcal{S}(\mathbb{R})$ such that $\operatorname{spt} \varphi \subset (d, 1 - d)$ and $\int \varphi = 1$. Let again $N = 1/d$ and define for $j = 1, 2, \ldots$,

$$\varphi_j(x) = \varphi([N^j x]) \quad \text{for } x \in [0, 1].$$

Then by (8.5) $\operatorname{spt} \varphi_j \cap C_d = \varnothing$, and by the Fourier inversion formula (3.67)

$$\varphi_j(x) = \sum_{k \in \mathbb{Z}} \widehat{\varphi}(k) e^{2\pi i x N^j k}, \quad x \in [0, 1],$$

so $\widehat{\varphi}_j(N^j k) = \widehat{\varphi}(k)$ and the other Fourier coefficients of φ_j vanish. Therefore by the Parseval formula (3.65) for any j and any $m > 1$,

$$
\begin{aligned}
0 = \int \varphi_j \, d\mu &= \sum_{k \in \mathbb{Z}} \overline{\widehat{\varphi}_j(k)} \widehat{\mu}(k) \\
&= \sum_{k \in \mathbb{Z}} \overline{\widehat{\varphi}_j(N^j k)} \widehat{\mu}(N^j k) = \sum_{k \in \mathbb{Z}} \overline{\widehat{\varphi}(k)} \widehat{\mu}(N^j k) \\
&= \overline{\widehat{\varphi}(0)} \widehat{\mu}(0) + \sum_{1 \le |k| \le m} \overline{\widehat{\varphi}(k)} \widehat{\mu}(N^j k) + \sum_{|k| > m} \overline{\widehat{\varphi}(k)} \widehat{\mu}(N^j k).
\end{aligned}
$$

The first term is $\mu(C_d) > 0$. For the last term we have

$$
\left| \sum_{|k| > m} \overline{\widehat{\varphi}(k)} \widehat{\mu}(N^j k) \right| \le \mu(C_d) \sum_{|k| > m} |\widehat{\varphi}(k)|,
$$

which we can make arbitrarily small choosing m large, since $\varphi \in \mathcal{S}(\mathbb{R})$. For any m we have for the middle term

$$
\left| \sum_{1 \le |k| \le m} \overline{\widehat{\varphi}(k)} \widehat{\mu}(N^j k) \right| \le 2m \sup_{|l| \ge N^j, l \in \mathbb{Z}} |\widehat{\mu}(l)|
$$

which goes to zero as $j \to \infty$. It follows that $\mu(C_d) = 0$, which is a contradiction. $\qquad\square$

All we needed in the above proof for $C = C_d$ is that there is a non-degenerate interval $I \subset [0, 1]$ and an increasing sequence (k_j) of positive integers such that

$$
[k_j x] \notin I \quad \text{for all } x \in C, j = 1, 2, \dots. \tag{8.6}
$$

Theorem 8.1 holds true for any compact set $C \subset [0, 1]$ with this property.

8.2 Pisot numbers and the corresponding measures

Next we will characterize the values of d for which $\widehat{\mu_d}(u)$ tends to 0 at infinity. For this we need the concept of a Pisot number.

Definition 8.2 A real number $\theta > 1$ is a *Pisot number* if there exists a real number $\lambda \ne 0$ such that

$$
\sum_{k=0}^{\infty} \sin^2(\lambda \theta^k) < \infty. \tag{8.7}
$$

This not a standard definition of Pisot numbers, but it is the form we shall use. Usually one defines Pisot numbers as algebraic integers whose conjugates have modulus less than 1. By a theorem of Pisot from 1938 these two definitions are equivalent. A proof can be found in Kahane and Salem [1963], Chapter VI, and in Salem [1963], Chapter I. The first indication that the above definition might be related to the number theoretic nature of θ is the following: write

$$\lambda\theta^k = \pi n_k + \delta_k \quad \text{where } n_k \in \mathbb{Z} \text{ and } -\pi/2 \leq \delta_k < \pi/2.$$

Then (8.7) is equivalent to $\sum_{k=0}^{\infty} \delta_k^2 < \infty$.

Algebraic integers are special types of algebraic numbers; they are solutions of polynomial equations with integer coefficients and with leading coefficient 1. That is, θ is an algebraic integer if there are integers $m_0, \ldots, m_{k-1}$ such that $P(\theta) = 0$ where $P(x) = x^k + m_{k-1}x^{k-1} + \cdots + m_0$. The conjugates of θ are the other complex solutions of $P(z) = 0$. For further information one can consult Appendix VI of Kahane and Salem [1963] and Salem [1963].

Obviously all integers greater than 1 are Pisot numbers. The smallest non-integral Pisot number is $1.3247\ldots$ It is a solution of $x^3 - x - 1 = 0$. Some quadratic equations giving Pisot numbers are $x^2 - x - 1 = 0$, which gives the golden ratio $\frac{1+\sqrt{5}}{2} = 1.618034\ldots$, and $x^2 - 2x - 1 = 0$, which gives $1 + \sqrt{2} = 2.414214\ldots$.

Theorem 8.3 *Let* $\mu_d, 0 < d < 1/2$, *be the Cantor measure as above. Then*

$$\lim_{u \to \infty} \widehat{\mu_d}(u) = 0$$

if and only if $1/d$ *is not a Pisot number.*

Proof Let $\theta = 1/d$. Suppose that $\widehat{\mu_d}(u)$ does not tend to 0 at infinity. Then there exist $\delta > 0$ and an increasing sequence (u_k) such that $u_k \to \infty$ and

$$|\widehat{\mu_d}(u_k)| > \delta$$

for all k. We can write

$$\pi(1-d)u_k = \lambda_k\theta^{m_k}$$

where $1 \leq \lambda_k < \theta$ and (m_k) is an increasing sequence of positive integers. Replacing the sequence (λ_k) by a subsequence if needed we can assume that $\lambda_k \to \lambda, 1 \leq \lambda \leq \theta$. By (8.4),

$$\delta < |\widehat{\mu_d}(u_k)| = |\Pi_{j=1}^{\infty} \cos(\pi(1-d)d^{j-1}u_k)|$$
$$= |\Pi_{j=1}^{\infty} \cos(\lambda_k\theta^{m_k-j+1})| \leq |\Pi_{j=0}^{m_k} \cos(\lambda_k\theta^j)|,$$

which gives

$$\Pi_{j=0}^{m_k}(1 - \sin(\lambda_k \theta^j)^2) \geq \delta^2.$$

Using the elementary inequality $x \leq -\log(1-x)$ for $0 < x < 1$ this yields

$$\sum_{j=0}^{m_k} \sin^2(\lambda_k \theta^j) \leq \log(1/\delta^2).$$

Hence for $l > k$,

$$\sum_{j=0}^{m_k} \sin^2(\lambda_l \theta^j) \leq \sum_{j=0}^{m_l} \sin^2(\lambda_l \theta^j) \leq \log(1/\delta^2).$$

Keeping k fixed and letting $l \to \infty$ we get

$$\sum_{j=0}^{m_k} \sin^2(\lambda \theta^j) \leq \log(1/\delta^2),$$

and letting $k \to \infty$,

$$\sum_{j=0}^{\infty} \sin^2(\lambda \theta^j) \leq \log(1/\delta^2).$$

Hence $\theta = 1/d$ is a Pisot number.

To prove the converse, suppose that $\theta = 1/d$ is a Pisot number. Then there exists a real number $\lambda \neq 0$ such that

$$\sum_{j=0}^{\infty} \sin^2(\lambda \theta^j) < \infty.$$

Reversing the above argument this implies that

$$p = \Pi_{j=0}^{\infty} |\cos(\lambda \theta^j)| > 0.$$

Using the formula (8.4) we get for $u_k = \lambda \theta^k/(\pi(1-d))$,

$$|\widehat{\mu_d}(u_k)| = |\Pi_{j=1}^{\infty} \cos(\lambda d^{j-1} \theta^k)| = |\Pi_{j=1}^{k} \cos(\lambda \theta^j)||\Pi_{j=0}^{\infty} \cos(\lambda \theta^{-j})|$$
$$\geq p|\Pi_{j=0}^{\infty} \cos(\lambda \theta^{-j})| = pq,$$

where $q > 0$ by similar calculus as above; $\sum_{j=0}^{\infty} \sin^2(\lambda \theta^{-j}) < \infty$ since $\theta > 1$. Hence $\widehat{\mu_d}(u)$ does not tend to 0 at infinity which proves the theorem. □

More is true: there is $\mu \in \mathcal{M}(C_d)$ such that $\lim_{u \to \infty} \widehat{\mu}(u) = 0$ if and only if $1/d$ is not a Pisot number. In fact, one can show that the Cantor sets C_d have a property similar to (8.6) if $1/d$ is a Pisot number; see Kahane and Salem [1963], Salem [1963] and Kechris and Louveau [1987].

The above results are related to the characterization of sets of uniqueness among the Cantor sets C_d, which is one of the main motivations for their study in Kahane and Salem [1963]: a compact set $C \subset [0, 1]$ is said to be a *set of uniqueness* if

$$\sum_{k \in \mathbb{Z}} c_k e^{-2\pi i k t} = 0 \; \forall t \in [0, 1] \setminus C \quad \text{implies} \quad c_k = 0 \; \forall k.$$

Otherwise C is said to be a *set of multiplicity*. The following result can be found on page 57 in Kahane and Salem [1963]:

If $C \subset [0, 1]$ is compact and there is $\mu \in \mathcal{M}(C)$ such that $\widehat{\mu}(u)$ tends to 0 at infinity, then C is a set of multiplicity.

This leads to, see Theorem IV, page 74, in Kahane and Salem [1963]:

C_d is a set of uniqueness if and only if $1/d$ is a Pisot number.

Here we again have a manifestation of the fact that the pointwise decay at infinity of the Fourier transform of a measure $\mu \in \mathcal{M}(\mathbb{R}^n)$ is a delicate matter which depends on other properties than size. Recall, however, that the average decay in the sense of the convergence of the integrals $\int |x|^{s-n} |\widehat{\mu}(x)|^2 \, dx$ depends solely on the size, that is, on the finiteness of the integrals $\iint |x - y|^{-s} \, d\mu x \, d\mu y$ which is determined by estimates on measures of balls.

8.2.1 Modified Cantor sets

We can easily modify the construction of the Cantor sets C_d and the corresponding measures μ_d to find for any $0 < s < 1$ measures $\mu \in \mathcal{M}(\mathbb{R})$ such that $I_s(\mu) < \infty$ and $\widehat{\mu}(u)$ does not tend to zero as $u \to \infty$. To see this choose positive integers $1 < N < M$ and set

$$I_{1,j} = [j/N, j/N + 1/M], \quad j = 0, 1, \ldots, N - 1.$$

The next level intervals $I_{2,j}, j = 1, 2 \ldots, N^2$, have length M^{-2} and each $I_{1,j}$ contains N of them in the same relative position as above. Continuing this yields the Cantor set $C_{M,N}$ of Hausdorff dimension $\log N / \log M$ and the natural uniformly distributed measure $\mu_{M,N} \in \mathcal{M}(C_{M,N})$. Now we can write $C_{M,N}$ as

$$C_{M,N} = \left\{ \sum_{j=0}^{\infty} \varepsilon_j M^{-j} / N : \varepsilon_j \in \{0, \ldots, N - 1\} \right\}.$$

Hence we can again get this measure as the weak limit of the discrete measures

$$\nu_k = N^{-k} \sum_{\varepsilon \in \mathcal{E}_k} \delta_{a(\varepsilon)}$$

where

$$\mathcal{E}_k = \{(\varepsilon_1, \ldots, \varepsilon_k) : \varepsilon_j \in \{0, \ldots, N-1\}\},$$

$$a(\varepsilon) = \sum_{j=0}^{k} \varepsilon_j M^{-j}/N \quad \text{for } \varepsilon = (\varepsilon_j) \in \mathcal{E}_k.$$

Then, as before for μ_d,

$$\widehat{v_k}(u) = N^{-k} \sum_{\varepsilon \in \mathcal{E}_k} \widehat{\delta_{a(\varepsilon)}}(u) = \Pi^k_{j=1} \frac{1}{N}(1 + e^{2\pi i u M^{-j}/N} + \cdots$$
$$+ e^{2\pi i u (N-1) M^{-j}/N}).$$

Thus

$$\widehat{\mu_{M,N}}(u) = \Pi^\infty_{j=1} \frac{1}{N}(1 + e^{2\pi i u M^{-j}/N} + \cdots + e^{2\pi i u (N-1) M^{-j}/N}).$$

If $u = NM^m$, $m \in \mathbb{N}$, the jth factor in this product is 1 if $j \leq m$ which implies that $\widehat{\mu_{M,N}}(NM^m)$ does not tend to zero as $m \to \infty$:

$$\liminf_{m \to \infty} |\widehat{\mu_{M,N}}(NM^m)| > 0. \tag{8.8}$$

We also have now with $s_{M,N} = \log N / \log M$ and with some positive constants a and b,

$$ar^{s_{M,N}} \leq \mu_{M,N}([x-r, x+r]) \leq br^{s_{M,N}} \quad \text{for } x \in C_{M,N}, 0 < r < 1. \tag{8.9}$$

Observe that $I_s(\mu) < \infty$ if $s < s_{M,N}$ and the numbers $s_{M,N}$ accumulate at 1.

8.3 Self-similar measures

The above Cantor measures are a subclass of more general self-similar measures which we now define. A mapping $S : \mathbb{R}^n \to \mathbb{R}^n$ is a (contractive) *similarity* if there is $0 < r < 1$ such that $|S(x) - S(y)| = r|x - y|$ for all $x, y \in \mathbb{R}^n$. This means that S has the representation

$$S(x) = rg(x) + a, \quad x \in \mathbb{R}^n,$$

for some $g \in O(n)$ and $a \in \mathbb{R}^n$.

A Borel measure $\mu \in \mathcal{M}(\mathbb{R}^n)$ is said to be *self-similar* if there are similarity maps $S_1, \ldots, S_N$ and numbers $p_1, \ldots, p_N \in (0, 1)$ such that $N \geq 2$,

$\sum_{j=1}^{N} p_j = 1$ and

$$\mu = \sum_{j=1}^{N} p_j S_{j\sharp}\mu. \tag{8.10}$$

Given any such S_j and p_j there exists a unique self-similar probability measure μ generated by this system by a theorem of Hutchinson [1981]. The proof is an elegant and simple application of the Banach fixed point theorem. It is also presented in the books Falconer [1985a] and Mattila [1995]. The support of μ is the unique non-empty compact invariant set K of the iterated function system (as it is generally called) S_j, $j = 1, \ldots, N$. This means that

$$K = \bigcup_{j=1}^{N} S_j(K).$$

Often one chooses $p_j = r_j^s$ where r_j is the contraction ratio of S_j and s is the *similarity dimension*, that is, the unique number such that $\sum_{j=1}^{N} r_j^s = 1$. If the pieces $S_j(K)$ are disjoint, or more generally if the S_j satisfy the so-called open set condition (see Hutchinson [1981], Falconer [1985a] or Mattila [1995]), then $\dim K = s$, moreover $0 < \mathcal{H}^s(K) < \infty$ and $\mathcal{H}^s \llcorner K$ is a self-similar measure.

The above classical Cantor sets C_d and measures μ_d fit into this setting with $S_1(x) = dx$ and $S_2(x) = dx + 1 - d$.

Suppose now that S_j, p_j, $j = 1, \ldots, N$, are as above and μ is the corresponding self-similar probability measure. Set

$$\mathcal{J}_m = \{(j_1, \ldots, j_m) : j_i \in \{1, \ldots, N\}\}, \quad m = 1, 2, \ldots,$$

$$S_J = S_{j_1} \circ \cdots \circ S_{j_m}, \quad p_J = p_{j_1} \cdot \cdots \cdot p_{j_m} \quad \text{for } J = (j_1, \ldots, j_m) \in \mathcal{J}_m.$$

If S_j is given by $S_j(x) = r_j g_j(x) + a_j, x \in \mathbb{R}^n$, then S_J, $J = (j_1, \ldots, j_m) \in \mathcal{J}_m$, is given by

$$S_J(x) = r_J g_J(x) + a_J, x \in \mathbb{R}^n, \quad \text{with } r_J = r_{j_1} \cdot \cdots \cdot r_{j_m},$$

$$g_J = g_{j_1} \circ \cdots \circ g_{j_m}, a_J \in \mathbb{R}^n.$$

Of course, the translation vectors a_J can easily be written explicitly in terms of the translations a_j, the dilations r_j and the rotations g_j, but that would not help us here.

Iterating the equation (8.10) we get for every $m = 1, 2, \ldots,$

$$\mu = \sum_{J \in \mathcal{J}_m} p_J S_{J\sharp}\mu. \tag{8.11}$$

We can now obtain the following limiting formula for the Fourier transform of μ:

Proposition 8.4 *For a self-similar measure $\mu \in \mathcal{M}(\mathbb{R}^n)$ as above*

$$\widehat{\mu}(x) = \lim_{m \to \infty} \sum_{J \in \mathcal{J}_m} p_J e^{-2\pi i a_J \cdot x} \quad \text{for } x \in \mathbb{R}^n$$

with uniform convergence on compact sets.

Proof If $x \in \mathbb{R}^n$ we have

$$\begin{aligned}
\widehat{\mu}(x) &= \int e^{-2\pi i x \cdot y} \, d\mu y = \sum_{J \in \mathcal{J}_m} p_J \int e^{-2\pi i x \cdot S_J(y)} \, d\mu y \\
&= \sum_{J \in \mathcal{J}_m} p_J \int e^{-2\pi i x \cdot (r_J g_J(y) + a_J)} \, d\mu y \\
&= \sum_{J \in \mathcal{J}_m} p_J e^{-2\pi i x \cdot a_J} \int e^{-2\pi i r_J g_J^{-1}(x) \cdot y} \, d\mu y \\
&= \sum_{J \in \mathcal{J}_m} p_J e^{-2\pi i x \cdot a_J} \widehat{\mu}(r_J g_J^{-1}(x)).
\end{aligned}$$

Since $r_J \leq (\max_{1 \leq j \leq N} r_j)^m$ for $J \in \mathcal{J}_m$, $\widehat{\mu}(r_J g_J^{-1}(x)) \to \widehat{\mu}(0) = 1$ as $J \in \mathcal{J}_m$ and $m \to \infty$ uniformly on compact sets. The proposition follows from this. $\square$

Strichartz [1990b], [1993a] and [1993b] studied the behaviour of Fourier transforms of self-similar measures systematically, in particular the asymptotic behaviour of L^2 averages over large balls.

8.4 Further comments

The main reference for this chapter is the classical book Kahane and Salem [1963]. In addition to the simplest Cantor sets C_d it discusses much wider classes, both deterministic and random, of Cantor sets and Fourier analytic questions related to them. Salem [1963] and Kechris and Louveau [1987] are also excellent references.

The product formula (8.4) of course immediately generalizes to product measures $\mu_{d_1} \times \cdots \times \mu_{d_n}$ in $\mathbb{R}^n$ because

$$\mathcal{F}(\mu_{d_1} \times \cdots \times \mu_{d_n})(x_1, \ldots, x_n) = \widehat{\mu_1}(x_1) \cdot \cdots \cdot \widehat{\mu_n}(x_n).$$

A product formula for a special but more general class of self-similar measures can be found in Chapter 4 of Strichartz [1990b].

9

Bernoulli convolutions

Problems on self-similar sets and measures become very delicate if no separation condition is assumed. We shall now investigate a very important class of self-similar measures, the Bernoulli convolutions, with emphasis on the overlapping case.

9.1 Absolute continuity of the Bernoulli convolutions

Let $0 < \lambda < 1$. The (infinite) *Bernoulli convolution* ν_λ with parameter λ is the probability distribution of

$$\sum_{j=0}^{\infty} \pm \lambda^j$$

where the signs are chosen independently with probability $1/2$. The term comes from the fact that this is the limit as $k \to \infty$ of the k-fold convolution product of the measures $(\delta_{-\lambda^j} + \delta_{\lambda^j})$, $j = 1, \ldots, k$. A formal definition of the above probabilistic description is the following.

Let

$$\Omega = \{-1, 1\}^{\mathbb{N}_0} = \{(\omega_j) : \omega_j = 1 \text{ or } \omega_j = -1, j = 0, 1, \ldots\},$$

and let μ be the infinite product of the probability measure $(\delta_{-1} + \delta_1)/2$ with itself. Then μ is determined by its values on the finite cylinder sets:

$$\mu(\{\omega : \omega_j = a_j \text{ for } j = 0, 1, \ldots, k\}) = 2^{-k-1} \; \forall a_j \in \{-1, 1\},$$
$$j = 0, 1, \ldots, k, k = 0, 1, \ldots.$$

120

Define the 'projection'

$$\Pi_\lambda : \Omega \to \mathbb{R}, \quad \Pi_\lambda(\omega) = \sum_{j=0}^{\infty} \omega_j \lambda^j.$$

Then ν_λ is defined as the image measure of μ under Π_λ:

$$\nu_\lambda(B) = \mu(\{\omega \in \Omega : \Pi_\lambda(\omega) \in B\}) \quad \text{for } B \subset \mathbb{R}.$$

We can also write ν_λ as the weak limit,

$$\nu_\lambda = \lim_{k \to \infty} 2^{-k-1} \sum \{\delta_{\sum_{j=0}^{k} \omega_j \lambda^n} : \omega_j \in \{-1, 1\}\}.$$

So the ν_λ have a fairly similar expression to the Cantor measures μ_λ of the previous chapter. In fact, for $0 < \lambda < 1/2$, ν_λ is just μ_λ but constructed on the interval $[-1/(1 - \lambda), 1/(1 - \lambda)]$ instead of $[0, 1]$. For $\lambda = 1/2$, $\nu_{1/2}$ is even simpler: it is the normalized Lebesgue measure on $[-2, 2]$. But when $\lambda > 1/2$ things become much more complicated. One can still think of the construction of ν_λ in the same spirit as the Cantor construction for $0 < \lambda < 1/2$, but now the construction intervals overlap. And when one continues the iterative construction the overlaps become very complicated and difficult to control by hand. Anyway, ν_λ is still a self-similar measure but without any separation conditions. To see this observe first that μ is shift invariant: defining the shift σ by $\sigma(\omega_0, \omega_1, \dots) = (\omega_1, \omega_2, \dots)$ we have $\mu(A) = \mu(\sigma^{-1}(A))$ for $A \subset \Omega$. Using this we compute for $B \subset \mathbb{R}$,

$$\nu_\lambda(B) = \mu(\{\omega : \Pi_\lambda(\omega) \in B\})$$
$$= \mu(\{\omega : \omega_0 = 1, 1 + \lambda\Pi_\lambda(\sigma(\omega)) \in B\})$$
$$\quad + \mu(\{\omega : \omega_0 = -1, -1 + \lambda\Pi_\lambda(\sigma(\omega)) \in B\})$$
$$= \frac{1}{2}\mu(\{\omega : 1 + \lambda\Pi_\lambda(\sigma(\omega)) \in B\}) + \frac{1}{2}\mu(\{\omega : -1 + \lambda\Pi_\lambda(\sigma(\omega)) \in B\})$$
$$= \frac{1}{2}\mu\left(\Pi_\lambda^{-1}\left(\frac{1}{\lambda}(B - 1)\right)\right) + \frac{1}{2}\mu\left(\Pi_\lambda^{-1}\left(\frac{1}{\lambda}(B + 1)\right)\right)$$
$$= \frac{1}{2}\nu_\lambda\left(\frac{1}{\lambda}(B - 1)\right) + \frac{1}{2}\nu_\lambda\left(\frac{1}{\lambda}(B + 1)\right).$$

It follows that ν_λ satisfies the equation

$$\nu_\lambda = \frac{1}{2}S_{1\sharp}\nu_\lambda + \frac{1}{2}S_{2\sharp}\nu_\lambda$$

with the similarities $S_1(x) = \lambda x + 1$ and $S_2(x) = \lambda x - 1$.

To find the Fourier transform of ν_λ we can go through the computation in Chapter 8 for the formula (8.4) and see that it is valid for all $0 < \lambda < 1$ and gives

$$\widehat{\nu_\lambda}(u) = \Pi_{j=0}^\infty \cos(2\pi \lambda^j u). \tag{9.1}$$

One of the main questions concerning the Bernoulli convolutions is: for which λ is ν_λ absolutely continuous with respect to Lebesgue measure? The complete answer is still unknown and we shall discuss some known partial results.

It is clear that ν_λ is singular for $0 < \lambda < 1/2$: its support is a Cantor set like C_λ. We have already observed that $\nu_{1/2}$ is just Lebesgue measure on $[-2, 2]$. For $1/2 < \lambda < 1$ the absolute continuity of ν_λ depends on the number theoretic nature of λ. The Pisot numbers appear again. Erdős [1939] proved that the Fourier transform of ν_λ does not tend to zero at infinity if $1/\lambda$ is a Pisot number; the fact we know for $0 < \lambda < 1/2$ from Theorem 8.3. This implies by the Riemann–Lebesgue lemma that ν_λ is not absolutely continuous (in fact, it is even singular) if $1/\lambda$ is a Pisot number. No other values of λ in $(1/2, 1)$ apart from the reciprocals of the Pisot numbers are known for which ν_λ is singular. Later Salem [1944] showed that also the converse of Erdős's results is true: the Fourier transform of ν_λ tends to zero at infinity if $1/\lambda$ is not a Pisot number. But of course this does not guarantee that ν_λ would be absolutely continuous. Erdős [1940] also proved that ν_λ is absolutely continuous for almost all λ in some interval $(a, 1), a < 1$. There were several other results but the real breakthrough was the following theorem of Solomyak [1995], part of whose proof we shall present.

Theorem 9.1 ν_λ *is absolutely continuous with respect to Lebesgue measure for almost all* $\lambda \in (1/2, 1)$. *Moreover,* $\nu_\lambda \in L^2$ *for almost all* $\lambda \in (1/2, 1)$.

The proof below is due to Peres and Solomyak [1996]. It does not use Fourier transform methods and was inspired by a non-Fourier proof of the second part of the projection theorem 4.1.

Proof We shall only give the proof for $\lambda \in [1/2, 2^{-2/3}]$ and make some comments for the rest at the end. We shall use the lower derivative of $\nu \in \mathcal{M}(\mathbb{R})$:

$$\underline{D}(\nu, x) = \liminf_{r \to 0} \frac{\nu([x - r, x + r])}{2r}, \quad x \in \mathbb{R}.$$

Due to Theorem 2.11 in order to prove that ν_λ is absolutely continuous for almost every λ on an interval J, it is enough to show that

$$I(J) = \int_J \int_\mathbb{R} \underline{D}(\nu_\lambda, x) \, d\nu_\lambda x \, d\lambda < \infty. \tag{9.2}$$

We shall prove this for suitable intervals J. In fact, it will then also show that $\nu_\lambda \in L^2$ for almost every λ on J; once one knows the absolute continuity, one can fairly easily show that $I(J) = \int_J \int_{\mathbb{R}} D(\nu_\lambda, x)^2 \, dx \, d\lambda$ (heuristically $(\frac{d\nu_\lambda}{dx})^2 dx = \frac{d\nu_\lambda}{dx} d\nu_\lambda$). We shall leave all measurability questions as exercises. By Fatou's lemma and the definition of ν_λ,

$$I(J) \leq \liminf_{r \to 0} (2r)^{-1} \int_J \int_\Omega \nu_\lambda([\Pi_\lambda(\omega) - r, \Pi_\lambda(\omega) + r])) \, d\mu\omega \, d\lambda.$$

Applying Fubini's theorem to the characteristic function of $\{(\omega, \tau, \lambda) : |\Pi_\lambda(\omega) - \Pi_\lambda(\tau)| \leq r\}$, we obtain

$$I(J) \leq \liminf_{r \to 0} (2r)^{-1} \int_\Omega \int_\Omega \mathcal{L}^1(\{\lambda \in J : |\Pi_\lambda(\omega) - \Pi_\lambda(\tau)| \leq r\}) \, d\mu\omega \, d\mu\tau.$$
(9.3)

Define

$$\varphi_{\omega,\tau}(\lambda) = \Pi_\lambda(\omega) - \Pi_\lambda(\tau) = \sum_{j=0}^\infty (\omega_j - \tau_j)\lambda^j.$$

We need to estimate

$$\mathcal{L}^1(\{\lambda \in J : |\varphi_{\omega,\tau}(\lambda)| \leq r\}).$$

To do this, observe that $\omega_n - \tau_n \in \{-2, 0, 2\}$ and write

$$\varphi_{\omega,\tau}(\lambda) = 2\lambda^{k(\omega,\tau)} g(\lambda),$$
(9.4)

where $k(\omega, \tau)$ is the smallest j such that $\omega_j \neq \tau_j$ and g is of the form (assuming without loss of generality that $\omega_{k(\omega,\tau)} > \tau_{k(\omega,\tau)}$)

$$g(x) = 1 + \sum_{j=1}^\infty b_j x^j \quad \text{with } b_j \in \{-1, 0, 1\}.$$
(9.5)

The essential ingredient in the proof is to find intervals J where the following δ *transversality* condition holds:

For any g as in (9.5), any $\delta > 0$ and any $x \in J$, $g(x) < \delta$ implies $g'(x) < -\delta$,
(9.6)

and to use this condition to estimate the integral $I(J)$.

We shall first show that the δ transversality on $J \subset (1/2, 1)$ implies that $I(J) < \infty$, consequently ν_λ is absolutely continuous for almost every $\lambda \in J$. So suppose that (9.6) holds for $J = [\lambda_0, \lambda_1] \subset (1/2, 1)$. We claim that then for g as in (9.5) and for all $\varrho > 0$,

$$\mathcal{L}^1(\{\lambda \in J : |g(\lambda)| \leq \varrho\}) \leq 2\varrho/\delta.$$
(9.7)

This is obvious if $\varrho \geq \delta$. Suppose that $\varrho < \delta$. Then $g'(\lambda) < -\delta$ whenever $|g(\lambda)| \leq \varrho$. Thus g is monotone on the set of (9.7) with $|g'| > \delta$, which implies (9.7).

By (9.4), $|\varphi_{\omega,\tau}(\lambda)| \leq r$ implies that $|g(\lambda)| \leq \lambda_0^{-k(\omega,\tau)}r/2$ for $\lambda \in J$. Applying (9.7) with $\varrho = \lambda_0^{-k(\omega,\tau)}r/2$, we obtain

$$\mathcal{L}^1(\{\lambda \in J : |\varphi_{\omega,\tau}(\lambda)| \leq r\}) \leq \delta^{-1}\lambda_0^{-k(\omega,\tau)}r.$$

Substituting this in (9.3) yields

$$I(J) \leq \liminf_{r\to 0}(2r)^{-1}\int_\Omega\int_\Omega \delta^{-1}\lambda_0^{-k(\omega,\tau)}r\,d\mu\omega\,d\mu\tau$$

$$= (2\delta)^{-1}\sum_{k=0}^{\infty}\lambda_0^{-k}\mu(\{\omega : k(\omega,\tau) = k\})\,d\mu\tau = (2\delta)^{-1}\sum_{k=0}^{\infty}\lambda_0^{-k}2^{-k-1} < \infty,$$

where we used $\lambda_0 > 1/2$ in the last step. So we have shown that the δ transversality on J implies that ν_λ is absolutely continuous for almost every $\lambda \in J$.

Transversality will be established by finding $\star$-functions:

Definition 9.2   A power series h is called a $\star$-function if for some $k \geq 1$ and $a_k \in [-1, 1]$,

$$h(x) = 1 - \sum_{j=1}^{k-1}x^j + a_k x^k + \sum_{j=k+1}^{\infty}x^j.$$

Lemma 9.3   *Suppose that $0 < \delta < 1, 0 < x_0 < 1$ and there is a $\star$-function h such that*

$$h(x_0) > \delta \quad and \quad h'(x_0) < -\delta.$$

Then the δ transversality (9.6) holds on $[0, x_0]$.

Proof We shall use the following elementary lemma.

Lemma 9.4   *Let*

$$f(x) = \sum_{j=1}^{k}c_j x^j - \sum_{j=k+1}^{\infty}c_j x^j, x \in [0, 1),$$

with $c_j \geq 0, j = 1, 2\dots$. If $x \in (0, 1)$ and $f(x) < 0$, then $f'(x) < 0$. Moreover, f can have at most one zero on $(0, 1)$.

Proof The first assertion follows from

$$\sum_{j=1}^{k}jc_j x^{j-1} \leq (k/x)\sum_{j=1}^{k}c_j x^j < (k/x)\sum_{j=k+1}^{\infty}c_j x^j \leq \sum_{j=k+1}^{\infty}jc_j x^{j-1}.$$

The second assertion follows from the first.   □

To prove Lemma 9.3, note that Lemma 9.4 gives that h'' has at most one zero on $[0, x_0]$. We have $h'(0) = -1 < -\delta$ if $k > 1$ and $h'(0) \leq h'(x_0) < -\delta$ otherwise. Since $\lim_{x \to 1^-} h'(x) = \infty$, we must have $h'(x) < -\delta$ for all $x \in (0, x_0)$, otherwise h'' would have at least two zeros. It follows that $h(x) > h(x_0) > \delta$ for $x \in (0, x_0)$.

Let g be as in (9.5) and $f(x) = g(x) - h(x)$. Then $f(x) = \sum_{j=1}^{l} c_j x^j - \sum_{j=l+1}^{\infty} c_j x^j$, where $c_j \geq 0$ and $l = k - 1$ or $l = k$. If $x \in [0, x_0]$ and $g(x) < \delta$, then $f(x) < 0$. So by Lemma 9.4 $f'(x) < 0$ which gives $g'(x) < -\delta$. This completes the proof of the lemma. $\qquad\square$

We return to the proof of the theorem. From (9.1) we see that by (3.24)

$$\widehat{v_\lambda}(u) = \widehat{v_{\lambda^2}}(u)\widehat{v_{\lambda^2}}(\lambda u) = \widehat{v_{\lambda^2} * \sigma_\lambda}(u),$$

where $\sigma_\lambda(A) = v_{\lambda^2}(\lambda^{-1} A)$. Hence $v_\lambda = v_{\lambda^2} * \sigma_\lambda$ and so v_λ is absolutely continuous if v_{λ^2} is. Therefore if we can prove the absolute continuity of v_λ for almost every $\lambda \in [1/2, 2^{-1/2}]$ we get it also in $[1/2, 2^{-1/4}]$, and then again in $[1/2, 2^{-1/8}]$, and so on. Consequently, it suffices to prove that v_λ is absolutely continuous for almost every $\lambda \in [1/2, 2^{-1/2}]$.

Here is a $\star$-function h with $h(2^{-2/3}) > 0.07$ and $h'(2^{-2/3}) < -0.09$ (which Peres and Solomyak have found by computer search):

$$h(x) = 1 - x - x^2 - x^3 + 0.5x^4 + \sum_{j=5}^{\infty} x^j.$$

So by Lemma 9.3 v_λ is absolutely continuous for almost all $\lambda \in [1/2, 2^{-2/3}]$. There is still a gap from $2^{-2/3}$ to $2^{-1/2}$. To fill this one can employ two more $\star$-functions and apply the above methods to some modified random sums. For the details, see Peres and Solomyak [1996]. $\qquad\square$

There really is a need for the additional tricks, because the whole interval $(1/2, 1)$ is not an interval of δ transversality. In fact, Solomyak [1995] found a power series as in (9.5) which has a double zero at some point of the interval $[0.649, 0.683]$.

9.2 Further comments

Bernoulli convolutions appear in a wide variety of topics; in Fourier analysis, probability, dynamical systems and number theory. An excellent survey on them is given by Peres, Schlag and Solomyak [2000]. The notion of transversality for power series was introduced and applied by Pollicott and Simon [1995]. We shall see it in action in a more general setting in Chapter 18, also with further applications to Bernoulli convolutions.

Kahane [1971] noticed that a method of Erdős [1940] gives that there is a set $E \subset (0, 1)$ of Hausdorff dimension 0 such that for every $\lambda \in (0, 1) \setminus E$ there is $\alpha > 0$ for which $|\widehat{\nu_\lambda}(u)| \lesssim |u|^{-\alpha}$. The proof of this with some extensions is also presented in Peres, Schlag and Solomyak [2000] and Shmerkin [2014]. Using this decay estimate Shmerkin proved that

$$\dim\{\lambda \in (1/2, 1) : \nu_\lambda \text{ is not absolutely continuous}\} = 0.$$

Shmerkin's proof essentially relied on the deep techniques developed by Hochman [2014] who had already proved that $\dim \nu_\lambda = 1$ outside a set of parameters λ of dimension zero. Later Shmerkin and Solomyak [2014] proved that $\nu_\lambda \in L^p$ for some $p > 1$ outside a zero-dimensional exceptional set of the numbers λ. We shall discuss other related exceptional set estimates in Sections 10.5 and 18.5.

The above results have natural analogues for asymmetric Bernoulli convolutions; the plus and minus are taken with probability p and $1 - p$, $0 < p < 1$. These are again examples of self-similar measures. Much more general self-similar measures have also been studied extensively. Hochman's [2014] theory deals with them and Shmerkin [2014] and Shmerkin and Solomyak [2014] proved absolute continuity and integrability results with zero-dimensional exceptional sets for a large class of such measures.

10

Projections of the four-corner Cantor set

In this chapter we study projections of a particular planar one-dimensional Cantor set. We shall give two proofs to show that almost all projections have length zero. This will be used in the next chapter for a construction of Besicovitch sets.

10.1 The Cantor sets $C(d)$

In this chapter we investigate orthogonal projections of the Cantor set

$$C(d) = C_d \times C_d, \quad 0 < d < 1/2,$$

where C_d is the linear Cantor set of Chapter 8.

The term four-corner Cantor set comes from the geometric construction in the plane, see Figure 10.1:

$$C(d) = \bigcap_{k=1}^{\infty} U_k^d, \quad U_k^d = \bigcup_{i=1}^{4^k} Q_{k,i}. \tag{10.1}$$

Here each $Q_{k,i}$ is a closed square of side-length d^k, and they are defined as follows. First the $Q_{1,i}$ are the four squares in the four corners of the unit square $[0, 1] \times [0, 1]$, that is, $[0, d] \times [0, d]$, $[0, d] \times [1 - d, 1]$, $[1 - d, 1] \times [0, d]$ and $[1 - d, 1] \times [1 - d, 1]$. If the squares $Q_{k,i}, i = 1, \ldots, 4^k$, have been constructed, the $Q_{k+1,j}$ are obtained in the same way inside and in the corners of the $Q_{k,i}$.

Defining s_d by

$$4d^{s_d} = 1, \quad \text{i.e.,} \quad s_d = \frac{\log 4}{\log(\frac{1}{d})},$$

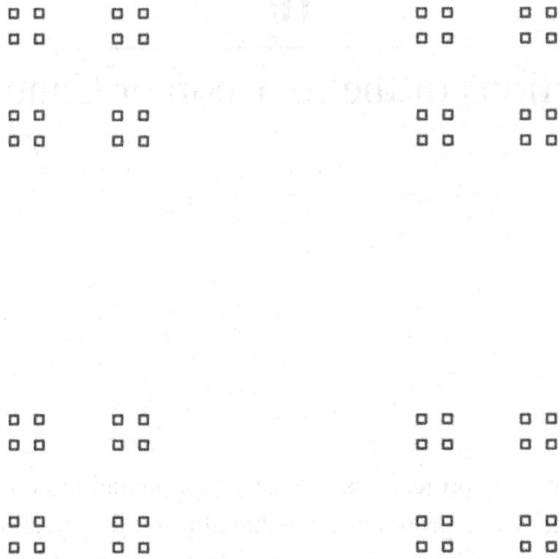

Figure 10.1 Four-corner Cantor set, more precisely the approximation $U_3^{1/4}$

we have

$$0 < \mathcal{H}^{s_d}(C(d)) < \infty \quad \text{and} \quad \dim C(d) = s_d.$$

This is easily derived directly from (10.1), for example as in Chapter 8 for the linear Cantor sets C_d.

10.2 Peres–Simon–Solomyak proof for the projections of $C(1/4)$

We shall now look at the projections of $C(d)$. Instead of parametrizing the orthogonal projections onto lines by the unit circle as before, we parametrize them with the angle the line makes with the x-axis; we set

$$p_\theta(x, y) = x \cos\theta + y \sin\theta, \quad (x, y) \in \mathbb{R}^2, \theta \in [0, \pi).$$

So with our earlier notation

$$p_\theta = P_\theta \quad \text{with } \boldsymbol{\theta} = (\cos\theta, \sin\theta).$$

We notice immediately that when $\theta = 0$ or $\theta = \frac{\pi}{2}$, that is, when we project into the coordinate axis, we get the Cantor sets C_d whose dimension is $\frac{\log 2}{\log(\frac{1}{d})} = \frac{1}{2}s_d$. Looking more carefully at these projections with different angles θ we easily find a countable dense set of angles θ for which $p_\theta(C(d))$ is a Cantor set in $\mathbb{R}$

with dimension strictly less than s_d. This happens always when p_θ maps two different squares $Q_{k,i}$ exactly onto the same interval. However, this behaviour is exceptional due to Marstrand's general projection theorem 4.1.

We now turn to the one-dimensional Cantor set $C(1/4)$. Observe that $p_\theta(C(1/4))$ is an interval when $\tan \theta = 1/2$. Soon we shall get precise information about other projections too. We have also $\mathcal{H}^1(C(1/4)) = \sqrt{2}$, see comments in Section 10.5. We shall now prove the following.

Theorem 10.1

$$\mathcal{L}^1(p_\theta(C(1/4))) = 0 \quad \text{for almost all } \theta \in [0, \pi).$$

The following elementary proof is due to Peres, Simon and Solomyak [2003]. Another proof is given by Kenyon [1997], which we shall also present below. It gives a sharper result, which in particular implies that there are only countably many directions θ for which $\mathcal{L}^1(p_\theta(C(1/4))) > 0$. The set of such directions is countably infinite and dense.

Set now $C = C(1/4)$. We can write

$$C = \bigcup_{i=1}^{4} \left(\frac{1}{4}C + c_i \right)$$

where $c_1 = (0, 0)$, $c_2 = \left(\frac{3}{4}, 0\right)$, $c_3 = \left(0, \frac{3}{4}\right)$, $c_4 = \left(\frac{3}{4}, \frac{3}{4}\right)$. Hence, writing again $\theta = (\cos\theta, \sin\theta)$,

$$p_\theta(C) = \bigcup_{i=1}^{4} \left(\frac{1}{4}p_\theta(C) + \theta \cdot c_i \right) \subset \mathbb{R}.$$

Let us first look more generally at this type of self-similar subset of $\mathbb{R}$. Let $K \subset \mathbb{R}$ be compact such that for some integer $m \geq 2$ and some $d_1, \ldots, d_m \in \mathbb{R}$ ($d_i \neq d_j$ for $i \neq j$),

$$K = \bigcup_{i=1}^{m} K_i \quad \text{with} \quad K_i = \frac{1}{m}K + d_i.$$

Lemma 10.2

(1) $\mathcal{L}^1(K_i \cap K_j) = 0$ *for* $i \neq j$.
(2) $K_i \cap K_j \neq \varnothing$ *for some* $i \neq j$.

Proof (1) follows easily from

$$\mathcal{L}^1(K) \leq \sum_{i=1}^{m} \mathcal{L}^1(K_i) = \sum_{i=1}^{m} \frac{1}{m}\mathcal{L}^1(K) = \mathcal{L}^1(K).$$

If $K_i \cap K_j = \emptyset$ for all $i \neq j$, then for some $\varepsilon > 0$ the open ε-neighbourhoods $K_i(\varepsilon)$ of the K_i are also disjoint. The ε-neighbourhood of $K_i = \frac{1}{m}K + d_i$ is $(\frac{1}{m}K)(\varepsilon) + d_i = \frac{1}{m}K(m\varepsilon) + d_i$, whence

$$\mathcal{L}^1(K_i(\varepsilon)) = \mathcal{L}^1\left(\frac{1}{m}K(m\varepsilon)\right) = \frac{1}{m}\mathcal{L}^1(K(m\varepsilon)).$$

It follows that

$$\mathcal{L}^1(K(\varepsilon)) = \sum_{i=1}^{m} \mathcal{L}^1(K_i(\varepsilon)) = \sum_{i=1}^{m} \frac{1}{m}\mathcal{L}^1(K(m\varepsilon)) = \mathcal{L}^1(K(m\varepsilon)).$$

This is a contradiction, since $K(\varepsilon)$ is a strict subset of $K(m\varepsilon)$ and both are bounded open sets. $\square$

Since

$$K_i = \frac{1}{m}K + d_i = \frac{1}{m}\left(\bigcup_{j=1}^{m}\left(\frac{1}{m}K + d_j\right)\right) + d_i = \bigcup_{j=1}^{m} K_{i,j},$$

where $K_{ij} = \frac{1}{m^2}K + \frac{1}{m}d_j + d_i$, we can write K also as the union of the m^2 sets K_{ij}. Set

$$\mathcal{I} = \{1, \ldots, m\},$$

$$\mathcal{I}^k = \{u : u = (i_1, \ldots, i_k), \; i_j \in I\}, \quad k = 1, 2, \ldots.$$

Then for each k,

$$K = \bigcup_{u \in \mathcal{I}^k} K_u, \quad \text{where} \quad K_u = m^{-k}K + d_u.$$

The translation numbers K_u were defined above for $k = 1, 2$, and the general case should be clear from this.

The following notion is due to Bandt and Graf [1992].

Definition 10.3 Let $\varepsilon > 0$. We say that K_u and K_v are ε-relatively close if $u, v \in \mathcal{I}^k$ for some k, $u \neq v$, and

$$|d_u - d_v| \leq \varepsilon d(K_u) = \varepsilon d(K)m^{-k}.$$

Observe that this means that

$$K_v = K_u + x$$

with $x = d_v - d_u$ and $|x| \leq \varepsilon d(K_u)$.

Lemma 10.4 *If for every $\varepsilon > 0$ there are k and $u, v \in \mathcal{I}^k$ with $u \neq v$ such that K_u and K_v are ε-relatively close, then $\mathcal{L}^1(K) = 0$.*

Proof To prove this suppose $\mathcal{L}^1(K) > 0$ and let $1/2 < t < 1$. Then there is some interval I such that $\mathcal{L}^1(K \cap I) > t\mathcal{L}^1(I)$. Pick small $\varepsilon > 0$ and K_u and K_v, $u, v \in I^k$, $u \neq v$, which are ε-relatively close. By an iteration of Lemma 10.2(1) $\mathcal{L}^1(K_u \cap K_v) = 0$. Setting $I_u = m^{-k}I + d_u$ and $I_v = m^{-k}I + d_v$, $\mathcal{L}^1(K_u \cap I_u) > t\mathcal{L}^1(I_u)$, $\mathcal{L}^1(K_v \cap I_v) > t\mathcal{L}^1(I_v)$ and $\mathcal{L}^1(I_v \setminus I_u) \leq \varepsilon d(K)m^{-k}$. It follows that

$$
\begin{aligned}
2tm^{-k}\mathcal{L}^1(I) &= t\mathcal{L}^1(I_u) + t\mathcal{L}^1(I_v) \\
&\leq \mathcal{L}^1(K_u \cap I_u) + \mathcal{L}^1(K_v \cap I_v) = \mathcal{L}^1((K_u \cap I_u) \cup (K_v \cap I_v)) \\
&\leq \mathcal{L}^1(I_u) + \mathcal{L}^1(I_v \setminus I_u) \leq (\mathcal{L}^1(I) + \varepsilon d(K))m^{-k}.
\end{aligned}
$$

This is a contradiction if ε is sufficiently small. $\qquad\square$

Proof of Theorem 10.1 We now return to the proof that $\mathcal{L}^1(p_\theta(C)) = 0$ for almost all θ. Let $p_\theta(C) = C^\theta$ to fit more conveniently with the notation C_u^θ above. For $\varepsilon > 0$ let

$$
\begin{aligned}
V_\varepsilon = \{\theta \in [0, \pi) : \exists\, k,\, u,\, v \text{ such that } u, v \in I^k,\, u \neq v \\
\text{and } C_u^\theta \text{ and } C_v^\theta \text{ are } \varepsilon\text{-relatively close}\}.
\end{aligned}
$$

It follows from Lemma 10.4 that it suffices to show that for every $\varepsilon > 0$,

$$
\mathcal{L}^1([0, \pi) \setminus V_\varepsilon) = 0.
$$

Then also $\mathcal{L}^1\big([0, \pi) \setminus \bigcap_{\varepsilon > 0} V_\varepsilon\big) = \mathcal{L}^1\big([0, \pi) \setminus \bigcap_{j=1}^\infty V_{\frac{1}{j}}\big) = 0$. So let $\varepsilon > 0$ and $\theta \in [0, \pi)$. By Lemma 10.2(2), $C_i^\theta \cap C_j^\theta \neq \varnothing$ for some $i \neq j$. This means that there are $x \in C_i$ and $y \in C_j$ such that $p_\theta x = p_\theta y$. Let $k > 1$ be an integer. Then $x \in C_u$ and $y \in C_v$ for some $u, v \in I^k$ with $u \neq v$. Let $\theta_0 \in [0, \pi)$ be such that $p_{\theta_0}(C_u) = p_{\theta_0}(C_v)$ (that is, p_{θ_0} maps the squares of side-length 4^{-k} which contain C_u and C_v onto the same interval). Then $|\theta - \theta_0| < c4^{-k}$ with some $c > 1$ independent of k. Moreover, $C_u^{\theta_0}$ and $C_v^{\theta_0}$ are '0-relatively close', and a simple geometric inspection shows that C_u^φ and C_v^φ are ε-relatively close when $|\varphi - \theta_0| < b\varepsilon 4^{-k}$, where $0 < b < 1$ is independent of k. Hence $[\theta - 2c4^{-k}, \theta + 2c4^{-k}] \cap V_\varepsilon$ contains an interval of length $b\varepsilon 4^{-k}$. Since this is true for every k, it follows that $\mathcal{L}^1([0, \pi) \setminus V_\varepsilon) = 0$ as required. $\qquad\square$

10.3 Kenyon's tilings and projections of $C(1/4)$

Here we shall give the proof of Kenyon for the fact that almost all projections of $C = C(1/4)$ have measure zero. In fact, we shall derive, following Kenyon [1997], much more precise information about the projections. Instead of projections we shall consider essentially the same mappings $\pi_t : \mathbb{R}^2 \to \mathbb{R}$, $t > 0$:

$$\pi_t(x, y) = tx + y.$$

Theorem 10.5 can of course be immediately turned into a statement for the projections p_θ.

For any positive integer m, let $m^* \in \{1, 2, 3\}$ be defined by

$$m^* = m4^{-j_0} \qquad \text{mod } 4$$

where j_0 is the largest integer j such that 4^j divides m. So $6^* = 2, 20^* = 1, 112^* = 3$, and so on. If $t = p/q$ is a positive rational in the reduced form, that is, p and q are positive integers having 1 as their greatest common divisor, then one quickly checks that p^* and q^* cannot both be even.

Theorem 10.5 *Let $t > 0$.*

(a) *If t is irrational, $\mathcal{L}^1(\pi_t(C)) = 0$.*

If $t = p/q$ is rational in the reduced form, then

(b) *$\mathcal{L}^1(\pi_t(C)) = 0$ and $\dim \pi_t(C) < 1$, provided both p^* and q^* are odd (that is, 1 or 3),*

(c) *$\pi_t(C)$ contains a non-degenerate interval and it is the closure of its interior, provided either p^* or q^* is even.*

Proof Kenyon's main idea is to study tilings of $\mathbb{R}$ with translates of certain self-similar subsets of $\mathbb{R}$. By a *tiling* of an open interval $I \subset \mathbb{R}$ we mean a covering of I with measurable sets $A_1, A_2, \cdots \subset \mathbb{R}$ such that $\mathcal{L}^1(A_j \cap I) > 0$ for all j and $\mathcal{L}^1(A_i \cap A_j) = 0$ for $i \neq j$.

For the proof we shall use the arithmetic expression (8.3) for $C(1/4)$:

$$C = \left\{ \sum_{j=1}^{\infty} 3 \cdot 4^{-j}(\varepsilon_{1,j}, \varepsilon_{2,j}) : \varepsilon_{k,j} \in \{0, 1\} \right\}.$$

Then

$$\pi_t \left(\tfrac{1}{3} C \right) = B_t := \left\{ \sum_{j=1}^{\infty} 4^{-j} \varepsilon_j : \varepsilon_j \in \{0, 1, t, t+1\} \right\}. \tag{10.2}$$

We shall now concentrate on the linear sets B_t forgetting about C. We restrict to $t \in (0, 1]$, which we may since $B_t = t B_{\frac{1}{t}}$ for $t > 0$. Then $B_t \subset [0, 2/3]$.

First we make some simple observations. Obviously B_0 is the same as the standard Cantor set $C_{1/4}$ (scaled to the interval $[0, 1/3]$), so it has Hausdorff dimension $1/2$. Similarly, B_1 is also a simple Cantor set of dimension

$\log 3/\log 4$. On the other hand,

$$B_2 = [0, 1] \quad \text{and} \quad B_{1/2} = [0, 1/2].$$

Observe also that B_t is self-similar. More precisely,

$$B_t = \bigcup_{\varepsilon_1 \in \{0,1,t,t+1\}} \left\{ \sum_{j=2}^{\infty} 4^{-j} \varepsilon_j + \varepsilon_1/4 : \varepsilon_j \in \{0, 1, t, t+1\} \right\}$$

$$= \bigcup_{\varepsilon_1 \in \{0,1,t,t+1\}} \tfrac{1}{4}(B_t + \varepsilon_1).$$

We shall also study dilations of B_t and we write the above formula as

$$4B_t = B_t + \{0, 1, t, t+1\} =: B_t + V_1. \tag{10.3}$$

Iterating this we have

$$4^m B_t = B_t + V_m \quad \text{for } m = 1, 2, \ldots, \tag{10.4}$$

where, for $t \in (0, 1]$,

$$V_m = \{0, 1, t, t+1\} + 4\{0, 1, t, t+1\} + \cdots + 4^{m-1}\{0, 1, t, t+1\} \subset [0, 4^m]. \tag{10.5}$$

From (10.3) we see as in Lemma 10.2 that

$$4\mathcal{L}^1(B_t) = \mathcal{L}^1(4B_t) \le \sum_{v \in V_1} \mathcal{L}^1(B_t + v) = 4\mathcal{L}^1(B_t),$$

so $\mathcal{L}^1((B_t + v_1) \cap (B_t + v_2)) = 0$ for $v_1, v_2 \in V_1, v_1 \ne v_2$. Similarly, the different translates of B_t in (10.4) intersect in measure zero:

$$\mathcal{L}^1((B_t + v_1) \cap (B_t + v_2)) = 0 \quad \text{for } v_1, v_2 \in V_m, v_1 \ne v_2, m = 1, 2, \ldots. \tag{10.6}$$

We shall now prove four lemmas. Part (a) of Theorem 10.5 obviously follows Lemmas 10.6 and 10.9.

Lemma 10.6 *If $\mathcal{L}^1(B_t) > 0$, then B_t contains a non-degenerate interval.*

Lemma 10.7 *If B_t contains a non-degenerate interval, it is the closure of its interior.*

Lemma 10.8 *If $\#V_m < 4^m$ for some $m = 1, 2, \ldots,$ then $\mathcal{L}^1(B_t) = 0$, and moreover,*

$$\dim B_t \le \frac{\log \#V_m}{m \log 4}.$$

Lemma 10.9 *If B_t contains a non-degenerate interval, then t is rational.*

Proof of Lemma 10.6 As $\mathcal{L}^1(B_t) > 0$, B_t has a point of density x. Then

$$\lim_{m \to \infty} \mathcal{L}^1(4^m(B_t - x) \cap [-1, 1]) = \lim_{m \to \infty} 4^m \mathcal{L}^1(B_t \cap [x - 4^{-m}, x + 4^{-m}]) = 2,$$

or equivalently by (10.4),

$$\lim_{m \to \infty} \mathcal{L}^1((B_t + V_m - 4^m x) \cap [-1, 1]) = 2,$$

and further,

$$\lim_{m \to \infty} \mathcal{L}^1((B_t + W_m) \cap [-1, 1]) = 2, \tag{10.7}$$

where W_m is the set of $v \in V_m - 4^m x$ for which $(B_t + v) \cap [-1, 1] \neq \varnothing$. As $B_t \subset [0, 1]$, we have $B_t + v \subset [-2, 2]$ and $v \in [-2, 2]$ for $v \in W_m$. Since the sets $B_t + v$ are pairwise almost disjoint, we obtain

$$\#W_m \leq 4/\mathcal{L}^1(B_t). \tag{10.8}$$

Thus the sets $W_m, m = 1, 2, \ldots$, are finite subsets of $[-2, 2]$ with uniformly bounded cardinality. It follows that some subsequence of them converges to a finite set $W \subset [-2, 2]$. By (10.7),

$$\mathcal{L}^1((B_t + W) \cap [-1, 1]) = 2.$$

Thus $(B_t + W) \cap [-1, 1]$ is a dense closed subset of $[-1, 1]$, so $(B_t + W) \cap [-1, 1] = [-1, 1]$. Hence the finite union of closed sets $B_t + w, w \in W$, has non-empty interior and so, as an easy topology exercise, some $B_t + w$ has non-empty interior and Lemma 10.6 follows. $\qquad \square$

Proof of Lemma 10.7 Let $x \in B_t$, let I be an open interval contained in B_t, and let $y \in I$. Then by (10.4) there are $v_m \in V_m, m = 1, 2 \ldots$, such that $x \in 4^{-m} B_t + 4^{-m} v_m$ and $y_m := 4^{-m} y + 4^{-m} v_m \in 4^{-m} I + 4^{-m} v_m \subset B_t$. Thus the points y_m are interior points of B_t converging to x. $\qquad \square$

Proof of Lemma 10.8 Let $N = \#V_m$. By (10.4), $B_t = 4^{-m} B_t + 4^{-m} V_m$, and so B_t is covered with N translates of $4^{-m} B_t$. Iterating this we see that for every $k = 1, 2, \ldots$, B_t is covered with N^k translates of $4^{-km} B_t$ each of them having diameter at most $4^{-km} d(B_t) < 4^{-km}$. Letting $s = \frac{\log N}{m \log 4}$, we have $N^k (4^{-km})^s = 1$ from which the lemma follows. $\qquad \square$

Proof of Lemma 10.9 Suppose B_t contains an open interval J. Then by Lemma 10.8, $\#V_m = 4^m$ for every $m = 1, 2, \ldots$, that is, the expressions defining V_m have no multiple points.

We have for all m,

$$|v_1 - v_2| \geq \mathcal{L}^1(J) \quad \text{for } v_1, v_2 \in V_m, v_1 \neq v_2; \tag{10.9}$$

otherwise two translates of J by elements of V_m would intersect in a non-degenerate interval which would contradict (10.6).

Fix a large integer N such that $d(4^{-N} B_t) < d(J)$. By (10.4) B_t is covered with 4^N almost disjoint translates, by elements of $4^{-N} V_N$, of $4^{-N} B_t$. Therefore for large enough N there are $v_1, v_2 \in 4^{-N} V_N$ with $v_1 < v_2$ such that

$$J_1 = 4^{-N} J + v_1 \subset J, \ J_2 = 4^{-N} J + v_2 \subset J \quad \text{and} \quad J_1 \cap J_2 = \varnothing.$$

Let us extract from the covering of B_t with $4^{-N} B_t + v$, $v \in 4^{-N} V_N$, a minimal covering, that is a tiling, for J:

$$J \subset \bigcup_{v \in V(J)} (4^{-N} B_t + v), \quad V(J) \subset 4^{-N} V_N.$$

Then for $i = 1, 2$,

$$J_i \subset \bigcup_{v \in V(J)} (4^{-2N} B_t + 4^{-N} v + v_i).$$

Set $V(J_i) = 4^{-N} V(J) + v_i$. Observe that $V(J_i) \subset 4^{-2N} V_N + v_i \subset 4^{-2N} V_{2N}$. We get the tilings of J_1 and J_2:

$$J_1 \subset \bigcup_{v \in V(J_1)} (4^{-2N} B_t + v),$$

$$J_2 \subset \bigcup_{v \in V(J_2)} (4^{-2N} B_t + v),$$

where the second is obtained from the first translating by $v_2 - v_1$. Such tilings are unique due to the almost disjointness of these translates. We shall use these to find a periodic tiling of $\mathbb{R}$ with period $v_2 - v_1$. First we shall construct a tiling of J extending the above tiling of J_1.

Let $I_1 = (a, b)$ be the largest open interval such that

$$J_1 \subset I_1 \subset \bigcup_{v \in V(J_1)} (4^{-2N} B_t + v).$$

Observe that $(4^{-2N} B_t + v) \cap I_1 = \varnothing$ for all $v \in 4^{-2N} V_{2N} \setminus V(J_1)$. Indeed, if $(4^{-2N} B_t + v) \cap I_1 \neq \varnothing$ for some $v \in 4^{-2N} V_{2N} \setminus V(J_1)$, then this inter-section would contain a non-degenerate interval by Lemma 10.7 and thus $4^{-2N} B_t + v$ would intersect some $4^{-2N} B_t + v'$, $v' \in V(J_1)$, in a positive mea-sure which would contradict (10.6). Since also $d(4^{-2N} B_t + v) < d(4^{-N} J) = d(J_1) \leq d(I_1)$, every set $4^{-2N} B_t + v$, $v \in 4^{-2N} V_{2N} \setminus V(J_1)$, lies either to the left or to the right of I_1. The sets $4^{-2N}B_t + v$, $v \in 4^{-2N}V_{2N}$, cover J but none of them for $v \in V(J_1)$ contains a neighbourhood of b (by the maximality of I_1). Hence, as the sets $4^{-2N}B_t + v$ are closed, there exists $v(b) \in 4^{-2N}V_{2N} \setminus V(J_1)$ such that $B_t + v(b)$ contains b, and then it contains b as its left extreme point. For any other $4^{-2N}B_t + v$, $v \in 4^{-2N}V_{2N} \setminus V(J_1)$, on the right of I_1, the left extreme point must be at least $b + 4^{-2N}\mathcal{L}^1(J)$ by (10.9). Therefore the sets $4^{-2N}B_t + v$, $v \in W_1 := V(J_1) \cup \{v(b)\}$, cover (a, c) for some $c > b$. Let b_2 be the largest of such numbers c, that is, let $I_2 = (a, b_2)$ be the maximal open interval for which

$$I_1 \subset I_2 \subset \bigcup_{v \in W_1} (4^{-2N}B_t + v).$$

Observe in passing that the above argument shows that any extension of a tiling of an open interval with translates of $4^{-2N}B_t$ is unique.

We can repeat the same procedure obtaining $W_2 := W_1 \cup \{v(b_2)\}$ and the maximal open interval

$$I_3 = (a, b_3) \subset \bigcup_{v \in W_2} (4^{-2N}B_t + v),$$

with $b_3 \geq b_2 + \mathcal{L}^1(J)$. We continue this until, after finitely many steps, we cover the right end-point d of J. On the way we find again the unique tiling of J_2 with the translates $4^{-2N}B_t + v$, $v \in V(J_2)$, which, as we already stated, is the translate by $r := v_2 - v_1$ of $4^{-2N}B_t + v$, $v \in V(J_1)$. Hence the tiling of (a, d) we have found is periodic with period r in the sense that if $4^{-2N}B_t + v$, $v \in V \subset 4^{-2N}V_{2N}$, tile an interval $I_0 \subset (a, d)$, then $4^{-2N}B_t + v + r$, $v \in V \subset 4^{-2N}V_{2N}$, tile the interval $I_0 + r$ provided it is contained in (a, d). Consequently we can extend this tiling to an r-periodic tiling of (a, ∞) by periodicity. Multiplying by 4^{2N} we get a $p = 4^{2N}r$-periodic tiling of $(4^{2N}a, \infty)$ with tiles $B_t + v$, with $v \in V_{2N}$ for the tiles meeting $(4^{2N}a, 4^{2N}b)$.

We can do the analogous construction to the left of J_1. Using again the uniqueness of these tilings we get a p-periodic tiling of the whole line $\mathbb{R}$:

$$\mathbb{R} = \bigcup_{w \in W} (B_t + w). \tag{10.10}$$

Here W is a discrete subset of $\mathbb{R}$. Moreover, by periodicity, $W = A + p\mathbb{Z}$ for some finite set A. The interval $(4^{2N}a, 4^{2N}b)$ is covered with tiles $B_t + v$ for which $v \in V_{2N}$.

Multiplying this tiling by 4^m and using (10.4) once more, we obtain

$$\mathbb{R} = \bigcup_{w \in W} \bigcup_{v \in V_m} (B_t + v + 4^m w). \tag{10.11}$$

For large $m > 2N$ only the tiles $B_t + v + 4^m w$ with $w = 0$ meet $(4^{2N}a, 4^{2N}b)$. Since $V_{2N} \subset V_m$, we conclude that (10.10) and (10.11) induce the same tiling for $(4^{2N}a, 4^{2N}b)$, whence these tilings of $\mathbb{R}$ must be identical by our previous observation about uniqueness.

Since (10.11) was obtained from (10.10) multiplying by 4^m, the interval $(4^{2N+m}a, 4^{2N+m}b)$ is covered with tiles of the form $B_t + v + 4^m w, v \in V_m, w \in V_{2N}$, which further are of the form $B_t + v, v \in V_{2N+m}$. Each $v \in V_{2N+m}$ can be written as

$$v = \sum_{j=0}^{\infty} (\varepsilon_j + t\eta_j) 4^j, \quad \varepsilon_j, \eta_j \in \{0, 1\}, \tag{10.12}$$

where $\varepsilon_j = \eta_j = 0$ for all but finitely many values of j. Using (10.4) and the fact, stated at the beginning of the proof, that there are no multiple points for V_m, we see that both terms of the form $v = \sum_{j=0}^{\infty} 4^j \varepsilon_j \in W, \varepsilon_j \in \{0, 1\}$, and of the form $v = \sum_{j=0}^{\infty} 4^j t\eta_j \in W, \eta_j \in \{0, 1\}$, are needed to cover $(4^{2N+m}a, 4^{2N+m}b) \subset 4^{2N+m} B_t$. Since $W = A + p\mathbb{Z}$ with A finite, we see, letting $m \to \infty$, that there are two different sequences (ε_j) and (ε'_j) and two different sequences (η_j) and (η'_j) such that both (finite sums) $\sum_{j=0}^{\infty} (\varepsilon_j - \varepsilon'_j) 4^j$ and $t \sum_{j=0}^{\infty} (\eta_j - \eta'_j) 4^j$ are integer multiples of p. This obviously implies that t is rational and completes the proof. $\qquad\square$

It remains to study the rational case. Let $t = p/q$ with the irreducible expression. Let us first check

Lemma 10.10 *If either p^* or q^* is even, then $\#V_m = 4^m$ for all $m = 1, 2, \ldots$.*

Proof We write again

$$V_m = \left\{ \sum_{j=0}^{m-1} (\varepsilon_j + t\eta_j) 4^j : \varepsilon_j, \eta_j \in \{0, 1\} \right\}.$$

The assertion of the lemma is that there are no multiple points for the sums above. This means that for any rational r the equation

$$\sum_{j=-\infty}^{\infty} \varepsilon_j 4^j + t \sum_{j=-\infty}^{\infty} \eta_j 4^j = r$$

has at most one solution among $\varepsilon_j, \eta_j \in \{0, 1\}$ which are non-zero only for finitely many j; the extension of the summation to negative values of j is no problem as the reader easily checks considering a suitable $4^k r$. With a bit of algebra this allows us to assume that $p = p^*$ and $q = q^*$.

The property of no multiple points means now that the equation

$$q \sum_{j=0}^{\infty} \varepsilon_j 4^j + p \sum_{j=0}^{\infty} \eta_j 4^j = q \sum_{j=0}^{\infty} \varepsilon_j' 4^j + p \sum_{j=0}^{\infty} \eta_j' 4^j$$

has only the solutions $\varepsilon_j = \varepsilon_j'$, $\eta_j = \eta_j'$ among $\varepsilon_j, \varepsilon_j', \eta_j, \eta_j' \in \{0, 1\}$ which are non-zero only for finitely many j, or equivalently that the equation

$$q \sum_{j=0}^{\infty} \gamma_j 4^j + p \sum_{j=0}^{\infty} \lambda_j 4^j = 0 \tag{10.13}$$

has only the trivial solutions $\gamma_j = \lambda_j = 0$ among $\gamma_j, \lambda_j \in \{-1, 0, 1\}$ which are non-zero only for finitely many j.

Suppose now that this last equation holds for $p = p^*$ and $q = q^*$, and either p^* or q^* is even. Then

$$q^* \gamma_0 + p^* \lambda_0 = 0 \quad \text{mod } 4.$$

Since $p^* = 2$ and $q^* = 1$ or 3, or $q^* = 2$ and $p^* = 1$ or 3, this is only possible if $\gamma_0 = \lambda_0 = 0$. Knowing this, we deduce from (10.13) that $\gamma_1 = \lambda_1 = 0$, and so on, $\gamma_j = \lambda_j = 0$ for all j. Thus there are no multiple points for the original sums, and the lemma follows. □

To finish the proofs of the statements (b) and (c) of Theorem 10.5 we introduce a measure μ on B_t. Let μ_m be the probability measure on $4^{-m} V_m$ giving equal measure $1/\#V_m$ to all of its points. We can extract a subsequence which converges weakly to a probability measure μ on B_t (since $4^{-m} V_m \subset B_t$).

Suppose now that $\#V_m = 4^m$ for all $m = 1, 2, \ldots$. By Lemma 10.10 this is the case if either p^* or q^* is even, but we shall use the statement obtained below also in the opposite case. Since $\mu_m(\{v\}) = 4^{-m}$ for all $v \in V_m$ and $|v_1 - v_2| \geq 1/q$ for all $v_1, v_2 \in V_m$, $v_1 \neq v_2$, we have for any interval J of length at least $1/(q4^m)$, $\mu_m(J) \leq 2q\mathcal{L}^1(J)$, whence also

$$\mu(J) \leq 2q\mathcal{L}^1(J). \tag{10.14}$$

This implies that μ is absolutely continuous and thus $\mathcal{L}^1(B_t) > 0$. By Lemmas 10.6 and 10.7 this finishes the proof of the statement (c) of Theorem 10.5.

We have left (b). We shall now show that if p^* and q^* are both odd, the Fourier transform of μ does not tend to zero at infinity, consequently μ is not absolutely continuous. Above we showed that this yields that $\#V_m < 4^m$ for some $m = 1, 2, \ldots$, and then appealing to Lemma 10.8, (b) follows.

So let us now proceed to study the Fourier transform of μ. Using (10.2) and calculating as in Chapter 8 we find that

$$\widehat{\mu}(u) = \Pi_{j=1}^{\infty} \tfrac{1}{4}(1 + e^{-2\pi i 4^{-j}u} + e^{-2\pi i 4^{-j}tu} + e^{-2\pi i 4^{-j}(1+t)u})$$
$$= \Pi_{j=1}^{\infty} \tfrac{1}{4}(1 + e^{-2\pi i 4^{-j}u})(1 + e^{-2\pi i 4^{-j}tu}).$$

Using again the formula $\frac{1+e^{ix}}{2} = e^{ix/2}\cos(x/2)$, we obtain

$$|\widehat{\mu}(u)| = |\Pi_{j=1}^{\infty} \cos(\pi 4^{-j}u)\Pi_{j=1}^{\infty}\cos(\pi 4^{-j}tu)|.$$

Recalling that $t = p/q$ and taking $u = q4^m$, $m \in \mathbb{N}$, we get

$$|\widehat{\mu}(u)| = |\Pi_{j=1}^{m} \cos(\pi 4^{m-j}q)\Pi_{j=m+1}^{\infty}\cos(\pi 4^{m-j}q)$$
$$\times \Pi_{j=1}^{m}\cos(\pi 4^{m-j}p)\Pi_{j=m+1}^{\infty}\cos(\pi 4^{m-j}p)|.$$

The products from 1 to m equal 1. The products from $m + 1$ to ∞ are independent of m, and the only way they could vanish is that at least one factor should be zero. But this is impossible when p^* and q^* are odd.

It follows that if p^* and q^* are both odd, the Fourier transform of μ does not tend to zero at infinity, which completes the proof of Theorem 10.5. $\square$

10.4 Average length of projections

Since $\mathcal{L}^1(p_\theta(C(1/4))) = 0$ for almost all $\theta \in (0, \pi)$, the integrals

$$I_k := \int_0^\pi \mathcal{L}^1\big(p_\theta(U_k^{1/4})\big)\, d\theta$$

tend to 0 when k tends to ∞; recall the definition of $U_k^{\frac{1}{4}}$ from (10.1). But how fast do they converge? Theorem 4.3 gives easily the lower bound

$$\int_0^\pi \mathcal{L}^1\big(p_\theta(U_k^{1/4})\big)\, d\theta \gtrsim k^{-1}. \tag{10.15}$$

To prove this it is enough to check that $I_1(\mu_k) \lesssim k$ when μ_k is the normalized Lebesgue measure on $U_k^{1/4}$ and then apply Theorem 4.3. Bateman and Volberg [2010] improved this to

$$\int_0^\pi \mathcal{L}^1\big(p_\theta(U_k^{1/4})\big)\, d\theta \gtrsim (\log k)k^{-1}. \tag{10.16}$$

Getting good upper bounds has turned out to be a very difficult problem. Using the notion of ε-relative closeness more effectively Peres and Solomyak [2002] derived a quantitative, but rather weak, upper bound. This was considerably improved by Nazarov, Peres and Volberg [2010] who proved with delicate

Fourier analytic and combinatorial arguments that for every $\delta > 0$,

$$\int_0^\pi \mathcal{L}^1\left(p_\theta\left(U_k^{1/4}\right)\right) d\theta \lesssim_\delta k^{\delta-1/6}. \tag{10.17}$$

Several authors developed this technique and result further. Łaba and Zhai [2010] proved similar upper bound estimates for more general Cartesian product Cantor sets. They also used the tiling methods of Kenyon [1997], which we discussed above, and of Lagarias and Wang [1996]. Bond, Łaba and Volberg [2014] extended these results to larger classes of product sets. Bond and Volberg [2010] managed without product structure proving an estimate of the type (10.17) for the one-dimensional Sierpinski gasket. Bond and Volberg [2012] proved the upper estimate $I_k \lesssim e^{-c\sqrt{\log k}}$ for rather general self-similar constructions with equal contraction ratios and without rotations. Bond and Volberg [2011] proved the lower bound (10.16) with orthogonal projections replaced by circular transformations. Bond, Łaba and Zhai [2013] studied the analogous question for radial projections from points. Peres and Solomyak [2002] showed that the estimate $I_k \approx k^{-1}$ holds almost surely for some random Cantor sets. A good survey on this topic was given by Łaba [2012].

A lower bound of the type (10.15) was proved in Mattila [1990] for a much larger class of sets, without any self-similarity assumptions. Tao [2009] proved an upper estimate for a very general class of sets.

The integral $\int_0^\pi \mathcal{L}^1(p_\theta(U)) d\theta$ gives the probability for a random line in the plane to meet the set U. Therefore it is often called Buffon's needle probability as a generalization of Count Buffon's famous eighteenth-century problem in geometric probability.

10.5 Further comments

As mentioned before, the proof we gave for Theorem 10.1 is due to Peres, Simon and Solomyak [2003]. It is also given in Bishop and Peres [2016], Section 9.5.

The presentation for the proof of Theorem 10.5 was based on Kenyon [1997] and influenced by Kahane [2013]. Very precise information about tilings of $\mathbb{R}$ was obtained by Lagarias and Wang [1996].

Theorem 10.5 left open what are the Hausdorff dimensions of the projections for the irrational values of t and for the rational t in the case (b). For the latter Kenyon [1997] gave a formula in the case of the Sierpinski gasket, Corollary 10 in his paper, but the same method works for our set C. We know by Marstrand's projection theorem that $\dim \pi_t(C) = 1$ for almost all

$t \in \mathbb{R}$. Furstenberg conjectured that this would hold for all irrational t. Recently Hochman [2014] verified this conjecture.

The four-corner Cantor set is an example of a self-similar set without rotations. That is, the generating similarities are composed only of dilations and translations, which makes it possible to check the dimension drop in many directions. Peres and Shmerkin [2009] proved that for many planar self-similar sets with rotations the situation is quite different; there are no exceptional directions for the dimension preservation. More precisely, let $K = \cup_{j=1}^{N} S_j(K)$ be a self-similar set such that $S_j(x) = r_j g_j(x) + a_j, 0 < r_j < 1, g_j \in O(2), a_j \in \mathbb{R}^2, j = 1, \ldots, N$. If the subgroup of $O(2)$ generated by $g_j, j = 1, \ldots, N$, is dense in $O(2)$, then

$$\dim p_\theta(K) = \min\{\dim K, 1\} \text{ for all } \theta \in [0, \pi).$$

Very roughly the idea is the following. By Marstrand's theorem 4.1 there are projections for which the dimension is preserved, assuming $\dim K \leq 1$. Thus an approximation of K at a small scale $\delta > 0$ satisfies a kind of discretized δ level dimension preservation. The self-similarity, the denseness assumption and an ergodic theorem imply that similar configurations appear at arbitrary small scales in every direction.

Nazarov, Peres and Shmerkin [2012] proved related results for convolutions of self-similar measures. Earlier Moreira [1998] had proven similar results for attractors of some non-linear dynamical systems. These are now included in a very general result of Hochman and Shmerkin [2012]. They proved the dimension preservation for a large class of sets and measures, and not only for projections, but for all non-singular C^1 maps. Other related results are due to Ferguson, Jordan and Shmerkin [2010] and Farkas [2014].

Fulfilling the above program of Peres and Shmerkin is far from trivial. But it does not seem to give an answer for the analogous question when $\dim K > 1$: is then $\mathcal{L}^1(p_\theta(K)) > 0$ for all $\theta \in (0, \pi)$? However, Shmerkin [2014] and Shmerkin and Solomyak [2014] have later proved that the exceptional set in this and many other similar settings has dimension zero.

Originally the fact that $C(1/4)$ projects into a set of measure zero in almost all directions is due to Besicovitch. It follows from his general theorem that any purely unrectifiable plane Borel set with finite one-dimensional Hausdorff measure has this property. The pure unrectifiability means that the set meets every rectifiable curve in zero length. For the proof and related matters, see, for example, Falconer [1985a], Section 6.4, or Mattila [1995], Chapter 18.

For the values of d other than $1/4$ there are several open problems about the Hausdorff dimension and measures of the projections $p_\theta(C(d))$. Again Theorem 4.1 tells us that if $d < 1/4$, that is, $\dim C(d) < 1$, then $\dim p_\theta(C(d)) = s_d$

for almost all θ. But what can be said about the measures? For example for what values of d is $\mathcal{H}^{s_d}(p_\theta(C(d))) > 0$ for almost all θ? This is true when $d < 1/9$ by an easy argument, see Mattila [2004]. This argument shows for all $0 < d < 1/4$ that $\mathcal{H}^{s_d}(p_\theta(C(d))) > 0$ for a non-empty open set of angles θ. Peres, Simon and Solomyak [2000] proved that when $1/6 < d < 1/4$, then also the set of θ with $\mathcal{H}^{s_d}(p_\theta(C(d))) = 0$ has positive measure. It is not known what happens when $1/9 < d < 1/6$.

The four-corner Cantor is sometimes called Garnett set or Garnett–Ivanov set. This is because Garnett and Ivanov showed in the 1970s that it has zero analytic capacity. Later many people studied it and related sets in connection with analytic capacity and the Cauchy transform. One can consult Chapter 19 of Mattila [1995] and in particular Tolsa's book [2014] for this.

Often it is not easy to compute the exact value of the Hausdorff measure even for fairly simple fractal sets. Davies gave in 1959 a simple elegant proof for the fact that $\mathcal{H}^1(C(1/4)) = \sqrt{2}$; this is unpublished, I am grateful to Kenneth Falconer for this information. Xiong and Zhou [2005] established formulas for the measures of a class of Sierpinski carpet type sets, including $C(1/4)$. These sets have dimension at most one. Computing the measure for sets of dimension bigger than one seems to be much harder. For example, the exact value for the von Koch snow-flake curve appears to be unknown.

11

Besicovitch sets

We say that a Borel set in $\mathbb{R}^n$, $n \geq 2$, is a *Besicovitch set*, or a Kakeya set, if it has zero Lebesgue measure and it contains a line segment of unit length in every direction. This means that for every $e \in S^{n-1}$ there is $b \in \mathbb{R}^n$ such that $\{te + b : 0 < t < 1\} \subset B$. It is not obvious that Besicovitch sets exist but they do in every $\mathbb{R}^n$, $n \geq 2$, as we shall now prove. We shall also show that their Hausdorff dimension is at least 2. Moreover, we shall discuss related Nikodym and Furstenberg sets.

11.1 Existence of Besicovitch sets

We show that Besicovitch sets exist using duality between points and lines.

Theorem 11.1 *For any $n \geq 2$ there exists a Borel set $B \subset \mathbb{R}^n$ such that $\mathcal{L}^n(B) = 0$ and B contains a whole line in every direction. Moreover, there exist compact Besicovitch sets in $\mathbb{R}^n$.*

Proof It is enough to find B in the plane since then we can take $B \times \mathbb{R}^{n-2}$ in higher dimensions.

Let $C \subset \mathbb{R}^2$ be a compact set such that $\pi(C) = [0, 1]$, where $\pi(x, y) = x$ for $(x, y) \in \mathbb{R}^2$, and $\mathcal{L}^1(p_\theta(C)) = 0$ for $\mathcal{L}^1$ almost all $\theta \in [0, \pi)$. Here p_θ is again the projection onto the line through the origin forming an angle θ with the x-axis. We can take as C a suitably rotated and dilated copy of $C(1/4)$ or we can modify the construction of $C(1/4)$ by placing the first four disjoint closed squares of side-length $\frac{1}{4}$ inside $[0, 1] \times [0, 1]$ so that their projections cover $[0, 1]$. Consider the lines

$$\ell(a, b) = \{(x, y) : y = ax + b\}, \quad (a, b) \in C,$$

and define B as their union:

$$B = \bigcup_{(a,b) \in C} \ell(a, b) = \{(x, ax + b) : x \in \mathbb{R}, (a, b) \in C\}.$$

143

From the latter representation it is easy to see that B is σ-compact and hence a Borel set. If we restrict x to $[0, 1]$, B will be compact, which will give us compact Besicovitch sets. Since $\pi(C) = [0, 1]$, B contains a line $\ell(a, b)$ for some b for all $0 \leq a \leq 1$. Taking a union of four rotated copies of B we get a Borel set that contains a line in every direction. It remains to show that $\mathcal{L}^2(B) = 0$.

We do this by showing that almost every vertical line meets B in a set of length zero and then using Fubini's theorem. For any $t \in \mathbb{R}$,

$$B \cap \{(x, y) : x = t\} = \{(t, at + b) : (a, b) \in C\}$$
$$= \{t\} \times \pi_t(C),$$
(11.1)

where $\pi_t(x, y) = tx + y$. The map π_t is essentially a projection p_θ for some θ, and hence we have $\mathcal{L}^1(\pi_t(C)) = 0$ for $\mathcal{L}^1$ almost all $t \in \mathbb{R}$. Thus $\mathcal{L}^2(B) = 0$. $\qquad\square$

11.2 Hausdorff dimension of Besicovitch sets

Reversing the above argument, we now use the projection theorems of Chapter 4 to prove that Besicovitch sets must have Hausdorff dimension 2 at least.

Theorem 11.2 *For every Besicovitch set B, $\dim B \geq 2$.*

Proof If B is a Besicovitch set in $\mathbb{R}^n$ and Π is the projection, $\Pi(x) = (x_1, x_2)$, then $\Pi(B)$ is contained in a G_δ set B' which contains a unit line segment in every direction and for which $\dim B' = \dim \Pi(B) \leq \dim B$. Thus we can assume that B is a G_δ Besicovitch set in the plane. For $a \in (0, 1), b \in \mathbb{R}$ and $q \in \mathbb{Q}$ denote by $I(a, b, q)$ the line segment $\{(q + t, at + b) : 0 \leq t \leq 1/2\}$ of length less than 1. Let C_q be the set of (a, b) such that $I(a, b, q) \subset B$. Then each C_q is a G_δ-set, because for any open set G, the set of (a, b) such that $I(a, b, q) \subset G$ is open. Since for every $a \in (0, 1)$, some $I(a, b, q) \subset B$, we have $\pi(\cup_{q \in \mathbb{Q}} C_q) = (0, 1)$, with $\pi(x, y) = x$, and so there is $q \in \mathbb{Q}$ for which $\mathcal{H}^1(C_q) > 0$. Then by Theorem 4.1, for almost all $t \in \mathbb{R}$, $\dim \pi_t(C_q) = 1$, where again $\pi_t(x, y) = tx + y$. We have now for $0 \leq t \leq 1/2$,

$$\{q + t\} \times \pi_t(C_q)$$
$$= \{(q + t, at + b) : (a, b) \in C_q\} \subset B \cap \{(x, y) : x = q + t\}.$$

Hence for a positive measure set of t, vertical t-sections of B have dimension 1. By Proposition 6.6 we obtain that $\dim B = 2$. $\qquad\square$

We give another proof for compact Besicovitch sets which shows more; even the Fourier dimension is at least 2. Recall from Section 3.6 the definition of the Fourier dimension $\dim_F$ and the fact that $\dim_F \leq \dim$.

Theorem 11.3 *For every compact Besicovitch set B, $\dim_F B \geq 2$.*

Proof We first skip the measurability problems and return to them at the end of the proof. For $e \in S^{n-1}$, let $a_e \in \mathbb{R}^n$ be such that $a_e + te \in B$ for all $0 \leq t \leq 1$. Fix a non-negative function $\varphi \in C_0^\infty(\mathbb{R})$ with $\text{spt}\,\varphi \subset [0, 1]$ and $\int \varphi = 1$. Define $\mu \in \mathcal{M}(B)$ by

$$\int g\,d\mu = \int_{S^{n-1}} \int_0^1 g(a_e + te)\varphi(t)\,dt\,d\sigma^{n-1}e$$

for continuous functions g on $\mathbb{R}^n$. Let $0 < \alpha < 1$ and $\xi \in \mathbb{R}^n$ with $|\xi| > 1$. The Fourier transform of μ at ξ is given by

$$\widehat{\mu}(\xi) = \int_{S^{n-1}} \int e^{-2\pi i \xi \cdot (a_e + te)} \varphi(t)\,dt\,d\sigma^{n-1}e,$$

which yields

$$|\widehat{\mu}(\xi)| \leq \int_{S^{n-1}} |\widehat{\varphi}(\xi \cdot e)|\,d\sigma^{n-1}e.$$

Let $\eta > 0$ and $S_{\xi,\eta} = \{e \in S^{n-1} : |\xi \cdot e| < \eta|\xi|\}$. Then $\sigma^{n-1}(S_{\xi,\eta}) \lesssim \eta$ and so for any $N > 1$,

$$|\widehat{\mu}(\xi)| \lesssim_N \eta + (\eta|\xi|)^{-N}.$$

Choosing $\eta = |\xi|^{-\alpha}$ and N such that $N/(N+1) = \alpha$, we have

$$|\widehat{\mu}(\xi)| \lesssim_\alpha |\xi|^{-\alpha}.$$

This yields $\dim_F B \geq 2$.

The measurability problem disappears if σ^{n-1} is replaced by a discrete measure $\sigma_k = \sum_{j=1}^{m_k} c_{k,j} \delta_{e_{k,j}}$. The above proof goes through if σ_k satisfies $\sigma_k(S^{n-1}) \lesssim 1$ and $\sigma_k(S_{\xi,\eta}) \lesssim \eta$ for $k > k_\eta$. We leave it as an easy exercise for the reader to check that σ^{n-1} can be written as a weak limit of such measures σ_k. Then for a given $\xi \in \mathbb{R}^n$ with $|\xi| > 1$, the corresponding measures μ_k satisfy $|\widehat{\mu_k}(\xi)| \lesssim_\alpha |\xi|^{-\alpha}$ for large k. Moreover, they converge weakly to a measure $\mu \in \mathcal{M}(B)$ with $|\widehat{\mu}(\xi)| \lesssim_\alpha |\xi|^{-\alpha}$, which completes the proof. □

Both proofs above give more. Let us consider this in the plane. Suppose that $B \subset \mathbb{R}^2$ is a Borel (compact in the case of Theorem 11.3) set and $E \subset S^1$ is a Borel set such that $\dim E = s$ and B contains a unit line segment in every direction $e \in E$. Then $\dim B \geq s + 1$ and $\dim_F B \geq 2s$. The first statement follows when one applies the generalization of Proposition 6.6 mentioned in Section 6.4. For the proof of the second statement one replaces σ^1 by a Frostman measure on E.

To find good lower bounds for the Hausdorff dimension of Besicovitch sets is an interesting problem to which we shall return extensively. The conjecture, usually called *Kakeya conjecture*, is:

Conjecture 11.4 Every Besicovitch set in $\mathbb{R}^n$ has Hausdorff dimension n.

This is true for $n = 2$ and open for $n \geq 3$. One can state the corresponding conjectures for the upper and lower Minkowski dimensions and for the packing dimension. In the plane they follow from the Hausdorff dimension version and for $n \geq 3$ they too are open.

Recall from the Introduction the connection to Stein's restriction conjecture. We shall return to this in Chapters 22 and 23.

Now we go back to orthogonal projections and use Besicovitch sets to show that in the plane there is no non-trivial analogue of Theorem 4.2:

Example 11.5 There is a Borel set $A \subset \mathbb{R}^2$ such that $\dim A = 2$, and even $\mathcal{L}^2(\mathbb{R}^2 \backslash A) = 0$, but the interior of $p_\theta(A)$, Int $p_\theta(A)$, is empty for all $\theta \in [0, \pi)$.

Proof Let B be the Besicovitch set of Theorem 11.1 and

$$A = \mathbb{R}^2 \backslash \bigcup_{q \in \mathbb{Q}^2} (B + q).$$

Then A has all the required properties. $\qquad\qquad\qquad\qquad\qquad\qquad\Box$

Let us still make a simple observation about the relations between different dimensions of Besicovitch sets:

Proposition 11.6 *If for all n every Besicovitch set in $\mathbb{R}^n$ has Hausdorff dimension at least $n - c(n)$, where $\lim_{n \to \infty} c(n)/n = 0$, then for all n every Besicovitch set in $\mathbb{R}^n$ has packing and upper Minkowski dimension n.*

Proof The packing, $\dim_P$, and upper Minkowski, $\overline{\dim}_M$, dimensions were defined in Section 2.3. Since $\dim_P \leq \overline{\dim}_M$, it is enough to consider the packing dimension. The only properties we need for it are the trivial inequality $\dim \leq \dim_P$ and the simple product inequality (see, e.g., Mattila [1995], Theorem 8.10):

$$\dim_P(A \times B) \leq \dim_P A + \dim_P B.$$

This holds for the upper Minkowski dimension, too, and is even simpler.

Suppose we have a Besicovitch set B in $\mathbb{R}^n$ of packing dimension less than n. Then for large $k \in \mathbb{N}$, $\dim_P B < n - c(kn)/(kn)$. But this gives for the k-fold product $B^k \subset \mathbb{R}^{kn}$,

$$\dim(B^k) \leq \dim_P(B^k) \leq k \dim_P B < kn - c(kn),$$

which is a contradiction, since B^k is a Besicovitch set in $\mathbb{R}^{kn}$. $\qquad\qquad\square$

11.3 Nikodym sets

In 1927 Nikodym [1927] constructed a kind of relative of Besicovitch sets; a Borel set A in the unit square $[0, 1] \times [0, 1]$ such that $\mathcal{L}^2(A) = 1$ and for every $x \in A$ there is a line L through x for which $L \cap (A \setminus \{x\}) = \varnothing$. Davies [1952a] simplified Nikodym's construction and also showed that it is possible to construct the set A so that there are uncountably many lines through every $x \in A$ which meet A only at x. Davies's construction of Nikodym sets is presented in de Guzmán's book [1981] too.

We shall call Nikodym sets the complements of sets like A. More precisely, we say that a Borel set $N \subset \mathbb{R}^n$ is a *Nikodym set* if $\mathcal{L}^n(N) = 0$ and for every $x \in \mathbb{R}^n$ there is a line L through x for which $L \cap N$ contains a unit line segment.

As Besicovitch sets, Nikodym sets allow a dual construction based on projections. We shall present it below in the plane following Falconer [1986]. This paper also contains higher dimensional formulations and other interesting and surprising related results and phenomena, see also Chapter 7 in Falconer [1985a] and Chapter 6 in Falconer [1990]. In particular, the following theorem is valid in $\mathbb{R}^n$ for any $n \geq 2$ with lines replaced by hyperplanes.

Theorem 11.7 *There is a Borel set $N \subset \mathbb{R}^2$ such that $\mathcal{L}^2(N) = 0$ and for every $x \in \mathbb{R}^2$ there is a line L through x for which $L \setminus \{x\} \subset N$.*

For an arc G in $G(2, 1)$ (identifying $G(2, 1)$ with S^1) we let $2G$ be the arc with the same centre as G and with double length. Recall that P_L is the orthogonal projection onto the line $L \in G(2, 1)$.

Lemma 11.8 *Let $Q \subset \mathbb{R}^2$ be a square, $\varepsilon_j > 0$, $j = 1, 2, \ldots$, and let G_j be subarcs of $G(2, 1)$ such that $G_1 \supset G_2 \supset \ldots$. Then there are compact sets $C_j \subset \mathbb{R}^2$ such that $C_1 \supset C_2 \supset \ldots$, and $Q \cap L \subset P_L(C_j)$ for $L \in G_j$ and $\mathcal{H}^1(P_L(C_j)) < \varepsilon_j$ for $L \in G(2, 1) \setminus 2G_j$.*

Proof This follows by the iterated Venetian blind construction. The idea of the construction is presented in Figure 11.1. There a line segment is replaced by many short parallel line segments. These project into small length in directions

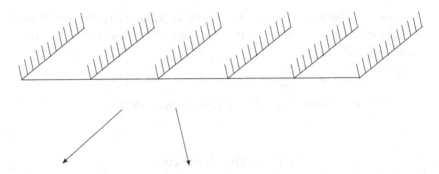

Figure 11.1 Venetian blinds

belonging to a small neighbourhood of the direction of these line segments, but outside a slightly bigger neighbourhood their projection contains that of the original segment. Next all these segments are replaced by many much shorter parallel line segments in a different direction. Then two intervals result where the union has small projection, but still nothing is lost in most directions. Iterating this in a suitable manner we find a finite union B_1 of line segments such that $Q \cap L \subset P_L(B_1)$ for $L \in G_1$ and $\mathcal{H}^1(P_L(B_1)) < \varepsilon_1$ for $L \in G(2, 1) \setminus 2G_1$. Enclosing each of these segments into a sufficiently narrow closed rectangle, the union of these will have the same properties as B_1. This is our first set C_1. Next we can perform a similar process inside each of the rectangles to get $C_2 \subset C_1$. Continuing this yields the lemma. We leave the details to the reader, or see Falconer [1986]. $\square$

Lemma 11.9 *For every $L \in G(2, 1)$ there is a Borel set $A_L \subset L$ such that $\mathcal{H}^1(A_L) = 0$ and if $x \in L$, then there is $y \in \mathbb{R}^2$ such that $P_L(y) = x$ and $P_{L'}(y) \in A_{L'}$ for every $L' \in G(2, 1), L' \neq L$.*

Proof Write $\mathbb{R}^2 = \cup_{m=1}^{\infty} Q_m$ where the Q_m are pairwise disjoint squares of sidelength 1. Let $G_{k,j} \subset G(2, 1)$, $j = 1, \ldots, 2^k$, $k = 1, 2 \ldots$, provide for each k a decomposition of $G(2, 1)$ into dyadic arcs such that $\cup_j G_{k,j} = G(2, 1)$, $\gamma_{2,1}(G_{k,j}) = 2^{-k}$ and, in the usual way, each $G_{k,j}$ splits into two disjoint arcs G_{k+1,j_1} and G_{k+1,j_2}. With the aid of Lemma 11.8 we find compact sets $C_{m,k,j} \subset \mathbb{R}^2$, $j = 1, \ldots, 2^k$, $k, m = 1, 2 \ldots$, such that

$$C_{m,k',j'} \subset C_{m,k,j} \quad \text{if } G_{k',j'} \subset G_{k,j}, \tag{11.2}$$

$$Q_m \cap L \subset P_L(C_{m,k,j}) \quad \text{for all } L \in G_{k,j}, \tag{11.3}$$

and

$$\mathcal{H}^1(P_L(C_{m,k,j})) < 2^{-2k-m} \quad \text{for all } L \in G(2, 1) \setminus 2G_{k,j}.$$

Set

$$A_L = \bigcap_{l=1}^{\infty} \bigcup_{1 \le j \le 2^k, k \ge l, m \ge 1, L \in G(2,1) \backslash 2G_{k,j}} P_L(C_{m,k,j}).$$

Then $\mathcal{H}^1(A_L) = 0$.

Let $L \in G(2, 1)$ and $x \in L$. Then $x \in Q_m \cap L$ for some m and there is a sequence (j_k) such that $L \in G_{k,j_k}$ for all $k \ge 1$. Thus by (11.3) $x \in Q_m \cap L \subset P_L(C_{m,k,j_k})$ and we find $y_k \in C_{m,k,j_k}$ such that $P_L(y_k) = x$. Using (11.2) we find a limit point y of the sequence (y_k) which belongs to C_{m,k,j_k} for all k. Clearly also $P_L(y) = x$.

Suppose then that $L' \in G(2, 1)$ and $L' \ne L$. Then for sufficiently large k, $L' \notin 2G_{k,j_k}$. Therefore $P_{L'}(y) \in A_{L'}$, and the lemma follows. $\square$

Proof of Theorem 11.7 We use the duality between lines and points induced by the reflexion in the unit circle. For $x \in \mathbb{R}^2 \setminus \{0\}$, let $L_x \in G(2, 1)$ be the line through x, let $x' = |x|^{-2}x$ and let M_x be the line orthogonal to L_x passing through x'. Then $y \in M_x$ if and only if the vectors $y - x'$ and x are orthogonal, that is, $(y - |x|^{-2}x) \cdot x = y \cdot x - 1 = 0$. Since this is symmetric in x and y, we have $y \in M_x$ if and only if $x \in M_y$. Observe also that $y \in M_x$ if and only if $P_{L_x}(y) = x'$.

For $L \in G(2, 1)$ let $A_L \subset L$ be as in Lemma 11.9 and define

$$N = \{x \in \mathbb{R}^2 \setminus \{0\} : x' \in A_{L_x}\}.$$

Then every line through the origin meets N in a set of length zero, so $\mathcal{L}^2(N) = 0$. Let $x \in \mathbb{R}^2 \setminus \{0\}$. Then by Lemma 11.9 there is $y \in \mathbb{R}^2$ such that $P_{L_x}(y) = x'$ and $P_L(y) \in A_L$ for every $L \in G(2, 1)$, $L \ne L_x$. Then $y \in M_x$ and thus $x \in M_y$. If $z \in M_y$ and $z \ne x$, then $y \in M_z$, so $z' = P_{L_z}(y) \in A_{L_z}$ and $z \in N$. This means that $M_y \setminus \{x\} \subset N$.

We only considered $x \ne 0$. But replacing N by $N \cup (N + a)$ for some $a \in \mathbb{R}^2 \setminus \{0\}$ we obtain the desired set. $\square$

Analogously to the Kakeya conjecture we have the *Nikodym conjecture*:

Conjecture 11.10 Every Nikodym set in $\mathbb{R}^n$ has Hausdorff dimension n.

Now we show using a projective transformation that every Nikodym set generates a Besicovitch set. Define

$$F(\widetilde{x}, x_n) = \frac{1}{x_n}(\widetilde{x}, 1) \quad \text{for } (\widetilde{x}, x_n) \in \mathbb{R}^n, x_n \ne 0. \tag{11.4}$$

If $e \in S^{n-1}$ with $e_n \ne 0$ and $a \in \mathbb{R}^{n-1}$, F maps the half-lines $\{te + (a, 0) : t \ne 0\}$ onto the half-lines $\{u(a, 1) + \frac{1}{e_n}(\widetilde{e}, 0) : u \ne 0\}$. Hence F maps every

Nikodym set, or even a set which contains a unit line segment in some line through $(a, 0)$ for all $a \in \mathbb{R}^{n-1}$, to a set which contains a line segment in every direction $(a, 1), a \in \mathbb{R}^{n-1}$. Taking the union of finitely many dilated and translated copies of these images one gets a Besicovitch set.

The following theorem is an immediate consequence of the above construction and Theorem 11.2:

Theorem 11.11 *If $1 \le s \le n$ and there is a Nikodym set in $\mathbb{R}^n$ of Hausdorff dimension s, then there is a Besicovitch set in $\mathbb{R}^n$ of Hausdorff dimension s. In particular, $\dim N \ge 2$ for every Nikodym set N in $\mathbb{R}^n$ and the Kakeya conjecture implies the Nikodym conjecture.*

Reversing the previous argument only gives partial Nikodym sets from Besicovitch sets, lines going through all points of a fixed hyperplane. I do not know if the Nikodym conjecture implies the Kakeya conjecture.

According to Lebesgue's theorem on differentiation of integrals

$$\lim_{B \to x} \frac{1}{\mathcal{L}^n(B)} \int_B f \, d\mathcal{L}^n = f(x) \quad \text{for almost all } x \in \mathbb{R}^n$$

for any locally integrable function f. Here $B \to x$ means that the limit is taken with balls B containing x and tending to x. The existence of Nikodym sets implies easily that balls cannot be replaced with arbitrary rectangular boxes even when f is a characteristic function. De Guzmán's books [1975] and [1981] discuss extensively differentiation theory of integrals and validity of such results with different classes of sets and functions.

11.4 Lines vs. line segments

We defined Besicovitch sets as sets of measure zero containing a unit line segment in every direction, but we showed in Theorem 11.1 that there exist sets of measure zero containing a whole line in every direction. In general, is there a difference in the sizes of these types of sets? This question was studied by Keleti [2014]. We present here some of his results.

First, as concerns Lebesgue measure there is a great difference. Let N be a Nikodym set of measure zero as in Theorem 11.7. Then for every $x \in \mathbb{R}^2$ there is an open half-line $L_x \subset N$ with end-point x. These half-lines cover a set of measure zero, but the corresponding lines, and even the corresponding closed half-lines, cover the whole plane.

For dimension the situation turns out to be different. Keleti posed the following *line segment extension conjecture*:

Conjecture 11.12 If A is the union of a family of line segments in $\mathbb{R}^n$ and B is the union of the corresponding lines, then dim A = dim B.

This is true in the plane:

Theorem 11.13 *Conjecture 11.12 is true in $\mathbb{R}^2$.*

We skip some measurability arguments, which are given in Keleti [2014], and consider only the case where B is a Borel set parametrized by another Borel set C as before. More precisely, we again let

$$l(a, b) = \{(x, y) : y = ax + b\},$$

and we set

$$\mathcal{L}(C) = \bigcup \{l(a, b) : (a, b) \in C\}.$$

Notice that $\mathcal{L}(C)$ is a Borel set, if C is a σ-compact. If C is a Borel set, then $\mathcal{L}(C)$ is a Suslin set, which also would suffice for the argument below.

For the proof of Theorem 11.13 we use the following lemma:

Lemma 11.14 *If $C \subset \mathbb{R}^2$ is a Borel set, then*

$$\dim \mathcal{L}(C) \cap \{(t, y) : y \in \mathbb{R}\} = \min\{\dim C, 1\} \quad \text{for almost all } t \in \mathbb{R}.$$

Proof As in the proof of Theorem 11.1, we have $\mathcal{L}(C) \cap \{(t, y) : y \in \mathbb{R}\} = \{t\} \times \pi_t(C)$, where $\pi_t(x, y) = tx + y$. The lemma follows then from Marstrand's projection theorem 4.1. □

Proof of Theorem 11.13 As already mentioned, we only handle the case where $B = \mathcal{L}(C)$ for some Borel set C. Let $J(a, b) \subset l(a, b)$ be the corresponding line segments composing A. We may assume that dim $B > 1$. Let $1 < s < \dim B$. Decomposing C into a countable union, we can suppose that for each $(a, b) \in C$, $J(a, b)$ meets two fixed line segments I and J which form the opposite sides of a rectangle.

Set $L_{v,t} = \{x \in \mathbb{R}^2 : v \cdot x = t\}$ for $v \in S^1$, $t \in \mathbb{R}$. By Theorem 6.7 we have for σ^1 almost all $v \in S^1$, $\dim L_{v,t} \cap B \geq s - 1$ for $t \in T_v$ where $T_v \subset \mathbb{R}$ with $\mathcal{L}^1(T_v) > 0$. Fix such a non-exceptional unit vector v in a way that there are parallel lines l_0 and l_1 which are orthogonal to v and which separate the line segments I and J. Rotating the whole picture we may assume that $v = (1, 0)$. Let L_t be the vertical line $\{(t, y) : y \in \mathbb{R}\}$. Then $l_0 = L_\alpha$ and $l_1 = L_\beta$ for some, say, $\alpha < \beta$.

We now have that for every $(a, b) \in C$ the line segment $J(a, b) \subset l(a, b)$ meets both lines L_α and L_β. Hence $L_{v,t} \cap A = L_{v,t} \cap B$ for all $t \in [\alpha, \beta]$.

As above,

$$\dim L_{v,t} \cap B \ge s - 1 \quad \text{for } t \in T_v.$$

Since T_v has positive measure we get by Lemma 11.14 that

$$\dim L_{v,t} \cap B \ge s - 1 \quad \text{for almost all } t \in \mathbb{R}.$$

Hence

$$\dim L_{v,t} \cap A \ge s - 1 \quad \text{for almost all } t \in [\alpha, \beta].$$

Therefore Proposition 6.6 yields that $\dim A \ge s$, from which the theorem follows. $\qquad\Box$

If true, the line segment extension conjecture would imply the Kakeya conjecture for the packing, and hence for upper Minkowski, dimension, and it would improve the known Hausdorff dimension estimates (discussed in Chapter 23) in dimensions $n \ge 5$:

Theorem 11.15 *(1) If Conjecture 11.12 is true for some n, then, for this n, every Besicovitch set in $\mathbb{R}^n$ has Hausdorff dimension at least $n - 1$.*

(2) If Conjecture 11.12 is true for all n, then every Besicovitch set in $\mathbb{R}^n$ has packing and upper Minkowski dimension n for all n.

Proof (2) follows from (1) by Proposition 11.6. To prove (1) we use the projective transformation F as in (11.4):

$$F(\widetilde{x}, x_n) = \frac{1}{x_n}(\widetilde{x}, 1) \quad \text{for } (\widetilde{x}, x_n) \in \mathbb{R}^n, x_n \ne 0.$$

For $e \in S^{n-1}$ with $e_n \ne 0$ and $a \in \mathbb{R}^{n-1}$, let $L(e, a)$ be the punctured line $\{te + (a, 0) : t \ne 0\}$. As already observed, F maps it onto the punctured line $\widetilde{L}(e, a) := \{u(a, 1) + \frac{1}{e_n}(\widetilde{e}, 0) : u \ne 0\}$. If B is a Besicovitch set in $\mathbb{R}^n$, it contains for every $e \in S^{n-1}$ for some $a_e \in \mathbb{R}^{n-1}$ a line segment $J_e \subset L(e, a_e)$. Thus $F(B)$ contains a line segment on $\widetilde{L}(e, a_e)$ for every $e \in S^{n-1}$ with $e_n \ne 0$. The union of the line extensions of these segments contains $\{\frac{1}{e_n}(\widetilde{e}, 0) : e \in S^{n-1}, e_n \ne 0\}$, which is the hyperplane $\mathbb{R}^{n-1} \times \{0\}$ of Hausdorff dimension $n - 1$. Hence, assuming Conjecture 11.12, $\dim F(B) \ge n - 1$, which implies that $\dim B \ge n - 1$. $\qquad\Box$

11.5 Furstenberg sets

The following question is in the spirit of Besicovitch sets: Let $0 < s < 1$ and suppose that $F \subset \mathbb{R}^2$ is a compact set with the property that for every $e \in S^1$

there is a line L_e in direction e for which dim $L_e \cap F \geq s$. What can be said about the dimension of F? Wolff [2003], Section 11.1, showed that dim $F \geq$ max$\{2s, s + 1/2\}$ and that there is such an F with dim $F = 3s/2 + 1/2$. The lower bound $2s$ is easier and its proof resembles the proof of Theorem 22.9. In Wolff [1999] he also connected this problem to the decay estimates of the L^1 spherical averages of the Fourier transform. When $s = 1/2$ Bourgain [2003] improved the lower bound 1 to dim $F \geq 1 + c$ for some absolute constant $c > 0$ using the work of Katz and Tao [2001]. Recall also Section 4.4 for the discrete level results in Katz and Tao [2001]. D. M. Oberlin [2014b] improved Wolff's lower bound for a class of sets related to the four-corner Cantor set. Some other recent results on this problem were obtained by Molter and Rela [2010], [2012] and [2013].

The above question comes from Furstenberg and the sets appearing in it are called Furstenberg sets. The origin seems to be the following remarkable result of Furstenberg [1970]:

For a positive integer p a closed subset $A \subset [0, 1]$ is called a p-set if $pA \subset A \cup (A + 1) \cup \cdots \cup (A + p - 1)$. Suppose that p and q are not powers of the same integer, A is a p-set, B is a q-set, $C = A \times B$, and $s > 0$ is an arbitrary positive number. If there is a line with positive, finite slope which intersects C in a set of Hausdorff dimension greater than s, then for almost every $u > 0$, there is a line of slope u intersecting C in a set of dimension greater than s.

11.6 Further comments

Besicovitch [1919] was the first to construct a set named after him solving a question on Riemann integrability. It was republished in Besicovitch [1928]. In doing this he also solved a problem of Kakeya [1917]: in how small (in terms of area) a plane domain a unit segment can be turned around continuously? The answer is: arbitrarily small. But in the plane it is impossible to turn around a unit segment continuously in a set of measure zero, as was shown by Tao [2008b]. However, in higher dimensions this is possible, as proven by E. Järvenpää, M. Järvenpää, Keleti and Máthé [2011].

Łaba [2008] has an interesting discussion on Besicovitch and his early work.

Fefferman [1971] was the first to apply these constructions to problems of Fourier transforms, to the ball multiplier problem. We shall return to this later as well as to other relations between Besicovitch sets and Fourier analysis.

Elementary geometric constructions of Besicovitch sets can be found, for example, in Falconer [1985a], de Guzmán [1981] and Stein [1993]. They are based on the *Perron tree*. This is a construction where a triangle is divided

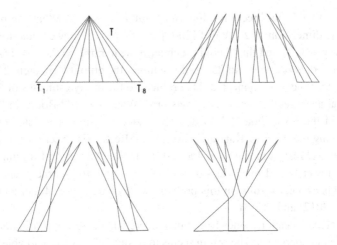

Figure 11.2 Perron tree

into many subtriangles and they are translated in order to have large overlap, whence a small area, but no directions of the unit line segments from the original triangle are lost. Such a construction where a triangle T is partitioned into eight subtriangles T_j is presented in Figure 11.2.

Simple and more analytic constructions are presented in Wolff [2003], Chapter 11 (due to Sawyer [1987]), and Bishop and Peres [2016], Section 9.1. Bishop and Peres also present a random construction of Besicovitch sets. In that book it is also shown, a result due to Keich [1999], that the Lebesgue measure of the δ-neighbourhood of a Besicovitch set B can be $\lesssim 1/\log(1/\delta)$. This is optimal, as follows from Córdoba's Kakeya maximal function estimate, Theorem 22.5. So this gives a sharp result on gauge functions with respect to which the Minkowski contents of Besicovitch sets are positive. Keich's paper also contains partial results for generalized Hausdorff measures, but sharp results for them seem to be unknown. Babichenko, Peres, Peretz, Sousi and Winkler [2014] constructed Besicovitch sets using games. They also gave a different proof for Keich's estimate.

Besicovitch [1964] used his general projection theorem and duality between points and lines to get a completely new way of finding Besicovitch sets. The construction presented above is in the same spirit but followed Falconer's modification in Falconer [1985a], Section 7.3. Kahane [1969] showed, see also Kahane [2013], that connecting $\{(\frac{1}{3}x, 0) : x \in C_{1/4}\}$ to $\{(\frac{1}{3}x + \frac{2}{3}, 1) : x \in C_{1/4}\}$ (recall Chapter 8 for the notation) with all possible line segments gives a Besicovitch set. Again the reason that this set has Lebesgue measure zero is that almost all projections of the four-corner Cantor set $C(1/4)$ have zero measure. Alexander [1975] related a compact plane set to every sequence $x_1, x_2, \cdots \in [0, 2/3]$, in a way somewhat similar to that of Kahane. He showed by an easy argument that for almost all such sequences it is a Besicovitch set. Moreover, using the above projection property of a set like the four-corner Cantor set, he showed that it is a Besicovitch set for every constant sequence $x_j = x \in [0, 2/3]$.

Körner [2003] showed that Besicovitch sets are generic; one can show their existence by the Baire category theorem. Fraser, Olson and Robinson [2014] proved some other category properties of Besicovitch sets.

Theorem 11.2 is due to Davies [1971] and Theorem 11.3 to D. M. Oberlin [2006a]. Later we shall give another proof for Theorem 11.2, due to Córdoba, and we shall obtain better lower bounds for the Hausdorff dimension of Besicovitch sets in dimensions bigger than two. Tao [1999a] used the projective transformation (11.4) to associate Nikodym sets to Besicovitch sets and thus the Kakeya conjecture to the Nikodym conjecture. The line segment extension conjecture and the related results in Section 11.4, as well as Proposition 11.6, are due to Keleti [2014]. Some further related results were obtained by Falconer and Mattila [2015].

Another interesting open question is: for which pairs of integers (k, n), $0 < k < n$, are there Borel sets $B \subset \mathbb{R}^n$ such that $\mathcal{L}^n(B) = 0$ and B contains a k-plane in every direction? We know that they exist when $k = 1$ for all n. They do not exist when $k > \frac{n}{2}$. This follows from Corollary 5.12 by the same argument we used for Example 11.5: if such a set B exists, then $A = \mathbb{R}^n \setminus \bigcup_{q \in \mathbb{Q}^n} (B + q)$ would contradict Corollary 5.12(c) because dim $A = n > 2(n - k)$ and $P_V(A)$ has empty interior for all $V \in G(n, n - k)$. We shall discuss in Chapter 24 some sharpenings of this by Marstrand [1979] and Falconer [1980a], in the case $k > n/2$, and by Bourgain [1991a] in the case $k \leq n/2$.

One could also ask about the existence of multi-line Besicovitch sets, sets of measure zero containing many line segments in every direction. Łaba and Tao [2001b] derived from their general results, which will be briefly discussed in Section 24.4, dimension estimates for such multi-line Besicovitch sets. In particular, if $B \subset \mathbb{R}^2$ contains a positive Hausdorff dimension collection of unit line segments in every direction, then $\mathcal{L}^2(B) > 0$. Orponen [2014b] gave an elegant direct proof for this and a related result.

The above Besicovitch's duality method can be adapted to many other curve packing problems. For example, there are circles in the plane centred at every point of a given line segment covering only a set of measure zero. To see this, let

$$B = \bigcup_{(a,b) \in C} \{(x, y) : (x - a)^2 + y^2 = a^2 + b\},$$

where C is as in the proof of Theorem 11.1, and modify the argument for that proof. However, if the centres form a set of positive Lebesgue measure, then the union of the circles must have positive Lebesgue measure. This was proved independently by Bourgain [1986] and Marstrand [1987]. In fact, Bourgain proved more; he showed that the circular maximal operator M_S,

$$M_S f(x) = \sup_{r>0} \int_{S^1} |f(x - ry)| \, d\sigma^1 y, \quad x \in \mathbb{R}^2,$$

is bounded from $L^p(\mathbb{R}^2)$ into $L^p(\mathbb{R}^2)$ for $p > 2$. The same result, with $p > n/(n-1)$ and due to Stein, for the spherical maximal operator, is valid and easier in higher dimensions. A consequence is the spherical differentiation theorem: if $f \in L^p(\mathbb{R}^n)$ and $p > n/(n-1)$, then

$$\lim_{r \to 0} \int_{S^{n-1}} f(x - ry) d\sigma^{n-1} y = f(x) \quad \text{for almost all } x \in \mathbb{R}^n.$$

See Stein [1993], Chapter XI, Grafakos [2008], Section 5.5, and de Guzmán [1981], Chapter 12, for these and other results on maximal and differentiation theorems along curves and surfaces.

The above circle and sphere packing result can be sharpened: if the centres of the spheres form a set of Hausdorff dimension bigger than one in $\mathbb{R}^n$, then the union of these spheres must have positive Lebesgue measure. This was proved by Mitsis [1999] for $n \geq 3$ and by Wolff [2000] for $n = 2$. Mitsis's argument, which worked only for $3/2$ in place of 1 in the plane, is geometric while Wolff's proof is very complicated involving geometric, combinatorial and Fourier analytic ideas. In fact, Wolff proved more: he showed that if $E \subset \mathbb{R}^n \times (0, \infty)$ and $F \subset \mathbb{R}^n, n \geq 2$, are Borel sets such that $\dim E > 1$ and $\mathcal{H}^{n-1}(\{y : |y - x| = r\} \cap F) > 0$ for $(x, r) \in E$, then $\mathcal{L}^n(F) > 0$. D. M. Oberlin [2006b] gave a simpler proof for this in dimensions $n \geq 3$.

On the other hand, one can again show by the duality method that there is a family of circles containing a circle of every radius and covering only a set of measure zero, see Falconer [1985a], Theorem 7.10. The same is true with spheres in $\mathbb{R}^n, n \geq 3$, as pointed out by Kolasa and Wolff [1999]. In that paper they proved for $n \geq 3$ that such a family of spheres must have Hausdorff dimension n. Wolff [1997] extended this to $n = 2$. More precisely, he proved that if the set of centres has Hausdorff dimension $s, 0 < s \leq 1$, the corresponding union has dimension at least $1 + s$. See also the discussion in Wolff [2003], Section 11.3. More generally, one would expect that if $E \subset \mathbb{R}^n \times (0, \infty)$ and $F \subset \mathbb{R}^n, n \geq 2$, are Borel sets such that $0 < s = \dim E \leq 1$ and $\mathcal{H}^{n-1}(\{y : |y - x| = r\} \cap F) > 0$ for $(x, r) \in E$, then $\dim F \geq n - 1 + s$. D. M. Oberlin [2007] proved this for $n \geq 3$.

Wisewell [2004] proved a very general result on packing curves and surfaces into a set of measure zero.

In analogy to the spherical average operator and the related maximal operator, Iosevich, Sawyer, Taylor and Uriarte-Tuero [2014] proved $L^p(\mu) \to L^q(\nu)$ inequalities for the operator $f \mapsto \lambda * (f\mu)$, where the measures μ and ν satisfy Frostman growth conditions and λ satisfies a Fourier decay condition.

Käenmäki and Shmerkin [2009] proved dimension results for Kakeya (or Besicovitch) type self-affine subsets of the plane.

12

Brownian motion

In this chapter we shall study subsets of Brownian trajectories and Fourier transforms of measures on them. In particular we shall see that they give us Salem sets of any dimension s, $0 < s < 2$.

12.1 Some facts on Brownian motion

We present first without proofs some basic facts about Brownian motion.

The n-dimensional *Brownian motion* (or one realization of it) is a probability measure on the space Ω_n of continuous functions $\omega : [0, \infty) \to \mathbb{R}^n$ such that $\omega(0) = 0$, the increments $\omega(t_2) - \omega(t_1)$ and $\omega(t_4) - \omega(t_3)$ are independent for $0 \le t_1 \le t_2 \le t_3 \le t_4$ and such that $\omega(t + h) - \omega(t)$ has Gaussian distribution with zero mean and variance h for $t \ge 0$ and $h > 0$. In particular,

$$P_n(\{\omega : |\omega(t + h) - \omega(t)| \le \varrho\}) = ch^{-n/2} \int_0^\varrho r^{n-1} e^{-r^2/(2h)}\, dr \quad (12.1)$$

for $t \ge 0$, $h > 0$ and $\varrho > 0$. Here c is chosen so that $c2^{n/2} \int_0^\infty r^{n-1} e^{-r^2}\, dr = 1$, which means $P_n(\Omega_n) = 1$. This gives

$$\int |\omega(t + h) - \omega(t)|^{-s}\, dP_n\omega = c_1 h^{-s/2} \quad (12.2)$$

for $t \ge 0$, $h > 0$ and $0 < s < n$. This formula is quite close to saying that the paths $\omega \in \Omega_n$ are almost surely Hölder continuous with exponent $1/2$. That is not quite true, but they are Hölder continuous with exponent s for any $0 < s < 1/2$, see Falconer [1985a], Lemma 8.22, for example. Hence for any $A \subset [0, \infty)$, $\dim \omega(A) \le 2 \dim A$ for P_n almost all $\omega \in \Omega_n$.

As usual, we shall denote by $\mathbf{E}$ the expectation:

$$\mathbf{E}(f) = \int f\, dP_n.$$

158

As a consequence of the Gaussian distribution and the fact that $e^{-\pi|x|^2}$ is its own Fourier transform (recall Section 3.1) we have the formula for the expectation of $e^{-2\pi ix\cdot\omega(t)}$ (the characteristic function):

$$\mathbf{E}(e^{-2\pi ix\cdot\omega(t)}) = e^{-2\pi|x|^2t}. \tag{12.3}$$

12.2 Dimension of trajectories

We introduce for any $\mu \in \mathcal{M}([0,\infty))$ and $\omega \in \Omega_n$ the image of μ under ω:

$$\mu_\omega = \omega_\sharp\mu \in \mathcal{M}(\mathbb{R}^n)$$

characterized by

$$\int g\,d\mu_\omega = \int g\circ\omega\,d\mu$$

for continuous functions g. In particular, when we take $\mu = \mathcal{L}^1 \llcorner [0,1]$ we get a natural probability measure on the trajectory from 0 to $\omega(1)$.

Theorem 12.1 *Let* $\mu \in \mathcal{M}([0,\infty))$. *If* $0 < s \le 1$ *and* $\mu([x-r,x+r]) \le r^s$ *for all* $x \in \mathbb{R}$ *and* $r > 0$, *then for* P_n *almost all* $\omega \in \Omega_n$ *and for all* $x \in \mathbb{R}^n$, $|x| \ge 2$,

$$|\widehat{\mu_\omega}(x)| \le C(\mu,\omega)(\log(|x|))^{1/2}|x|^{-s}.$$

Proof By the definition of μ_ω,

$$\widehat{\mu_\omega}(x) = \int e^{-2\pi ix\cdot\omega(t)}\,d\mu t.$$

Let us compute $\mathbf{E}(|\widehat{\mu_\omega}(x)|^{2q})$ for positive integers q. We have by Fubini's theorem

$$|\widehat{\mu_\omega}(x)|^{2q} = \int\cdots\int \exp(2\pi ix\cdot(\omega(t_1)+\cdots+\omega(t_q)-\omega(u_1)-\omega(u_q)))$$
$$\times d\mu t_1\cdots d\mu t_q\,d\mu u_1\cdots d\mu u_q.$$

Since the integrand is symmetric with respect to $t_1,\ldots,t_q$, the t-integrals over $t_{\sigma(1)} < \cdots < t_{\sigma(q)}$ are equal for all permutations σ of $1,\ldots,q$ and their sum is the full t-integral, and similarly for the u-integrals. Since there are $q!$ such permutations we obtain (the integrand is as above),

$$|\widehat{\mu_\omega}(x)|^{2q} = (q!)^2\int_{0<t_1<\cdots<t_q}\int_{0<u_1<\cdots<u_q}\exp(\ldots)\,d\mu t_1\ldots d\mu t_q\,d\mu u_1\ldots d\mu u_q.$$

It is enough to integrate over variables $t_1,\ldots,t_q,u_1,\ldots,u_q$ such that $t_i \ne u_j$ for all i and j, because singletons have zero μ measure. Then for any given

$t_1, \ldots, t_q, u_1, \ldots, u_q$, we can write them in the increasing order $v_1 < v_2 < \cdots < v_{2q}$ and we can write

$$\omega(t_1) + \cdots + \omega(t_q) - \omega(u_1) - \omega(u_q) = \varepsilon_1 \omega(v_1) + \cdots + \varepsilon_{2q} \omega(v_{2q}),$$

where $\varepsilon_j \in \{-1, 1\}$ are such that $\sum_{j=1}^{2q} \varepsilon_j = 0$, $\varepsilon_j = 1$ if $v_j = t_i$ for some i and $\varepsilon_j = -1$ if $v_j = u_i$ for some i. Conversely, every sequence $\varepsilon_1, \ldots, \varepsilon_{2q} \in \{-1, 1\}$ with $\sum_{j=1}^{2q} \varepsilon_j = 0$ determines uniquely the order of the variables $t_1, \ldots, t_q, u_1, \ldots, u_q$ in this manner. It follows that we can write the above integral summing over all sequences (ε_j), $\varepsilon_j \in \{-1, 1\}$, such that $\sum_{j=1}^{2q} \varepsilon_j = 0$:

$$|\widehat{\mu_\omega}(x)|^{2q} = (q!)^2 \sum_{(\varepsilon_j)} \int_{0 < t_1 < \cdots < t_{2q}} \exp(2\pi i x \cdot (\varepsilon_1 \omega(t_1) + \cdots + \varepsilon_{2q} \omega(t_{2q})))$$

$$\times \, d\mu t_1 \cdots d\mu t_{2q}.$$

Next we write

$$\varepsilon_1 \omega(t_1) + \cdots + \varepsilon_{2q} \omega(t_{2q})$$
$$= (\varepsilon_1 + \cdots + \varepsilon_{2q}) \omega(t_1) + (\varepsilon_2 + \cdots + \varepsilon_{2q})(\omega(t_2) - \omega(t_1))$$
$$+ \cdots + \varepsilon_{2q}(\omega(t_{2q}) - \omega(t_{2q-1})).$$

For $0 < t_1 < \cdots < t_{2q}$, $\omega(t_1), \omega(t_2) - \omega(t_1), \ldots, \omega(t_{2q}) - \omega(t_{2q-1})$ are independent random variables. Thus using (12.3),

$$\mathbf{E}(\exp(2\pi i x \cdot (\varepsilon_1 \omega(t_1) + \cdots + \varepsilon_{2q} \omega(t_{2q}))))$$
$$= \mathbf{E}(\exp(2\pi i x \cdot (\varepsilon_1 + \cdots + \varepsilon_{2q}) \omega(t_1))) \cdots$$
$$\times \mathbf{E}(\exp(2\pi i x \cdot \varepsilon_{2q}(\omega(t_{2q}) - \omega(t_{2q-1}))))$$
$$= \exp(-2\pi |x|^2 ((\varepsilon_1 + \cdots + \varepsilon_{2q})^2 t_1 + \cdots + \varepsilon_{2q}^2 (t_{2q} - t_{2q-1}))).$$

Set $a_j = 2\pi |x|^2 (\varepsilon_j + \cdots + \varepsilon_{2q})^2$. Then $a_j \geq 0$ for all j and $a_j \geq |x|^2$ for even j. We have now

$$\mathbf{E}(|\widehat{\mu_\omega}(x)|^{2q}) = (q!)^2 \sum_{(\varepsilon_j)} \int_{0 < t_1 < \cdots < t_{2q}} \exp(-a_1 t_1 - a_2(t_2 - t_1) - \ldots$$

$$- a_{2q}(t_{2q} - t_{2q-1})) \, d\mu t_1 \cdots d\mu t_{2q}.$$

For any $a > 0$ we have by our assumption on μ,

$$\int_{t > t_0} e^{-a(t - t_0)} \, d\mu t = \int_0^1 \mu(\{t > t_0 : e^{-a(t-t_0)} > r\}) \, dr$$

$$= \int_0^1 \mu(\{t : t_0 < t < t_0 + a^{-1} \log(1/r)\}) \, dr$$

$$\leq a^{-s} \int_0^1 (\log(1/r))^s \, dr = C(s) a^{-s}.$$

Using this and integrating the above first over t_j with even j we obtain

$$\mathbf{E}(|\widehat{\mu_\omega}(x)|^{2q})$$
$$\leq (q!)^2 \sum_{(\varepsilon_j)} (C(s)|x|^{-2s})^q \int \cdots \int_{0 < t_1 < t_3 < \cdots < t_{2q-1}} 1 \, d\mu t_1 \, d\mu t_3 \cdots d\mu t_{2q-1}$$
$$= \frac{(2q)!}{q!} \mu(\mathbb{R})^q (C(s)|x|^{-2s})^q,$$

the last equation comes from the facts that there are $\frac{(2q)!}{(q!)^2}$ (binomial coefficient) sequences (ε_j) to consider and the last multiple integral is $\mu(\mathbb{R})^q/q!$ by the same symmetry reasons as before. Since $\frac{(2q)!}{q!} \leq 2^q q^q$, we have with some constant C, independent of q,

$$\mathbf{E}(|\widehat{\mu_\omega}(x)|^{2q}) \leq (Cq|x|^{-2s})^q. \tag{12.4}$$

Choose a set $Z \subset \mathbb{R}^n \setminus B(0,1)$ such that $|z_1 - z_2| \geq \min\{|z_1|^{-s}, |z_2|^{-s}\}/2$ for $z_1, z_2 \in Z$, $z_1 \neq z_2$, and that for every $x \in \mathbb{R}^n$ there is $z \in Z$ for which $|x - z| < |x|^{-s}$. Then for $k = 1, 2 \ldots$ the number of points $z \in Z$ such that $2^{k-1} \leq |z| \leq 2^k$ is about $2^{n(s+1)k}$ so that

$$\sum_{z \in Z} |z|^{-n(s+2)} < \infty.$$

For $z \in Z$ let q_z be the integer for which $\log|z| < q_z \leq \log|z| + 1$. Then by (12.4)

$$\mathbf{E}\left(\sum_{z \in Z} |z|^{-n(s+2)} \frac{|\widehat{\mu_\omega}(z)|^{2q_z}}{(Cq_z|z|^{-2s})^{q_z}}\right) < \infty.$$

Thus for P_n almost all ω the term in the series tends to zero as $z \in Z$, $|z| \to \infty$, which gives that

$$|\widehat{\mu_\omega}(z)| \leq C(\omega)(\log|z|)^{1/2}|z|^{-s},$$

that is, our claim for the points in Z is proven. Recall from (3.19) that $\widehat{\mu_\omega}$ is Lipschitz. By the choice of Z, for every $x \in \mathbb{R}^n$ there is $z \in Z$ for which $|x - z| < |x|^{-s}$, whence, when $|x| \geq 2$,

$$|\widehat{\mu_\omega}(x)| \lesssim |\widehat{\mu_\omega}(z)| + |x|^{-s} \lesssim (\log|x|)^{1/2}|x|^{-s}. \qquad \square$$

Combining this theorem with the fact that $\dim \omega(A) \leq 2 \dim A$ and using Frostman's lemma we obtain

Theorem 12.2 *Let $A \subset [0, \infty)$ be a Borel set. For P_n almost all $\omega \in \Omega$:*

(a) if $n \geq 2$, then $\dim \omega(A) = 2 \dim A$,
(b) if $n = 1$, then $\dim \omega(A) = 2 \dim A$ provided $\dim A \leq 1/2$, otherwise, $\mathcal{L}^1(\omega(A)) > 0$.

Proof If $0 < s < \dim A$, Frostman's lemma gives us a measure $\mu \in \mathcal{M}(A)$ which satisfies the assumptions of Theorem 12.1. If $n \geq 2$ or $s < 1/2$, then for P_n almost all $\omega \in \Omega$, $I_t(\mu_\omega) < \infty$ for $0 < t < 2s$ by Theorem 3.10, which implies $\dim \omega(A) \geq 2t$. This proves the case $n \geq 2$ and the first part for $n = 1$. In the second part $\dim A > 1/2$ and we can take $s > 1/2$, which yields that $\widehat{\mu_\omega} \in L^2(\mathbb{R})$ proving that $\mathcal{L}^1(\omega(A)) > 0$ by Theorem 3.3. $\square$

Recall Salem sets from Definition 3.11.

Corollary 12.3 *If $A \subset [0, \infty)$ is a Borel set, then for P_n almost all $\omega \in \Omega$, $\omega(A)$ is a Salem set.*

12.3 Further comments

The results of this chapter are due to Kahane from 1966. The presentation here follows very closely Kahane [1985], Chapter 17. In that book Kahane presents various results on Fourier transforms, the Hausdorff dimension and many other stochastic processes. See also the notes on pages 288–289 and the references given to Kahane's papers, and to many interesting papers of Kaufman related to this topic. Xiao [2013], Section 4.2, discusses many recent results and references on Fourier dimension and stochastics processes. Other good references for Brownian motion and its relations to Hausdorff dimension are Mörters and Peres [2010], Falconer [1985a], [1990], and Bishop and Peres [2016].

The almost sure Hausdorff dimension of the graph $\{(t, \omega(t)) : t > 0\}$ is $3/2$, if $n = 1$, and 2, if $n \geq 2$, see Mörters and Peres [2010] or Falconer [1990]. Recall however from Section 6.3 that the graphs are not Salem sets.

When $n = 1$ Kaufman [1975] proved that almost surely $\mathrm{Int}\, \omega(A) \neq \varnothing$ if $A \subset [0, \infty)$ is a Borel set with $\dim A > 1/2$. For a more general result, see Theorem 2 in Chapter 18 of Kahane [1985].

13

Riesz products

In this chapter we introduce an important class of measures on the real line, called Riesz products. We study their absolute continuity, singularity and relations to the Hausdorff dimension. We shall also use them to construct a singular measure which locally behaves very much like a Lebesgue measure.

13.1 Definition of Riesz products

Formally the *Riesz product* induced by the sequences (a_j), $a_j \in [-1, 1]$, and (λ_j), $\lambda_j \in \mathbb{N}$, is the infinite product

$$\Pi_{j=1}^{\infty}(1 + a_j \cos(2\pi\lambda_j x)), \quad x \in [0, 1].$$

We shall always assume that

$$\lambda_{j+1} \geq 3\lambda_j \quad \text{for all } j.$$

This guarantees that every integer m has at most one representation in the form $m = \sum_j \varepsilon_j \lambda_j$ where $\varepsilon_j \in \{-1, 0, 1\}$. To check this, observe first that for all k,

$$\lambda_{k+1} - \lambda_k - \cdots - \lambda_1 > \lambda_k + \cdots + \lambda_1$$

and reduce the claim to this.

The first question is: in what sense does the Riesz product exist? If $\sum_j |a_j| < \infty$ this product converges pointwise to a continuous function. In general pointwise convergence may fail and we should consider the product as a measure. Observe that we get Lebesgue measure if all $a_j = 0$.

Define

$$f_N(x) = \Pi_{j=1}^{N}(1 + a_j \cos(2\pi\lambda_j x)), \quad x \in [0, 1].$$

Using the formula

$$\cos\alpha\cos\beta = (\cos(\alpha+\beta)+\cos(\alpha-\beta))/2$$

we can expand f_N as a trigonometric polynomial of the form

$$f_N(x) = 1 + \sum_m b_m \cos(2\pi m x) \tag{13.1}$$

where m attains the values

$$m = \sum_{j=1}^{N} \varepsilon_j \lambda_j, \quad \varepsilon_j \in \{-1, 0, 1\},$$

and

$$b_m = \Pi_{\varepsilon_j \neq 0}(a_j/2) \quad \text{when} \quad m = \sum_{j=1}^{N} \varepsilon_j \lambda_j. \tag{13.2}$$

By (13.1) $\int_0^1 f_N(x)\,dx = 1$ for all N and we can think of f_N as a probability measure on $[0, 1]$. By the general weak compactness Theorem 2.4 we can extract weakly converging subsequences of (f_N). But we want to show that the whole sequence converges. This follows if we can show that the limit measure is the same for every converging subsequence. To achieve this we use Fourier analysis on $[0, 1]$. Then instead of the Fourier transform we consider the Fourier coefficients $\widehat{\mu}(k)$, $k \in \mathbb{Z}$. Recall from (3.66) that we have for $\mu, \nu \in \mathcal{M}([0, 1])$, $\mu = \nu$ if and only if $\widehat{\mu}(k) = \widehat{\nu}(k)$ for all $k \in \mathbb{Z}$.

We can compute the Fourier coefficients of f_N from the formulas (13.1) and (13.2). For $|k| \leq \lambda_N$ we get

$$\widehat{f_N}(k) = \Pi_{\varepsilon_j \neq 0}(a_j/2) \quad \text{if} \quad k = \sum_j \varepsilon_j \lambda_j \quad \text{with} \quad \varepsilon_j \in \{-1, 0, 1\}$$

with the interpretation $\widehat{f_N}(0) = 1$, and

$$\widehat{f_N}(k) = 0 \quad \text{if } k \text{ does not have such a representation.}$$

From this we see that for every $k \in \mathbb{Z}$ there is N_k such that

$$\widehat{f_N}(k) = \widehat{f_{N_k}}(k) \quad \text{for all } N \geq N_k.$$

This gives that for any weak limit measure μ,

$$\widehat{\mu}(k) = \widehat{f_{N_k}}(k) \quad \text{for all } k \in \mathbb{Z},$$

which implies the uniqueness of μ and the weak convergence of the Riesz product.

Definition 13.1 For the sequences $a = (a_j)$, $a_j \in [-1, 1]$, and $\lambda = (\lambda_j)$, $\lambda_j \in \mathbb{N}$ with $\lambda_{j+1} \geq 3\lambda_j$, the Riesz product

$$\mu_{a,\lambda} = \mu_a = \Pi_{j=1}^{\infty}(1 + a_j \cos(2\pi\lambda_j x))$$

is the weak limit of the partial products f_N, $f_N(x) = \Pi_{j=1}^{N}(1 + a_j \times \cos(2\pi\lambda_j x))$, $x \in [0, 1]$, as $N \to \infty$.

We shall use the notation μ_a, when the sequence λ will be kept fixed and a will vary.

We still record from the above calculations the Fourier coefficients of $\mu_{a,\lambda}$:

$$\widehat{\mu_{a,\lambda}}(k) = \Pi_{\varepsilon_j \neq 0}(a_j/2) \quad \text{if} \quad k = \sum_j \varepsilon_j \lambda_j \quad \text{with} \quad \varepsilon_j \in \{-1, 0, 1\}, \quad (13.3)$$

$$\widehat{\mu_{a,\lambda}}(k) = 0 \quad \text{if } k \text{ does not have such a representation.} \quad (13.4)$$

13.2 Absolute continuity of Riesz products

Riesz products are continuous measures, that is, the singletons have measure 0, see for example Zygmund [1959], Section V.7. Some Riesz products are singular and some absolutely continuous.

Theorem 13.2 *The Riesz product μ_a is absolutely continuous with respect to Lebesgue measure if and only if $\sum_j a_j^2 < \infty$. In that case $\mu_a \in L^2([0, 1])$. If $\sum_j a_j^2 = \infty$, μ_a and $\mathcal{L}^1$ are mutually singular.*

Proof Assume first that $\sum_j a_j^2 < \infty$. We shall prove that the L^2 norms of f_N are uniformly bounded. It is then an exercise to show using Hölder's inequality and basic properties of weak convergence that $\mu_a \in L^2$. We estimate the squares of the factors in the product by

$$(1 + a_j \cos(2\pi\lambda_j x))^2 \leq (1 + a_j^2)\left(1 + \frac{2a_j}{1 + a_j^2}\cos(2\pi\lambda_j x)\right).$$

Hence we can conclude

$$\int_0^1 f_N(x)^2\, dx = \int_0^1 \Pi_{j=1}^{N}(1 + a_j\cos(2\pi\lambda_j x))^2\, dx$$

$$\leq \Pi_{j=1}^{N}(1 + a_j^2)\int_0^1 \Pi_{j=1}^{N}\left(1 + \frac{2a_j}{1 + a_j^2}\cos(2\pi\lambda_j x)\right)dx$$

$$= \Pi_{j=1}^{N}(1 + a_j^2) \leq \Pi_{j=1}^{\infty}(1 + a_j^2) < \infty.$$

Suppose then that $\sum_j a_j^2 = \infty$. We see from (13.3) that

$$\int e^{\pm 2\pi i \lambda_j x} \, d\mu_a x = \widehat{\mu}_a(\pm \lambda_j) = a_j/2,$$

and

$$\int e^{\pm 2\pi i \lambda_j x \pm 2\pi i \lambda_l x} \, d\mu_a x = \widehat{\mu}_a(\pm \lambda_j \pm \lambda_l) = a_j a_l/4 \quad \text{for} \quad j \neq l.$$

These relations yield that the functions

$$x \mapsto e^{2\pi i \lambda_j x} - a_j/2$$

form a bounded orthogonal sequence in $L^2(\mu_a)$. As $\sum_j a_j^2 = \infty$, we can find c_j such that

$$\sum_j c_j^2 < \infty, c_j a_j \geq 0 \quad \text{and} \quad \sum_j c_j a_j = \infty.$$

An easy way to see this is to apply the Banach–Steinhaus theorem to the linear functionals $L_k : l^2 \to \mathbb{R}$, $L_k(c_j) = \sum_{j=1}^{k} a_j c_j$, with norms $\|L_k\| = (\sum_{j=1}^{k} a_j^2)^{1/2} \to \infty$ as $k \to \infty$.

Now the series $\sum_j c_j(e^{2\pi i \lambda_j x} - a_j/2)$ is an orthogonal series converging in $L^2(\mu_a)$. On the other hand the series $\sum_j c_j e^{2\pi i \lambda_j x}$ is an orthogonal series converging in $L^2(\mathcal{L}^1 \, \llcorner \, [0,1])$. Any series converging in L^2 has almost everywhere converging subsequences. So there is a sequence (N_m) such that the sequence $\sum_{j=1}^{N_m} c_j(e^{2\pi i \lambda_j x} - a_j/2)$ converges μ_a almost everywhere to a finite value. Next there is a subsequence (N_m') of (N_m) such that the sequence $\sum_{j=1}^{N_m'} c_j e^{2\pi i \lambda_j x}$ converges $\mathcal{L}^1$ almost everywhere to a finite value. Then also $\sum_{j=1}^{N_m'} c_j(e^{2\pi i \lambda_j x} - a_j/2)$ converges μ_a almost everywhere. If μ_a and $\mathcal{L}^1$ were not mutually singular there would be a point x where both of these sequences converge. Then also their difference would converge. But the difference is $\sum_j c_j a_j/2 = \infty$. This contradiction completes the proof of the theorem. $\qquad \Box$

13.3 Riesz products and Hausdorff dimension

We shall now show that sets with sufficiently small Hausdorff dimension have zero μ_a measure for certain Riesz products.

Theorem 13.3 *Suppose that there is $C < \infty$ such that $\lambda_{j+1} \leq C \lambda_j$ for all j. Then for $0 < s < 1$, $I_s(\mu_{a,\lambda}) < \infty$ if and only if*

$$\sum_{k=1}^{\infty} \lambda_k^{s-1} \Pi_{j=1}^{k} \left(1 + a_j^2/2\right) < \infty.$$

Proof By Theorem 3.21 the condition $I_s(\mu_{a,\lambda}) < \infty$ means that

$$\sum_{m \in \mathbb{Z}} |\widehat{\mu_{a,\lambda}}(m)|^2 |m|^{s-1} < \infty.$$

From (13.3) we see that this means that

$$\sum_{\varepsilon} \left(\Pi_{\varepsilon_j \neq 0} \frac{a_j}{2} \right)^2 \left| \sum_j \varepsilon_j \lambda_j \right|^{s-1} < \infty,$$

where the summation is over the sequences $\varepsilon = (\varepsilon_j)$, $\varepsilon_j \in \{-1, 0, 1\}$, $j = 1, 2, \ldots$ such that $\varepsilon_j \neq 0$ for some but only for finitely many indices j. For any such ε denote by k_ε the largest j for which $\varepsilon_j \neq 0$. Then due to the condition $\lambda_{j+1} \geq 3\lambda_j$,

$$\left| \sum_j \varepsilon_j \lambda_j \right| \approx \lambda_{k_\varepsilon}.$$

For any $j_1 < j_2 < \cdots < j_l$ there are 2^l sequences ε such that $\{j : \varepsilon_j \neq 0\} = \{j_1, \ldots, j_l\}$. Writing $b_j = \frac{a_j}{\sqrt{2}}$ it follows that

$$\sum_{\varepsilon} \left(\Pi_{\varepsilon_j \neq 0} \frac{a_j}{2} \right)^2 \left| \sum_j \varepsilon_j \lambda_j \right|^{s-1} \approx \sum_{k=1}^{\infty} \sum_{\varepsilon : k_\varepsilon = k} \left(\Pi_{\varepsilon_j \neq 0} \frac{a_j}{2} \right)^2 \lambda_k^{s-1}$$

$$= \sum_{k=1}^{\infty} \sum_{j_1 < \cdots < j_l = k} 2^l \left(\Pi_{i=1}^{l} \frac{a_{j_i}}{2} \right)^2 \lambda_k^{s-1} = \sum_{k=1}^{\infty} \sum_{j_1 < \cdots < j_l = k} \left(\Pi_{i=1}^{l} b_{j_i} \right)^2 \lambda_k^{s-1}$$

$$= b_1^2 \lambda_1^{s-1} + \left(b_2^2 + (b_1 b_2)^2 \right) \lambda_2^{s-1} + \left(b_3^2 + (b_1 b_3)^2 + (b_2 b_3)^2 \right.$$

$$\left. + (b_1 b_2 b_3)^2 \right) \lambda_3^{s-1} + \cdots = \sum_{k=1}^{\infty} b_k^2 (1 + b_1^2) \cdots \cdots (1 + b_{k-1}^2) \lambda_k^{s-1}$$

$$= \sum_{k=1}^{\infty} \left(\Pi_{j=1}^{k} (1 + b_j^2) - \Pi_{j=1}^{k-1} (1 + b_j^2) \right) \lambda_k^{s-1},$$

with the interpretation $\Pi_{j=1}^{k-1}(1 + b_j^2) = 1$ when $k = 1$. Using $\lambda_k^{s-1} - \lambda_{k+1}^{s-1} \approx \lambda_k^{s-1}$ one sees that the finiteness of the last sum is equivalent to

$$\sum_{k=1}^{\infty} \lambda_k^{s-1} \Pi_{j=1}^{k} (1 + b_j^2) < \infty$$

and the theorem follows. $\qquad\qquad\square$

The above theorem immediately yields sufficient conditions for the absolute continuity of $\mu_{a,\lambda}$ with respect to Hausdorff measures. For example, we have

Corollary 13.4 *If* $\lambda_j = \lambda_0^j \geq 3$ *for all* j *and* $a_j = a_0 \in (-1, 1)$ *for all* j, *then for every Borel set* $A \subset \mathbb{R}$,

$$\dim A < 1 - \frac{\log\left(1 + a_0^2/2\right)}{\log \lambda_0} \quad \text{implies} \quad \mu_{a,\lambda}(A) = 0.$$

Proof Let $\dim A < s < 1 - \frac{\log(1+a_0^2/2)}{\log \lambda_0}$. Then by Theorem 13.3 $I_s(\mu_{a,\lambda}) < \infty$. If $\mu_{a,\lambda}(A)$ were positive, we could find by a standard approximation theorem (for example, Mattila [1995], Theorem 1.10) a compact set $C \subset A$ with $\mu_{a,\lambda}(C) > 0$. Since also $I_s(\mu_{a,\lambda} \,\llcorner\, C) < \infty$, this would lead to the contradiction $\dim C \geq s$ by Theorem 2.8. $\qquad\qquad\square$

13.4 Uniformly locally uniform measures

We shall now use Riesz products to construct measures on $[0, 1]$ which locally look very much like Lebesgue measure but are singular. To express this more precisely we consider blow-ups of $\mu \in \mathcal{M}([0, 1])$: if $I = [a, b] \subset [0, 1], a < b$, we define

$$\mu_I(B) = \mu(\{x : (x - a)/(b - a) \in B\})/\mu([a, b]) \quad \text{for } B \subset \mathbb{R}.$$

This means that μ_I is the image of μ under the map $x \mapsto (x - a)/(b - a)$, which maps $[a, b]$ onto $[0, 1]$, normalized to a probability measure. In particular, if $I = [a, b]$ then

$$\mu_I([0, y]) = \frac{\mu([a, a + y(b - a)])}{\mu([a, b])} \quad \text{for } y \in [0, 1].$$

If μ_I is close to Lebesgue measure on $[0, 1]$ whenever $d(I)$ is small, μ itself is in a sense nearly uniformly distributed over $[0, 1]$. Clearly, μ is then a continuous measure; it cannot have point masses. We shall now use Riesz products to show that μ could still be singular.

Two useful metrics which metrize weak convergence are given by,

$$d_1(\mu, \nu) = \sup\{|\mu([a, b]) - \nu([a, b])| : [a, b] \subset [0, 1]\},$$
$$d_2(\mu, \nu) = \sup\{|\mu([0, y]) - \nu([0, y])| : y \in [0, 1]\}.$$

One checks easily that $d_2(\mu, \nu) \leq d_1(\mu, \nu) \leq 2d_2(\mu, \nu)$.

Following Freedman and Pitman [1990] we give the following definition.

Definition 13.5 A measure $\mu \in \mathcal{M}([0, 1])$ is called *locally uniform* at $x \in [0, 1]$ if

$$\mu_I \to \mathcal{L}^1 \,\llcorner\, [0, 1] \quad \text{weakly when} \quad x \in I, d(I) \to 0.$$

If this convergence is uniform in $[0, 1]$ with respect to the above metrics d_1 and d_2, we say that μ is *uniformly locally uniform*.

Thus μ is uniformly locally uniform if and only if

$$\lim_{\delta \to 0} \sup_{d(I)<\delta} \sup\{|\mu_I([a, b]) - (b - a)| : [a, b] \subset [0, 1]\} = 0,$$

or, equivalently,

$$\lim_{\delta \to 0} \sup_{d(I)<\delta} \sup\{|\mu_I([0, y]) - y| : y \in [0, 1]\} = 0,$$

which means that for any $\varepsilon > 0$,

$$\left| \frac{\mu([a, a + y(b - a)])}{\mu([a, b])} - y \right| < \varepsilon \quad \text{for all } y \in [0, 1]$$

whenever $0 \le a < b \le 1$ and $b - a$ is sufficiently small.

These conditions can be translated to the behaviour of μ on adjacent dyadic intervals. We say that the subintervals I and J of $[0, 1]$ are adjacent dyadic intervals if they are of the form

$$I = [j2^{-k}, (j + 1)2^{-k}), J = [(j + 1)2^{-k}, (j + 2)2^{-k}),$$
$$j = 0, 1, \ldots, 2^k - 2, \quad k \in \mathbb{N}.$$

In the case $(j + 2)2^{-k} = 1$ we take J to be the closed interval $[1 - 2^{-k}, 1]$.

Lemma 13.6 *A measure $\mu \in \mathcal{M}([0, 1])$ is uniformly locally uniform if and only if*

$$\mu(I)/\mu(J) \to 1 \quad \text{as } k \to \infty$$

uniformly over all pairs of adjacent dyadic intervals of length 2^{-k}.

Proof The 'only if' is easier and not needed here, so we skip it. Assume that the above condition holds. Let us say for a positive integer N that the dyadic intervals I and J of the same length are N-adjacent if there are dyadic intervals $I_0, I_1, \ldots, I_M$ such that $M \le N$, I_{j-1} and I_j are adjacent for all j, $I_0 = I$ and $I_M = J$. Then for a fixed N, $\mu(I)/\mu(J) \to 1$ uniformly as $k \to \infty$ for all N-adjacent dyadic intervals I and J of length 2^{-k}.

Let $t > 1$ and let N be a large integer. Let $0 \le a < b \le 1$ with $b - a$ so small that $\mu(I)/\mu(J) < t$ for all $(N + 1)$-adjacent dyadic intervals I and J with $d(I) = d(J) \le b - a$. We can choose $(N + 1)$-adjacent intervals I_j, $j = 0, 1, \ldots, N + 1$, such that $d(I_j) \le b - a$ and

$$\bigcup_{j=1}^{N} I_j \subset [a, b] \subset \bigcup_{j=0}^{N+1} I_j.$$

Then for any $i \in \{0, 1, \ldots, N + 1\}$,

$$\mu([a, b]) \le \sum_{j=0}^{N+1} \mu(I_j) \le (N + 2)t\mu(I_i).$$

In the same way, using the inclusion $\cup_{j=1}^{N} I_j \subset [a, b]$,

$$\mu([a, b]) \ge (N/t)\mu(I_i).$$

Let $0 \le y \le 1$ and let N_y be the number of the intervals I_j contained in $[a, a + y(b - a)]$. Then $[a, a + y(b - a)]$ is contained in $N_y + 2$ intervals I_j and by the above argument for all i such that $I_i \subset [a, a + y(b - a)]$,

$$(N_y/t)\mu(I_i) \le \mu([a, a + y(b - a)]) \le (N_y + 2)t\mu(I_i).$$

Denoting by d the length of the intervals I_j we have

$$(b - a)/d - 2 \le N \le (b - a)/d, \quad y(b - a)/d - 2 \le N_y \le y(b - a)/d,$$

so

$$yN - 2 \le N_y \le y(N + 2).$$

Putting all these estimates together we find, when $I_i \subset [a, a + y(b - a)]$,

$$\mu_{[a,b]}([0, y]) = \frac{\mu([a, a + y(b - a)])}{\mu([a, b])} \le \frac{(N_y + 2)t\mu(I_i)}{(N/t)\mu(I_i)} \le t^2 \frac{N + 2}{N} y + \frac{2t^2}{N}$$

and

$$\mu_{[a,b]}([0, y]) \ge t^{-2} \frac{N}{N + 2} y - \frac{2t^{-2}}{N + 2}.$$

Since we can choose t arbitrarily close to 1 and N arbitrarily large, the lemma follows. $\square$

Theorem 13.7 *There exists a singular uniformly locally uniform Borel measure on $[0, 1]$.*

Proof We choose the sequence $a = (a_j)$ such that $a_j \to 0$ and $\sum_j a_j^2 = \infty$, for example, $a_j = 1/\sqrt{j}$ will do. Then by Theorem 13.2 $\mu_{a,\lambda}$ is singular for any sequence $\lambda = (\lambda_j)$ as before. Now we choose the integers λ_j so that the ratios λ_{j+1}/λ_j are integers greater than or equal to 3 and they are of the form $\lambda_j = 2^{k(j)}$. The sequence $k(j)$ of positive integers will be defined inductively. Making it very rapidly increasing will make $\mu_{a,\lambda}$ uniformly locally uniform.

From now on we shall write $\mu = \mu_{a,\lambda}$. We continue to use the notation f_N as before. The first observation is that μ is invariant under translation by $1/\lambda_1$. This means that $\int \varphi(x + 1/\lambda_1) d\mu x = \int \varphi(x) d\mu x$ for continuous functions φ.

This follows from the definition of μ and the fact that λ_1 divides every λ_m. In particular,

$$\mu([i/\lambda_1, (i+1)/\lambda_1)) = 1/\lambda_1 \quad \text{for all } i = 0, 1, \ldots, \lambda_1 - 1. \tag{13.5}$$

Lemma 13.8 *Fix N. Let j, k and m be non-negative integers with $j < k$ and $m(k+1) \leq \lambda_{N+1}$. Set*

$$I = [mj/\lambda_{N+1}, m(j+1)/\lambda_{N+1}), \quad J = [mk/\lambda_{N+1}, m(k+1)/\lambda_{N+1}).$$

If

$$b < f_N(x)/f_N(y) < c \quad \text{for all } x \in I, y \in J,$$

then

$$b < \mu(I)/\mu(J) < c.$$

Proof Let

$$\mu_N = \Pi_{j=N+1}^{\infty}(1 + a_j \cos(2\pi\lambda_j x)).$$

Then μ can be written as $\mu(A) = \int_A f_N \, d\mu_N$, so in particular,

$$\mu(I) = \int_I f_N \, d\mu_N \quad \text{and} \quad \mu(J) = \int_J f_N \, d\mu_N.$$

Applying (13.5) to μ_N we see that $\mu_N(I) = \mu_N(J)$; notice that I and J are composed of m intervals of the type $[i/\lambda_{N+1}, (i+1)/\lambda_{N+1}), i = 0, 1, \ldots, \lambda_{N+1} - 1$. The lemma follows immediately from these facts. $\qquad\square$

Now we shall define the sequence $k(j)$ inductively. We take k(1)=1. Suppose that $k(1) < \cdots < k(j)$ have been defined. Then f_j is strictly positive and uniformly continuous, whence we can choose $k(j+1) \geq k(j) + 2$ so large that

$$1 - a_{j+1} < \frac{f_j(x)}{f_j(y)} < 1 + a_{j+1} \quad \text{provided} \quad |x - y| \leq 2^{1-k(j+1)}. \tag{13.6}$$

This completes the inductive definition.

Let $j \geq 2$. Then

$$f_j(x) = f_{j-1}(x)(1 + a_j \cos(2\pi\lambda_j x)).$$

Thus

$$(1 - a_j)f_{j-1}(x) \leq f_j(x) \leq (1 + a_j)f_{j-1}(x).$$

Let I and J be adjacent dyadic intervals of length 2^{-k} with $k(j) \leq k \leq k(j+1)$. If $x \in I$ and $y \in J$, then $|x - y| \leq 2^{1-k} \leq 2^{1-k(j)}$. Thus we can apply (13.6) with j replaced by $j - 1$ to obtain

$$\frac{f_j(x)}{f_j(y)} \leq \frac{(1 + a_j)f_{j-1}(x)}{(1 - a_j)f_{j-1}(y)} \leq \frac{(1 + a_j)^2}{1 - a_j}.$$

The lower bound $\frac{(1-a_j)^2}{1+a_j}$ is obtained in the same way. Hence we have by Lemma 13.8

$$\frac{(1 - a_j)^2}{1 + a_j} \leq \frac{\mu(I)}{\mu(J)} \leq \frac{(1 + a_j)^2}{1 - a_j}.$$

Appealing to Lemma 13.6 completes the proof of the theorem.   □

13.5 Further comments

F. Riesz [1918] introduced the measures carrying his name.

Theorem 13.2 is classical, Peyriére [1975] and Brown and Moran [1974] proved more generally that μ_a and μ_b are mutually singular if $\sum_j |a_j - b_j|^2 = \infty$. The results for the Hausdorff dimension are due to Peyriére from the same paper, in fact, Peyriére proves considerably better estimates than Corollary 13.4, see also Kahane [2010]. Notice that these do not follow by the same energy method, since Theorem 13.3 is sharp. Hare and Roginskaya [2002] used the energy method to obtain results relaxing on the condition $\lambda_{j+1} \geq 3\lambda_j$.

Freedman and Pitman [1990] introduced the concept of uniformly locally uniform measure with a probabilistic motivation. Locally uniform measures are essentially the same as the measures whose only micromeasure in the sense of Furstenberg [2008] is Lebesgue measure. They are also essentially the same as the measures whose tangent measures in the sense of Preiss [1987] are constant multiples of Lebesgue measure. Such an example with tangent measures was constructed by Preiss [1987]. The main difference between tangent measures and the above blow-ups is that for tangent measures one blows up the whole measure, and not only locally, whence they have usually unbounded support. This is important in the applications of Preiss [1987]; see also Mattila [1995], Chapters 14 and 17.

The locally uniform measures and measures whose tangent measures are constant multiples of Lebesgue measure possess some regularity properties, for example some doubling properties. However, Orponen and Sahlsten [2012] constructed a very badly non-doubling measure all of whose tangent measures are equivalent (but not equal) to Lebesgue measure.

Various aspects of Riesz products have been discussed in many books, see Zygmund [1959], Kahane [1985], Katznelson [1968], Havin and Jöricke

[1995], Graham and McGehee [1970] and Grafakos [2008]. Much more than discussed here has been done on Riesz products in different settings and their relations to Hausdorff dimension; see, for example, Fan and Zhang [2009] and Shieh and Zhang [2009] and the references given there.

14

Oscillatory integrals (stationary phase) and surface measures

In this chapter we study integrals of the type

$$I(\lambda) = \int e^{i\lambda\varphi(x)}\psi(x)\,dx, \quad \lambda > 0, \tag{14.1}$$

and in particular their behaviour as $\lambda \to \infty$. As a standing assumption the functions φ and ψ defined on $\mathbb{R}^n$ will be infinitely differentiable and ψ will have compact support, φ is real valued and ψ complex valued. The reader will easily see that often much less smoothness suffices. As special cases we obtain the estimates for the Bessel functions and the Fourier transform of the surface measure on the sphere presented in Chapter 3.

14.1 One-dimensional case

In this section φ and ψ will be defined on $\mathbb{R}$.

Theorem 14.1 *If $\varphi'(x) \neq 0$ when $x \in \mathrm{spt}\,\psi$, then for every $N \in \mathbb{N}$,*

$$I(\lambda) \leq C(\varphi, \psi, N)\lambda^{-N} \quad for\ \lambda > 0.$$

When $N = 1$, we can take

$$C(\varphi, \psi) = \int \left| \frac{d}{dx}\left(\frac{\psi(x)}{\varphi'(x)}\right) \right| dx.$$

Proof Integrating by parts,

$$|I(\lambda)| = \left| \int \frac{1}{i\lambda\varphi'(x)} \frac{d}{dx}\left(e^{i\lambda\varphi(x)}\right)\psi(x)\,dx \right|$$

$$= \left| -\int e^{i\lambda\varphi(x)} \frac{d}{dx}\left(\frac{\psi(x)}{i\lambda\varphi'(x)}\right) dx \right| \leq C(\varphi, \psi)/\lambda.$$

The cases $N \geq 2$ follow by similar calculations. $\qquad\qquad\square$

If $\varphi'(x) = 0$ but some higher order derivative does not vanish, the following *van der Corput's lemma* is useful:

Theorem 14.2 *Suppose $k \in \{1, 2, \dots\}$ is such that $|\varphi^{(k)}(x)| \geq 1$ for $x \in [a, b]$. Then with $C_k = 5 \cdot 2^{k-1} - 2$,*

$$\left| \int_a^b e^{i\lambda\varphi(x)} \, dx \right| \leq C_k \lambda^{-1/k} \quad \text{for } \lambda > 0, \tag{14.2}$$

(i) if $k = 1$ and φ' is monotone, or
(ii) if $k \geq 2$.

Proof Suppose first (i). Integrating by parts

$$\left| \int_a^b e^{i\lambda\varphi(x)} \, dx \right| = \left| \frac{e^{i\lambda\varphi(b)}}{i\lambda\varphi'(b)} - \frac{e^{i\lambda\varphi(a)}}{i\lambda\varphi'(a)} - \int_a^b e^{i\lambda\varphi(x)} \frac{d}{dx} \left(\frac{1}{i\lambda\varphi'(x)} \right) dx \right|$$

$$\leq 2\lambda^{-1} + \lambda^{-1} \int_a^b \left| \frac{d}{dx} \left(\frac{1}{\varphi'(x)} \right) \right| dx$$

$$= 2\lambda^{-1} + \lambda^{-1} \left| \varphi'(b)^{-1} - \varphi'(a)^{-1} \right| \leq 3\lambda^{-1},$$

where in the last equality and inequality we used the facts that $\frac{d}{dx}(\frac{1}{\varphi'(x)})$ and $\varphi'(x)$ do not change sign on $[a, b]$.

Suppose then that $k \geq 2$. We use induction on k and assume that (14.2) holds for $k - 1$. We may assume that $\varphi^{(k)}(x) \geq 1$ for $x \in [a, b]$, since $\varphi^{(k)}$ does not change sign on $[a, b]$. Then $\varphi^{(k-1)}$ is strictly increasing and there is a unique $c \in [a, b]$ such that $|\varphi^{(k-1)}(x)|$ has its minimum at c. Either $\varphi^{(k-1)}(c) = 0$ or $c = a$ or $c = b$. Suppose $\varphi^{(k-1)}(c) = 0$ and let $\delta > 0$. Then $|\varphi^{(k-1)}(x)| \geq \delta$ when $x \in [a, b] \setminus [c - \delta, c + \delta]$ and the induction hypothesis gives

$$\left| \int_a^{c-\delta} e^{i\lambda\varphi(x)} \, dx \right| \leq C_{k-1}(\lambda\delta)^{-1/(k-1)}$$

and

$$\left| \int_{c+\delta}^b e^{i\lambda\varphi(x)} \, dx \right| \leq C_{k-1}(\lambda\delta)^{-1/(k-1)}.$$

Since

$$\left| \int_{c-\delta}^{c+\delta} e^{i\lambda\varphi(x)} \, dx \right| \leq 2\delta,$$

we obtain

$$\left| \int_a^b e^{i\lambda\varphi(x)} \, dx \right| \leq 2C_{k-1}(\lambda\delta)^{-1/(k-1)} + 2\delta.$$

(Here we only consider the case $a \le c - \delta, c + \delta \le b$; the reader can easily check the remaining cases.) Choosing $\delta = \lambda^{-1/k}$ we get

$$\left| \int_a^b e^{i\lambda\varphi(x)} dx \right| \le (2C_{k-1} + 2)\lambda^{-1/k}.$$

If $c = a$ or $c = b$, a similar argument gives

$$\left| \int_a^b e^{i\lambda\varphi(x)} dx \right| \le C_{k-1}(\lambda\delta)^{-1/(k-1)} + \delta,$$

and we take again $\delta = \lambda^{-1/k}$.

As $2C_{k-1} + 2 = 5 \cdot 2^{k-1} - 2 = C_k$, the proof is complete. $\square$

Corollary 14.3 *Under the assumptions of Theorem 14.2, for any C^∞-function $\psi : \mathbb{R} \to \mathbb{C}$,*

$$\left| \int_a^b e^{i\lambda\varphi(x)}\psi(x)\, dx \right| \le C_k \lambda^{-1/k} \left(|\psi(b)| + \int_a^b |\psi'(x)|\, dx \right) \quad \text{for } \lambda > 0.$$

Proof Let

$$F(x) = \int_a^x e^{i\lambda\varphi(t)} dt.$$

Then

$$\int_a^b e^{i\lambda\varphi(x)}\psi(x)\, dx = \int_a^b F'(x)\psi(x)\, dx = F(b)\psi(b) - \int_a^b F(x)\psi'(x)\, dx,$$

and by Theorem 14.2, $|F(x)| \le C_k \lambda^{-1/k}$ for all $x \in [a, b]$, from which the theorem follows. $\square$

We now discuss applications to Bessel functions and the surface measure σ^{n-1} on the sphere S^{n-1}. We defined in (3.32)

$$J_m(t) = \frac{(t/2)^m}{\Gamma(m + 1/2)\Gamma(1/2)} \int_{-1}^1 e^{its}(1 - s^2)^{m-1/2}\, ds,$$

for $m > -1/2$. For the formula for radial functions, (3.33), and for $\widehat{\sigma^{n-1}}$, (3.41), we only need the integral and half integral values of m. When $m + 1/2$ is a positive integer, we already saw in Section 3.3 that the estimate (3.35)

$$|J_m(t)| \le C(m)t^{-1/2} \quad \text{for } t > 0,$$

holds. This case was almost trivial; then J_m is a linear combination of simple elementary functions. Now we derive (3.35) from Corollary 14.3 when m is a non-negative integer. First when $m \in \mathbb{N}_0$, we have the alternative formula:

$$J_m(t) = \frac{1}{2\pi} \int_0^{2\pi} e^{it \sin \theta} e^{-im\theta} \, d\theta.$$

This is easily checked for $m = 0$, and for $m > 0$ it follows by induction from the recursion relation (3.38):

$$\frac{d}{dt}(t^{-m} J_m(t)) = -t^m J_{m+1}(t).$$

We leave the details for the reader or one can consult Grafakos [2008], Appendix B. This alternative formula combined with Corollary 14.3 yields (3.35): we apply Corollary 14.3 with $\lambda = t$ and $\varphi(x) = \sin x$. Then $\varphi'(x) = 0$ when x is $\pi/2$ or $3\pi/2$ and $\varphi''(x) = \pm 1$ for these values of x. We can find smooth non-negative functions ψ_1, ψ_2 and ψ_3 such that $\psi_1 + \psi_2 + \psi_3 = 1$, ψ_1 has support in a small neighbourhood of $\pi/2$ and it equals 1 in a smaller neighbourhood of $\pi/2$, and similarly ψ_2 has support and it equals 1 near $3\pi/2$. Then we can write $J_m(t)$ as a sum of three terms; to two of them we apply Corollary 14.3 with $k = 2$, and to one of them we apply Theorem 14.1.

Thus we get the decay estimate (3.42) for the spherical surface measure. This argument is heavily based on the radial symmetry of the sphere; via integration in polar coordinates we could employ the estimates for the one-dimensional integrals (14.1). For other surface measures we need analogous estimates for higher dimensional integrals, which we now investigate.

14.2 Higher dimensional case

For the rest of this chapter φ and ψ will be smooth functions in $\mathbb{R}^n$, and as before ψ has compact support, φ is real valued and ψ complex valued. We let again

$$I(\lambda) = \int e^{i\lambda\varphi(x)} \psi(x) \, dx \quad \text{for } \lambda > 0.$$

Theorem 14.4 *If $\nabla\varphi(x) \neq 0$ when $x \in$ spt ψ, then for every $N \in \mathbb{N}$,*

$$I(\lambda) \leq C(\varphi, \psi, N)\lambda^{-N} \quad \text{for } \lambda > 0. \tag{14.3}$$

Proof Suppose first that for some j, $\partial_j\varphi(x) \neq 0$ for $x \in$ spt ψ. Then by Fubini's theorem, writing $\widetilde{x} = (x_1, \ldots, x_{j-1}, x_{j+1}, \ldots, x_n)$, and $C = \{\widetilde{x} : x \in$ spt $\psi\}$,

$$I(\lambda) = \int_C \left(\int_{\mathbb{R}} e^{i\lambda\varphi(x)} \psi(x) dx_j \right) d\widetilde{x}.$$

An application of Theorem 14.1 to the inner integral yields (14.3); obviously the proof of Theorem 14.1 shows that the constants involved depending on $\tilde{x}$ are uniformly bounded.

In the general case we can cover spt ψ with finitely many balls B_k such that some $\partial_{j_k} \varphi(x) \neq 0$ for $x \in B_k$. Writing $\psi = \sum_k \psi_k$ with spt $\psi_k \subset B_k$, the theorem follows. $\square$

Next we consider points where the gradient vanishes. We call such points *critical*. A point x_0 is called a *non-degenerate critical point* of φ if $\nabla \varphi(x_0) = 0$ and the *Hessian determinant*

$$h_\varphi(x_0) := \det(\partial_j \partial_k \varphi(x_0)) \neq 0. \tag{14.4}$$

The corresponding *Hessian matrix* is denoted by

$$H_\varphi(x_0) := (\partial_j \partial_k \varphi(x_0)).$$

Theorem 14.5 *If all critical points of φ in* spt ψ *are non-degenerate, then*

$$|I(\lambda)| \leq C(\varphi, \psi) \lambda^{-n/2} \quad for \ \lambda > 0. \tag{14.5}$$

Proof We may assume that spt $\psi \subset B(0,1)$, $\|\psi\|_\infty \leq 1$, $\|\nabla \psi\|_\infty \leq 1$ and $\|H_\varphi\|_\infty \leq 1$. We first consider the case where φ is a special quadratic polynomial, $\varphi = Q$:

$$Q(x) = x_1^2 + \cdots + x_k^2 - x_{k+1}^2 - \cdots - x_n^2$$

for some $k = 1, \ldots, n$. We use induction on n to prove that for any special quadratic polynomial Q in $\mathbb{R}^n$ as above and for any smooth ψ in $\mathbb{R}^n$ with spt $\psi \subset B(0,1)$ and with $\|\partial^\alpha \psi\|_\infty \lesssim_n 1$ for any partial derivative $\partial^\alpha \psi$ of order $|\alpha| \leq n$,

$$\left| \int e^{i\lambda Q(x)} \psi(x) \, dx \right| \lesssim_n \lambda^{-n/2}.$$

The case $n = 1$ follows from Corollary 14.3. Suppose the result holds for $n - 1$ and let $\psi \in C^\infty(\mathbb{R}^n)$ be as above. By Fubini's theorem

$$I(\lambda) = \lambda^{-1/2} \int e^{i\lambda(x_2^2 + \cdots + x_k^2 - x_{k+1}^2 \cdots - x_n^2)} \psi_\lambda(x_2, \ldots, x_n) \, d(x_2, \ldots, x_n),$$

where

$$\psi_\lambda(x_2, \ldots, x_n) = \lambda^{1/2} \int e^{i\lambda x_1^2} \psi(x_1, \ldots, x_n) \, dx_1.$$

Corollary 14.3 applied to ψ and its partial derivatives tells us that $\|\partial^\alpha \psi_\lambda\|_\infty \lesssim_n 1$ for $|\alpha| \leq n - 1$. The induction hypothesis gives that

$$\left| \int e^{i\lambda(x_2^2 + \cdots + x_k^2 - x_{k+1}^2 - \cdots - x_n^2)} \psi_\lambda(x_2, \ldots, x_n) \, d(x_2, \ldots, x_n) \right| \lesssim \lambda^{(1-n)/2}.$$

The theorem follows from these for such quadratic polynomials.

For the general case we use the following calculus lemma, called *Morse's lemma*:

Lemma 14.6 *Let $\varphi : U \to \mathbb{R}$ be a C^∞ function with $U \subset \mathbb{R}^n$ open, and let $x_0 \in U$ be such that $\varphi(x_0) = 0$, $\nabla\varphi(x_0) = 0$ and $h_\varphi(x_0) \neq 0$. Then there exists a diffeomorphism $G : V \to W$ with $V, W \subset \mathbb{R}^n$ open, $0 \in V$, $x_0 \in W \subset U$, $G(0) = x_0$, and for some $k \in \{1, \ldots, n\}$,*

$$\varphi \circ G(x) = \sum_{j=1}^{k-1} x_j^2 - \sum_{j=k}^{n} x_j^2 \quad \text{for } x \in V.$$

Proof We may assume $x_0 = 0$. We may also assume that the matrix $H_\varphi(0)$ is diagonal with all diagonal elements non-zero. This is achieved by first diagonalizing $H_\varphi(0)$ by an orthogonal transfomation O so that $S = O^{-1} \circ H_\varphi(0) \circ O$ is diagonal. By direct computation using the chain rule $H_{\varphi \circ O}(0) = O^T \circ H_\varphi(0) \circ O$. Since the transpose O^T is O^{-1}, we have $H_{\varphi \circ O}(0) = S$, which justifies our assumption.

Under this assumption, $\partial_1 \varphi(0) = 0$ and $\partial_1^2 \varphi(0) \neq 0$. By the implicit function theorem there is a smooth function $g : W_1 \to \mathbb{R}$, $W_1 \subset \mathbb{R}^{n-1}$ open, $0 \in W_1$, such that $g(0) = 0$ and

$$\partial_1 \varphi(g(\tilde{x}), \tilde{x}) = 0, \ \partial_1^2 \varphi(g(\tilde{x}), \tilde{x}) \neq 0 \quad \text{for } \tilde{x} = (x_2, \ldots, x_n) \in W_1,$$

and $\partial_1 \varphi(x_1, \tilde{x}) \neq 0$ when $(x_1, \tilde{x}) \in U, \tilde{x} \in W_1$ and $x_1 \neq g(\tilde{x})$. Let $\psi = \varphi \circ F$, $F(x) = (x_1 + g(\tilde{x}), \tilde{x})$. Then by the chain rule $\partial_1 \psi(0, \tilde{x}) = 0$ and $\partial_1^2 \psi(0, \tilde{x}) \neq 0$ for $\tilde{x} \in W_1$ and by Taylor's theorem, taking W_1 sufficiently small, we can write

$$\psi(x) = \psi(0, \tilde{x}) \pm h(x) x_1^2$$

where h is a strictly positive smooth function. Define $E(x) = (\frac{x_1}{\sqrt{h(x)}}, \tilde{x})$. Then

$$\psi \circ E(x) = \pm x_1^2 + \psi(0, \tilde{x})$$

and so

$$\varphi \circ F \circ E(x) = \pm x_1^2 + \psi(0, \tilde{x}).$$

We leave it to the reader to check that $F \circ E$ is a diffeomorphism in a neighbourhood of the origin. The lemma follows by repeating this with $\psi(0, \tilde{x})$ in place of $\varphi(x)$ and so on.   □

We can now complete the proof of Theorem 14.5. Each point of spt ψ has a ball neighbourhood where either $\nabla \psi \neq 0$ or we can perform the change of variable by a diffeomorphism G provided by Morse's lemma. Covering the whole spt ψ with a finite number of such balls B_j and using a partition unity to write $\psi = \sum_j \psi_j$ with spt $\psi_j \subset B_j$, we can write $I(\lambda) = \sum_j I_j(\lambda)$ with $I_j(\lambda) = \int e^{i\lambda\varphi(x)}\psi_j(x)\,dx$. If j corresponds to a non-critical point, $|I_j(\lambda)| \lesssim \lambda^{-n/2}$ by Theorem 14.4. For j corresponding to non-degenerate critical points we have

$$I_j(\lambda) = \int e^{i\lambda Q_j(x)}\psi(G_j(x))J_{G_j}(x)\,dx,$$

where G_j and $Q_j = \varphi \circ G_j$ are given by Morse's lemma. For these $|I_j(\lambda)| \lesssim \lambda^{-n/2}$ by the special case considered above.   □

14.3 Surface measures

We shall consider Fourier transforms of measures on smooth hypersurfaces of $\mathbb{R}^n$. If σ is the surface measure on such a surface S, we shall consider measures μ of the type $d\mu = \zeta\,d\sigma$ where ζ is a smooth function with sufficiently small compact support. Moreover, we shall assume that spt $\zeta \cap S$ is a graph of a smooth function φ over its tangent plane at a point $p \in S$. Without loss of generality we assume that $p = 0$ and the tangent plane is $\mathbb{R}^{n-1} = \mathbb{R}^{n-1} \times \{0\}$. The reader can of course easily deduce various generalizations from this basic case.

So let $U \subset \mathbb{R}^{n-1}$ be bounded and open, and let $0 \in U$, $\varphi: U \to \mathbb{R}$ and $\zeta: \mathbb{R}^n \to \mathbb{R}$ be smooth functions, ζ with compact support, such that

$$S = \{(x, \varphi(x)) : x \in U\},$$
$$\varphi(0) = 0, \nabla\varphi(0) = 0,$$
$$\text{spt}\,\zeta \subset \{(x, t) : x \in U, t \in \mathbb{R}\}.$$

Then the measure $\mu = \zeta\sigma$ is given by

$$\int g\,d\mu = \int_U g(x, \varphi(x))\psi(x)\,dx$$

for $g \in C_0(\mathbb{R}^n)$ where

$$\psi(x) = \zeta(x, \varphi(x))\sqrt{1 + |\nabla\varphi(x)|^2}.$$

Thus the Fourier transform of μ is, writing $\xi = (\tilde{\xi}, \xi_n)$,

$$\widehat{\mu}(\xi) = \int e^{-2\pi i(\tilde{\xi}\cdot x + \xi_n \varphi(x))} \psi(x) \, dx.$$

In order to obtain the optimal decay $|\xi|^{(1-n)/2}$ as in the case of the sphere, we need to make curvature assumptions. The *Gaussian curvature* of S at $(x, \varphi(x))$ is the Hessian determinant $h_\varphi(x)$, which is the the product of the principal curvatures, that is, the eigenvalues of $H_\varphi(x)$.

Theorem 14.7 *With the above assumptions, if $h_\varphi(x) \neq 0$ for $x \in U$, then*

$$|\widehat{\mu}(\xi)| \leq C(\varphi, \zeta)|\xi|^{(1-n)/2} \quad \text{for } \xi \in \mathbb{R}^n.$$

Proof Let $\xi = \lambda \eta$ with $\lambda = |\xi| > 0$ and $|\eta| = 1$, and

$$\varphi_\eta(x) = -2\pi(\eta_1 x_1 + \cdots \eta_{n-1} x_{n-1} + \eta_n \varphi(x)), \quad x \in U.$$

Then we need to show that

$$|\widehat{\mu}(\xi)| = \left| \int e^{i\lambda \varphi_\eta(x)} \psi(x) \, dx \right| \lesssim_\eta \lambda^{(1-n)/2}.$$

The implicit constant may a priori depend on η, since the integral is a continuous function of η and hence attains a maximum on S^{n-1}.

We have

$$\nabla \varphi_\eta(x) = -2\pi((\eta_1, \ldots, \eta_{n-1}) + \eta_n \nabla \varphi(x))$$

and

$$H_{\varphi_\eta}(x) = -2\pi \eta_n H_\varphi(x).$$

If $\eta_n = 0$, $\nabla \varphi_\eta(x) \neq 0$ for all $x \in U$, and the required estimate follows from Theorem 14.4. If $\eta_n \neq 0$, the assumption $h_\varphi(x) \neq 0$ for $x \in U$ implies that $h_{\varphi_\eta}(x) \neq 0$ for $x \in U$, and the required estimate follows from Theorem 14.5. $\qquad\square$

14.4 Further comments

The contents of this chapter are classical and they are discussed in the books of Grafakos [2008], Muscalu and Schlag [2013], Stein [1993], Sogge [1993] and Wolff [2003]. In particular Stein [1993] includes many historical comments and goes much further. For example, it gives rather precise asymptotic formulas, not only decay estimates, and studies also surfaces for which some of the principal curvatures may vanish.

PART III

Deeper applications of the
Fourier transform

15

Spherical averages and distance sets

15.1 The Wolff–Erdoğan distance set theorem

In this chapter we show how the spherical averages of Fourier transforms of measures, $\sigma(\mu)(r) = \int_{S^{n-1}} |\widehat{\mu}(rv)|^2 \, d\sigma^{n-1}v$, can be used for distance set estimates. From Theorem 4.6 we see that for a Borel set $A \subset \mathbb{R}^n$,

$$\dim A > (n+1)/2 \quad \text{implies} \quad \mathcal{L}^1(D(A)) > 0,$$

where $D(A)$ is the distance set:

$$D(A) = \{|x - y| : x, y \in A\}.$$

The conjecture is

$$\dim A > n/2 \quad \text{implies} \quad \mathcal{L}^1(D(A)) > 0.$$

This will remain open but we shall be able to improve Falconer's result above to the following theorem of Wolff [1999] for $n = 2$ and Erdoğan [2005] for general n:

Theorem 15.1 *Let $A \subset \mathbb{R}^n$ be a Borel set and $n \geq 2$.*

(a) If $\dim A > n/2 + 1/3$, then $\mathcal{L}^1(D(A)) > 0$.
(b) If $n/2 \leq \dim A \leq n/2 + 1/3$, then $\dim D(A) \geq \frac{6 \dim A + 2 - 3n}{4}$.

In this chapter we show that certain decay estimates for the spherical averages imply this theorem; these estimates will be proven in the next chapter. In fact, when $n > 2$ the proof relies on Tao's bilinear restriction theorem which will be proven only in Chapter 25.

The lower bound in (b) holds also for $\dim A < n/2$, but then the bound $\dim A - (n-1)/2$ given by Theorem 4.6 is better.

As mentioned before, in $\mathbb{R}$ there is no such result: one can construct compact sets $C \subset \mathbb{R}$ such that $\dim C = 1$ and $\mathcal{L}^1(D(C)) = 0$.

15.2 Spherical averages and distance measures

Recall from Section 3.7 the quadratic spherical averages of $\mu \in \mathcal{M}(\mathbb{R}^n)$: for $r > 0$,

$$\sigma(\mu)(r) = \int_{S^{n-1}} |\widehat{\mu}(rv)|^2 \, d\sigma^{n-1}v = r^{1-n} \int_{S^{n-1}(r)} |\widehat{\mu}(v)|^2 \, d\sigma_r^{n-1}v.$$

For the energy-integrals of μ we have as before in (3.50):

$$I_s(\mu) = \gamma(n, s) \int_0^\infty \sigma(\mu)(r) r^{s-1} \, dr, \quad 0 < s < n. \tag{15.1}$$

Recall also from Section 4.2 the distance measure $\delta(\mu) \in \mathcal{M}(D(A))$ of a measure $\mu \in \mathcal{M}(A)$ defined by

$$\int \varphi \, d\delta(\mu) = \iint \varphi(|x - y|) \, d\mu x \, d\mu y$$

for continuous functions φ on $\mathbb{R}$. We shall also consider the weighted distance measure $\Delta(\mu)$ defined by

$$\int \varphi \, d\Delta(\mu) = \int u^{(1-n)/2} \varphi(u) \, d\delta(\mu)u$$

and the weighted spherical averages $\Sigma(\mu)$;

$$\Sigma(\mu)(r) = r^{(n-1)/2} \sigma(\mu)(r).$$

Suppose now that $I_{(n+1)/2}(\mu) < \infty$. Notice that then $\Sigma(\mu) \in L^1$ because of (15.1). By integration in spherical coordinates and by the formulas (3.41) and (4.10), for $u > 0$ these two are related by

$$\Delta(\mu)(u) = c(n)\sqrt{u} \int_0^\infty \sqrt{r} J_{(n-2)/2}(2\pi r u) \Sigma(\mu)(r) \, dr. \tag{15.2}$$

Here again $J_{(n-2)/2}$ is a Bessel function for which we have the asymptotic formula (3.37),

$$J_m(u) = \frac{\sqrt{2}}{\sqrt{\pi u}} \cos(u - \pi m/2 - \pi/4) + O(u^{-3/2}), \quad u \to \infty.$$

Hence

$$c(n) J_{(n-2)/2}(2\pi u) = \frac{1}{\sqrt{u}}(a_1 \cos(2\pi u) + b_1 \sin(2\pi u)) + K(u),$$

where

$$|K(u)| \lesssim \min\{u^{-3/2}, u^{-1/2}\}. \tag{15.3}$$

Here and below a_j and b_j are complex constants depending only on n. The local estimate for K with $u^{-1/2}$ holds since $J_{(n-2)/2}$ is bounded. Thus for $u > 0$,

$$\Delta(\mu)(u) = S(\mu)(u) + L(\mu)(u), \tag{15.4}$$

$$S(\mu)(u) = a_2 \int_0^\infty \cos(2\pi r u)\Sigma(\mu)(r)\,dr + b_2 \int_0^\infty \sin(2\pi r u)\Sigma(\mu)(r)\,dr,$$

$$L(\mu)(u) = \sqrt{u} \int_0^\infty \sqrt{r}\,\Sigma(\mu)(r)K(ru)\,dr.$$

Let $\Sigma_1(\mu)$ be the even extension of $\Sigma(\mu)$ to the negative reals; $\Sigma_1(\mu)(r) = \Sigma(\mu)(|r|)$, and $\Sigma_2(\mu)$ the odd extension; $\Sigma_2(\mu)(r) = -\Sigma(\mu)(|r|)$ for $r < 0$. Then for $u > 0$ we can write S as

$$S(\mu)(u)$$
$$= (a_2/2) \int_{-\infty}^\infty \cos(2\pi r u)\Sigma_1(\mu)(r)\,dr + (b_2/2) \int_{-\infty}^\infty \sin(2\pi r u)\Sigma_2(\mu)(r)\,dr$$
$$= (a_2/2) \int_{-\infty}^\infty e^{-2\pi i r u}\Sigma_1(\mu)(r)\,dr + (ib_2/2) \int_{-\infty}^\infty e^{-2\pi i r u}\Sigma_2(\mu)(r)\,dr.$$

As $\Delta(\mu)(u) = 0$ for $u < 0$, (15.4) stays in force for $u < 0$ when we define $S(u) = L(\mu)(u) = 0$ for $u < 0$. Then for all $u \in \mathbb{R}, u \neq 0$,

$$S(\mu)(u) = (a_2/4) \int_{-\infty}^\infty e^{-2\pi i r u}\Sigma_1(\mu)(r)\,dr + (a_2/4)\operatorname{sgn}(u)$$
$$\times \int_{-\infty}^\infty e^{-2\pi i r u}\Sigma_1(\mu)(r)\,dr + (ib_2/4) \int_{-\infty}^\infty e^{-2\pi i r u}\Sigma_2(\mu)(r)\,dr$$
$$+ (ib_2/4)\operatorname{sgn}(u) \int_{-\infty}^\infty e^{-2\pi i r u}\Sigma_2(\mu)(r)\,dr,$$

where the sign function is defined by $\operatorname{sgn}(u) = 1$ if $u > 0$ and $\operatorname{sgn}(u) = -1$ if $u < 0$. Since the Fourier transform maps L^2 onto itself isometrically we can define the Hilbert transform

$$H : L^2(\mathbb{R}) \to L^2(\mathbb{R}), \quad \widehat{Hf} = -i\operatorname{sgn}\widehat{f} \quad \text{with}$$
$$\|Hf\|_2 = \|f\|_2 \quad \text{for } f \in L^2(\mathbb{R}).$$

If $\Sigma(\mu) \in L^2(\mathbb{R})$, we can now write S as

$$S(\mu) = \mathcal{F}(a_3(\Sigma_1(\mu) + i H(\Sigma_1(\mu))) + b_3(\Sigma_2(\mu) + i H(\Sigma_2(\mu)))). \tag{15.5}$$

To estimate $L(\mu)$ suppose $(n-1)/2 \le s \le (n+1)/2$ and set $a = (n+1)/2 - s \in [0, 1]$. Then for $u > 0$ by (15.3) and (15.1),

$$
\begin{aligned}
|L(\mu)(u)| &= \left| \sqrt{u} \int_0^\infty \sqrt{r} \Sigma(\mu)(r) K(ru) \, dr \right| \\
&\lesssim \int_0^{1/u} r^{(n-1)/2} \sigma(\mu)(r) \, dr + u^{-1} \int_{1/u}^\infty r^{(n-3)/2} \sigma(\mu)(r) \, dr \\
&= u^{-a} \int_0^{1/u} (ru)^a r^{(n-1)/2-a} \sigma(\mu)(r) \, dr \qquad\qquad (15.6) \\
&\quad + u^{-a} \int_{1/u}^\infty (ru)^{a-1} r^{(n-1)/2-a} \sigma(\mu)(r) \, dr \\
&\le u^{-a} \int_0^\infty r^{s-1} \sigma(\mu)(r) \, dr = \gamma(n, s)^{-1} u^{s-(n+1)/2} I_s(\mu).
\end{aligned}
$$

Recall that we have worked under the assumption $I_{(n+1)/2}(\mu) < \infty$.

Proposition 15.2 *Suppose $\mu \in \mathcal{M}(\mathbb{R}^n)$, $n \ge 2$, $s > 0$ and $I_s(\mu) < \infty$.*

(a) *If $s > n/2$ and $\int_1^\infty \sigma(\mu)(r)^2 r^{n-1} \, dr < \infty$, then $\Delta(\mu) \in L^2(\mathbb{R})$. In particular, $\delta(\mu) \ll \mathcal{L}^1$.*

(b) *If $0 < t < 1$, $s > (n+t-1)/2$ and $\int_1^\infty \sigma(\mu)(r)^2 r^{n+t-2} \, dr < \infty$, then $I_t(\Delta(\mu)) < \infty$.*

Proof To prove (a) we may assume $s < (n+1)/2$, because $I_s(\mu) < \infty$ implies $I_{s'}(\mu) < \infty$ for $s' < s$. Let $f_\varepsilon = \psi_\varepsilon * \mu$ where ψ_ε, $\varepsilon > 0$, is an approximate identity as in Section 3.2 with $\widehat{f_\varepsilon}(x) = \widehat{\psi}(\varepsilon x) \widehat{\mu}(x) \to \widehat{\mu}(x)$ as $\varepsilon \to 0$ and $|\widehat{f_\varepsilon}(x)| \le |\widehat{\mu}(x)|$. First we have by (15.5) and Plancherel's theorem,

$$
\|S(f_\varepsilon)\|_2 \lesssim \|\Sigma(f_\varepsilon)\|_2 \le \|\Sigma(\mu)\|_2 < \infty,
$$

because by the assumption in (a), $\Sigma(\mu) \in L^2(\mathbb{R})$. Secondly by (15.6)

$$
\int_0^1 L(f_\varepsilon)(u)^2 du \lesssim \int_0^1 u^{2s-n-1} \, du \, I_s(f_\varepsilon)^2 \lesssim I_s(\mu)^2,
$$

as $2s > n$, and, applying (15.6) with $s = (n-1)/2$,

$$
\int_1^\infty L(f_\varepsilon)(u)^2 du \lesssim \int_1^\infty u^{-2} \, du \, I_s(f_\varepsilon)^2 \lesssim I_s(\mu)^2.
$$

Thus by (15.4) the norms $\|\Delta(f_\varepsilon)\|_2$, $\varepsilon > 0$, are uniformly bounded from which one easily concludes that $\Delta(\mu) \in L^2(\mathbb{R})$.

Notice that in the above proof for part (a) we did not need any information about the Hilbert transform beyond Plancherel's formula, which we used to define it. For part (b) we shall use the following inequality for $0 < t \le 1$:

$$\int_{-\infty}^{\infty} |r|^{t-1} |Hf(r)|^2 \, dr \lesssim_t \int_{-\infty}^{\infty} |r|^{t-1} |f(r)|^2 \, dr. \tag{15.7}$$

We do not prove this here. It follows from standard weighted inequalities for singular integrals using the fact that $|r|^{t-1}$ is a so-called A_2-weight; see, for example, Duoandikoetxea [2001], Theorem 7.11.

To prove (b) we assume first that μ is a smooth non-negative function f with compact support. Recall from (3.46) the mutual energy

$$I_t(g, h) = \iint |x - y|^{-t} g(x) h(y) \, dx \, dy$$
$$= \gamma(1, t) \int |x|^{t-1} \widehat{g}(x) \overline{\widehat{h}(x)} \, dx, \quad 0 < t < 1,$$

for $g, h \in L^\infty(\mathbb{R}) \cap L^2(\mathbb{R})$ such that $I_s(|f|, |g|) < \infty$. Recall also, see (3.47), that $I_t(g) = I_t(g, g) \ge 0$. The functions $S(f)$ and $L(f)$ are bounded with sufficient decay at infinity so that we can apply this to them. For instance, the estimate $|L(f)(u)| \lesssim |u|^{-1}$ follows easily from the definition of $L(f)$, or from (15.6), and then the same estimate holds for $S(f)$ by (15.4) since $\Delta(f)$ has compact support. In particular, $L(f), S(f) \in L^2(\mathbb{R})$. From the identity

$$\Delta(f)(u)\Delta(f)(v) = S(f)(u)S(f)(v) + \Delta(f)(u)L(f)(v)$$
$$+ \Delta(f)(v)L(f)(u) - L(f)(u)L(f)(v),$$

we get

$$I_t(\Delta(f)) = I_t(S(f)) + 2I_t(\Delta(f), L(f)) - I_t(L(f))$$
$$\le I_t(S(f)) + 2I_t(\Delta(f), L(f)).$$

Using (15.5), (15.7) and the obvious fact that the energy can be written in terms of the inverse Fourier transform in place of the Fourier transform, we obtain

$$I_t(S(f)) = \gamma(1, t) \int |r|^{t-1} |\mathcal{F}^{-1}(S(f))(r)|^2 \, dr \lesssim \int_0^\infty r^{t-1} \Sigma(f)(r)^2 \, dr$$
$$= \int_0^\infty r^{t+n-2} \sigma(f)(r)^2 \, dr.$$

To estimate $I_t(\Delta(f), L(f))$ we use (15.6) and the elementary fact

$$\int_0^\infty v^{-a} |u - v|^{-b} \, dv \approx u^{1-a-b} \quad \text{for } u > 0, a, b \in (0, 1).$$

Thus

$$|I_t(\Delta(f), L(f))| = \left| \int_0^\infty \int_0^\infty \Delta(f)(u)L(f)(v)|u - v|^{-t} \, du \, dv \right|$$

$$\lesssim I_s(f) \int_0^\infty \Delta(f)(u) \int_0^\infty v^{s-(n+1)/2}|u - v|^{-t} \, du \, dv$$

$$\approx I_s(f) \int_0^\infty \delta(f)(u)u^{-(n-1)/2+s-(n+1)/2-t+1} \, du$$

$$= I_s(f) \int_0^\infty \delta(f)(u)u^{s-t-n+1} \, du$$

$$= I_s(f)I_{n-1-s+t}(f) \lesssim I_s(f)^2,$$

where the last equality follows from the definition of $\delta(f)$ and the last inequality from the fact that $n - 1 - s + t \leq s$. Combining these estimates we have established

$$I_t(\Delta(f)) \leq I_t(S(f)) + 2I_t(\Delta(f), L(f)) \lesssim \int_0^\infty r^{t+n-2}\sigma(f)(r)^2 \, dr + I_s(f)^2.$$

We apply this with $f = \psi_\varepsilon * \mu$ as above. The above inequality remains valid in the limit which completes the proof. $\qquad\square$

Proposition 15.2 immediately leads to the following proposition:

Proposition 15.3 *Suppose that C, s and t are positive numbers, $t \leq s$, and $\mu \in \mathcal{M}(\mathbb{R}^n)$, $n \geq 2$, is such that $I_s(\mu) < \infty$ and*

$$\sigma(\mu)(r) \leq Cr^{-t} \tag{15.8}$$

for all $r > 0$.

(a) If $s + t \geq n$, then $\mathcal{L}^1(D(\operatorname{spt} \mu)) > 0$.
(b) If $s + t < n$, then $\dim D(\operatorname{spt} \mu) \geq s + t + 1 - n$.

Proof In case (a),

$$\int_1^\infty \sigma(\mu)(r)^2 r^{n-1} \, dr \leq C \int_1^\infty \sigma(\mu)(r)r^{n-1-t} \, dr$$

$$= \gamma(n, n - t)CI_{n-t}(\mu) \lesssim I_s(\mu) < \infty.$$

By Proposition 15.2, $\delta(\mu) \ll \mathcal{L}^1$ and so $\mathcal{L}^1(D(\operatorname{spt} \mu)) > 0$.

In case (b), set $u = s + t + 1 - n$. Then $u < 1$ and we may of course assume that $u > 0$. We may also assume that $t < s$, which gives $s > (n + u - 1)/2$, and we can apply Proposition 15.2(b) with t replaced by u. We have

$$\int_1^\infty \sigma(\mu)(r)^2 r^{n+u-2}\,dr = \int_1^\infty \sigma(\mu)(r)^2 r^{s+t-1}\,dr$$

$$\leq C \int_1^\infty \sigma(\mu)(r) r^{s-1}\,dr = \gamma(n,s)^{-1} C I_s(\mu) < \infty.$$

By Proposition 15.2, $I_u(\Delta(\mu)) < \infty$ and so dim $D(\mathrm{spt}\,\mu) \geq u$. $\qquad\square$

Recall the definition of Salem sets from 3.11.

Corollary 15.4 *Let $A \subset \mathbb{R}^n, n \geq 2$, be a Borel Salem set.*

(a) If dim $A > n/2$*, then $\mathcal{L}^1(D(A)) > 0$.*
(b) If dim $A > (n-1)/2$*, then* dim $D(A) \geq 2\dim A + 1 - n$.

Proof For any $0 < s < \dim A$ there is $\mu \in \mathcal{M}(A)$ with $I_s(\mu) < \infty$ and $|\widehat{\mu}(x)|^2 \leq C(\mu)|x|^{-s}$. Then also $\sigma(\mu)(r) \leq C(\mu)r^{-s}$. We apply the above proposition with $s = t$: if $s > n/2$ we get (a), and if $s > (n-1)/2$, we get (b). $\qquad\square$

15.3 The decay of spherical averages

In view of the previous results, good answers to the following question are likely to give improvements for the distance set problem.

For what pairs (s, t) of positive numbers is the estimate

$$\sigma(\mu)(r) \lesssim r^{-t} I_s(\mu) \tag{15.9}$$

valid for all $\mu \in \mathcal{M}(B(0, 1))$ and for all $r > 1$? Unless we are in the optimal case $s = t$, we have to restrict to $\mu \in \mathcal{M}(B(0, 1))$ for scaling reasons. We should emphasize that although getting such estimates is easily reduced to getting them for non-negative smooth functions with compact support, the non-negativity is essential: there is no hope of getting the same estimates for general real valued smooth functions with compact support.

We already derived in Lemma 3.15 the easy estimate which says that (15.9) holds with $t = s$ if $s \leq (n-1)/2$. Obviously this implies

$$\sigma(\mu)(r) \leq C(s,t) I_s(\mu) r^{-(n-1)/2} \quad \text{for } s \geq (n-1)/2.$$

The following estimate was proved by Wolff for $n = 2$ and by Erdoğan for general n. Combined with Proposition 15.3 it gives immediately Theorem 15.1. We shall prove this estimate in the next chapter.

Theorem 15.5 *For all* $(n - 2)/2 \leq s < n, n \geq 2, \varepsilon > 0$ *and* $\mu \in \mathcal{M}(\mathbb{R}^n)$ *with* $\operatorname{spt} \mu \subset B(0, 1)$,

$$\sigma(\mu)(r) \leq C(n, s, \varepsilon) r^{\varepsilon - (n + 2s - 2)/4} I_s(\mu) \quad \text{for } r > 1.$$

Up to ε this is the best possible for $1 \leq s < 2$ in the plane, but it is not known if it is the best possible when $n > 2$ for the relevant interval $n/2 \leq s \leq (n + 2)/2$. Outside these intervals of s the estimates can be improved, as we shall see below, but the known improvements do not improve the distance set results. In addition to $s \leq (n - 1)/2$, which we have already settled, we now discuss these estimates for other values of s.

Define

$$t_n(s)$$

$$= \sup\{t : \sigma(\mu)(r) \lesssim_{n,s,t} r^{-t} I_s(\mu) \text{ for all } \mu \in \mathcal{M}(\mathbb{R}^n) \text{ with } \operatorname{spt} \mu \subset B(0, 1)\}.$$

It is clear by the formula (3.50) that $t_n(s) \leq s$ for any s. By Lemma 3.15 $t_n(s) = s$ for $0 < s \leq (n - 1)/2$.

In $\mathbb{R}^2$ the exact value of $t_2(s)$ is known for all $0 < s < 2$:

Theorem 15.6

$$t_2(s) = s \quad for \quad 0 < s \leq 1/2,$$

$$= 1/2 \quad for \quad 1/2 \leq s \leq 1,$$

$$= s/2 \quad for \quad 1 \leq s < 2.$$

In order to complete the proof of this after what already has been said, we need to give examples showing that $t_2(s) \leq 1/2$ for $1/2 \leq s \leq 1$ and $t_2(s) \leq s/2$ for $1 \leq s < 2$. This will be done below.

In $\mathbb{R}^n, n > 2$, the best known estimates are the following:

Theorem 15.7

$$t_n(s) = s \quad for \quad 0 < s \leq (n - 1)/2,$$

$$\geq (n - 1)/2 \quad for \quad (n - 1)/2 \leq s \leq n/2,$$

$$\geq (n + 2s - 2)/4 \quad for \quad n/2 \leq s \leq (n + 2)/2,$$

$$\geq s - 1 \quad for \quad (n + 2)/2 \leq s < n,$$

$$\leq s \quad for \quad (n - 1)/2 \leq s \leq n - 2,$$

$$\leq s/2 + n/2 - 1 \quad for \quad 1 \leq n - 2 \leq s < n.$$

We already know the first three lines and the fifth one. Now we verify the fourth.

Proposition 15.8 *If* $1 < s < n$, *then* $t_n(s) \geq s - 1$.

Proof Let $\mu \in \mathcal{M}(\mathbb{R}^n)$ and choose a radial C^∞ function φ on $\mathbb{R}^n$ with $\mathrm{spt}\,\varphi \subset B(0, 1)$ such that $\widehat{\varphi}$ is non-negative and $\widehat{\varphi} \geq 1$ on $\mathrm{spt}\,\mu$. Let $\nu = \widehat{\varphi}^{-1}\mu \in \mathcal{M}(\mathbb{R}^n)$. Then $\widehat{\mu} = \widehat{\widehat{\varphi}\nu} = \varphi * \widehat{\nu}$. Thus we have for $r > 2$ using Schwartz's inequality, Fubini's theorem, the fact $\mathrm{spt}\,\varphi \subset B(0, 1)$ and (3.45),

$$\sigma(\mu)(r) = \int |\varphi * \widehat{\nu}(rv)|^2 \, d\sigma^{n-1}v = \int \left| \int \varphi(x - rv)\widehat{\nu}(x) \, dx \right|^2 d\sigma^{n-1}v$$

$$\leq \int \left(\int |\varphi(x - rv)| \, dx \int |\varphi(x - rv)||\widehat{\nu}(x)|^2 \, dx \right) d\sigma^{n-1}v$$

$$\leq \|\varphi\|_1 \|\varphi\|_\infty \int \int_{|x-rv|\leq 1} |\widehat{\nu}(x)|^2 \, dx \, d\sigma^{n-1}v$$

$$\lesssim \int_{||x|-r|\leq 1} \sigma^{n-1}(\{v : |x - rv| \leq 1\})|\widehat{\nu}(x)|^2 \, dx$$

$$\lesssim r^{1-n} \int_{||x|-r|\leq 1} |\widehat{\nu}(x)|^2 \, dx \approx r^{1-s} \int_{||x|-r|\leq 1} |x|^{s-n}|\widehat{\nu}(x)|^2 \, dx$$

$$\leq r^{1-s} \int |x|^{s-n}|\widehat{\nu}(x)|^2 \, dx = \gamma(n, s)^{-1} r^{1-s} I_s(\nu) \leq \gamma(n, s)^{-1} r^{1-s} I_s(\mu).$$

The last inequality follows since $I_s(\nu) \leq I_s(\mu)$ by the definition of ν and the fact that $\widehat{\varphi} \geq 1$ on $\mathrm{spt}\,\mu$. We have also used the obvious estimate $\sigma^{n-1}(\{v : |y - v| \leq \varrho\}) \lesssim \varrho^{n-1}$ for all $y \in \mathbb{R}^n, \varrho > 0$. $\square$

We shall now discuss the counter-examples giving the upper bounds in Theorems 15.6 and 15.7. First we prove the missing part in $\mathbb{R}^2$:

Proposition 15.9 *We have $t_2(s) \leq 1/2$ for $0 < s \leq 1$ and $t_2(s) \leq s/2$ for $1 \leq s < 2$.*

Proof To prove the first statement we show that if $0 < s < 1$ there is $\mu \in \mathcal{M}(\mathbb{R}^2)$ such that $I_s(\mu) < \infty$ and

$$\sigma(\mu)(r) \gtrsim 1/\sqrt{r} \quad \text{for some arbitrarily large } r > 0.$$

For this we can take as μ one of the measures $\mu_{M,N} \in \mathcal{M}(\mathbb{R})$ constructed in 8.2.1 considered as a measure in the x-axis in $\mathbb{R}^2$. As we observed there we have such a μ for which $I_s(\mu) < \infty$ and $\widehat{\mu}(x, 0)$ does not tend to zero as $x \to \infty$. Notice that as $\mathrm{spt}\,\mu \subset \{(x, 0) : x \in \mathbb{R}\}$,

$$\widehat{\mu}(x, y) = \widehat{\mu}(x, 0) \quad \text{for all } y \in \mathbb{R}.$$

Since $\widehat{\mu}$ is Lipschitz continuous we can then find a positive number a and arbitrarily large values $r > 0$ such that

$$|\widehat{\mu}(x, y)| > a \quad \text{for all } x \in [r - a, r], \quad y \in \mathbb{R}.$$

The length of the arc $C(r) = \{(x, y) : |(x, y)| = r, r - a \le x \le r\}$ is at least $b\sqrt{r}$ for some $b > 0$ independent of r. Therefore,

$$\sigma(\mu)(r) = r^{-1} \int_{S(r)} |\widehat{\mu}(v)|^2 \, d\sigma_r^1 v \ge a^2 b / \sqrt{r}$$

as desired.

For the second statement we need to show that if $1 < s < 2$ and $t > s/2$, then there is $\mu \in \mathcal{M}(\mathbb{R}^2)$ such that $I_s(\mu) < \infty$ and

$$\sigma(\mu)(r) \gtrsim r^{-t} \quad \text{for some arbitrarily large } r > 0.$$

We shall again make use of the measures $\mu_{M,N} \in \mathcal{M}(\mathbb{R})$ and also of the standard Cantor measures μ_d. Choose positive numbers $s_1, s_2, s_2', \varepsilon \in (0, 1)$ such that

$$s < s_1 + s_2 < s_1 + s_2' < 2t \quad \text{and} \quad (1 + s_2')/2 + \varepsilon < t.$$

Let μ_1 be one of the measures $\mu_{M,N}$ with the properties

$$\mu_1([x - \varrho, x + \varrho]) \lesssim \varrho^{s_1} \quad \text{for } x \in \mathbb{R}, \quad \varrho > 0,$$

and for some $a > 0, L > 1$ and $L^k \le r_k < L^{k+1}$,

$$|\widehat{\mu_1}(x)| > a \text{ for } r_k - a \le x \le r_k, \quad k = 1, 2, \ldots.$$

The existence of such a measure follows from (8.8) and (8.9) using again the Lipschitz continuity of $\widehat{\mu_1}$. Let μ_2 be the measure μ_d with $s_2 = \log 2/\log(1/d)$; it has by (8.2) the properties

$$\mu_2([x - \varrho, x + \varrho]) \lesssim \varrho^{s_2} \quad \text{for } x \in \mathbb{R}, \quad \varrho > 0$$

and

$$I_{s_2'}(\mu_2) = \infty.$$

As in the proof for the first part, we see again that the length of the arc $C(r_k) = \{(x, y) : |(x, y)| = r_k, r_k - a \le x \le r_k\}$ is at least $b\sqrt{r_k}$. By (3.45) with $r_0 = 0$,

$$I_{s_2'}(\mu_2)$$

$$= \gamma(1, s_2') \sum_{k=1}^{\infty} \int_{b\sqrt{r_{k-1}}}^{b\sqrt{r_k}} |y|^{s_2'-1} |\widehat{\mu_2}(y)|^2 \, dy \lesssim \sum_{k=1}^{\infty} r_k^{(s_2'-1)/2} \int_0^{b\sqrt{r_k}} |\widehat{\mu_2}(y)|^2 \, dy.$$

As $I_{s_2'}(\mu_2) = \infty$ there are arbitrarily large values of k for which

$$r_k^{(s_2'-1)/2} \int_0^{b\sqrt{r_k}} |\widehat{\mu_2}(y)|^2 \, dy > L^{-k\varepsilon} \ge r_k^{-\varepsilon}. \tag{15.10}$$

Let μ be the product measure $\mu = \mu_1 \times \mu_2$. Then $I_s(\mu) < \infty$ and $\widehat{\mu}(x, y) = \widehat{\mu_1}(x)\widehat{\mu_2}(y)$. For those k for which (15.10) is valid we get

$$
\begin{aligned}
\sigma(\mu)(r_k) &= r_k^{-1} \int_{S(r_k)} |\widehat{\mu_1}(x)\widehat{\mu_2}(y)|^2 \, d\sigma_{r_k}^1(x, y) \\
&\geq a^2 r_k^{-1} \int_0^{b\sqrt{r_k}} |\widehat{\mu_2}(y)|^2 \, dy \geq a^2 r_k^{-1-(s_2'-1)/2-\varepsilon} \geq a^2 r_k^{-t}. \qquad \square
\end{aligned}
$$

It remains to verify the last line of Theorem 15.7:

Proposition 15.10 *We have* $t_n(s) \leq s/2 + n/2 - 1$ *for* $1 \leq s < n$.

Proof Observe first that if $0 < t < t_n(s)$, then

$$
\int |\widehat{f}(rv)|^2 \, d\sigma^{n-1}v \lesssim r^{-t} I_s(|f|) \tag{15.11}
$$

for all (not only non-negative) smooth functions f with spt $f \subset B(0, 1)$. Choose non-negative C^∞ functions φ on $\mathbb{R}$ and ψ on $\mathbb{R}^{n-1}$ with spt $\varphi \subset (-1/4, 1/4)$, $\widehat{\varphi}(0) > 0$, spt $\psi \subset B(0, 1/2)$ and $\widehat{\psi}(0) > 0$. Fix $R > 1$ and define

$$
f(x) = e^{2\pi i R x_1} \varphi(x_1) R^{(n-1)/2} \psi(R^{1/2}x') \quad \text{for } x = (x_1, x') \in \mathbb{R} \times \mathbb{R}^{n-1}.
$$

Then

$$
\widehat{f}(\xi_1, \xi') = \widehat{\varphi}(\xi_1 - R)\widehat{\psi}(R^{-1/2}\xi'),
$$

from which one checks that for some positive constants a and b independent of R, $|\widehat{f}(Rv)| > a$ when $v \in S^{n-1} \cap B((1, 0, \ldots, 0), b/\sqrt{R})$. This yields

$$
\int |\widehat{f}(Rv)|^2 \, d\sigma^{n-1}v \gtrsim R^{-(n-1)/2}. \tag{15.12}
$$

Set also $g = |f|$, for which we have

$$
\widehat{g}(\xi_1, \xi') = \widehat{\varphi}(\xi_1)\widehat{\psi}(R^{-1/2}\xi').
$$

Thus

$$
I_s(g) = \gamma(n, s) \int |\xi|^{s-n} |\widehat{\varphi}(\xi_1)|^2 |\widehat{\psi}(R^{-1/2}\xi')|^2 \, d\xi.
$$

Let $\varepsilon > 0$ and

$$
D = \{(\xi_1, \xi') \in \mathbb{R}^n : |\xi_1| \leq R^\varepsilon, |\xi'| \leq R^{1/2+\varepsilon}\}.
$$

We split the integration in the formula for $I_s(g)$ over D and its complement. For any positive integer N the fast decay of $\widehat{\varphi}$ and $\widehat{\psi}$ gives

$$
\int_{\mathbb{R}^n \setminus D} |\xi|^{s-n} |\widehat{\varphi}(\xi_1)|^2 |\widehat{\psi}(R^{-1/2}\xi')|^2 \, d\xi \lesssim R^{-N}.
$$

For the integral over D we get

$$\int_D |\xi|^{s-n} |\widehat{\varphi}(\xi_1)|^2 |\widehat{\psi}(R^{-1/2}\xi')|^2 \, d\xi$$

$$\lesssim \int_{|\xi_1|<R^\varepsilon, |\xi'|<R^\varepsilon} |\xi|^{s-n} \, d\xi + R^\varepsilon \int_{R^\varepsilon}^{R^{1/2+\varepsilon}} u^{s-2} \, du.$$

The first of the integrals on the right hand side is $\lesssim R^{s\varepsilon}$. The bound on the second depends on the range of s. If $0 < s < 1$, it is $\lesssim R^\varepsilon$. If $s = 1$, it is $\lesssim R^\varepsilon \log R$. If $s > 1$, it is $\lesssim R^{(s-1)/2+s\varepsilon}$. Combining this with (15.11) and (15.12), we get when $s > 1$,

$$R^{-(n-1)/2} \lesssim R^{-t+(s-1)/2+s\varepsilon},$$

which yields $(n-1)/2 \geq t - (s-1)/2 - s\varepsilon$, and leads to the last estimate $t(s) \leq s/2 + n/2 - 1$ of Theorem 15.7. The case $s \leq 1$ does not lead to anything new, but when $n = 2$, it gives another way to see that $t_2(s) \leq 1/2$ for $0 < s \leq 1$. $\qquad\square$

Let us give a dual characterization of $t_n(s)$:

Proposition 15.11 *For any $0 < s < n$ we have that $t_n(s)$ is the supremum of the numbers t such that*

$$\int |\widehat{g\sigma^{n-1}}(rx)| \, d\mu x \leq C(\mu) r^{-t/2} \|g\|_{L^2(S^{n-1})} \quad \text{for } r > 1,$$

for all $\mu \in \mathcal{M}(B(0,1))$ with $I_s(\mu) < \infty$ and for all $g \in L^2(S^{n-1})$.

Proof Let $\mu \in \mathcal{M}(B(0,1))$. By duality in $L^2(S^{n-1})$ and the product formula for the Fourier transform,

$$\sigma(\mu)(r) = \int |\widehat{\mu}(rv)|^2 \, d\sigma^{n-1}v = \sup_{\|g\|_{L^2(S^{n-1})}\leq 1} \left| \int \widehat{\mu}(rv)g(v) \, d\sigma^{n-1}v \right|^2$$

$$= \sup_{\|g\|_{L^2(S^{n-1})}\leq 1} \left| \int \widehat{g\sigma^{n-1}}(rx) \, d\mu x \right|^2.$$

We still need to get absolute values inside the integral sign in the last integral. This can be done by observing that

$$\int |\widehat{g\sigma^{n-1}}(rx)| \, d\mu x = \sup_{\|h\|_{L^\infty(\mu)}\leq 1} \int \widehat{g\sigma^{n-1}}(rx)h(x) \, d\mu x,$$

writing $h = h_1 - h_2 + i(h_3 - h_4)$ with non-negative functions h_j and investigating the measures $h_j\mu$. $\qquad\square$

This proposition is from Wolff [1999] and Barceló, Bennett, Carbery and Rogers [2011]. The authors of the latter paper also proved an L^2 version: $t_n(s)$ is the supremum of the numbers t such that

$$\int |\widehat{g\sigma^{n-1}}(rx)|^2 \, d\mu x \leq C(\mu) r^{-t} \|g\|^2_{L^2(S^{n-1})} \quad \text{for } r > 1$$

for all $\mu \in \mathcal{M}(B(0, 1))$ with $I_s(\mu) < \infty$ and for all $g \in L^2(S^{n-1})$.

15.4 Distance sets in finite fields

Iosevich and Rudnev [2007c] have developed analogues of some of the above methods and results in finite fields. Here we shall give a very brief sketch, for more details and related references, see Iosevich and Rudnev [2007c], Hart, Iosevich, Koh and Rudnev [2011] and Chapman, Erdoğan, Hart, Iosevich and Koh [2012]. Let $\mathbb{F}$ be a finite field of q elements and n a positive integer. For $A \subset \mathbb{F}^n$ we take now

$$\Delta(A) = \left\{ \sum_{j=1}^{n} (x_j - y_j)^2 : (x_1, \ldots, x_n), (y_1, \ldots, y_n) \in A \right\}$$

for the distance set, and we ask about its size (cardinality) as compared to the size of A. In particular, when is $\Delta(A)$ all of $\mathbb{F}$, or has cardinality $\approx q$? Iosevich and Rudnev [2007c] proved, among other things, that there is a constant C (any $C > 2$ works) such that

$$\Delta(A) = \mathbb{F} \quad \text{if } \#A \geq Cq^{(n+1)/2}.$$

This can be considered as the analogue of Theorem 4.6(a). This is fairly sharp, at least when n is odd: Hart, Iosevich, Koh and Rudnev [2011] showed that then the condition $\#A \geq cq^{(n+1)/2}$ with some positive constant c is not sufficient to guarantee that $\Delta(A) = \mathbb{F}$. However, for $n = 2$, Chapman, Erdoğan, Hart, Iosevich and Koh [2012] showed that $\#A \geq q^{4/3}$ suffices, which is analogous to Wolff's distance set Theorem 15.1(a) in the plane. Also for subsets of the sphere $\{x \in \mathbb{F}^n : \sum_{j=1}^{n} x_j^2 = 1\}$ better estimates hold, see Hart, Iosevich, Koh and Rudnev [2011].

To prove such results with Fourier methods one needs the Fourier transform. For $f : \mathbb{F}^n \to \mathbb{C}$ it is

$$\widehat{f}(\xi) = q^{-n} \sum_{x \in \mathbb{F}^n} \chi(-x \cdot \xi) f(x), \quad \xi \in \mathbb{F}^n,$$

where χ is a non-trivial additive character, that is, a homomorphism of the additive group $\mathbb{F}$ into the multiplicative group of non-zero complex numbers, which is not identically 1. By $x \cdot \xi$ we mean the usual inner product in $\mathbb{F}^n$. Sets now play the role of measures and the spherical averages $\sigma(\mu)(r) = \int |\widehat{\mu}(rv)|^2 \, d\sigma^{n-1} v$ are replaced by

$$\sigma_A(r) = \sum_{|\xi|^2=r} |\widehat{\chi_A}(\xi)|^2.$$

The sums

$$S(A, q) = \frac{q^{3n+1}}{(\#A)^4} \sum_{r \in \mathbb{F}} \sigma_A(r)^2$$

take the role of the integrals $\int_0^\infty \sigma(\mu)(r)^2 r^{n-1} \, dr$. Iosevich and Rudnev proved, in analogy to Proposition 15.2(a), that if $n \geq 2$, $\#A \gtrsim q^{n/2}$ and $S(A, q) \lesssim 1$, then $\#\Delta(A) \gtrsim q$.

15.5 Further comments

Theorems 15.1 and 15.5 were proven by Wolff [1999] for $n = 2$ and by Erdoğan [2005] for $n > 2$. The underlying ideas in the method come from developments in the study of Fourier restriction and related problems, which we shall investigate in the last part of the book. They were first used for the distance set problem by Bourgain [1994], who proved a weaker result. Theorems 15.1 and 15.5, with the proofs we shall present, hold for general norms for which the unit ball is a convex set whose smooth boundary has non-vanishing Gaussian curvature.

The method of application of spherical averages to distance sets was developed by Mattila [1987] where Propositions 15.2, 15.3 and Corollary 15.4 were obtained. Greenleaf, Iosevich, Liu and Palsson [2013] gave a different proof for the first part of Proposition 15.2 by the methods discussed in Section 4.4 avoiding the calculations with Bessel functions. Shayya [2011] proved the following extension: if $n/2 < s < n$, $v \in \mathcal{M}(\mathbb{R}^n)$, $\int |x|^{-s} \, dx < \infty$ and

$$\int_0^\infty |\int |\widehat{v}(rv) \, d\sigma^{n-1} v|^2 r^{n-1} \, dr < \infty,$$

then $\mathcal{L}^1(\{|x| : x \in \text{spt} \, v\}) > 0$. Proposition 15.2(a) follows from this with $v = \mu * \widetilde{\mu}$ where $\widetilde{\mu}(A) = \mu(-A)$. The proof follows similar lines as the one above combined with another result from Shayya [2011]: if σ is a finite complex Borel measure on $\mathbb{R}^n$ with compact support and $\widehat{\sigma} \in L^2(\{x : v \cdot x \geq 0\})$ for some half-space $\{x : v \cdot x \geq 0\}$, then σ is absolutely continuous.

The first two decay estimates in Theorem 15.7 and the planar examples discussed above also originated in Mattila [1987]. The fourth and last estimates of Theorem 15.7 were proven by Sjölin [1993]. The simple proof of Proposition 15.8 is due to Wolff [2003].

Wolff [1999] investigated also L^p, $1 \le p < \infty$, averages $\int_{S^1} |\widehat{\mu}(rv)|^p \, d\sigma^1 v$ in the plane. The estimates for $p > 2$ obtained from $p = 2$ by interpolation are sharp, Wolff constructed examples to show this. For $1 \le p < 2$ sharp estimates are not known. Wolff related L^1 estimates to dimension estimates for Furstenberg sets, recall Section 11.5.

Bennett and Vargas [2003] improved Wolff's L^1 estimates. The method uses random sums, which Wolff also used, and some of the interesting estimates for them are sharp.

Many variants of spherical averages have been studied. Sjölin [1997] proved estimates for radial functions and for averages over the boundaries of cubes. In Sjölin [2002] he investigated linear combinations of products of radial functions and spherical harmonics. Sjölin and Soria [2003] studied averages with respect to very general measures in place of the surface measure on the sphere. In all these three papers the authors looked at both general and non-negative functions. The behaviour differs considerably for these two classes. The upper bounds for $t_n(s)$ given in Theorems 15.6 and 15.7 are valid for much more general surfaces than the sphere. Barceló, Bennett, Carbery, Ruiz and Vilela [2007] constructed worse counter-examples for the paraboloid than what are known for the sphere. Worse behaviour can also occur if one considers signed measures in place of positive ones, see Iosevich and Rudnev [2007b]. Iosevich and Rudnev [2009] showed that certain bad estimates of Fourier averages of measures imply some structural properties of these measures.

In Proposition 15.2 we saw that $\Delta(\mu) \in L^2$ if $\Sigma(\mu) \in L^2$. There is a much more precise relation between the L^2 norms, namely

$$\int_0^\infty \Delta(\mu)(u)^2 \, du = c(n) \int_0^\infty \Sigma(\mu)(r)^2 \, dr.$$

The key to this is the formula (15.2) which gives $\Delta(\mu)$ as the so-called Hankel transform of $\Sigma(\mu)$. Then the above identity is a Plancherel type formula for this transform which is proved for example in Watson [1944].

From Theorem 7.4 it follows that if A and B are Borel sets in $\mathbb{R}^n$ with $\dim B > (n+1)/2$ and $\dim A + \dim B - n > u$, then for θ_n almost all $g \in O(n)$,

$$\mathcal{L}^n(\{z \in \mathbb{R}^n : \dim A \cap (\tau_z \circ g)(B) \ge u\}) > 0. \tag{15.13}$$

Combining the results in Section 6 of Mattila [1987] with the Wolff–Erdoğan estimate in Theorem 15.5 gives some further information for the case when the assumption $\dim B > (n+1)/2$ is not valid. Namely, if $\dim A \le (n+1)/2$, dim $B \leq (n + 1)/2$ and dim A + dim $B/2 - (3n + 2)/4 > u > 0$, then for θ_n almost all $g \in O(n)$, (15.13) holds. Notice that these conditions imply that dim $A > n/2 + 1/3$ or dim $B > n/2 + 1/3$.

As mentioned in Section 5.4 D. M. Oberlin and R. Oberlin [2013b] used Erdoğan's estimate in Theorem 15.5 to prove certain projection theorems in $\mathbb{R}^3$.

A note added in proof

Very recently Lucà and Rogers [2015] improved some of the estimates for $t_n(s)$ in the range $n/2 + 2/3 + 1/n \leq s < n$ with applications to partial differential equations in the spirit of Chapter 17.

16

Proof of the Wolff–Erdoğan Theorem

We now begin the proof of Theorem 15.5 and we will prove the following almost equivalent

Theorem 16.1 *For all* $(n-2)/2 < s < n, n \geq 2, \varepsilon > 0$ *and every* $\mu \in \mathcal{M}(\mathbb{R}^n)$ *with* $\operatorname{spt} \mu \subset B(0, 1)$ *and*

$$\mu(B(x, \varrho)) \leq \varrho^s \quad \text{for all } x \in \mathbb{R}^n, \varrho > 0, \tag{16.1}$$

the spherical averages satisfy

$$\sigma(\mu)(r) \leq C(n, s, \varepsilon)\mu(\mathbb{R}^n)r^{\varepsilon - \frac{n+2s-2}{4}} \quad \text{for } r > 1. \tag{16.2}$$

This gives Theorem 15.1 by Frostman's lemma, Proposition 15.3 and the fact that (16.1) implies $I_t(\mu) < \infty$ for $0 < t < s$. It is not quite enough to get Theorem 15.5, because a measure with finite energy need not satisfy any uniform growth condition on measures of balls. At the end of this chapter we shall explain how to repair this. Briefly the idea is that the proof of Theorem 16.1 goes through assuming $\mu(B(x, \varrho)) \leq \varrho^s$ only for $\varrho \geq 1/r$, and given $r > 1$ any $\mu \in \mathcal{M}(B(0, 1))$ with $I_s(\mu) < \infty$ can be written as a sum of roughly $\log r$ measures satisfying (16.1) for $\varrho \geq 1/r$.

Since the proof of Theorem 16.1 is rather complicated, I first give a sketch. It is fairly easy to see that instead of integrating over the sphere $S(r)$ it suffices to estimate the integral of $|\widehat{\mu}|^2$ over the annulus $A_r = \{x : r - 1 < |x| < r + 1\}$, and then by duality this is reduced to proving

$$\int |\widehat{f}|^2 \, d\mu \lesssim r^{\varepsilon + (3n - 2s - 2)/4} \tag{16.3}$$

for functions f with $\|f\|_2 = 1$ and $\operatorname{spt} f \subset A_r$. Assume that $\operatorname{spt} f$ is contained in a part of A_r above a cube $I_0 \subset \mathbb{R}^{n-1}$ with $d(I_0) \approx r$. Consider a Whitney decomposition of $I_0 \times I_0 \setminus \{(x, x) : x \in I_0\}$ with cubes $I \times J$ and let $A_r(I)$ be

201

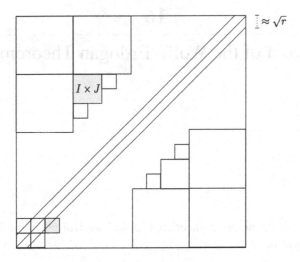

Figure 16.1 The Whitney decomposition

the part of A_r above I. Then we can, just from the definition of the Fourier transform, write

$$\widehat{f}(\xi)^2 = \sum_{k=k_n}^{k_r} \sum_{I \times J \in \mathcal{E}_k} \widehat{f_I}(\xi)\widehat{f_J}(\xi) + \sum_{I \times J \in \mathcal{E}} \widehat{f_I}(\xi)\widehat{f_J}(\xi),$$

where $f_I = f \chi_{A_r(I)}$ and for $I \times J \in \mathcal{E}_k$ both $A_r(I)$ and $A_r(J)$ are dyadic subregions of A_r of diameter roughly $2^{-k}r$ and the distance between $A_r(I)$ and $A_r(J)$ is also roughly $2^{-k}r$. Here $2^{-k_r} \approx 1/\sqrt{r}$, see Figures 16.1 and 16.2. The second sum consists of similar terms with $k = k_r$ and it is much easier to estimate. To estimate the contribution of the first sum we first observe that since there are $\lesssim \log r$ values of k to consider, it is enough to get the upper bound of (16.3) for each of them separately. This rather easily reduces our problem to getting the estimate

$$\int |\widehat{f_I}\widehat{f_J}| \, d\mu \lesssim r^{\varepsilon + (3n - 2s - 2)/4} \|f_I\|_2 \|f_J\|_2,$$

for any fixed $I \times J \in \mathcal{E}_k$.

A simple geometric observation is that the support of $f_I * f_J$ is contained in a rectangular box $R_{I,J}$ with $n - 1$ side-lengths $c2^{-k}r$ and one side-length $c2^{-2k}r$. Let $\varphi_{R_{I,J}}$ be a smooth approximation of the characteristic function of

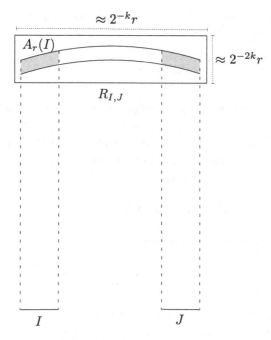

Figure 16.2 The dyadic subregions

$R_{I,J}$ as in Lemma 3.16 such that $\varphi_{R_{I,J}} = 1$ on $R_{I,J}$. Then we find that

$$\int |\widehat{f_I}\,\widehat{f_J}|\,d\mu \leq \int |\widehat{f_I}\,\widehat{f_J}|\mu_{R_{I,J}},$$

where $\mu_{R_{I,J}} = |\widehat{\varphi_{R_{I,J}}}| * \mu$ is as in Lemma 3.17. This lemma gave us estimates on $\mu_{R_{I,J}}$ based on the growth condition $\mu(B(x,\varrho)) \leq \varrho^s$. In order to use these effectively we decompose the space into rectangular boxes $P \in \mathcal{P}$ with $n-1$ side-lengths $c2^{-k}$ and one side-length c. Letting ψ_P be a suitable smooth approximation of the characteristic function of P we find by simple estimation

$$\int |\widehat{f_I}\,\widehat{f_J}|\,d\mu = \int |\mathcal{F}(\varphi_{R_{I,J}} \cdot (f_I * f_J))|\,d\mu$$

$$\lesssim \sum_{P \in \mathcal{P}} \left(\int |\widehat{f_{I,P}}\,\widehat{f_{J,P}}|^2 \right)^{\frac{1}{2}} \left(\int (\mu_{R_{I,J}} \psi_P)^2 \right)^{\frac{1}{2}},$$

where $f_{I,P}$ is defined by localizing the Fourier transform of f_I to P; $\widehat{f_{I,P}} = \psi_P \widehat{f_I}$.

Let us now assume that $n = 2$. The second factor in the above sum is estimated by Lemma 3.17 which gives

$$\int (\mu_{R_{I,J}} \psi_P)^2 \lesssim r^{2-s} 2^{-k}.$$

The first factor is estimated by the support properties of the functions $f_{I,P}$. More precisely, the support of $f_{I,P}$ is contained in a slight fattening $\widetilde{A}_r(I)$ of $A_r(I)$, which again has diameter about $2^{-k}r$ and the distance between $\widetilde{A}_r(I)$ and $\widetilde{A}_r(J)$ is about $2^{-k}r$. Since the 'angle' between $\widetilde{A}_r(I)$ and $\widetilde{A}_r(J)$ is about 2^{-k}, it follows by simple geometry that

$$\mathcal{L}^2((-\widetilde{A}_r(I) + x) \cap \widetilde{A}_r(J)) \lesssim 2^k \qquad (16.4)$$

for all $x \in \mathbb{R}^n$. Using this one can estimate

$$\int |\widehat{f_{I,P}} \, \widehat{f_{J,P}}|^2 \lesssim 2^k \int |f_{I,P}|^2 \int |f_{J,P}|^2.$$

These estimates lead to (16.3) when $n = 2$ but the corresponding inequalities in higher dimensions would not be good enough. When $n \geq 3$ deep bilinear restriction estimates come to the rescue and they can be used to complete the proof.

We now begin the detailed proof of Theorem 16.1. The following proposition allows us to estimate integrals over annuli instead of spheres:

Proposition 16.2 *Let* $\mu \in \mathcal{M}(B(0, 1))$ *and let* $\alpha > 0$. *Let*

$$A_r = \{x \in \mathbb{R}^n : r - 1 < |x| < r + 1\} \quad \text{for } r > 1.$$

If $\mu(\mathbb{R}^n) \leq 1$ *and for all* $0 < \varepsilon < 1, r > 1$,

$$r^{1-n} \int_{A_r} |\widehat{\mu}(x)|^2 \, dx \lesssim_{\alpha,\varepsilon} \mu(\mathbb{R}^n) r^{\varepsilon-\alpha} \qquad (16.5)$$

then for all $0 < \varepsilon < 1, r > 1$,

$$\sigma(\mu)(r) \lesssim_{\alpha,\varepsilon} \mu(\mathbb{R}^n) r^{\varepsilon-\alpha}.$$

Proof Choose $\varphi \in \mathcal{S}(\mathbb{R}^n)$, $\varphi \geq 0$, such that $\varphi = 1$ on spt μ. Let $N \in \mathbb{N}$. Then as $\widehat{\varphi} \in \mathcal{S}(\mathbb{R}^n)$,

$$|\widehat{\varphi}(y)| \leq C_N |y|^{-N} \quad \text{for all } y \in \mathbb{R}^n.$$

We may assume that $r^{1-\varepsilon} > 2$ so that $r - r^\varepsilon > 1$. Using Schwartz's inequality we derive,

$$\sigma(\mu)(r) = \int_{S^{n-1}} |\hat{\mu}(rv)|^2 \, d\sigma^{n-1}v = \int_{S^{n-1}} |\widehat{\varphi\mu}(rv)|^2 \, d\sigma^{n-1}v$$

$$= \int_{S^{n-1}} |\hat{\varphi} * \hat{\mu}(rv)|^2 \, d\sigma^{n-1}v = \int_{S^{n-1}} \left| \int \hat{\varphi}(rv - x)\hat{\mu}(x) \, dx \right|^2 d\sigma^{n-1}v$$

$$\leq \int_{S^{n-1}} \int |\hat{\varphi}(rv - x)| \, dx \int |\hat{\varphi}(rv - x)||\hat{\mu}(x)|^2 \, dx \, d\sigma^{n-1}v$$

$$= \int_{S^{n-1}} \|\hat{\varphi}\|_1 \int |\hat{\varphi}(rv - x)||\hat{\mu}(x)|^2 \, dx \, d\sigma^{n-1}v$$

$$\lesssim \int_{S^{n-1}} \int |\hat{\varphi}(rv - x)||\hat{\mu}(x)|^2 \, dx \, d\sigma^{n-1}v$$

$$\lesssim \iint_{\{(x,v):|rv-x|<r^\varepsilon\}} d\sigma^{n-1}v|\hat{\mu}(x)|^2 \, dx$$

$$+ \sum_{j=1}^{\infty} \iint_{\{(x,v):(r^\varepsilon)^j \leq |rv-x| < (r^\varepsilon)^{j+1}\}} |\hat{\varphi}(rv - x)||\hat{\mu}(x)|^2 \, dx \, d\sigma^{n-1}v.$$

Let

$$I_1(r) = \iint_{\{(x,v):|rv-x|<r^\varepsilon\}} d\sigma^{n-1}v|\hat{\mu}(x)|^2 \, dx$$

and

$$I_2(r) = \sum_{j=1}^{\infty} \iint_{\{(x,v):(r^\varepsilon)^j \leq |rv-x| < (r^\varepsilon)^{j+1}\}} |\hat{\varphi}(rv - x)||\hat{\mu}(x)|^2 \, dx \, d\sigma^{n-1}v.$$

First we estimate $I_1(r)$. For (x, v) satisfying $|rv - x| < r^\varepsilon$ we have $|r - |x|| < r^\varepsilon$ and,

$$\left| v - \frac{x}{|x|} \right| \leq \left| v - \frac{x}{r} \right| + \left| \frac{x}{r} - \frac{x}{|x|} \right| = r^{-1}(|rv - x| + ||x| - r|) < 2r^{\varepsilon-1},$$

so

$$\sigma^{n-1}(\{v \in S^{n-1} : |rv - x| < r^\varepsilon\}) \leq \sigma^{n-1}\left(\left\{ v \in S^{n-1} : \left| v - \frac{x}{|x|} \right| < 2r^{\varepsilon-1} \right\} \right)$$

$$\lesssim r^{(\varepsilon-1)(n-1)}.$$

Covering the interval $(r - r^\varepsilon, r + r^\varepsilon)$ with $k \leq 2r^\varepsilon$ intervals of length 2, we find $r_i \in (r - r^\varepsilon, r + r^\varepsilon)$, $i = 1, \ldots, k$, such that

$$\{x : r - r^\varepsilon < |x| < r + r^\varepsilon\} \subset \bigcup_{i=1}^{k} A_{r_i}.$$

Therefore, applying (16.5)

$$I_1(r) \lesssim r^{(\varepsilon-1)(n-1)} \int_{\{x: r-r^\varepsilon < |x| < r+r^\varepsilon\}} |\widehat{\mu}(x)|^2 \, dx$$

$$\leq r^{(\varepsilon-1)(n-1)} \sum_{i=1}^k \int_{A_{r_i}} |\widehat{\mu}(x)|^2 \, dx$$

$$\lesssim r^{(\varepsilon-1)(n-1)} \sum_{i=1}^k r_i^{n-1+\varepsilon-\alpha} \mu(\mathbb{R}^n)$$

$$\lesssim r^{(\varepsilon-1)(n-1)+\varepsilon+n-1+\varepsilon-\alpha} \mu(\mathbb{R}^n) = r^{(n+1)\varepsilon-\alpha} \mu(\mathbb{R}^n).$$

For $I_2(r)$ we have, as $|\widehat{\mu}(x)|^2 \leq \mu(\mathbb{R}^n)^2 \leq \mu(\mathbb{R}^n)$,

$$I_2(r) \leq \sum_{j=1}^\infty C_N r^{-\varepsilon N j} \int_{S^{n-1}} \mathcal{L}^n(B(rv, r^{\varepsilon(j+1)})) \, d\sigma^{n-1} v \mu(\mathbb{R}^n)$$

$$= C_N \sigma^{n-1}(S^{n-1}) \alpha(n) \sum_{j=1}^\infty r^{-\varepsilon N j + n\varepsilon(j+1)} \mu(\mathbb{R}^n)$$

$$= C_N \sigma^{n-1}(S^{n-1}) \alpha(n) r^{n\varepsilon} \sum_{j=1}^\infty r^{j\varepsilon(n-N)} \mu(\mathbb{R}^n)$$

$$\lesssim_{\varepsilon, N} r^{n\varepsilon + (n-N)\varepsilon} \mu(\mathbb{R}^n) \leq r^{-\alpha} \mu(\mathbb{R}^n),$$

when N is chosen big enough so that $(N - 2n)\varepsilon > \alpha$. $\qquad\qquad\square$

Proposition 16.3 *Let $\mu \in \mathcal{M}(B(0, 1))$ and $\alpha > 0$. If*

$$\int |\widehat{f}|^2 \, d\mu \lesssim_{\alpha, \varepsilon} r^{\varepsilon+n-1-\alpha}, \qquad\qquad (16.6)$$

for all $r > 1$ and $f \in L^2(\mathbb{R}^n)$ such that $\|f\|_2 = 1$ and $\operatorname{spt} f \subset A_r$, then the estimate (16.5) follows.

Proof By duality, the product formula and Schwartz's inequality,

$$\int_{A_r} |\widehat{\mu}|^2 = \sup_{\substack{\|f\|_2=1 \\ \operatorname{spt} f \subset A_r}} \left| \int f \widehat{\mu} \right|^2 = \sup_{\substack{\|f\|_2=1 \\ \operatorname{spt} f \subset A_r}} \left| \int \widehat{f} \, d\mu \right|^2 \leq \sup_{\substack{\|f\|_2=1 \\ \operatorname{spt} f \subset A_r}} \int |\widehat{f}|^2 \, d\mu \mu(\mathbb{R}^n).$$

$\qquad\qquad\qquad\qquad\qquad\qquad\qquad\qquad\qquad\qquad\qquad\qquad\qquad\qquad\square$

Proof of Theorem 16.1 With Propositions 16.2 and 16.3 in mind it is enough to prove that if $r > 1$ and $0 < \varepsilon < 1$, then

$$\int |\widehat{f}|^2 \, d\mu \lesssim_{s, \varepsilon} r^{\varepsilon+n-1-(n+2s-2)/4} = r^{\varepsilon+(3n-2s-2)/4}, \qquad\qquad (16.7)$$

provided

$$\|f\|_2 = 1, \text{spt } f \subset A_r, \mu \in \mathcal{M}(B(0,1)) \quad \text{and}$$

$$\mu(B(x, \varrho)) \le \varrho^s \text{ for all } x \in \mathbb{R}^n, \varrho > 0. \tag{16.8}$$

The rest of the proof of Theorem 16.1, and hence of Theorem 15.1, consists of proving that (16.8) implies (16.7). We shall do this with a bilinear approach, that is, we write $f^2 = \sum_{I,J} f_I f_J$ for suitable functions f_I and estimate the integrals $\int |f_I f_J| \, d\mu$. The functions f_I are supported in spherical dyadic annuli. To get the needed spherical decomposition of A_r more formally we may and shall assume that f lives above some cube $[0, r2^{-k_n})^{n-1}$ with $2^{-k_n} \approx 1/n$. Then the standard dyadic decomposition of $[0, r2^{-k_n})^{n-1}$ gives the desired decomposition of the part of A_r above it.

So let f and μ be as in (16.8), $r > 1$ and $0 < \varepsilon < 1$. Let $I_0 = [0, r2^{2-k_n})^{n-1}$ with $2^{2-k_n} < 1/n$, and for any cube $I \subset I_0$ let

$$A_r(I) = \{x \in \mathbb{R}^n : (x_1, \ldots, x_{n-1}) \in I, x_n > 0, r - 1 < |x| < r + 1\}.$$

By easy geometry

$$\overline{A_r(I)} \subset R_I \tag{16.9}$$

where R_I is a rectangular box with $n - 1$ side-lengths $c_1 d(I)$ and one side-length $c_1 r(d(I)/r)^2 = c_1 d(I)^2/r$. Here and later in this chapter $c_1, c_2, \ldots$ will be positive constants depending only on n. Clearly, $d(A_r(I)) \approx d(I)$.

We may now assume that spt $f \subset A_r(I_0)$. Consider the dyadic partition $\mathcal{D}_k$ of I_0 into disjoint subcubes of side-length $r2^{-k}, k = k_n, k_n + 1, \ldots$. We express $I_0 \times I_0 \setminus \{(x, x) : x \in I_0\}$ as a Whitney type decomposition by cubes $I \times J$ with disjoint interiors;

$$I_0 \times I_0 \setminus \{(x, x) : x \in I_0\} = \bigcup_{k=k_n}^{\infty} \bigcup_{I \times J \in \mathcal{E}_k} I \times J$$

such that $I, J \in \mathcal{D}_k$ and

$$r2^{-k} \le d(I, J) \le 2n \cdot r2^{-k} \quad \text{whenever} \quad I \times J \in \mathcal{E}_k. \tag{16.10}$$

There are several easy ways to achieve this. One way to define $\mathcal{E}_k$ is to declare that $I \times J \in \mathcal{E}_k$ if $I, J \in \mathcal{D}_k$ and I and J are not adjacent but their parents are. Adjacent means that the cubes are disjoint but their closures intersect. The parent of $I \in \mathcal{D}_k$ is the unique cube in $\mathcal{D}_{k-1}$ which contains I. Let k_r be the largest integer such that $2^{-k_r} \ge 1/\sqrt{r}$. So $k_r \approx \log \sqrt{r}$ and we have for some

$c_2 > 0$,

$$\{(x, y) \in I_0 \times I_0 : |x - y| > c_2\sqrt{r}\} \subset \bigcup_{k=k_n}^{k_r} \bigcup_{I \times J \in \mathcal{E}_k} I \times J.$$

Then $I_0 \times I_0 \setminus \bigcup_{k=k_n}^{k_r} \bigcup_{I \times J \in \mathcal{E}_k} I \times J$ is the union of disjoint cubes $I \times J$, $I, J \in \mathcal{D}_{k_r}$ of side-length $\approx \sqrt{r}$; call this family $\mathcal{E}$. Hence

$$I_0 \times I_0 = \bigcup_{k=k_n}^{k_r} \bigcup_{I \times J \in \mathcal{E}_k} I \times J \cup \bigcup_{I \times J \in \mathcal{E}} I \times J,$$

the whole union being disjoint. Let $\mathcal{E}'$ be the family of those cubes $I \in \mathcal{D}_{k_r}$ for which there is $J \in \mathcal{D}_{k_r}$ such that $I \times J \in \mathcal{E}$, and for $I \in \mathcal{E}'$ set $\mathcal{E}(I) = \{J : I \times J \in \mathcal{E}\}$. Clearly $I \in \mathcal{E}(I)$ and $\#\mathcal{E}(I)$ is bounded by a constant depending only on n. We also have the disjoint union

$$A_r(I_0) \times A_r(I_0) = \bigcup_{k=k_n}^{k_r} \bigcup_{I \times J \in \mathcal{E}_k} A_r(I) \times A_r(J) \cup \bigcup_{I \times J \in \mathcal{E}} A_r(I) \times A_r(J).$$

Note that $d(A_r(I)) \approx d(A_r(J)) \approx 2^{-k}r$ when $I \times J \in \mathcal{E}_k$.

For $\xi \in \mathbb{R}^n$ we now have

$$\widehat{f}(\xi)^2 = \left(\int e^{-2\pi i \xi \cdot x} f(x)\, dx \right)^2$$

$$= \int_{A_r(I_0)} \int_{A_r(I_0)} e^{-2\pi i \xi \cdot (x+y)} f(x) f(y)\, dx\, dy$$

$$= \sum_{k=k_n}^{k_r} \sum_{I \times J \in \mathcal{E}_k} \int_{A_r(I)} \int_{A_r(J)} e^{-2\pi i \xi \cdot (x+y)} f(x) f(y)\, dx\, dy$$

$$+ \sum_{I \times J \in \mathcal{E}} \int_{A_r(I)} \int_{A_r(J)} e^{-2\pi i \xi \cdot (x+y)} f(x) f(y)\, dx\, dy$$

$$= \sum_{k=k_n}^{k_r} \sum_{I \times J \in \mathcal{E}_k} \widehat{f_I}(\xi)\widehat{f_J}(\xi) + \sum_{I \times J \in \mathcal{E}} \widehat{f_I}(\xi)\widehat{f_J}(\xi),$$

where

$$f_I = f \chi_{A_r(I)}.$$

Set

$$S_1(\xi) = \sum_{k=k_n}^{k_r} \sum_{I \times J \in \mathcal{E}_k} \widehat{f_I}(\xi)\widehat{f_J}(\xi) \tag{16.11}$$

and

$$S_2(\xi) = \sum_{I \times J \in \mathcal{E}} \widehat{f_I}(\xi) \widehat{f_J}(\xi). \tag{16.12}$$

For $I \times J \in \mathcal{E}_k, k_n \leq k \leq k_r$, $\widehat{f_I}\widehat{f_J} = \widehat{f_I * f_J}$. Recalling (16.9) it follows by elementary geometry that for some c_3,

$$\operatorname{spt} f_I * f_J \subset \overline{A_r(I)} + \overline{A_r(J)} \subset R_I + R_J \subset R_{I,J}, \tag{16.13}$$

where $R_{I,J}$ is a rectangular box with $n-1$ side-lengths $c_3 r 2^{-k} \geq \sqrt{r}$ and one side-length $c_3 r 2^{-2k}$. Thus $R_{I,J}$ is a $(c_3 r 2^{-2k}, c_3 r 2^{-k}, \ldots, c_3 r 2^{-k})$-box according to the terminology of Section 3.9. For $I \times J \in \mathcal{E}$, we have also (16.13) with $R_{I,J}$ a $(c_3, c_3 \sqrt{r}, \ldots, c_3 \sqrt{r})$-box. Let us recall some facts about such boxes from Section 3.9.

The dual $\widetilde{R}$ of an $(r_1, \ldots, r_n)$-box R is the $(\frac{1}{r_n}, \ldots, \frac{1}{r_1})$-box centred at the origin with the $\frac{1}{r_j}$ side parallel to the r_j side of R. We associated an affine mapping A_R and a smooth function φ_R to each $(r_1, \ldots, r_n)$-box R such that, denoting $Q_0 = [0, 1]^n$, A_R is of the form

$$A_R(x) = g(Lx) + a, \quad g \in O(n), a \in \mathbb{R}^n, \quad Lx = (r_1 x_1, \ldots, r_n x_n) \quad \text{for } x \in \mathbb{R}^n,$$

$$R = A_R(Q_0), \quad \widetilde{R} = g(L^{-1}(Q_0)), \tag{16.14}$$

and

$$\varphi_R = \varphi \circ A_R^{-1} \quad \text{with} \quad \varphi_R = 1 \quad \text{on} \quad R \quad \text{and} \quad \operatorname{spt} \varphi_R \subset 2R. \tag{16.15}$$

Here $\varphi \in \mathcal{S}(\mathbb{R}^n)$ is such that $\varphi = 1$ on Q_0 and $\operatorname{spt} \varphi \subset 2Q_0$. Then by Lemma 3.16

$$\widehat{\varphi_R}(x) = r_1 \cdots r_n e^{-2\pi i x \cdot a} \widehat{\varphi}(L(g^{-1}(x))), \tag{16.16}$$

$$|\widehat{\varphi_R}(x)| \lesssim_M r_1 \cdots r_n \sum_{j=1}^{\infty} 2^{-Mj} \chi_{2^j \widetilde{R}}(x) \quad \text{for } x \in \mathbb{R}^n, M \in \mathbb{N}, \tag{16.17}$$

and

$$\|\widehat{\varphi_R}\|_1 = \|\widehat{\varphi}\|_1. \tag{16.18}$$

We shall also use the estimates for the function

$$\mu_R = |\widehat{\varphi_R}| * \mu$$

given in Lemma 3.17.

First we estimate $\int |S_2(\xi)| \, d\mu\xi$. Recalling (16.12), $S_2 = \sum_{I \times J \in \mathcal{E}} \widehat{f_I}\widehat{f_J}$ and $\operatorname{spt} f_I \subset R_I$ where R_I is a $(c_3, c_3 \sqrt{r}, \ldots, c_3 \sqrt{r})$-box. Thus, as $\varphi_{R_I} = 1$ on R_I,

$$\widehat{f_I} = \widehat{(\varphi_{R_I} f_I)} = \widehat{\varphi_{R_I}} * \widehat{f_I}.$$

Applying Schwartz's inequality and (16.18), we deduce

$$|\widehat{f_I}(x)| = \left| \int \widehat{\varphi_{R_I}}(x-y)\widehat{f_I}(y)\,dy \right|$$

$$\leq \left(\int |\widehat{\varphi_{R_I}}(x-y)|\,dy \right)^{\frac{1}{2}} \left(\int |\widehat{\varphi_{R_I}}(x-y)||\widehat{f_I}(y)|^2\,dy \right)^{\frac{1}{2}}$$

$$= \|\widehat{\varphi}\|_1^{\frac{1}{2}} \left(\int |\widehat{\varphi_{R_I}}(x-y)||\widehat{f_I}(y)|^2\,dy \right)^{\frac{1}{2}}.$$

Hence by (3.60) and Plancherel's theorem,

$$\int |\widehat{f_I}|^2\,d\mu \lesssim \int (|\widehat{\varphi_{R_I}}| * \mu)|\widehat{f_I}|^2 \lesssim (\sqrt{r})^{n-s} \int |\widehat{f_I}|^2 = r^{\frac{n}{2}-\frac{s}{2}} \int |f_I|^2. \tag{16.19}$$

Recall that $\mathcal{E}'$ is the set of I such that $I \times J \in \mathcal{E}$ for some J and $\mathcal{E}(I) = \{J : I \times J \in \mathcal{E}\}$. For every $I \in \mathcal{E}'$ there are only boundedly many elements, say at most N, in $\mathcal{E}(I)$. Here N depends only on n. Denote by $J(I)$ a cube $J \in \mathcal{E}(I)$ for which the integral $\int |\widehat{f_J}|^2\,d\mu$ is largest. Then, as $I \in \mathcal{E}(I)$,

$$\int |\widehat{f_I}|^2\,d\mu \leq \int |\widehat{f_{J(I)}}|^2\,d\mu.$$

Moreover, again any J can appear as $J(I)$ for at most N different cubes I, so any $x \in \mathbb{R}^n$ can belong to at most N sets $A_R(J(I))$, which gives

$$\sum_{I \in \mathcal{E}'} |f_{J(I)}|^2 = \sum_{I \in \mathcal{E}'} \chi_{A_R(J(I))}|f|^2 \leq N|f|^2.$$

Using these facts, Schwartz's inequality, the fact that $\|f\|_2 = 1$ and (16.19), we get

$$\int |S_2(\xi)|\,d\mu\xi \leq \sum_{I \times J \in \mathcal{E}} \int |\widehat{f_I}\widehat{f_J}|\,d\mu$$

$$\leq \sum_{I \times J \in \mathcal{E}} \left(\int |\widehat{f_I}|^2\,d\mu \right)^{1/2} \left(\int |\widehat{f_J}|^2\,d\mu \right)^{1/2}$$

$$= \sum_{I \in \mathcal{E}'} \left(\int |\widehat{f_I}|^2\,d\mu \right)^{1/2} \sum_{J \in \mathcal{E}(I)} \left(\int |\widehat{f_J}|^2\,d\mu \right)^{1/2}$$

$$\leq \sum_{I \in \mathcal{E}'} \left(\int |\widehat{f_I}|^2\,d\mu \right)^{1/2} N \left(\int |\widehat{f_{J(I)}}|^2\,d\mu \right)^{1/2}$$

$$\leq \sum_{I \in \mathcal{E}'} N \int |\widehat{f_{J(I)}}|^2\,d\mu \lesssim \sum_{I \in \mathcal{E}'} r^{\frac{n}{2}-\frac{s}{2}} \int |\widehat{f_{J(I)}}|^2$$

$$= \sum_{I \in \mathcal{E}'} r^{\frac{n}{2}-\frac{s}{2}} \int |f_{J(I)}|^2 \leq N r^{\frac{n}{2}-\frac{s}{2}} \int |f|^2 \leq r^{\frac{3n-2s-2}{4}},$$

since $\frac{n}{2} - \frac{s}{2} \leq \frac{3n-2s-2}{4}$, which is better than the desired estimate (16.7).

We are left with estimating $\int |S_1(\xi)| \, d\mu\xi$. Recall from (16.11) and (16.13) that

$$S_1 = \sum_{k=k_n}^{k_r} \sum_{I \times J \in \mathcal{E}_k} \widehat{f_I} \widehat{f_J}$$

with

$$\text{spt } f_I * f_J \subset R_{I,J}$$

where $R_{I,J}$ is a $(c_3 r 2^{-2k}, c_3 r 2^{-k}, \ldots c_3 r 2^{-k})$-box. It is enough to prove for every $I \times J \in \mathcal{E}_k$ that

$$\int |\widehat{f_I} \widehat{f_J}| \, d\mu \lesssim r^{\varepsilon + (3n-2s-2)/4} \|f_I\|_2 \|f_J\|_2, \tag{16.20}$$

because if (16.20) holds, using the facts that $k_r \lesssim r^\varepsilon$ and that for every $I \in \mathcal{D}_k$ there are at most N cubes $J \in \mathcal{D}_k$ such that $I \times J \in \mathcal{E}_k$, we obtain

$$
\begin{aligned}
\int |S_1| \, d\mu &\lesssim r^\varepsilon \sup_{\{k:k \leq k_r\}} \sum_{I \times J \in \mathcal{E}_k} \int |\widehat{f_I} \widehat{f_J}| \, d\mu \\
&\lesssim r^{\varepsilon + (3n-2s-2)/4} \sup_{\{k:k \leq k_r\}} \sum_{I \times J \in \mathcal{E}_k} \|f_I\|_2 \|f_J\|_2 \\
&\lesssim r^{\varepsilon + (3n-2s-2)/4} \sup_{\{k:k \leq k_r\}} \sum_{I \in \mathcal{D}_k} \|f_I\|_2^2 \\
&= r^{\varepsilon + (3n-2s-2)/4} \|f\|_2^2 \\
&= r^{\varepsilon + (3n-2s-2)/4},
\end{aligned}
$$

yielding the required estimate (16.7).

Let $\mathcal{P}$ be a partition of $\mathbb{R}^n$ into a union of disjoint $(c_4 \cdot 2^{-k}, \ldots, c_4 \cdot 2^{-k}, c_4)$-boxes P such that their longer sides are parallel to v where v is an arbitrarily fixed vector in $A_R(I) \cup A_R(J)$ with $|v| = r$. Choose $\psi \in \mathcal{S}(\mathbb{R}^n)$ such that

$$\int \psi = 1, \psi \geq 0 \quad \text{and} \quad \text{spt} \, \check{\psi} \subset Q_0.$$

We can get such a ψ as $\psi = \widehat{(\eta * \widetilde{\eta})} = |\widehat{\eta}|^2$ where η satisfies

$$\eta \geq 0, \text{spt} \, \eta \subset \frac{1}{2} Q_0 \quad \text{and} \quad \widetilde{\eta}(x) = \eta(-x) \quad \text{for all } x \in \mathbb{R}^n.$$

For any $P \in \mathcal{P}$ let $\psi_P = \psi \circ A_P^{-1}$. Then as every ψ_P is bounded and decays quickly off P,

$$\sum_{P \in \mathcal{P}} \psi_P^p \approx_p 1 \quad \text{for } 1 \leq p < \infty. \tag{16.21}$$

Let

$$f_{I,P} = \mathcal{F}^{-1}(\psi_P \widehat{f_I}) = \widecheck{\psi}_P * f_I \quad \text{for } P \in \mathcal{P},$$

so that $\widehat{f_{I,P}} = \psi_P \widehat{f_I}$. Set $S_{I,P} = \text{spt } f_{I,P}$. Then

$$S_{I,P} \subset \text{spt } f_I + \text{spt } \widecheck{\psi}_P.$$

As in (16.16) and (16.14) $\widecheck{\psi}_P(x) = \lambda_P \widecheck{\psi}(L(g^{-1}(x)))$, $\lambda_P \in \mathbb{C}$, and $\widetilde{P} = g(L^{-1}(Q_0))$. Thus the fact that spt $\widecheck{\psi} \subset Q_0$ implies that spt $\widecheck{\psi}_P \subset \widetilde{P}$, hence

$$S_{I,P} \subset \text{spt } f_I + \widetilde{P} \subset \overline{A_r(I)} + \widetilde{P}.$$

Define $\pi(x_1, \ldots, x_n) = (x_1, \ldots, x_{n-1})$. We shall now check that if c_4 is chosen large enough, but depending only on n,

$$
\begin{aligned}
S_{I,P} &\subset \overline{A_r(I)} + \widetilde{P} \\
&\subset \{x \in \mathbb{R}^n : \pi(x) \in \tfrac{5}{4}I, r - 2 < |x| < r + 2\} =: \widetilde{A}_r(I),
\end{aligned}
\tag{16.22}
$$

and similarly $S_{J,P} \subset \widetilde{A}_r(J)$. Notice that by (16.10)

$$d(\widetilde{A}_r(I), \widetilde{A}_r(J)) \geq 2^{-k-1}r \approx d(\widetilde{A}_r(I)) \approx d(\widetilde{A}_r(J)) \quad \text{for } I \times J \in \mathcal{E}_k.
\tag{16.23}$$

Recall that $\widetilde{P}$ is a $(1/c_4, 2^k/c_4, \ldots, 2^k/c_4)$-box centred at the origin whose shorter sides are parallel to v and

$$A_r(I) = \{x \in \mathbb{R}^n : \pi(x) \in I, x_n > 0, r - 1 < |x| < r + 1\},$$

where I is a cube in $\mathbb{R}^{n-1}$ with side-length $2^{-k}r$. We restricted k to $2^k \leq \sqrt{r}$, so $2^k \leq 2^{-k}r$ and $d(\widetilde{P}) \leq n2^k/c_4 \leq 2^{-k}r/4$ if $c_4 \geq 4n$. Then $\pi(x) \in \tfrac{5}{4}I$ for $x \in A_r(I) + \widetilde{P}$.

To verify $r - 2 < |x| < r + 2$ for the points of $A_r(I) + \widetilde{P}$ we may assume that $r > 2$ and $v = (0, \ldots, 0, r)$. Then $A_r(I) \subset A_r \cap B(v, 5n2^{-k}r)$ by (16.10) and we are reduced to showing that $r - 2 < |x| < r + 2$ whenever $x = y + z$ with $y \in A_r \cap B(v, 5n2^{-k}r)$ and $|z_j| \leq 2^k/c_4$ for $j = 1, \ldots, n-1$, $|z_n| \leq 1/c_4$. We have

$$|x|^2 = |y + z|^2 = |y|^2 + |z|^2 + 2y \cdot z,$$

where

$$r - 1 < |y| < r + 1, |y_j| \leq 5n2^{-k}r \quad \text{for } j = 1, \ldots, n-1, |y_n| \leq 2r,$$
$$|z_j| \leq 2^k/c_4 \leq \sqrt{r}/c_4 \quad \text{for } j = 1, \ldots, n-1, |z_n| \leq 1/c_4,$$

whence

$$|x|^2 = |y|^2 + |z|^2 + 2y \cdot z \le (r+1)^2 + nr/c_4^2 + 10nr/c_4 + 2r/c_4$$
$$\le (r+1)^2 + 15nr/c_4,$$
$$|x|^2 = |y|^2 + |z|^2 + 2y \cdot z \ge (r-1)^2 - 10nr/c_4 - 2r/c_4$$
$$\ge (r-1)^2 - 14nr/c_4,$$

and so

$$|x| \le \sqrt{(r+1)^2 + 15nr/c_4} \le r + 1 + (15nr/c_4)/(r+1) < r + 2,$$

and

$$|x| \ge \sqrt{(r-1)^2 - 14nr/c_4} \ge r - 1 - (14nr/c_4)/(r-1) > r - 2,$$

provided c_4 is sufficiently large; here we used the inequalities $\sqrt{1+a} \le 1 + a$ and $\sqrt{1-a} \ge 1 - a$ valid for $0 < a < 1$.

We shall first finish the proof of (16.20) for $n = 2$. We have $\varphi_{R_{I,J}} \cdot (f_I * f_J) \equiv f_I * f_J$, because spt $f_I * f_J \subset R_{I,J}$ and by (16.15) $\varphi_{R_{I,J}} \equiv 1$ on $R_{I,J}$. Hence using (16.21) with $p = 3$ and Schwartz's inequality,

$$\int |\widehat{f_I}\widehat{f_J}|\,d\mu = \int |\mathcal{F}(\varphi_{R_{I,J}} \cdot (f_I * f_J))|\,d\mu = \int |\widehat{\varphi_{R_{I,J}}} * \mathcal{F}(f_I * f_J)|\,d\mu$$

$$\le \iint |\widehat{\varphi_{R_{I,J}}}(x - y)\widehat{f_I * f_J}(y)|\,dy\,d\mu x$$

$$= \iint |\widehat{\varphi_{R_{I,J}}}(x - y)|\,d\mu x|\widehat{f_I}(y)\widehat{f_J}(y)|\,dy$$

$$= \int |\widehat{f_I}\widehat{f_J}|\left(|\widehat{\varphi_{R_{I,J}}}| * \mu\right) \approx \sum_{P \in \mathcal{P}} \int |\widehat{f_I}\widehat{f_J}|\mu_{R_{I,J}}\psi_P^3$$

$$= \sum_{P \in \mathcal{P}} \int |\widehat{f_I}\widehat{f_J}\psi_P^2|\mu_{R_{I,J}}\psi_P$$

$$\le \sum_{P \in \mathcal{P}} \left(\int |\widehat{f_I}\widehat{f_J}\psi_P^2|^2\right)^{\frac{1}{2}} \left(\int (\mu_{R_{I,J}}\psi_P)^2\right)^{\frac{1}{2}}$$

$$= \sum_{P \in \mathcal{P}} \left(\int |\widehat{f_{I,P}}\widehat{f_{J,P}}|^2\right)^{\frac{1}{2}} \left(\int (\mu_{R_{I,J}}\psi_P)^2\right)^{\frac{1}{2}}.$$

$$(16.24)$$

We have the geometric estimate:

$$\mathcal{L}^2((-S_{I,P} + x) \cap S_{J,P}) \le \mathcal{L}^2((-\widetilde{A}_r(I) + x) \cap \widetilde{A}_r(J)) \lesssim 2^k \quad \text{for all } x \in \mathbb{R}^2.$$
$$(16.25)$$

This follows from (16.22) and the following simple plane geometry lemma scaling by r and taking $l = 2^{-k}$, $\varrho = 2/r$. Recall that $A(\varrho)$ is the open ϱ-neighbourhood of A.

Lemma 16.4 *Let* $l, \varrho \in (0, 1)$ *and let* $S_1, S_2 \subset \{(x, y) \in S^1 : y > 0\}$ *be arcs with* $d(S_1, S_2) \geq l$. *Then*

$$\mathcal{L}^2(S_1(\varrho) \cap (S_2(\varrho) + z)) \leq C\varrho^2/l \quad \text{for all } z \in \mathbb{R}^2,$$

where C is an absolute constant.

Proof Observe that any two unit tangent vectors of S_1 and S_2 form an an angle $\gtrsim l$. Therefore, the length of the arc $S_z = S_1(\varrho) \cap (S_2 + z)$ is $\lesssim \varrho/l$. Since $S_1(\varrho) \cap (S_2(\varrho) + z)$ is essentially the ϱ-neighbourhood of S_z, the lemma follows from this. $\qquad\square$

Now we can estimate the right hand side of (16.24). For the first integral we have by Plancherel's theorem and Schwartz's inequality,

$$\int |\widehat{f_{I,P}} \widehat{f_{J,P}}|^2 = \int |(f_{I,P} * f_{J,P})|^2 = \int |(f_{I,P} * f_{J,P})|^2$$

$$= \int \left| \int_{(-S_{I,P}+x) \cap S_{J,P}} f_{I,P}(x - y) f_{J,P}(y) \, dy \right|^2 dx$$

$$\leq \int \left(\mathcal{L}^2((-S_{I,P}+x) \cap S_{J,P}) \int |f_{I,P}(x-y)|^2 |f_{J,P}(y)|^2 \, dy \right) dx.$$

Hence by (16.25),

$$\int |\widehat{f_{I,P}} \widehat{f_{J,P}}|^2 \lesssim 2^k \int |f_{I,P}|^2 \int |f_{J,P}|^2. \tag{16.26}$$

The second integral on the right hand side of (16.24) is estimated by using Lemma 3.17. To simplify notation let $R = R_{I,J}$. Recall that R is a $(c_3 r 2^{-2k}, c_3 r 2^{-k})$-rectangle, $\widetilde{R}$ is a $(c_3^{-1} r^{-1} 2^k, c_3^{-1} r^{-1} 2^{2k})$-rectangle centred at the origin and every $P \in \mathcal{P}$ is a $(c_4 2^{-k}, c_4)$-rectangle. Hence denoting the centre of P by c_P,

$$P \subset K\widetilde{R} + c_P \quad \text{with } K = c_3 c_4 r 2^{-2k}.$$

By (3.60),

$$\int \mu_R^2 \psi_P^2 \leq \int \|\mu_R\|_\infty \mu_R \psi_P^2 \lesssim (r 2^{-k})^{2-s} \int \mu_R \psi_P^2. \tag{16.27}$$

Furthermore by the definition of ψ_P,

$$\psi_P(x)^2 \lesssim 2^{-j(s+1)} \quad \text{for } x \in 2^j P \setminus 2^{j-1} P, \quad j \in \mathbb{N}.$$

Thus by (3.62)

$$
\int \mu_R \psi_P^2 \lesssim \int_P \mu_R + \sum_{j=1}^{\infty} 2^{-j(s+1)} \int_{2^j P \setminus 2^{j-1} P} \mu_R
$$

$$
\leq \int_{K\tilde{R}} \mu_R(x - c_P)\, dx + \sum_{j=1}^{\infty} 2^{-j(s+1)} \int_{2^j K\tilde{R}} \mu_R(x - c_P)\, dx
$$

$$
\lesssim K^s (r2^{-2k})^{-1}(r2^{-k})^{1-s} + \sum_{j=1}^{\infty} 2^{-j(s+1)} (2^j K)^s (r2^{-2k})^{-1}(r2^{-k})^{1-s}
$$

$$
= 2K^s r^{-s} 2^{k(1+s)} = 2(c_3 c_4)^s 2^{k(1-s)}.
$$

Combining this last estimate with (16.27) we get

$$
\int \mu_R^2 \psi_P^2 \lesssim r^{2-s} 2^{-k}. \tag{16.28}
$$

Finally using (16.24), (16.26),(16.28), Schwartz's inequality, Plancherel's theorem and (16.21), we obtain

$$
\int |\widehat{f_I}\, \widehat{f_J}|\, d\mu \lesssim \sum_{P \in \mathcal{P}} 2^{\frac{k}{2}} \left(\int |f_{I,P}|^2 \int |f_{J,P}|^2 \right)^{\frac{1}{2}} r^{1-\frac{s}{2}} 2^{-\frac{k}{2}}
$$

$$
\leq r^{1-\frac{s}{2}} \left(\sum_{P \in \mathcal{P}} \int |f_{I,P}|^2 \right)^{\frac{1}{2}} \left(\sum_{P \in \mathcal{P}} \int |f_{J,P}|^2 \right)^{\frac{1}{2}}
$$

$$
= r^{1-\frac{s}{2}} \left(\sum_{P \in \mathcal{P}} \int |\widehat{f_I}|^2 \psi_P^2 \right)^{\frac{1}{2}} \left(\sum_{P \in \mathcal{P}} \int |\widehat{f_J}|^2 \psi_P^2 \right)^{\frac{1}{2}}
$$

$$
\lesssim r^{1-\frac{s}{2}} \left(\int |\widehat{f_I}|^2 \right)^{\frac{1}{2}} \left(\int |\widehat{f_J}|^2 \right)^{\frac{1}{2}}
$$

$$
= r^{1-\frac{s}{2}} \|f_I\|_2 \|f_J\|_2,
$$

which proves (16.20) in the case $n = 2$.

It remains to consider $n \geq 3$. For that we need to use the following bilinear estimate which will be proven at the end of Chapter 25. The proof below works also for $n = 2$, but taking into account the difficulty of the proof of the bilinear estimate we have preferred to give a separate proof for $n = 2$.

Theorem 16.5 *Let* $(n-2)/2 < s < n, n \geq 2, q > \frac{4s}{n+2s-2}, c > 0$ *and* $\mu \in \mathcal{M}(\mathbb{R}^n)$ *such that*

$$
\mu(B(x, \varrho)) \leq \varrho^s \quad \text{for all } x \in \mathbb{R}^n, \quad \varrho > 0.
$$

There is a constant $\eta_n \in (0, 1)$ *depending only on* n *such that if* $0 < \eta < \eta_n, r > 1/\eta, f_j \in L^2(\mathbb{R}^n),$ spt $f_j \subset A_r \cap B(v_j, \eta r), |v_j| = r, j = 1, 2, c\eta r \leq d(A_r \cap B(v_1, \eta r)), A_r \cap B(v_2, \eta r)) \leq \eta r,$ *then*

$$\|\widehat{f_1}\widehat{f_1}\|_{L^q(\mu)} \leq C(n, s, q, c)\eta^{-1/q}(\eta r)^{n-1-s/q}\|f_1\|_2\|f_2\|_2.$$

Notice that $s > (n - 2)/2$ means that $\frac{4s}{n+2s-2} > 1$. To continue the proof of (16.20), let q be as in the above theorem. We first estimate by (16.21) and Hölder's inequality with $q' = q/(q - 1)$,

$$\int |\widehat{f_I}\widehat{f_J}|\,d\mu \approx \sum_{P \in \mathcal{P}} \int |\widehat{f_I}\widehat{f_J}|\psi_P^{2+1/q'}\,d\mu$$

$$= \sum_{P \in \mathcal{P}} \int |\widehat{f_{I,P}}\,\widehat{f_{J,P}}|\psi_P^{1/q'}\,d\mu \leq \sum_{P \in \mathcal{P}} \|\widehat{f_{I,P}}\,\widehat{f_{J,P}}\|_{L^q(\mu)}\|\psi_P\|_{L^1(\mu)}^{1/q'}.$$

As spt $f_{I,P} \subset \widetilde{A}_r(I)$ and spt $f_{J,P} \subset \widetilde{A}_r(J)$ by (16.22), $d(\mathrm{spt}\, f_{I,P}) \lesssim 2^{-k}r, d(\mathrm{spt}\, f_{J,P}) \lesssim 2^{-k}r$ and $d(\mathrm{spt}\, f_{I,P}, \mathrm{spt}\, f_{J,P}) \approx 2^{-k}r$ by (16.23), we can apply Theorem 16.5 with $\eta \approx 2^{-k}$ to estimate

$$\|\widehat{f_{I,P}}\,\widehat{f_{J,P}}\|_{L^q(\mu)} \lesssim 2^{k/q}(2^{-k}r)^{n-1-s/q}\|f_{I,P}\|_2\|f_{J,P}\|_2;$$

we can assume that r is big enough so that $\eta < \eta_n$. By the fast decay of ψ_P outside P we have

$$\int \psi_P\,d\mu \lesssim \sum_j 2^{-2sj} \int \chi_{2^jP}\,d\mu = \sum_j 2^{-2sj}\mu(2^jP).$$

The $(c_42^{j-k}, \ldots, c_42^{j-k}, c_42^j)$-box 2^jP can be covered with roughly 2^k balls of radius c_42^{j-k}. Hence $\mu(2^jP) \lesssim 2^k(c_42^{j-k})^s$ and so

$$\int \psi_P\,d\mu \lesssim \sum_j 2^{-2sj}2^k(2^{j-k})^s = 2^{k(1-s)}\sum_j 2^{-sj} \approx 2^{k(1-s)}.$$

Putting these estimates together we get

$$\int |\widehat{f_I}\widehat{f_J}|\,d\mu \lesssim \sum_{P \in \mathcal{P}} 2^{-k(n-1-s/q-1/q+(s-1)/q')}r^{n-1-s/q}\|f_{I,P}\|_2\|f_{J,P}\|_2.$$

The factor of $-k$ in the exponent of 2 is $n - 1 - s/q - 1/q + (s - 1)/q' = n - 2 + s - 2s/q > 0$ by the assumption on q. Furthermore, given $\varepsilon > 0$ we can choose q such that $1/q = \frac{n+2s-2}{4s} - \frac{\varepsilon}{s}$ which means that $n - 1 - s/q =$

$\varepsilon + \frac{3n-2s-2}{4}$. This leads to

$$\int |\widehat{f_I}\widehat{f_J}|\,d\mu \lesssim \sum_{P \in \mathcal{P}} r^{\varepsilon + \frac{3n-2s-2}{4}} \|f_{I,P}\|_2 \|f_{J,P}\|_2$$

$$\leq r^{\varepsilon + \frac{3n-2s-2}{4}} \left(\sum_{P \in \mathcal{P}} \|f_{I,P}\|_2^2 \right)^{1/2} \left(\sum_{P \in \mathcal{P}} \|f_{J,P}\|_2^2 \right)^{1/2}$$

$$= r^{\varepsilon + \frac{3n-2s-2}{4}} \|f_I\|_2 \|f_J\|_2 = r^{\varepsilon + \frac{3n-2s-2}{4}}.$$

This completes the proof of (16.20) and thus also of Theorems 15.1 and 16.1. □

We still have to explain how Theorem 15.5 follows from Theorem 16.1. Going through the proof of Theorem 16.1 one finds that once a big enough r is fixed the growth condition $\mu(B(x, \varrho)) \leq \varrho^s$ is only used for $\varrho \gtrsim 1/r$.

Let $\mu \in \mathcal{M}(B(0,1))$ with $I_s(\mu) < \infty$ and let $r > 1$. Replacing μ with $\mu(\mathbb{R}^n)^{-1}\mu$ we may assume that $\mu(\mathbb{R}^n) = 1$. Set

$$\theta(\mu, x) = \sup_{\varrho \geq 1/r} \varrho^{-s}\mu(B(x, \varrho))$$

for $x \in B(0, 1)$. Then

$$2^{-s} \leq \theta(\mu, x) \lesssim r^s.$$

Hence there are $\lesssim \log r$ values of $k \in \mathbb{Z}$ for which the set

$$E_k = \{x \in B(0, 1) : 2^{k-1} \leq \theta(\mu, x) < 2^k\}$$

is non-empty. Let $\mu_k = 2^{-k}\mu \llcorner E_k$. Then $\mu = \sum_k 2^k \mu_k$ and

$$\mu_k(B(x, \varrho)) \leq 2^s \varrho^s \quad \text{for } x \in \mathbb{R}^n, \quad \varrho \geq 1/r.$$

This inequality follows observing that if $\mu_k(B(x, \varrho)) > 0$, then $B(x, \varrho) \subset B(y, 2\varrho)$ for some $y \in E_k$. Moreover, $\mu_k(\mathbb{R}^n) \lesssim 2^{-2k}I_s(\mu)$. To see this cover E_k with balls $B_j = B(x_j, \varrho_j)$, $\varrho_j \geq 1/r$, such that $\sum_j \chi_{B_j} \lesssim 1$ and $\mu(B_j) \geq 2^{k-2}\varrho_j^s$. The existence of such balls follows from Besicovitch's covering theorem, see, for example, Mattila [1995], Theorem 2.7. Then

$$\int_{B_j} |x - y|^{-s}\,d\mu y \geq (2\varrho_j)^{-s}\mu(B_j) \geq 2^{k-2-s}.$$

for $x \in B_j$, whence

$$\mu_k(\mathbb{R}^n) \leq \sum_j \mu_k(B_j) \leq 2^{-k} \sum_j \mu(B_j)$$

$$\lesssim 2^{-2k} \sum_j \int_{B_j} \int_{B_j} |x - y|^{-s} \, d\mu y \, d\mu x \lesssim 2^{-2k} I_s(\mu).$$

Therefore, applying Theorem 16.1 to μ_k, we get

$$\sigma(\mu)(r) \lesssim \log r \sup_k \sigma(2^k \mu_k)(r) = 2^{2k} \log r \sup_k \sigma(\mu_k)(r) \lesssim r^{\varepsilon - \frac{n+2s-2}{4}} I_s(\mu).$$

Theorem 15.5 follows from this.

16.1 Further comments

As mentioned before Theorem 16.1 is due to Wolff [1999], for $n = 2$ and due to Erdoğan [2005], for $n > 2$. Erdoğan [2004] first gave a different proof from that of Wolff, and some related results, when $n = 2$ and in Erdoğan [2006] he gave a weaker estimate than the one in Erdoğan [2005].

Erdoğan and D. M. Oberlin [2013] used methods similar to those presented in this chapter to investigate L^2 averages of Fourier transforms of measures over certain polynomial curves in the plane.

17

Sobolev spaces, Schrödinger equation and spherical averages

We begin this chapter by giving a Fourier analytic definition of Sobolev spaces. Then we study the Hardy–Littlewood maximal function and convergence of ball averages of Sobolev functions. One of the main goals of this chapter is to show, using results of Barceló, Bennett, Carbery and Rogers [2011], how estimates of spherical averages can be used to derive information on the behaviour of some integral operators related to partial differential equations.

17.1 Sobolev spaces and the Hardy–Littlewood maximal function

For $\sigma \geq 0$ we define the fractional order *Sobolev space* $H^\sigma(\mathbb{R}^n)$ as

$$H^\sigma(\mathbb{R}^n) = \{f \in L^2(\mathbb{R}^n) : \|f\|_{H^\sigma(\mathbb{R}^n)} < \infty\},$$

where

$$\|f\|_{H^\sigma(\mathbb{R}^n)} = \left(\int |\widehat{f}(\xi)|^2 (1 + |\xi|^2)^\sigma \, d\xi \right)^{1/2}$$

is the Sobolev norm. Recall that in Chapter 5 we already introduced this norm for measures. There we also observed that for integers σ these spaces are classical Sobolev spaces with the weak derivatives of order σ in L^2. Of course, $H^0(\mathbb{R}^n) = L^2(\mathbb{R}^n)$.

For $\alpha \in \mathbb{R}$ we can define, at least as a tempered distribution, the *Bessel kernel* G_α by

$$\widehat{G_\alpha}(\xi) = (1 + |\xi|^2)^{-\alpha/2}, \quad \xi \in \mathbb{R}^n.$$

219

Then $H^\sigma(\mathbb{R}^n)$ consists of all *Bessel potentials* $G_\sigma * g$ of functions $g \in L^2(\mathbb{R}^n)$. These and related potential spaces have been extensively studied for example by Adams and Hedberg [1996].

The case when $\sigma > n/2$ is simple, for the questions we are interested in. Then, if $f \in H^\sigma(\mathbb{R}^n)$ and, for example, $f \in L^1(\mathbb{R}^n)$, it follows from Schwartz's inequality that $\widehat{f} \in L^1(\mathbb{R}^n)$ and thus f is continuous (recall Corollary 3.1). Therefore we shall only consider $\sigma \leq n/2$. We shall prove estimates using measures with finite energy, in particular $(n - 2\sigma)$-energy. Hence we shall often assume $\sigma < n/2$.

Notice that

$$H^{\sigma_2}(\mathbb{R}^n) \subset H^{\sigma_1}(\mathbb{R}^n) \quad \text{if} \quad \sigma_1 < \sigma_2. \tag{17.1}$$

Another easy observation is that smooth functions are dense in Sobolev spaces:

Lemma 17.1 $C_0^\infty(\mathbb{R}^n)$ *is dense in* $H^\sigma(\mathbb{R}^n)$.

Proof Let $\{\psi_\varepsilon : \varepsilon > 0\}$ be a smooth approximate identity; $\psi_\varepsilon(x) = \varepsilon^{-n}\psi(x/\varepsilon)$ where ψ is a C^∞ function with spt $\psi \subset B(0, 1)$ and $\int \psi = 1$. If $f \in H^\sigma(\mathbb{R}^n)$, then $f_\varepsilon = \psi_\varepsilon * f$ is infinitely differentiable, $\widehat{f_\varepsilon}(\xi) = \widehat{\psi}(\varepsilon\xi)\widehat{f}(\xi)$ and

$$\|f - f_\varepsilon\|_{H^\sigma(\mathbb{R}^n)} = \left(\int |(1 - \widehat{\psi}(\varepsilon\xi))\widehat{f}(\xi)|^2 (1 + |\xi|^2)^\sigma \, d\xi \right)^{1/2} \to 0$$

as $\varepsilon \to 0$. Thus C^∞-functions are dense in $H^\sigma(\mathbb{R}^n)$. This would be enough for us and we leave it as an exercise to get the approximation by compactly supported C^∞-functions. $\qquad\square$

The following lemma is useful for studying various integral operators.

Lemma 17.2 Let $0 < \sigma < n/2, \mu \in \mathcal{M}(B(0, 1))$, *and let* $\eta : \mathbb{R}^n \to \mathbb{C}, \varphi : \mathbb{R}^n \to \mathbb{R}$ *and* $\varrho : \mathbb{R}^n \to (0, \infty)$ *be Borel functions such that* $\eta \in L^2(\mathbb{R}^n), |\widehat{\eta}(x)| \lesssim (1 + |x|)^{-n-1}$ *and for some* $\alpha > n/2 - \sigma, |\eta(x)| \lesssim (1 + |x|)^{-\alpha}$ *for* $x \in \mathbb{R}^n$. *Then*

$$\int \left| \int \eta(\varrho(x)\xi)\widehat{f}(\xi)e^{2\pi i(x\cdot\xi + \varphi(\xi))} \, d\xi \right| d\mu x \leq C(n, \sigma, \eta)\sqrt{I_{n-2\sigma}(\mu)}\|f\|_{H^\sigma(\mathbb{R}^n)} \tag{17.2}$$

for all $f \in H^\sigma(\mathbb{R}^n)$. *In particular, the constant depends neither on* φ *nor on* ϱ.

Proof We may assume that ϱ is bounded below by some positive constant c; consider the measures $\mu \llcorner \{x : \varrho(x) > 1/j\}, j = 1, 2, \ldots$. By duality it suffices to show that for any Borel function $w \in L^\infty(\mu)$ with $\|w\|_{L^\infty(\mu)} \leq 1$

we have

$$\left| \iint \eta(\varrho(x)\xi)\widehat{f}(\xi)e^{2\pi i(x\cdot\xi+\varphi(\xi))}\,d\xi\,w(x)\,d\mu x \right| \lesssim \sqrt{I_{n-2\sigma}(\mu)}\|f\|_{H^\sigma(\mathbb{R}^n)}.$$
(17.3)

We would like to apply Fubini's theorem. This is legitimate because

$$\iint |\eta(\varrho(x)\xi)\widehat{f}(\xi)|\,d\xi\,d\mu x$$

$$\leq \left(\iint |\eta(\varrho(x)\xi)|^2\,d\xi\,d\mu x \right)^{1/2} \left(\iint |\widehat{f}(\xi)|^2\,d\xi\,d\mu x \right)^{1/2}$$

$$= \left(\iint \varrho(x)^{-n}|\eta(\zeta)|^2\,d\zeta\,d\mu x \right)^{1/2} \|f\|_2 \mu(\mathbb{R}^n)^{1/2}$$

$$\leq c^{-n/2}\mu(\mathbb{R}^n)\|\eta\|_2\|f\|_2 < \infty.$$

Writing $e^{2\pi i(x\cdot\xi+\varphi(\xi))} = e^{2\pi i\varphi(\xi)}e^{2\pi ix\cdot\xi}$ and using Schwartz's inequality, we see then that the square of the left hand side of (17.3) is bounded by

$$\int |\widehat{f}(\xi)|^2|\xi|^{2\sigma}\,d\xi \int \left| \int \eta(\varrho(x)\xi)e^{2\pi ix\cdot\xi}w(x)\,d\mu x \right|^2 |\xi|^{-2\sigma}\,d\xi.$$

Expressing the inner integral squared as a double integral and using again Fubini's theorem, which we can by our assumptions on η, we find that it suffices to show that

$$\iiint \eta(\varrho(x)\xi)\overline{\eta(\varrho(y)\xi)}e^{2\pi i(x-y)\cdot\xi}|\xi|^{-2\sigma}\,d\xi\,w(x)\overline{w(y)}\,d\mu x\,d\mu y \lesssim I_{n-2\sigma}(\mu).$$

For this it is enough that for $0 < \sigma < n/2$,

$$\left| \int \eta(r_1\xi)\overline{\eta(r_2\xi)}|\xi|^{-2\sigma}e^{2\pi i(x-y)\cdot\xi}\,d\xi \right| \lesssim |x-y|^{2\sigma-n}$$
(17.4)

for $x, y \in \mathbb{R}^n$ and for all positive numbers r_1 and r_2. In the left hand side we have the value at $y - x$ of the Fourier transform of $r_1^{-n}r_2^{-n}\eta_{r_1}\overline{\eta_{r_2}}k_{2\sigma}$ where $\eta_r(\xi) = r^n\eta(r\xi)$ and $k_{2\sigma}$ is the Riesz kernel. By Lemma 3.9 the inequality (17.4) reduces to

$$|(r_1^{-n}\widehat{\eta_{r_1}}) * (r_2^{-n}\widehat{\overline{\eta_{r_2}}}) * k_{n-2\sigma}(z)| \lesssim |z|^{2\sigma-n}$$

for $z \in \mathbb{R}^n$ and for all positive numbers r_1 and r_2. For this we only need that

$$|(r^{-n}\widehat{\eta_r}) * k_{n-2\sigma}(x)| = \left| r^{-n} \int \widehat{\eta}((x-y)/r)|y|^{2\sigma-n}\,dy \right| \lesssim |x|^{2\sigma-n}$$

for $x \in \mathbb{R}^n$ and $r > 0$. By change of variable $z = y/r$ this reduces to

$$\left| \int \widehat{\eta}(x-z)|z|^{2\sigma-n}\,dz \right| \lesssim |x|^{2\sigma-n}$$

for $x \in \mathbb{R}^n$. This is easily checked by our decay assumption on $\widehat{\eta}$:

$$\left| \int \widehat{\eta}(x-z)|z|^{2\sigma-n} \, dz \right| \lesssim \sum_{j=1}^{\infty} 2^{-(n+1)j} \int_{B(x,2^j)} |z|^{2\sigma-n} \, dz \lesssim |x|^{2\sigma-n},$$

where we used the elementary inequality

$$r^{-n} \int_{B(x,r)} |y|^{2\sigma-n} \, dy \lesssim |x|^{2\sigma-n},$$

valid for $x \in \mathbb{R}^n$ and and $r > 0$, whose easy verification we leave to the reader.
$\square$

We define a *Hardy–Littlewood maximal function* $\widetilde{M} f$ by

$$\widetilde{M} f(x) = \sup_{r>0} \left| \alpha(n)^{-1} r^{-n} \int_{B(x,r)} f \right|, \quad x \in \mathbb{R}^n,$$

where $\alpha(n) = \mathcal{L}^n(B(0,1))$. Notice that this is not the usual definition; we have absolute value signs outside the integral. This is because we have to be a little careful in the coming estimates, since the absolute value of a function in $H^\sigma(\mathbb{R}^n)$ need not always be in $H^\sigma(\mathbb{R}^n)$. Lemma 17.2 leads to information about the boundedness of $\widetilde{M} f$ and further to a refinement (not a generalization) of Lebesgue's differentiation theorem:

Theorem 17.3 *Let $0 < \sigma < n/2$ and $\mu \in \mathcal{M}(B(0,1))$ with $I_{n-2\sigma}(\mu) < \infty$. Then*

$$\|\widetilde{M} f\|_{L^1(\mu)} \le C(n,\sigma)\sqrt{I_{n-2\sigma}(\mu)} \|f\|_{H^\sigma(\mathbb{R}^n)}$$

for all $f \in H^\sigma(\mathbb{R}^n)$. Moreover, the finite limit $\lim_{r \to 0} \alpha(n)^{-1} r^{-n} \int_{B(x,r)} f \, d\mathcal{L}^n$ exists for μ almost all $x \in \mathbb{R}^n$.

Proof Let η be the inverse transform of the characteristic function of the unit ball; $\widehat{\eta} = \chi_{B(0,1)}$, and define $\eta_r(x) = r^n \eta(rx)$ for $x \in \mathbb{R}^n, r > 0$. Then $\widehat{\eta_r}(x) = \widehat{\eta}(x/r)$ and

$$\int_{B(x,r)} f \, d\mathcal{L}^n = \widehat{\eta_r} * f(x).$$

Thus we can easily find a Borel function $r : \mathbb{R}^n \to (0, \infty)$ such that

$$\widetilde{M} f(x) \lesssim \sup_{r>0} |r^{-n} \widehat{\eta_r} * f(x)| \le |2r(x)^{-n} \widehat{\eta_{r(x)}} * f(x)| \quad \text{for all } x \in \mathbb{R}^n.$$

Observe that by the Fourier inversion formula

$$\int \eta(r(x)\xi) \widehat{f}(\xi) e^{2\pi i x \cdot \xi} \, d\xi = r(x)^{-n} \mathcal{F}^{-1}(\eta_{r(x)} \widehat{f})(x) = r(x)^{-n} \widehat{\eta_{r(x)}} * f(x).$$

It follows that

$$\|\widetilde{M}f\|_{L^1(\mu)} \lesssim \int \left| \int \eta(r(x)\xi)\widehat{f}(\xi)e^{2\pi i x\cdot\xi}\, d\xi \right| d\mu x. \qquad (17.5)$$

The function η satisfies the decay condition of Lemma 17.2 by (3.40). Hence the first part of the theorem follows now immediately from Lemma 17.2. The second part follows with a standard argument. Let $\lambda > 0$, $\varepsilon > 0$ and choose, by Lemma 17.1, a smooth function φ for which $\|f - \varphi\|_{H^\sigma(\mathbb{R}^n)} < \lambda\varepsilon$. Since $\lim_{r\to 0}\alpha(n)^{-1}r^{-n}\int_{B(x,r)}\varphi\,d\mathcal{L}^n = \varphi(x)$, we have

$$\limsup_{r_1 < r_2, r_2 \to 0} \left| \alpha(n)^{-1}r_2^{-n}\int_{B(x,r_2)} f\,d\mathcal{L}^n - \alpha(n)^{-1}r_1^{-n}\int_{B(x,r_1)} f\,d\mathcal{L}^n \right|$$

$$= \limsup_{r_1 < r_2, r_2 \to 0} \left| \alpha(n)^{-1}r_2^{-n}\int_{B(x,r_2)} (f - \varphi)\,d\mathcal{L}^n - \alpha(n)^{-1}r_1^{-n}\int_{B(x,r_1)} (f - \varphi)\,d\mathcal{L}^n \right|$$

$$\leq 2\widetilde{M}(f - \varphi)(x).$$

Hence by the first part of the theorem we see that

$$\mu\left(\left\{ x : \limsup_{r_1 < r_2, r_2 \to 0} \left| \alpha(n)^{-1}r_2^{-n}\int_{B(x,r_2)} f\,d\mathcal{L}^n - \alpha(n)^{-1}r_1^{-n}\int_{B(x,r_1)} f\,d\mathcal{L}^n \right| > \lambda \right\}\right)$$

$$\leq \mu(\{x : 2\widetilde{M}(f - \varphi)(x) > \lambda\}) \lesssim \sqrt{I_{n-2\sigma}(\mu)}\|f - \varphi\|_{H^\sigma(\mathbb{R}^n)}/\lambda$$

$$\leq \sqrt{I_{n-2\sigma}(\mu)}\varepsilon.$$

Thus

$$\limsup_{r_1 < r_2, r_2 \to 0} \left| \alpha(n)^{-1}r_2^{-n}\int_{B(x,r_2)} f\,d\mathcal{L}^n - \alpha(n)^{-1}r_1^{-n}\int_{B(x,r_1)} f\,d\mathcal{L}^n \right| = 0$$

for μ almost all $x \in \mathbb{R}^n$ and the Cauchy criterion implies that $\alpha(n)^{-1}r^{-n} \times \int_{B(x,r)} f\,d\mathcal{L}^n$ converges for such points x. $\qquad\square$

Corollary 17.4 *If $0 < \sigma \leq n/2$, then for all $f \in H^\sigma(\mathbb{R}^n)$,*

$$\dim\left(\mathbb{R}^n \setminus \left\{ x : \exists \lim_{r\to 0}\alpha(n)^{-1}r^{-n}\int_{B(x,r)} f\,d\mathcal{L}^n \in \mathbb{C} \right\}\right) \leq n - 2\sigma.$$

Proof Because of (17.1) we may assume that $\sigma < n/2$. It is easy to see that the set B where $\alpha(n)^{-1}r^{-n}\int_{B(x,r)} f$ fails to converge is a Borel set. If we had $\dim B > n - 2\sigma$ we could find by Theorem 2.8 $\mu \in \mathcal{M}(B)$ with $I_{n-2\sigma}(\mu) < \infty$ which would contradict Theorem 17.3. $\qquad\square$

The upper bound $n - 2\sigma$ is sharp. This follows from the relations between Bessel potentials, Bessel capacities and Hausdorff dimension, see Adams and Hedberg [1996], in particular Proposition 2.3.7 and Corollary 5.1.14.

Of course in the above corollary we cannot have that the averages converge to f outside a set of Hausdorff dimension at most $n - 2\sigma$, because f is only defined almost everywhere. But this corollary gives a way to define f more precisely. We shall now turn to similar questions for integral operators related to partial differential equations.

17.2 Schrödinger equation and related integral operators

Consider the Schrödinger equation

$$2\pi i \partial_t u(x, t) + \Delta_x u(x, t) = 0, \quad x \in \mathbb{R}^n, t \in \mathbb{R}.$$

Here t represents time and $u_0 = u(\cdot, 0)$ gives the initial values of the solution. When $u_0 \in \mathcal{S}(\mathbb{R}^n)$ the solution can be written explicitly as

$$u(x, t) = \int \widehat{u_0}(\xi) e^{2\pi i (x \cdot \xi - t|\xi|^2)} \, d\xi. \tag{17.6}$$

When $u_0 \in L^2(\mathbb{R}^n)$, this integral can be interpreted in the L^2 sense to mean that for a fixed t the Fourier transform of $u(\cdot, t)$ is $e^{-2\pi i t|\xi|^2} \widehat{u_0}(\xi)$. Then also $u(x, t) \to u_0(x)$ in L^2 as $t \to 0$. We shall be interested in pointwise convergence.

More generally, let $m \geq 1$ and

$$u(x, t) = \int \widehat{u_0}(\xi) e^{2\pi i (x \cdot \xi - t|\xi|^m)} \, d\xi. \tag{17.7}$$

When $m = 1$, these integrals are closely related to the solutions of the wave equation

$$\partial_t^2 u(x, t) - c\Delta_x u(x, t) = 0, \quad x \in \mathbb{R}^n, t \in \mathbb{R}.$$

For more details, see, for example, Section 4.3 of the book Evans [1998].

Below we shall consider these integral operators for general $m \geq 1$. We now fix such an m for the rest of this chapter.

One of the main questions is: when and in what sense does $u(x, t)$ converge to $u(x, 0)$ when t approaches 0? The natural setting beyond smooth initial values to study this and related questions is that of Sobolev spaces.

Again, the case $\sigma > n/2$ is simple: in that case if $u_0 \in H^\sigma(\mathbb{R}^n)$, then $\widehat{u_0} \in L^1(\mathbb{R}^n)$ and the function u defined by (17.6) is continuous in $\mathbb{R}^{n+1}$.

It is natural to approach the convergence and boundedness properties of the integrals in (17.6) by introducing the approximating operators S_t^N defined for

$t \in \mathbb{R}$ and $N = 1, 2, \ldots$, by

$$S_t^N f(x) = \int \psi(\xi/N)\widehat{f}(\xi)e^{2\pi i(x\cdot\xi - t|\xi|^m)} d\xi, \quad x \in \mathbb{R}^n.$$

Here most of the time ψ could be any real valued radial Schwartz function in $\mathbb{R}^n$ with $\psi(0) = 1$, but for slight convenience let us fix it to be

$$\psi(x) = \psi(|x|) = e^{-|x|^2}.$$

If $u_0 \in S(\mathbb{R}^n)$, it is obvious that

$$\lim_{N\to\infty} S_t^N u_0(x) = u(x, t) \quad \text{for all } x \in \mathbb{R}^n, \quad t \in \mathbb{R}, \tag{17.8}$$

where $u(x, t)$ is given by (17.7). The point here is that if u_0 only is in some Sobolev space $H^\sigma(\mathbb{R}^n)$, one can still get, depending on σ, almost everywhere convergence, or even almost everywhere convergence with respect to a certain Hausdorff measure. If $\widehat{u_0}$ is not integrable, we cannot define $u(x, t)$ by the formula (17.7), but we can take (17.8) as the definition of $u(x, t)$ whenever the limit exists.

Theorem 17.5 *Let* $0 < \sigma < n/2, t \in \mathbb{R}$ *and* $\mu \in \mathcal{M}(B(0, 1))$ *with* $I_{n-2\sigma}(\mu) < \infty$. *Then*

$$\| \sup_{N\geq 1} |S_t^N f| \|_{L^1(\mu)} \leq C(n, \sigma)\sqrt{I_{n-2\sigma}(\mu)}\|f\|_{H^\sigma(\mathbb{R}^n)}$$

for all $f \in H^\sigma(\mathbb{R}^n)$. *Moreover, the finite limit* $\lim_{N\to\infty} S_t^N f(x)$ *exists for* μ *almost all* $x \in \mathbb{R}^n$.

Proof We can easily find a Borel function $\varrho : \mathbb{R}^n \to (0, \infty)$ such that

$$\sup_{N\geq 1} |S_t^N f(x)| \leq 2 \left| \int \psi(\varrho(x)|\xi|)\widehat{f}(\xi)e^{2\pi i(x\cdot\xi - t|\xi|^m)} d\xi \right| \quad \text{for all } x \in \mathbb{R}^n.$$

The first part of the theorem follows now immediately from Lemma 17.2 choosing $\varphi(\xi) = -t|\xi|^m$.

The second part follows as the second part of Theorem 17.3. □

Corollary 17.6 *If* $0 < \sigma \leq n/2$ *and* $t \in \mathbb{R}$, *then for all* $f \in H^\sigma(\mathbb{R}^n)$,

$$\dim \left(\mathbb{R}^n \setminus \left\{ x : \exists \lim_{N\to\infty} S_t^N f(x) \in \mathbb{C} \right\} \right) \leq n - 2\sigma.$$

This follows by the same argument as Corollary 17.4.

We shall now consider boundedness over $t \in \mathbb{R}$ and convergence when $t \to 0$. Recall from Section 15.3 that $t_n(s)$ is the supremum of the numbers τ

such that

$$\sigma(\mu)(r) = \int |\widehat{\mu}(rv)|^2 \, d\sigma^{n-1}v \lesssim r^{-\tau} I_s(\mu)$$

for all $\mu \in \mathcal{M}(B(0,1))$.

Lemma 17.7 *If $f \in L^2(\mathbb{R}^n)$ and $x \in \mathbb{R}^n$, then*

$$\sup_{t \in \mathbb{R}} \sup_{N \geq 1} S_t^N f(x) = \sup_{t \in \mathbb{Q}} \sup_{N \geq 1} S_t^N f(x).$$

In particular, the function $\sup_{t \in \mathbb{R}} \sup_{N \geq 1} S_t^N f$ is a Borel function.

Proof If $\sup_{t \in \mathbb{R}} \sup_{N \geq 1} S_t^N f(x) > a$, then $S_t^N f(x) > a$ for some $N \geq 1$ and $t \in \mathbb{R}$, whence also for some $t \in \mathbb{Q}$. Hence $\sup_{t \in \mathbb{Q}} \sup_{N \geq 1} S_t^N f(x) > a$ and the lemma follows. $\qquad\square$

Theorem 17.8 *Let $n \geq 2, 0 < s < n, 2\sigma > n - t_n(s)$ and $\mu \in \mathcal{M}(B(0,1))$ with $I_s(\mu) < \infty$. Then*

$$\left\| \sup_{t \in \mathbb{R}} \sup_{N \geq 1} |S_t^N f| \right\|_{L^1(\mu)} \leq C(n, \sigma, s)\sqrt{I_s(\mu)} \| f \|_{H^\sigma(\mathbb{R}^n)}$$

for all $\mu \in \mathcal{M}(B(0,1))$ and $f \in H^\sigma(\mathbb{R}^n)$. Moreover, for μ almost all $x \in \mathbb{R}^n$ the finite limit $\lim_{N \to \infty} S_t^N f(x)$ exists for all $t \in \mathbb{R}$. That is, the μ exceptional set is independent of t.

Proof Choose $0 < \tau < t_n(s)$ such that $2\sigma > n - \tau > 0$. Using polar coordinates we have

$$|S_t^N f(x)| = \left| \int \psi(|\xi|/N) \widehat{f}(\xi) e^{2\pi i(x \cdot \xi - t|\xi|^m)} \, d\xi \right|$$

$$= \left| \int_0^\infty \psi(r/N) r^{n-1} e^{-2\pi i t r^m} \int_{S^{n-1}} \widehat{f}(rv) e^{2\pi i r x \cdot v} \, d\sigma^{n-1}v \, dr \right|$$

$$\lesssim \int_0^\infty r^{n-1} \left| \int_{S^{n-1}} \widehat{f}(rv) e^{2\pi i r x \cdot v} \, d\sigma^{n-1}v \right| \, dr.$$

Thus by Fubini's theorem

$$\left\| \sup_{t \in \mathbb{R}} \sup_{N \geq 1} |S_t^N f| \right\|_{L^1(\mu)} \leq \int \int_0^\infty r^{n-1} \left| \int_{S^{n-1}} \widehat{f}(rv) e^{2\pi i r x \cdot v} \, d\sigma^{n-1}v \right| \, dr \, d\mu x$$

$$= \int_0^\infty r^{n-1} \int |\widehat{f_r \sigma^{n-1}}(x)| \, d\mu x \, dr,$$

where we have written $f_r(v) = \widehat{f}(-rv)$. Using the dual characaterization of $t_n(s)$ in Proposition 15.11 and Schwartz's inequality we get

$$\left\| \sup_{t \in \mathbb{R}} \sup_{N \geq 1} |S_t^N f| \right\|_{L^1(\mu)} \lesssim \sqrt{I_s(\mu)} \int_0^\infty r^{n-1-\tau/2} \|f_r\|_{L^2(S^{n-1})} \, dr$$

$$\leq \sqrt{I_s(\mu)} \left(\int_0^\infty \frac{r^{n-1-\tau}}{(1+r^2)^\sigma} \, dr \right)^{1/2} \left(\int_0^\infty \|f_r\|_{L^2(S^{n-1})}^2 (1+r^2)^\sigma r^{n-1} \, dr \right)^{1/2}.$$

Here the first integral is finite by the choice of τ. The second integral is

$$\int_0^\infty \|f_r\|_{L^2(S^{n-1})}^2 (1+r^2)^\sigma r^{n-1} \, dr = \int_0^\infty \int |\widehat{f}(rv)|^2 \, d\sigma^{n-1} v (1+r^2)^\sigma r^{n-1} \, dr$$

$$= \int |\widehat{f}(\xi)|^2 (1+|\xi|^2)^\sigma \, d\xi = \|f\|_{H^\sigma(\mathbb{R}^n)}^2.$$

Thus

$$\left\| \sup_{t \in \mathbb{R}} \sup_{N \geq 1} |S_t^N f| \right\|_{L^1(\mu)} \lesssim \sqrt{I_s(\mu)} \|f\|_{H^\sigma(\mathbb{R}^n)}.$$

The proof of the second statement is similar to the proof of the second part of Theorem 17.3. Set now

$$S^* f(x) = \sup_{t \in \mathbb{R}} \sup_{N \geq 1} |S_t^N f(x)|$$

and

$$S^{**} f(x) = \sup_{t \in \mathbb{R}} \limsup_{N_1 < N_2, N_1 \to \infty} |S_t^{N_2} f(x) - S_t^{N_1} f(x)|.$$

Then $S^{**} f(x) \leq 2S^* f(x)$. Let $\lambda > 0$, $\varepsilon > 0$ and choose $\varphi \in \mathcal{S}(\mathbb{R}^n)$ for which $\|f - \varphi\|_{H^\sigma(\mathbb{R}^n)} < \lambda\varepsilon$. Then $S^{**}\varphi(x) = 0$ for all $x \in \mathbb{R}^n$. Therefore

$$\mu(\{x : S^{**} f(x) > \lambda\}) = \mu(\{x : S^{**}(f - \varphi)(x) > \lambda\})$$

$$\leq \mu(\{x : 2S^*(f - \varphi)(x) > \lambda\})$$

$$\lesssim \lambda^{-1} \sqrt{I_s(\mu)} \|f - \varphi\|_{H^\sigma(\mathbb{R}^n)} \leq \sqrt{I_s(\mu)}\varepsilon.$$

It follows that $\mu(B) = 0$ where $B = \{x : S^{**} f(x) > 0\}$. By the definition of $S^{**} f(x)$, for all $x \in \mathbb{R}^n \setminus B$ the sequence $(S_t^N f(x))_{N \geq 1}$ is a Cauchy sequence for all $t \in \mathbb{R}$, from which the second statement follows. $\qquad \square$

Corollary 17.9 *Let* $n \geq 2, 0 < s < n, 2\sigma > n - t_n(s)$ *and* $f \in H^\sigma(\mathbb{R}^n)$. *Then there exists a Borel set* $B \subset \mathbb{R}^n$ *such that* $\dim B \leq s$ *and the limit* $\lim_{N \to \infty} S_t^N f(x) \in \mathbb{C}$ *exists for all* $x \in \mathbb{R}^n \setminus B$ *and all* $t \in \mathbb{R}$.

Proof Define $S^{**} f(x)$ and $B = \{x : S^{**} f(x) > 0\}$ as above. Since, as before, the supremum over t can be taken over rationals, B is a Borel set. Moreover, dim $B \leq s$, because otherwise we could find $\mu \in \mathcal{M}(B)$ with $I_s(\mu) < \infty$ and this would contradict Theorem 17.8. The corollary follows from this.   □

Let us now consider the convergence as t tends to 0 of $u(x, t)$ as in (17.6) with Sobolev initial values u_0. More precisely, as mentioned earlier, we define $u(x, t)$ as the limit in (17.8) whenever it exists. If $n \geq 2, 0 < s < n, 2\sigma > n - t_n(s)$ and $u_0 \in H^\sigma(\mathbb{R}^n)$, we then have by Corollary 17.9 that $u(x, t)$ exists for every $t \in \mathbb{R}$ for x outside a set of Hausdorff dimension at most s, and also for μ almost $x \in \mathbb{R}^n$, if $\mu \in \mathcal{M}(\mathbb{R}^n)$ with $I_s(\mu) < \infty$.

Theorem 17.10 *If $n \geq 2, 0 < s < n$ and $2\sigma > n - t_n(s)$, then for all $u_0 \in H^\sigma(\mathbb{R}^n)$,*

$$\dim\{x : u(x, t) \not\to u_0(x) \text{ as } t \to 0\} \leq s.$$

Proof If $v_0 \in \mathcal{S}(\mathbb{R}^n)$, then for the corresponding function v, $v(x, t)$ tends to $v_0(x)$ as $t \to 0$ for all $x \in \mathbb{R}^n$. Moreover, if x is such that $u(x, t)$ is defined for all t,

$$\limsup_{t \to 0} |u(x, t) - u_0(x)|$$

$$= \limsup_{t \to 0} |(u(x, t) - v(x, t)) - (u_0(x) - v_0(x))| \leq 2S^*(u_0 - v_0)(x)$$

with the notation of the proof of Theorem 17.8. The theorem follows from these observations with same arguments as above.   □

We somewhat reformulate the previous results. Recall that S_t^N depends also on $m \geq 1$. For $0 < \sigma < n/2$ denote by $s_{m,n}(\sigma)$ the infimum of the positive numbers s such that

$$\left\| \sup_{t \in \mathbb{R}} \sup_{N \geq 1} |S_t^N f| \right\|_{L^1(\mu)} \lesssim \sqrt{I_s(\mu)} \|f\|_{H^\sigma(\mathbb{R}^n)}$$

for all $\mu \in \mathcal{M}(B(0, 1))$, $f \in H^\sigma(\mathbb{R}^n)$. Then, as before, if $u_0 \in H^\sigma(\mathbb{R}^n)$,

$$\dim\{x : u(x, t) \not\to u_0(x) \text{ as } t \to 0\} \leq s_{m,n}(\sigma).$$

Taking into account Theorem 17.8 and the lower bounds we had for $t_n(s)$ in Theorem 15.7, we obtain

Theorem 17.11 *Let $n \geq 2$ and $0 < \sigma < n/2$. Then*

$$s_{m,n}(\sigma) \leq n + 1 - 2\sigma \quad \text{if} \quad 1/2 < \sigma \leq n/4,$$

$$s_{m,n}(\sigma) \leq 3n/2 + 1 - 4\sigma \quad \text{if} \quad n/4 < \sigma \leq (n+1)/4,$$

$$s_{m,n}(\sigma) \leq n - 2\sigma \quad \text{if} \quad (n+1)/4 < \sigma \leq n/2.$$

Proof By Theorem 17.10 $s_{m,n}(\sigma) \le s$ when $2\sigma \ge n - t_n(s)$. All we have to do is to substitute $t_n(s)$ from Theorem 15.7, solve s in terms of σ, and check the ranges of σ corresponding to the ranges of s in Theorem 15.7. $\square$

We now prove that the last estimate is sharp.

Theorem 17.12 *We have for all n, m and σ,*

$$s_{m,n}(\sigma) \ge n - 2\sigma.$$

In particular,

$$s_{m,n}(\sigma) = n - 2\sigma \text{ if } n \ge 2 \text{ and } (n+1)/4 \le \sigma < n/2.$$

Proof Let f_N be the Fourier transform of the characteristic function of the ball $B(0, N)$ and $\mu = N^n \mathcal{L}^n \llcorner B(0, 1/N)$. Then

$$S_t^N f_N(x) = \int_{B(0,N)} \psi(|\xi|/N) e^{2\pi i(x \cdot \xi - t|\xi|^m)} \, d\xi, \quad x \in \mathbb{R}^n.$$

Taking $t = N^{-m}/12$ we have that the real part of $e^{2\pi i(x \cdot \xi - t|\xi|^m)}$ is at least $\cos(\pi/3) = 1/2$ when $\xi \in B(0, N)$ and $x \in B(0, 1/(12N))$. Hence

$$\int \sup_{0 < t < 1} |S_t^N f_N| d\mu \gtrsim N^n \int_{B(0,1/(12N))} \int_{B(0,N)} \psi(|\xi|/N) \, d\xi dx \approx N^n.$$

On the other hand, $I_s(\mu) \approx N^s$ and $\|f_N\|_{H^\sigma(\mathbb{R}^n)}^2 \approx N^{n+2\sigma}$ so that the inequality

$$\left\| \int \sup_{0 < t < 1} |S_t^N f_N| \right\|_{L^1(\mu)} \lesssim \sqrt{I_s(\mu)} \|f_N\|_{H^\sigma(\mathbb{R}^n)}$$

is possible only if $N^n \lesssim N^{s/2 + n/2 + \sigma}$, which for large N yields $s \ge n - 2\sigma$, as required. $\square$

We shall now use another, more traditional, method to obtain estimates in the one-dimensional case.

Theorem 17.13 *Suppose that $n = 1$ and $m > 1$. Then for $1/4 \le \sigma < 1/2$,*

$$\left\| \sup_{t \in \mathbb{R}} \sup_{N \ge 1} |S_t^N f| \right\|_{L^1(\mu)} \le C(m, \sigma) \sqrt{I_{1-2\sigma}(\mu)} \|f\|_{H^\sigma(\mathbb{R})}$$

for all $\mu \in \mathcal{M}([-1, 1])$ and $f \in H^\sigma(\mathbb{R})$. In other words,

$$s_{m,1}(\sigma) \le 1 - 2\sigma \quad \text{if} \quad m > 1 \quad \text{and} \quad 1/4 < \sigma \le 1/2.$$

We need the following lemma due to Sjölin [2007].

Lemma 17.14 *Suppose that* $m > 1, 1/2 \leq \gamma < 1$ *and* $\eta \in S(\mathbb{R})$. *Then*

$$\left| \int \eta(\xi/N) e^{i(x\xi - t|\xi|^m)} |\xi|^{-\gamma} \, d\xi \right| \leq C(m, \gamma, \eta) |x|^{\gamma - 1}$$

for all $x, t \in \mathbb{R}$ *and all* $N \geq 1$.

Before discussing the proof of the lemma we show how Theorem 17.13 follows from it.

Proof of Theorem 17.13 We again find Borel functions $t : [-1, 1] \to \mathbb{R}$, $N : [-1, 1] \to [1, \infty)$ and $w : [-1, 1] \to \mathbb{C}$ with $\|w\|_\infty \leq 1$ such that

$$\left\| \sup_{t \in \mathbb{R}} \sup_{N \geq 1} |S_t^N f| \right\|_{L^1(\mu)} \leq 2 \int S_{t(x)}^{N(x)} f(x) w(x) \, d\mu x.$$

Using the definition of $S_t^N f$, Fubini's theorem twice and Schwartz's inequality, we obtain

$$\left| \int S_{t(x)}^{N(x)} f(x) w(x) \, d\mu x \right|^2 = \left| \iint \psi(|\xi|/N(x)) \widehat{f}(\xi) e^{2\pi i (x\xi - t(x)|\xi|^m)} \, d\xi \, w(x) \, d\mu x \right|^2$$

$$\leq \int |\widehat{f}(\xi)|^2 |\xi|^{2\sigma} \, d\xi \int \left| \int \psi(|\xi|/N(x)) e^{2\pi i (x\xi - t(x)|\xi|^m)} w(x) \, d\mu x \right|^2 |\xi|^{-2\sigma} \, d\xi$$

$$\leq \|f\|_{H^\sigma(\mathbb{R}^n)}^2 \iiint \psi(|\xi|/N(x)) \psi(|\xi|/N(y)) e^{2\pi i ((x-y)\xi - t(x)|\xi|^m)} |\xi|^{-2\sigma}$$

$$\times \, d\xi \, w(x) \overline{w(y)} \, d\mu x \, d\mu y.$$

We now apply Lemma 17.14 with $\eta = \psi$ which we can do since by our choice $\psi(\xi) = e^{-\xi^2}$. We then have $\psi(|\xi|/N(x)) \psi(|\xi|/N(y)) = \eta(\xi/\sqrt{N(x)^2 + N(y)^2})$. This gives

$$\left| \int \psi(|\xi|/N(x)) \psi(|\xi|/N(y)) e^{2\pi i ((x-y)\xi - t(x)|\xi|^m)} |\xi|^{-2\sigma} \, d\xi \right| \lesssim |x - y|^{2\sigma - 1}.$$

The double μ integration completes the proof of the theorem. $\qquad\square$

The proof of Lemma 17.14 consists of several applications of van der Corput's lemma, more precisely of Corollary 14.3.

Proof of Lemma 17.14 We may assume that $x \neq 0$ and $t > 0$. We write

$$\int \eta(\xi/N) e^{i(x\xi - t|\xi|^m)} |\xi|^{-\gamma} \, d\xi = I_1 + I_2,$$

where

$$I_1 = \int_{|\xi| \leq 1/|x|} \eta(\xi/N) e^{i(x\xi - t|\xi|^m)} |\xi|^{-\gamma} \, d\xi$$

and

$$I_2 = \int_{|\xi| > 1/|x|} \eta(\xi/N) e^{i(x\xi - t|\xi|^m)} |\xi|^{-\gamma} \, d\xi.$$

The estimate for I_1 is trivial:

$$|I_1| \le 2\|\eta\|_\infty \int_0^{1/|x|} \xi^{-\gamma} \, d\xi = \frac{2\|\eta\|_\infty}{1-\gamma} |x|^{\gamma-1}.$$

To estimate I_2 suppose first that $|x|^m \le t/2$. We apply van der Corput's lemma with $\varphi(\xi) = x\xi - t\xi^m$ for $\xi > 1/|x|$. Then

$$|\varphi'(\xi)| = |x| \left| 1 - \frac{mt}{x} \xi^{m-1} \right| \ge |x| \left(\left| \frac{mt}{x} \xi^{m-1} \right| - 1 \right)$$

$$\ge |x| \left(\left| \frac{2|x|^m}{|x|} |x|^{1-m} \right| - 1 \right) = |x|.$$

Obviously, φ' is monotonic on $[1/|x|, \infty)$. For ψ in van der Corput's lemma we choose $\psi(\xi) = \xi^{-\gamma} \eta(\xi/N)$. Then

$$\psi'(\xi) = \xi^{-\gamma} \eta'(\xi/N)/N - \gamma \xi^{-\gamma-1} \eta(\xi/N)$$

and

$$\int_{1/|x|}^\infty |\psi'(\xi)| \, d\xi \le N^{-1} |x|^\gamma \int_{1/|x|}^\infty |\eta'(\xi/N)| \, d\xi + \gamma \|\eta\|_\infty \int_{1/|x|}^\infty \xi^{-\gamma-1} \, d\xi \lesssim |x|^\gamma.$$

Therefore Corollary 14.3 with $k = 1$ yields

$$\left| \int_{1/|x|}^\infty \eta(\xi/N) e^{i(x\xi - t|\xi|^m)} |\xi|^{-\gamma} \, d\xi \right| = \left| \int_{1/|x|}^\infty \psi(\xi) e^{i\varphi(\xi)} \, d\xi \right| \lesssim |x|^{\gamma-1}.$$

The integral over $(-\infty, -1/|x|]$ has the same estimate, so the estimate in the case $|x|^m \le t/2$ is complete.

We have left the case $|x|^m > t/2$. The proof proceeds along similar lines: the integral is split into three subintegrals and van der Corput's lemma is applied to each of them, twice with $k = 1$ and once with $k = 2$. We skip the details and refer the reader to Sjölin [2007]. □

Corollary 17.15 *If $m > 1$, then*

$$s_{m,1}(\sigma) = 1 - 2\sigma \quad \text{for} \quad 1/4 \le \sigma \le 1/2.$$

In particular, for all $u_0 \in H^\sigma(\mathbb{R})$ with $1/4 \le \sigma \le 1/2$,

$$\dim\{x : u(x,t) \not\to u_0(x) \text{ as } t \to 0\} \le 1 - 2\sigma.$$

This follows immediately from Theorems 17.12 and 17.13.

One can rather easily modify the above argument to get the following result, for the details, see Barceló, Bennett, Carbery and Rogers [2011].

Theorem 17.16 *Suppose that $n \geq 2$ and $m = 2$. Then for $n/4 \leq \sigma < n/2$,*

$$\left\| \sup_{t \in \mathbb{R}} \sup_{N \geq 1} |S_t^N f| \right\|_{L^1(\mu)} \leq C(n, \sigma) \sqrt{I_{n-2\sigma}(\mu)} \|f\|_{H^\sigma(\mathbb{R}^n)}$$

for all $\mu \in \mathcal{M}(B(0, 1))$ and $f \in H^\sigma(\mathbb{R}^n)$. In other words,

$$s_{2,n}(\sigma) \leq n - 2\sigma \text{ if } n \geq 2 \quad \text{and} \quad n/4 \leq \sigma < n/2,$$

and hence for all $u_0 \in H^\sigma(\mathbb{R}^n)$ with $n/4 \leq \sigma < n/2$,

$$\dim\{x : u(x, t) \not\to u_0(x) \text{ as } t \to 0\} \leq n - 2\sigma.$$

Let us see briefly how the above relates to some classical results on the one-dimensional Schrödinger equation. Carleson [1980] proved that, in the case $m = 2, n = 1$, if $u_0 \in H^{1/4}(\mathbb{R})$, then $\lim_{t \to 0} u(x, t) = u_0(x)$ for almost all $x \in \mathbb{R}$. Dahlberg and Kenig [1982] showed that this is false for any $\sigma < 1/4$. So combining with Corollary 17.15 we see that somewhat surprisingly there is a jump from no estimate for $\sigma < 1/4$ to a dimension estimate $1/2$ for $\sigma = 1/4$. We shall now give an example that confirms this.

Example 17.17 Let $0 < \sigma < 1/4$. Then there is $u_0 \in H^\sigma(\mathbb{R})$ such that

$$S_t u_0(x) = u(x, t) = \int \widehat{u_0}(\xi) e^{2\pi i (x\xi - t\xi^2)} \, d\xi$$

(defined in the L^2 sense, recall (17.6)) fails to converge as $t \to 0$ in a set of positive Lebesgue measure. More precisely,

$$\mathcal{L}^1 \left(\left\{ x \in \mathbb{R} : \limsup_{t \to 0} |S_t u_0(x)| = \infty \right\} \right) > 0.$$

Proof We do not give the full proof. We shall only show that the weak type inequality

$$\mathcal{L}^1 \left(\left\{ x : \sup_{0 < t < 1} |S_t f(x)| \geq \lambda \right\} \right) \lesssim \lambda^{-2} \|f\|_{H^\sigma(\mathbb{R})}^2 \tag{17.9}$$

fails when $0 < \sigma < 1/4$. Then we shall discuss how the failure of almost everywhere convergence follows essentially from this.

Choose $N > 1$ and let f_N be the inverse Fourier transform of the characteristic function of $[N, N + \sqrt{N}]$. Then

$$\|f_N\|_{H^\sigma(\mathbb{R})} = \left(\int_N^{N+\sqrt{N}} (1 + \xi^2)^\sigma \, d\xi \right)^{1/2} \approx N^{1/4+\sigma}. \tag{17.10}$$

On the other hand with the change of variable $\xi = \sqrt{N}\zeta + N$ we compute

$$\begin{aligned}
|S_t f_N(x)| &= \left| \int_N^{N+\sqrt{N}} e^{2\pi i(x\xi - t\xi^2)} \, d\xi \right| \\
&= \sqrt{N} \left| \int_0^1 e^{2\pi i(x(\sqrt{N}\zeta + N) - Nt(\zeta^2 + 2\sqrt{N}\zeta + N))} \, d\zeta \right| \\
&= \sqrt{N} \left| \int_0^1 e^{2\pi i(\sqrt{N}(x - 2Nt)\zeta - Nt\zeta^2)} \, d\zeta \right|.
\end{aligned}$$

For any $x \in [0, 1/3]$ we choose $t = t(x) = x/(2N)$ and have

$$|S_{t(x)} f_N(x)| = \sqrt{N} \left| \int_0^1 e^{-\pi i x \zeta^2} \, d\zeta \right| \geq \sqrt{N} \int_0^1 \cos(\pi x \zeta^2) \, d\zeta \geq \sqrt{N}/2.$$

Consequently, if (17.9) holds, we have by (17.10) for all $N > 1$,

$$\begin{aligned}
1/3 &\leq \mathcal{L}^1 \left(\left\{ x : \sup_{0<t<1} |S_t f_N(x)| \geq \sqrt{N}/2 \right\} \right) \\
&\lesssim N^{-1} \|f_N\|_{H^\sigma(\mathbb{R})}^2 \approx N^{-1+1/2+2\sigma} = N^{2\sigma-1/2},
\end{aligned}$$

which forces $\sigma \geq 1/4$.

To conclude the proof one can apply a very general result of Nikishin [1972] for certain families of operators $T_t, 0 < t < 1$, on $L^p(\nu)$ spaces with $1 \leq p \leq 2$. Roughly speaking it says that the almost everywhere finiteness of $\limsup_{t \to 0} |T_t f(x)|$ for every in $f \in L^p(\nu)$ implies a weak type inequality. We can apply such a result with

$$T_t f(x) = \int f(\xi) e^{2\pi i(x\xi - t\xi^2)} \, d\xi, \quad x \in \mathbb{R}, f \in L^2((1 + \xi^2)^\sigma \, d\xi).$$

We saw above that the weak type inequality (17.9) fails if $\sigma < 1/4$, so almost everywhere convergence fails too. For more precise details, see Dahlberg and Kenig [1982]. $\square$

Nikishin [1972] extended a theorem of Stein [1961]; Stein considered translation invariant operators, Nikishin's operators are much more general. De Guzmán's book [1981] presents both of these results and others with proofs.

17.3 Further comments

Theorem 17.3 is classical. Description of fine behaviour of Sobolev functions can be expressed by various capacities more precisely than with Hausdorff measures. See, for example, Adams and Hedberg [1996], Chapter 6, Evans and Gariepy [1992], Section 4.8, and Ziemer [1989], Section 3.3.

The results of this chapter are mainly from Barceló, Bennett, Carbery and Rogers [2011]. That paper gives a thorough discussion on earlier related results. The authors developed there also a third method, based on Tao's bilinear restriction theorem 25.3, to improve Theorems 17.11 and 17.16 for certain ranges. Namely, they proved that

$$s_{m,n}(\sigma) \leq \frac{n+3}{n+1}(n - 2\sigma) \quad \text{if} \quad m > 1 \text{ and } n \geq 2.$$

In addition to the example giving Theorem 17.12, the paper of Barceló, Bennett, Carbery and Rogers contains other examples. They show that $\sigma \geq 1/4$ is a necessary condition to get any non-trivial estimates for any values of m and n. Moreover,

$$s_{m,n}(\sigma) \geq n + 1 - 4\sigma \quad \text{if} \quad \sigma < n/4,$$

and

$$s_{1,n}(\sigma) \geq n + 2 - 4\sigma \quad \text{if} \quad \sigma < (n + 1)/4.$$

It follows that Theorem 17.11 is sharp when $n = 2$ and $m = 1$: $s_{1,2}(\sigma) = 4 - 4\sigma$ for $1/2 < \sigma \leq 3/4$ and $s_{1,2}(\sigma) = 2 - 2\sigma$ for $3/4 < \sigma \leq 1$. In higher dimensions it probably is not sharp, because the estimates for spherical averages probably are not sharp.

Bennett and Rogers [2012] proved the sharp estimate

$$\dim\{x : u(x, t) \not\to u_0(x) \text{ as } t \to 0\} \leq n - 2\sigma$$

for all radial functions $u_0 \in H^\sigma(\mathbb{R}^n)$ with $1/4 \leq \sigma < 1/2$.

Example 17.17 is due to Dahlberg and Kenig [1982]. I am grateful to Keith Rogers for a simplified presentation.

There are many recent results on boundedness and almost everywhere convergence (with respect to Lebesgue measure) in higher dimensions, see, for example, Lee [2006b], Lee, Rogers and Seeger [2013], Bourgain [2013], Sjölin [2013] and Sjölin and Soria [2014] (sharp one-dimensional results were discussed above). Bourgain [2013] proved that when $m = 2$ almost everywhere convergence takes place for $f \in H^\sigma(\mathbb{R}^n)$ if $\sigma > 1/2 - 1/(4n)$; for $n = 2$ this was previously shown by Lee [2006b]. Bourgain's proof relies on recent results and methods developed by Bourgain and Guth [2011] on multilinear restriction theory, see Section 25.13. Bourgain also constructed examples to show that at least $\sigma > 1/2 - 1/n$ is needed for almost everywhere convergence when $n \geq 5$.

18

Generalized projections of Peres and Schlag

In Chapter 5 we proved several estimates on dimensions of exceptional sets related to orthogonal projections, many of them sharp. For example in Theorem 5.6

$$\dim\{e \in S^{n-1} : \dim_S \mu_e < t\} \leq \max\{0, n - 1 + t - \dim_S \mu\}.$$

In this chapter we present a setting, due to Peres and Schlag [2000], where such results can be established for much more general parametrized families of mappings π_λ. The crucial property required for such mappings, in addition to standard regularity properties, is transversality. This means, roughly speaking, that if $|\pi_{\lambda_0}(x) - \pi_{\lambda_0}(y)|$ is too small relative to $d(x, y)$ for some $\lambda_0 \in J$, then the mapping $\lambda \mapsto |\pi_\lambda(x) - \pi_\lambda(y)|$ is rapidly growing in a neighbourhood of λ_o. Orthogonal projections obviously possess such a property (we say a bit more below).

18.1 Tranversality of degree 0 in the one-dimensional case

First we shall consider one-dimensional families of mappings and later explain how the results can be extended to cases where the parameter space is higher dimensional. The setting in this section and Section 18.2 will be the following:

Setting: Let (Ω, d) be a compact metric space, and let $J \subset \mathbb{R}$ be a bounded open interval. Let

$$\pi_\lambda : \Omega \to \mathbb{R}, \quad \lambda \in J,$$

be a family of mappings such that the function $\lambda \mapsto \pi_\lambda(x)$ is in $C^\infty(J)$ for every fixed $x \in \Omega$, and to every compact interval $I \subset J$ and $l = 0, 1, \ldots,$

236

there corresponds a finite positive constant $C_{l,I}$ such that

$$|\partial_\lambda^l \pi_\lambda(x)| \leq C_{l,I} \quad \text{for all } \lambda \in I, x \in \Omega. \tag{18.1}$$

We shall always assume these derivative bounds of all orders. Peres and Schlag also formulate results on restricted regularity, where for some positive integer N (18.1) is only assumed for $0 \leq l \leq N$.

Definition 18.1 Set

$$\Phi_\lambda(x, y) = \frac{\pi_\lambda(x) - \pi_\lambda(y)}{d(x, y)} \quad \text{for } \lambda \in J, \ x, y \in \Omega, x \neq y.$$

The family π_λ, $\lambda \in J$, is *transversal*, if there exists a positive constant C_0 such that

$$|\Phi_\lambda(x, y)| \leq C_0 \quad \Longrightarrow \quad |\partial_\lambda \Phi_\lambda(x, y)| \geq C_0 \tag{18.2}$$

for $\lambda \in J$ and $x, y \in \Omega, x \neq y$. The family π_λ, $\lambda \in J$, is *regular*, if to every $l \in \mathbb{N}$ there corresponds a positive constant C_l such that

$$|\Phi_\lambda(x, y)| \leq C_0 \quad \Longrightarrow \quad |\partial_\lambda^l \Phi_\lambda(x, y)| \leq C_l \tag{18.3}$$

for $\lambda \in J$ and $x, y \in \Omega, x \neq y$.

Often the bounds on derivatives as in (18.3) hold for all x, y and l, but they will be needed only for the critical values for which $|\Phi_\lambda(x, y)|$ is small.

Later on we shall consider a generalization, transversality of degree $\beta \geq 0$; then the above definition corresponds to the case $\beta = 0$. We shall first present the detailed proofs in this special case in order to have the main ideas less obscured by technicalities. Then we shall sketch the changes required to deal with the general case of β transversality. The case $\beta = 0$ is enough for geometric applications, such as orthogonal projections and pinned distance sets, but $\beta > 0$ is needed for some other applications, in particular for Bernoulli convolutions.

Notation: For $\pi_\lambda \colon \Omega \to \mathbb{R}$ as in Definition 18.1, we write

$$\mu_\lambda = \pi_{\lambda\sharp}\mu \quad \text{for } \mu \in \mathcal{M}(\Omega).$$

Then $\mu_\lambda \in \mathcal{M}(\mathbb{R})$ and $\mu_\lambda(B) = \mu(\pi_\lambda^{-1}(B))$ for $B \subset \mathbb{R}$. The s-energy, $s > 0$, of $\mu \in \mathcal{M}(\Omega)$ is as before

$$I_s(\mu) = \iint d(x, y)^{-s} \, d\mu x \, d\mu y,$$

and the Sobolev t-energy, $t \in \mathbb{R}$, of $v \in \mathcal{M}(\mathbb{R}^m)$,

$$\mathcal{I}_t(v) = \int |\widehat{v}(x)|^2 |x|^{t-m} |\, dx$$

with, due to (3.45),

$$\gamma(m, t)\mathcal{I}_t(v) = I_t(v) \quad \text{for } 0 < t < m.$$

Recall also from (5.4) the Sobolev dimension $\dim_S \mu$,

$$\dim_S v = \sup\{t : \mathcal{I}_t(v) < \infty\}. \tag{18.4}$$

Our goal is to show that the finiteness of the energy $I_s(\mu)$ implies that the Sobolev dimensions $\dim_S \mu_\lambda$ are large. The obstacle is that the π_λ could be badly non-injective. The transversality puts obstacles to this obstacle: if π_λ maps x and y close to each other, then $\pi_{\lambda'}$ does not map them too close to each other when λ' moves away from λ.

Example 18.2 A basic example comes from orthogonal projections in $\mathbb{R}^2$ which we now write as

$$\pi_\lambda(x_1, x_2) = x_1 \cos \lambda + x_2 \sin \lambda, \quad \lambda \in J = [0, \pi).$$

For Ω we can take any closed disc containing the support of μ. Then

$$\Phi_\lambda(x, y) = u \cos \lambda + v \sin \lambda \quad \text{where} \quad (u, v) = \frac{x - y}{|x - y|},$$

and

$$\frac{d}{d\lambda} \Phi_\lambda(x, y) = -u \sin \lambda + v \cos \lambda.$$

All the conditions of Definition 18.1 are clearly satisfied with $\beta = 0$. Other examples will be discussed at the end of this chapter.

The main general theorem in this setting (for $\beta = 0$) is:

Theorem 18.3 *Let $\mu \in \mathcal{M}(\Omega)$ with $I_s(\mu) < \infty$ for some $s > 0$. Assume that the mappings π_λ, $\lambda \in J$, satisfy the transversality and regularity conditions of Definition 18.1. Then*

$$\int_I \mathcal{I}_s(\mu_\lambda) \, d\lambda \leq C(s, I) I_s(\mu) \tag{18.5}$$

for compact intervals $I \subset J$. Therefore,

$$\dim_S \mu_\lambda \geq s \quad \text{for almost all } \lambda \in J. \tag{18.6}$$

Furthermore, for any $t \in (0, s]$ we have the estimates

$$\dim\{\lambda \in J : \dim_S \mu_\lambda < t\} \leq t, \tag{18.7}$$
$$\dim\{\lambda \in J : \dim_S \mu_\lambda < t\} \leq 1 + t - s \quad if\ 1 + t - s \geq 0, \tag{18.8}$$

and

$$\dim_S \mu_\lambda \geq s - 1 \quad for\ all\ \lambda \in J. \tag{18.9}$$

As in the case of orthogonal projections this gives us the corollary:

Corollary 18.4 *Let $A \subset \Omega$ be a Borel set and $s = \dim A$.*

(a) If $s \leq 1$ and $t \in (0, s]$, then

$$\dim\{\lambda \in J : \dim \pi_\lambda(A) < t\} \leq t.$$

(b) If $1 < s \leq 2$, then

$$\dim\{\lambda \in J : \dim \pi_\lambda(A) < t\} \leq 1 + t - s \quad if\ 1 + t - s \geq 0,$$
$$\dim \pi_\lambda(A) \geq s - 1 \quad for\ all\ \lambda \in J,$$

and

$$\dim\{\lambda \in J : \mathcal{L}^1(\pi_\lambda(A)) = 0\} \leq 2 - s.$$

(c) If $2 < s \leq 3$, then

$$\dim\{\lambda \in J : the\ interior\ of\ \pi_\lambda(A)\ is\ empty\} \leq 3 - s.$$

As remarked above, our enemies in the proof of Theorem 18.3 will be triples (x, y, λ) such that $|\Phi_\lambda(x, y)|$ is small. The following lemma gives some control over what is happening around them.

Lemma 18.5 *Fix $x, y \in \Omega, x \neq y$, write $r = d(x, y)$ and suppose that (18.2) and (18.3) hold. Then*

$$\{\lambda \in J : |\Phi_\lambda(x, y)| < C_0\} = \bigcup_{j=1}^{N} I_j,$$

where the I_j are disjoint open subintervals of J such that:

(i) $\mathcal{L}^1(I_j) \leq 2$ for all j, $\mathcal{L}^1(I_j) \geq 2C_0/C_1$ for all but at most two indices j, and $N \leq C_1 \mathcal{L}^1(J)/(2C_0) + 2$.
(ii) The function $\lambda \mapsto \Phi_\lambda(x, y)$ is strictly monotone on any interval I_j.
(iii) There exist points $\lambda_j \in \overline{I_j}$, which satisfy: if $\lambda \in \overline{I_j}$, then $|\Phi_\lambda(x, y)| \geq |\Phi_{\lambda_j}(x, y)|$ and $|\Phi_\lambda(x, y)| \geq C_0|\lambda - \lambda_j|$.

(iv) *There exists a constant* $\delta > 0$ *depending only on* C_0 *with the following properties. Except for at most two exceptional values of* j, $\Phi_{\lambda_j}(x, y) = 0$ *and* $|\Phi_\lambda(x, y)| \leq C_0/2$ *for* $|\lambda - \lambda_j| \leq \delta$. *For each of the two possible exceptional values of* j, *either* $|\Phi_\lambda(x, y)| \geq C_0/4$ *for all* $\lambda \in I_j$ *or* $|\Phi_\lambda(x, y)| \leq C_0/2$ *for all* $\lambda \in J$ *for which* $|\lambda - \lambda_j| \leq \delta$. *For all* j, $J \cap (\lambda_j - \delta, \lambda_j + \delta) \subset I_j$.

Proof Since $\{\lambda \in J : |\Phi_\lambda(x, y)| < C_0\}$ is open, it can be written in a unique way as a union of disjoint open intervals I_j. On each of these by (18.2) either $\partial_\lambda \Phi_\lambda(x, y) \geq C_0$ or $\partial_\lambda \Phi_\lambda(x, y) \leq -C_0$. The item (ii) follows immediately from this. The first inequality of (i) is also easy: for $\lambda_1, \lambda_2 \in I_j$ with $\lambda_1 < \lambda_2$,

$$2C_0 \geq |\Phi_{\lambda_2}(x, y) - \Phi_{\lambda_1}(x, y)| = \left| \int_{\lambda_1}^{\lambda_2} \partial_\lambda \Phi_\lambda(x, y) \, d\lambda \right| \geq C_0(\lambda_2 - \lambda_1).$$

For all but at most two intervals I_j we have $\overline{I_j} \subset J$, so $\Phi_\lambda(x, y) = \pm C_0$ at the end-points, with both values attained. Hence by (18.3)

$$2C_0 = \left| \int_{I_j} \partial_\lambda \Phi_\lambda(x, y) \, d\lambda \right| \leq C_1 \mathcal{L}^1(I_j).$$

The last two statements of (i) follow from this.

If $\overline{I_j} \subset J$, then I_j contains the unique zero λ_j of $\Phi_\lambda(x, y)$. For the possible one or two other values of j for which $\Phi_\lambda(x, y)$ does not have zero, we take λ_j as the end-point of $\overline{I_j}$ which gives the minimum for $|\Phi_\lambda(x, y)|$ on $\overline{I_j}$ (extending $\Phi_\lambda(x, y)$ in the obvious way to the end-points of $\overline{I_j}$ which are not in J).

To prove (iii), let $\lambda \in I_j$. The inequalities

$$|\Phi_\lambda(x, y)| \geq |\Phi_\lambda(x, y) - \Phi_{\lambda_j}(x, y)| \geq C_0|\lambda - \lambda_j|$$

follow from the monotonicity and the mean-value theorem. The last item (iv) is also easy and we leave its checking to the reader. $\qquad\square$

We begin the proof of Theorem 18.3 with (18.7). It is considerably simpler than the rest. In particular the proof does not involve the use of Fourier transforms.

Proof of (18.7) If (18.7) fails for some $t \in (0, s]$, then for some $\tau < t$,

$$\mathcal{H}^t(\{\lambda \in J : \dim_S \mu_\lambda < \tau\}) > 0.$$

The set $S_\tau = \{\lambda \in J : \dim_S \mu_\lambda < \tau\}$ is a Borel set. This is easily proven: check first that $\lambda \mapsto \widehat{\mu_\lambda}(u)$ is continuous for a fixed $u \in J$, whence $\mathcal{I}_\sigma(\mu_\lambda)$ is lower semicontinuous for every $\sigma \in \mathbb{R}$, then use the definition of the Sobolev dimension. Hence Frostman's lemma gives us a measure $\nu \in \mathcal{M}(S_\tau)$ with

$v(B(x,r)) \leq r^t$ for $x \in \mathbb{R}$ and $r > 0$. Given $x, y \in \Omega, x \neq y$, write $B_{x,y} = \{\lambda \in J : |\Phi_\lambda(x,y)| < C_0\}$, and split

$$\int_J I_\tau(\mu_\lambda) \, dv\lambda = \int_J \int_\Omega \int_\Omega |\pi_\lambda(x) - \pi_\lambda(y)|^{-\tau} \, d\mu x \, d\mu y \, dv\lambda$$

$$= \int_\Omega \int_\Omega \left(\int_{J \setminus B_{x,y}} |\Phi_\lambda(x,y)|^{-\tau} \, dv\lambda \right) d(x,y)^{-\tau} \, d\mu x \, d\mu y$$

$$(18.10)$$

$$+ \int_\Omega \int_\Omega \left(\int_{B_{x,y}} |\Phi_\lambda(x,y)|^{-\tau} \, dv\lambda \right) d(x,y)^{-\tau} d\mu x \, d\mu y.$$

$$(18.11)$$

The integral on line (18.10) is easily estimated:

$$\int_\Omega \int_\Omega \left(\int_{J \setminus B_{x,y}} |\Phi_\lambda(x,y)|^{-\tau} \, dv\lambda \right) d(x,y)^{-\tau} \, d\mu x \, d\mu y$$

$$\leq C_0^{-\tau} v(J) \int_\Omega \int_\Omega d(x,y)^{-\tau} \, d\mu x \, d\mu y \lesssim I_\tau(\mu) \lesssim I_s(\mu) < \infty,$$

because $\tau \leq t \leq s$. To estimate the integral on line (18.11), let $B_{x,y} = \cup_{j=1}^N I_j$ and $\lambda_j \in I_j$ be as in Lemma 18.5. The estimate $|\Phi_\lambda(x,y)| \geq C_0 |\lambda - \lambda_j|$ for $\lambda \in I_j$ gives

$$\int_\Omega \int_\Omega \left(\int_{B_{x,y}} |\Phi_\lambda(x,y)|^{-\tau} \, dv\lambda \right) d(x,y)^{-\tau} \, d\mu x \, d\mu y$$

$$= \int_\Omega \int_\Omega \left(\sum_{j=1}^N \int_{I_j} |\Phi_\lambda(x,y)|^{-\tau} \, dv\lambda \right) d(x,y)^{-\tau} \, d\mu x \, d\mu y$$

$$\leq C_0^{-\tau} \int_\Omega \int_\Omega \left(\sum_{j=1}^N \int_{I_j} |\lambda - \lambda_j|^{-\tau} \, dv\lambda \right) d(x,y)^{-\tau} \, d\mu x \, d\mu y.$$

Here

$$\int_{I_j} |\lambda - \lambda_j|^{-\tau} \, dv\lambda = \int_0^\infty v(\{\lambda \in I_j : |\lambda - \lambda_j|^{-\tau} \geq r\}) \, dr$$

$$= \int_0^\infty v(B(\lambda_j, r^{-1/\tau})) \, dr$$

$$\leq \int_0^1 v(\mathbb{R}) \, dr + \int_1^\infty r^{-t/\tau} \, dr \lesssim 1,$$

since $t > \tau$. As the number of the intervals I_j is bounded independently of x and y, we obtain

$$\int_\Omega \int_\Omega \left(\int_{B_{x,y}} |\Phi_\lambda(x,y)|^{-\tau} dv\lambda \right) d(x,y)^{-\tau} d\mu x \, d\mu y \lesssim \int_\Omega \int_\Omega d(x,y)^{-\tau} d\mu x \, d\mu y.$$

The expression on the right is again finite, since $\tau \leq t \leq s$ and $I_s(\mu) < \infty$. So we have shown that $\int_J I_\tau(\mu_\lambda) dv\lambda < \infty$, and, in particular, $I_\tau(\mu_\lambda) < \infty$ for v almost all $\lambda \in J$. This implies that $\dim_S \mu_\lambda \geq \tau$ for v almost all $\lambda \in J$. This contradicts $v \in \mathcal{M}(S_\tau)$ and finishes the proof. $\square$

A central tool in the rest of the proof of Theorem 18.3 is the following Littlewood–Paley (dyadic) decomposition of the Sobolev norm:

Lemma 18.6 *There exists a Schwartz function $\psi \in \mathcal{S}(\mathbb{R}^m)$ with the following properties:*

(i) $\widehat{\psi} \geq 0$ and spt $\widehat{\psi} \subset \{x \in \mathbb{R}^m : 1 \leq |x| \leq 4\}$,
(ii) $\sum_{j \in \mathbb{Z}} \psi(2^{-j}x) = 1$ for $x \in \mathbb{R}^m \setminus \{0\}$,
(iii) for all $v \in \mathcal{M}(\mathbb{R}^m)$ and $t \in \mathbb{R}$ the following decomposition of the Sobolev norm holds:

$$\mathcal{I}_t(v) \approx_{t,m} \sum_{j \in \mathbb{Z}} 2^{j(t-m)} \int_{\mathbb{R}^m} (\psi_{2^{-j}} * v)(x) \, dvx,$$

where $\psi_{2^{-j}}(x) = 2^{jm} \psi(2^j x)$.

Proof Take a radial function $\eta \in \mathcal{S}(\mathbb{R}^m)$, $\eta(x) = h(|x|)$, with h non-increasing, $0 \leq \eta \leq 1$ and $\eta(x) = 1$ for all $x \in B(0,1)$ and spt $\eta \subset B(0,2)$. Since also $x \mapsto \eta(x/2) - \eta(x)$ is a Schwartz function, we may choose $\psi \in \mathcal{S}(\mathbb{R}^m)$ so that

$$\widehat{\psi}(x) = \eta(x/2) - \eta(x), \quad x \in \mathbb{R}^m.$$

The function ψ has the desired properties:

(i) The claim $\widehat{\psi} \geq 0$ follows directly from the fact that $\eta(x)$ is non-increasing in $|x|$. The properties $\eta = 1$ on $B(0,1)$ and $\eta = 0$ outside $B(0,2)$ imply that spt $\widehat{\psi} \subset \{x \in \mathbb{R}^m : 1 \leq |x| \leq 4\}$.

(ii) Fix $x \in \mathbb{R}^m \setminus \{0\}$ and let $k \in \mathbb{Z}$ be such that $2^{-(k+1)}|x| < 1 \leq 2^{-k}|x|$. Then $|2^{2-k}x| \geq 4$, whence

$$\sum_{j \in \mathbb{Z}} \widehat{\psi}(2^{-j}x) = \widehat{\psi}(2^{-k}x) + \widehat{\psi}(2^{1-k}x) = \eta(2^{-(k+1)}x) - \eta(2^{1-k}x) = 1 - 0 = 1.$$

(iii) Let $v \in \mathcal{M}(\mathbb{R}^m)$ and $t \in \mathbb{R}$. For $x \in \mathbb{R}^m \setminus \{0\}$ let $k \in \mathbb{Z}$ be as in the proof of (ii). Then $2^k \approx |x|$, whence

$$\sum_{j \in \mathbb{Z}} 2^{j(t-m)} \widehat{\psi}(2^{-j}x) = 2^{k(t-m)} \widehat{\psi}(2^{-k}x) + 2^{(k-1)(t-m)} \widehat{\psi}(2^{1-k}x)$$

$$\approx |x|^{t-m}\big(\widehat{\psi}(2^{-k}x) + \widehat{\psi}(2^{1-k}x)\big) = |x|^{t-m}.$$

Finally, note that $\widehat{\psi_{2^{-j}}}(x) = \widehat{\psi}(2^{-j}x)$ and apply Parseval's theorem in the form of (3.27) to get

$$\mathcal{I}_t(v) = \int_{\mathbb{R}^m} |x|^{t-m} |\widehat{v}(x)|^2 \, dx$$

$$\approx \sum_{j \in \mathbb{Z}} 2^{j(t-m)} \int_{\mathbb{R}^m} \widehat{\psi}(2^{-j}x) |\widehat{v}(x)|^2 \, dx$$

$$= \sum_{j \in \mathbb{Z}} 2^{j(t-m)} \int_{\mathbb{R}^m} (\psi_{2^{-j}} * v)(x) \, dvx. \qquad \square$$

To get a better feeling of the proof of (18.5) we first prove a simple variant under the *strong transversality condition*:

$$|\partial_\lambda \Phi_\lambda(x, y)| \geq c > 0 \quad \text{for all } \lambda \in J, x, y \in \Omega, x \neq y. \qquad (18.12)$$

Theorem 18.7 *If (18.12) holds and $\mu \in \mathcal{M}(\Omega)$ with $I_s(\mu) < \infty$, then for any compact interval $I \subset J$,*

$$\int_I |\widehat{\mu_\lambda}(u)|^2 \, d\lambda \leq C(s, I) I_s(\mu) |u|^{-s} \quad \text{for all } u \in \mathbb{R}. \qquad (18.13)$$

In particular, for $0 < t < s$,

$$\int_I \mathcal{I}_t(\mu_\lambda) \, d\lambda \leq C(s, t, I) I_s(\mu). \qquad (18.14)$$

Proof Let $\varrho \in C_0^\infty(\mathbb{R})$ with $\operatorname{spt} \varrho \subset J$ and $\varrho = 1$ on I. Then

$$\int_I |\widehat{\mu_\lambda}(u)|^2 \, d\lambda = \int_{-\infty}^\infty |\widehat{\mu_\lambda}(u)|^2 \varrho(\lambda) \, d\lambda$$

$$= \int_{-\infty}^\infty \int e^{-2\pi i u v} \, d\mu_\lambda v \int e^{2\pi i u w} \, d\mu_\lambda w \varrho(\lambda) \, d\lambda$$

$$= \int_{-\infty}^\infty \iint e^{-2\pi i u(\pi_\lambda(x) - \pi_\lambda(y))} \, d\mu x \, d\mu y \varrho(\lambda) \, d\lambda$$

$$= \iint \int_{-\infty}^\infty e^{-2\pi i u d(x,y) \Phi_\lambda(x,y)} \varrho(\lambda) \, d\lambda \, d\mu x \, d\mu y.$$

Let $u \in \mathbb{R}, u \neq 0$, and $x, y \in \Omega$ with $x \neq y$; note that since $I_s(\mu) < \infty$, the singletons have μ measure 0 and so it is enough to consider $x \neq y$. Let

$r = d(x, y)$. Because of (18.12) we can apply the estimate of Theorem 14.1 to obtain

$$\left| \int_{-\infty}^{\infty} e^{-2\pi i u d(x,y)\Phi_\lambda(x,y)} \varrho(\lambda) \, d\lambda \right| \lesssim (|u| d(x, y))^{-s},$$

and the first estimate follows integrating twice with respect to μ. The second estimate follows from the first, since $|\widehat{\mu_\lambda}(u)|^2 \leq \mu(\Omega)^2 \leq d(\Omega)^s I_s(\mu)$:

$$\int_I \mathcal{I}_t(\mu_\lambda) \, d\lambda = \int_I \int_{-\infty}^{\infty} |u|^{t-1} |\widehat{\mu_\lambda}(u)|^2 \, du \, d\lambda$$

$$\lesssim I_s(\mu) \left(\int_0^1 |u|^{t-1} \, du + \int_1^\infty |u|^{t-1-s} \, du \right) \approx I_s(\mu).$$

$\square$

Above (18.14) also holds for $t = s$, as we shall prove even under the weaker hypothesis of Theorem 18.3.

Proof of (18.5) This will be based on the following inequality: Let $\psi \in \mathcal{S}(\mathbb{R})$ be the function provided by Lemma 18.6, let $\varrho \in C_0^\infty(\mathbb{R})$ be any function with support in J, and let $q \in \mathbb{N}$. Then for all $x, y \in \Omega$ and $j \in \mathbb{Z}$,

$$\left| \int_{\mathbb{R}} \varrho(\lambda) \psi(2^j (\pi_\lambda(x) - \pi_\lambda(y))) \, d\lambda \right| \lesssim (1 + 2^j d(x, y))^{-q}, \tag{18.15}$$

with the implicit constant independent of x, y and j.

Before we start proving (18.15), let us see how the estimate (18.5) follows from it. Suppose I is a compact interval with $I \subset J$. Choose $\varrho \in C_0^\infty(\mathbb{R})$ so that spt $\varrho \subset J$, $0 \leq \varrho \leq 1$ and $\varrho|I \equiv 1$. Furthermore, let $q \in \mathbb{N}$ with $q > s$. Then, by applying the assertion (iii) of Lemma 18.6 to the measures $\mu_\lambda \in \mathcal{M}(\mathbb{R})$ we obtain by (18.15)

$$\int_I \mathcal{I}_s(\mu_\lambda) \, d\lambda \leq \int_{\mathbb{R}} \mathcal{I}_s(\mu_\lambda) \varrho(\lambda) \, d\lambda$$

$$\approx \int_{\mathbb{R}} \sum_{j \in \mathbb{Z}} 2^{j(s-1)} \left(\int_{\mathbb{R}} (\psi_{2^{-j}} * \mu_\lambda)(u) \, d\mu_\lambda u \right) \varrho(\lambda) \, d\lambda$$

$$= \int_{\mathbb{R}} \sum_{j \in \mathbb{Z}} 2^{js} \left(\int_{\mathbb{R}} \int_{\mathbb{R}} \psi(2^j (u - v)) \, d\mu_\lambda v \, d\mu_\lambda u \right) \varrho(\lambda) \, d\lambda$$

$$= \int_\Omega \int_\Omega \sum_{j \in \mathbb{Z}} 2^{js} \left(\int_{\mathbb{R}} \psi(2^j (\pi_\lambda(x) - \pi_\lambda(y))) \varrho(\lambda) \, d\lambda \right) d\mu y \, d\mu x$$

$$\lesssim \int_\Omega \int_\Omega \sum_{j \in \mathbb{Z}} 2^{js} (1 + 2^j d(x, y))^{-q} \, d\mu y \, d\mu x. \tag{18.16}$$

Fix $x, y \in \Omega$, $x \neq y$, and write $r = d(x, y) > 0$. Then the series inside the integral is

$$\sum_{j \in \mathbb{Z}} 2^{js} (1 + 2^j r)^{-q} \leq \sum_{2^j \leq 1/r} 2^{js} + r^{-q} \sum_{2^j > 1/r} 2^{j(s-q)}$$

$$\approx r^{-s} + r^{-q-s+q} = 2r^{-s} = 2d(x, y)^{-s},$$

since the value of a geometric series is roughly its dominating term. Plugging this into (18.16) gives

$$\int_I \mathcal{I}_s(\mu_\lambda) \, d\lambda \lesssim \int_\Omega \int_\Omega d(x, y)^{-s} \, d\mu y \, d\mu x = I_s(\mu)$$

as required.

Now we shall deal with (18.15). Fix $x, y \in \Omega$, $x \neq y$, and write again $r = d(x, y)$. Then $\pi_\lambda(x) - \pi_\lambda(y) = r \Phi_\lambda(x, y) =: r \Phi(\lambda)$. We may assume that

$$2^j r > 1. \tag{18.17}$$

Choose an auxiliary function $\varphi \in C_0^\infty(\mathbb{R})$ with $0 \leq \varphi \leq 1$, $\varphi|[-1/2, 1/2] \equiv 1$, spt $\varphi \subset [-1, 1]$, and split the integration in (18.15) into two parts:

$$\int_\mathbb{R} \varrho(\lambda) \psi(2^j (\pi_\lambda(x) - \pi_\lambda(y))) \, d\lambda$$

$$= \int_\mathbb{R} \varrho(\lambda) \psi(2^j r \Phi(\lambda)) \varphi(C_0^{-1} \Phi(\lambda)) \, d\lambda \tag{18.18}$$

$$+ \int_\mathbb{R} \varrho(\lambda) \psi(2^j r \Phi(\lambda))(1 - \varphi(C_0^{-1} \Phi(\lambda))) \, d\lambda. \tag{18.19}$$

Here $C_0 > 0$ is the transversality constant of Definition 18.1. The integral of line (18.19) is easy to bound: since the integrand vanishes whenever $|\Phi(\lambda)| \leq C_0/2$. But if $|\Phi(\lambda)| \geq C_0/2$, we have for all $q \in \mathbb{N}$ the estimate

$$|\psi(2^j r \Phi(\lambda))| \lesssim (1 + C_0 2^{j-1} r)^{-q} \lesssim (1 + 2^j r)^{-q},$$

whence

$$\left| \int_\mathbb{R} \varrho(\lambda) \psi(2^j r \Phi(\lambda))(1 - \varphi(C_0^{-1} \Phi(\lambda))) \, d\lambda \right| \lesssim (1 + 2^j r)^{-q}. \tag{18.20}$$

Moving on to line (18.18), let the intervals $I_1, \ldots, I_N$, the points $\lambda_i \in I_i$ and the constant $\delta > 0$ be as in Lemma 18.5. Choose another auxiliary function $\chi \in C_0^\infty(\mathbb{R})$ with $0 \leq \chi \leq 1$, $\chi|(-\delta/2, \delta/2) \equiv 1$, spt $\chi \subset (-\delta, \delta)$, and split the integration on line (18.18) into $N + 1$ parts:

$$\int_{\mathbb{R}} \varrho(\lambda)\psi(2^j r \Phi(\lambda))\varphi(C_0^{-1}\Phi(\lambda))\,d\lambda$$

$$= \sum_{i=1}^{N} \int_{\mathbb{R}} \varrho(\lambda)\chi(\lambda - \lambda_i)\psi(2^j r \Phi(\lambda))\varphi(C_0^{-1}\Phi(\lambda))\,d\lambda \qquad (18.21)$$

$$+ \int_{\mathbb{R}} \varrho(\lambda)\left(1 - \sum_{i=1}^{N} \chi(\lambda - \lambda_i)\right)\psi(2^j r \Phi(\lambda))\varphi(C_0^{-1}\Phi(\lambda))\,d\lambda. \quad (18.22)$$

With the aid of parts (iii) and (iv) of Lemma 18.5, the integral on line (18.22) is easy to handle. If the integrand is non-vanishing at some point $\lambda \in J$, we must have $\varphi(C_0^{-1}\Phi(\lambda)) \neq 0$, which gives that $|\Phi(\lambda)| < C_0$: in particular $\lambda \in I_k$ for some $1 \le k \le N$. Then $\chi(\lambda - \lambda_i) = 0$ for $i \neq k$ by Lemma 18.5(iv). But, since the integrand is non-vanishing at λ, this enables us to conclude that $\chi(\lambda - \lambda_k) < 1$: in particular, $|\lambda - \lambda_k| \ge \delta/2$. Then Lemma 18.5(iii) shows that $|\Phi(\lambda)| \ge C_0|\lambda - \lambda_k| \ge C_0\delta/2$, and using the rapid decay of ψ as in (18.20) one obtains

$$\left|\int_{\mathbb{R}} \varrho(\lambda)\left(1 - \sum_{i=1}^{N} \chi(\lambda - \lambda_i)\right)\psi(2^j r \Phi(\lambda))\varphi(C_0^{-1}\Phi(\lambda))\,d\lambda\right| \lesssim (1 + 2^j r)^{-q}.$$

Now we turn our attention to the N integrals on the line (18.21). Since N is bounded, it is enough to get the required estimate for each of them separately. They are of the form

$$\int f(\lambda)\psi(ag(\lambda))\,d\lambda$$

where $f \in C_0^\infty(\mathbb{R})$, $\|f\|_\infty \lesssim 1$, $a = 2^j r > 1$ and $g(\lambda) = \Phi_\lambda(x, y)$. We have also that $|g'(\lambda)| \ge C_0$ on an interval containing spt f. We need to show that

$$\left|\int f(\lambda)\psi(ag(\lambda))\,d\lambda\right| \lesssim a^{-q}. \qquad (18.23)$$

By Lemma 18.6, spt $\widehat{\psi} \subset \{u : 1 \le |u| \le 4\}$, so all the derivatives of $\widehat{\psi}$ vanish at 0. But the Fourier transform of $u \mapsto u^l \psi(u)$ is $(-2\pi i)^{-l}\widehat{\psi}^{(l)}$, whence

$$\int_{\mathbb{R}} u^l \psi(u)\,du = (-2\pi i)^{-l}\widehat{\psi}^{(l)}(0) = 0$$

for $l \ge 0$. Since g is strictly monotone, say, strictly increasing, on an interval I containing spt f, we can change variable; $\lambda = h(\eta)$ with $h = (g|I)^{-1}$, to write

$$\int f(\lambda)\psi(ag(\lambda))\,d\lambda = \int f(h(\eta))h'(\eta)\psi(a\eta)\,d\eta.$$

Let

$$F(\eta) = f(h(\eta))h'(\eta) = \frac{f(h(\eta))}{g'(h(\eta))}.$$

By Taylor's formula

$$F(\eta) = \sum_{l=0}^{2(q-1)} \frac{F^{(l)}(0)}{l!}\eta^l + O(F^{(2q-1)}(\eta)\eta^{2q-1}).$$

Thus

$$\int f(\lambda)\psi(ag(\lambda))\,d\lambda$$
$$= \int_{|\eta|<1/\sqrt{a}} \psi(a\eta)\left(\sum_{l=0}^{2(q-1)} \frac{F^{(l)}(0)}{l!}\eta^l + O(F^{(2q-1)}(\eta)\eta^{2q-1})\right) d\eta$$
$$+ \int_{|\eta|\geq 1/\sqrt{a}} O((a|\eta|)^{-2q-1})|F(\eta)|\,d\eta,$$

where in the second integral we have used the rapid decay of ψ. The second integral is easy: since $|g'(\lambda)| \geq C_0$ on spt f, $|h'(\eta)| \leq C_0^{-1}$ and so $\|F\|_\infty \lesssim 1$, whence

$$\left|\int_{|\eta|\geq 1/\sqrt{a}} O((a|\eta|)^{-2q-1})|F(\eta)|\,d\eta\right| \lesssim \int_{|\eta|\geq 1/\sqrt{a}} (a|\eta|)^{-2q-1})\,d\eta = q^{-1}a^{-q-1},$$

which is what we want. In the first integral we use

$$\int_{\mathbb{R}} u^l \psi(u)\,du = 0.$$

This gives

$$\int_{|\eta|<1/\sqrt{a}} \psi(a\eta)\left(\sum_{l=0}^{2(q-1)} \frac{F^{(l)}(0)}{l!}\eta^l + O(F^{(2q-1)}(\eta)\eta^{2q-1})\right) d\eta$$
$$= -\int_{|\eta|\geq 1/\sqrt{a}} \psi(a\eta)\sum_{l=0}^{2(q-1)} \frac{F^{(l)}(0)}{l!}\eta^l\,d\eta + \int_{|\eta|<1/\sqrt{a}} O(F^{(2q-1)}(\eta)\eta^{2q-1})\,d\eta.$$

As the derivative of $h = (g|I)^{-1}$ is bounded below by a positive constant and all the derivatives of f and g are bounded by constants depending on the degree, it follows by routine calculus that $\|F^{(l)}\|_\infty \lesssim_l 1$ for $l = 0, 1, 2, \ldots$

and so

$$\int_{|\eta|<1/\sqrt{a}} |F^{(2q-1)}(\eta)\eta^{2q-1}|\, d\eta \lesssim a^{-q}.$$

Finally by the rapid decay of ψ and our assumption $a = 2^j r \geq 1$,

$$\left| \int_{|\eta|\geq 1/\sqrt{a}} \psi(a\eta)\eta^l\, d\eta \right| \lesssim \int_{1/\sqrt{a}}^{\infty} a^{-2q-l-1}\eta^{-2q-1}\, d\eta \leq a^{-q-1}.$$

These estimates give (18.23), and hence also (18.15) and (18.5).

Both (18.8) and (18.9) are derived from (18.5) with the help of the following Hausdorff dimension result. We present it in higher dimensions since it will also be used in Section 18.3. □

Lemma 18.8 *Let $U \subset \mathbb{R}^n$ be an open set, and let $h_j \in C^\infty(U)$, $j \in \mathbb{N}$. Suppose there exist finite constants $B > 1$, $R > 1$, $C > 0$ and $C_\eta > 0$ for all multi-indices $\eta \in \mathbb{N}_0^n$ such that (a) $\|\partial^\eta h_j\|_\infty \leq C_\eta B^{j|\eta|}$ and (b) $\|h_j\|_1 \leq C R^{-j}$ for $j \in \mathbb{N}$. Let $1 \leq r < R$.*

(i) If $B^n < R/r$, then $\sum_j r^j |h_j(\lambda)| < \infty$ for all $\lambda \in U$.
(ii) If $\sigma \in (0, n)$ is such that $B^\sigma \leq R/r$, then

$$\dim \left\{ \lambda \in U : \sum_{j\in\mathbb{N}} r^j |h_j(\lambda)| = \infty \right\} \leq n - \sigma.$$

Proof Let $0 < \beta < \sigma$, let Q be a compact cube in U, and set

$$E_j = \{\lambda \in Q : |h_j(\lambda)| \geq (j^2 r^j)^{-1}\}, \quad E = \bigcap_{k=1}^{\infty} \bigcup_{j=k}^{\infty} E_j.$$

Then

$$\left\{ \lambda \in Q : \sum_j r^j |h_j(\lambda)| = \infty \right\} \subset \bigcap_{k=1}^{\infty} \bigcup_{j=k}^{\infty} E_j.$$

Thus to prove (i) it suffices to show that $E_j = \varnothing$ for large enough j, and to prove (ii) it suffices to show that $\dim E_j \leq n - \beta$ for large enough j.

If $N \in \mathbb{N}$ and $u_1 + ty \in U$ for all $0 \leq t \leq N$, then N applications of the fundamental theorem of calculus together with the formula $\binom{N}{i} =$

$\binom{N-1}{i} + \binom{N-1}{i-1}$, $1 \le i \le N - 1$, yield

$$\left| \sum_{i=0}^{N} \binom{N}{i} (-1)^i h_j(u_1 + iy) \right|$$

$$= \left| \sum_{i=0}^{N-1} \binom{N-1}{i} (-1)^i (h_j(u_1 + (i+1)y) - h_j(u_1 + iy)) \right|$$

$$= \left| \sum_{i=0}^{N-1} \left(\binom{N-1}{i} (-1)^i \int_{u_1+iy}^{u_1+(i+1)y} h_j'(u_2) \, du_2 \right) \right|$$

$$\le \int_{u_1}^{u_1+y} \left| \sum_{i=0}^{N-1} \binom{N-1}{i} (-1)^i h_j'(u_2 + iy) \right| du_2$$

$$\le \cdots \le \int_{u_1}^{u_1+y} \int_{u_2}^{u_2+y} \cdots \int_{u_N}^{u_N+y} |h_j^{(N)}(u_{N+1})| \, du_{N+1} \cdots du_2$$

$$\le \|h_j^{(N)}\|_\infty |y|^N \le C_N (B^j |y|)^N,$$

where, of course, we are considering the h_j as one-dimensional functions on subline-segments of $[u_1, u_1 + Ny]$. So if the closed cube $Q(\lambda, NL)$ with centre λ and side-length $2NL$ is contained in U, we have for $y \in Q(0, NL)$,

$$|h_j(\lambda)| \le C_N B^{jN} |y|^N + \sum_{i=1}^{N} \binom{N}{i} |h_j(\lambda + iy)|.$$

Integrating gives

$$2^n L^n |h_j(\lambda)| \le C_N B^{jN} \int_{Q(0,L)} |y|^N \, dy + \sum_{i=1}^{N} \binom{N}{i} \int_{Q(0,L)} |h_j(\lambda + iy)| \, dy$$

$$\le 2^n n^N C_N B^{jN} L^{N+n} + \sum_{i=1}^{N} \binom{N}{i} \int_{Q(0,NL)} |h_j(\lambda + y)| \, dy$$

$$\le 2^n n^N C_N B^{jN} L^{N+n} + 2^N \int_{Q(0,NL)} |h_j(\lambda + y)| \, dy.$$

Suppose now that $\lambda \in E_j$ so that $|h_j(\lambda)| \ge (j^2 r^j)^{-1}$. We choose $L = L_{j,N}$ in order to have

$$2^n L_{j,N}^n |h_j(\lambda)| \le \frac{1}{2} 2^n L_{j,N}^n (j^2 r^j)^{-1} + 2^N \int_{Q(0,NL_{j,N})} |h_j(\lambda + y)| \, dy,$$

so that

$$\int_{Q(0,NL_{j,N})} |h_j(\lambda + y)| \, dy \ge \frac{1}{2} 2^{-N} (j^2 r^j)^{-1} (2L_{j,N})^n \ge 2^{-N} (j^2 r^j)^{-1} L_{j,N}^n.$$

$$(18.24)$$

This is achieved by $L = L_{j,N} := B^{-j}(C'_N j^2 r^j)^{-1/N}$, where $C'_N = 2n^N C_N$.

To prove (i), assume $B^n < R/r$ and choose $N \in \mathbb{N}$ so large that $r^{n/N} B^n < R/r$. If $\lambda \in E_j$, assumption (b), the inequality (18.24) and the definition of $L_{j,N}$ imply

$$
\begin{aligned}
C \geq R^j \int_{Q(0, NL_{j,N})} |h_j(\lambda + y)| \, dy &\geq R^j 2^{-N} (j^2 r^j)^{-1} L^n_{j,N} \\
&= R^j 2^{-N} (j^2 r^j)^{-1} \left(B^{-j}(C'_N j^2 r^j)^{-1/N} \right)^n \\
&= 2^{-N} C'^{-n/N}_N j^{-2(1+n/N)} \left(\frac{R}{B^n r^{1+n/N}} \right)^j .
\end{aligned}
$$

Since $\frac{R}{B^n r^{1+n/N}} > 1$, this shows that $j \in \mathbb{N}$ cannot be arbitrarily large, and E_j is thus empty for large $j \in \mathbb{N}$.

To deal with (ii), fix $N \in \mathbb{N}$ for a moment. For every $j \in \mathbb{N}$ choose closed cubes $Q_{j,1}, \ldots, Q_{j,m_j}$, of side-length $NL_{j,N}$ such that every $Q_{j,i}$ meets E_j, the cubes $Q_{j,k}$ and $Q_{j,l}$ have disjoint interiors for $k \neq l$, and $E_j \subset Q_{j,1} \cup \cdots \cup Q_{j,m_j}$. Pick points $\lambda_{j,i} \in E_j \cap Q_{j,i}$ and apply (18.24):

$$
m_j 2^{-N} (j^2 r^j)^{-1} L^n_{j,N} \leq \sum_{i=1}^{m_j} \int_{Q(\lambda_{j,i}, NL_{j,N})} |h_j(y)| \, dy \lesssim \|h_j\|_1 \leq C R^{-j},
$$

because any point in $\mathbb{R}^n$ can belong to at most boundedly many cubes $Q(\lambda_{j,i}, NL_{j,N}), i = 1, \ldots, m_j$. This gives

$$
m_j \leq 2^N j^2 r^j L^{-n}_{j,N} C R^{-j} = 2^N C C'^{n/N}_N j^{2(1+n/N)} \left(\frac{B^n r^{1+n/N}}{R} \right)^j .
$$

Now we choose N appropriately: suppose, as in (ii), that $B^\sigma \leq R/r$ and choose numbers $\beta \in (0, \sigma)$ and $N \in \mathbb{N}$ such that $B^\beta r^{\beta/N} < R/r$. Since $L_{j,N} \leq L_{k,N}$ for $j \geq k$, we have $d(Q_{j,i}) \leq nNL_{j,N} \leq nNL_{k,N}$ for $j \geq k$, and so

$$
\begin{aligned}
\mathcal{H}^{n-\beta}_{nNL_{k,N}} \left(\bigcup_{j=k}^{\infty} E_j \right) &\leq \sum_{j=k}^{\infty} \sum_{i=1}^{m_j} d(Q_{j,i})^{n-\beta} \leq \sum_{j=k}^{\infty} m_j (nNL_{j,N})^{n-\beta} \\
&\lesssim \sum_{j=k}^{\infty} j^{2(1+n/N)} \left(\frac{B^n r^{1+n/N}}{R} \right)^j (nNB^{-j}(C'_N j^2 r^j)^{-1/N})^{n-\beta} \\
&\approx \sum_{j=k}^{\infty} j^{2(1+\beta/N)} \left(\frac{B^\beta r^{1+\beta/N}}{R} \right)^j < \infty,
\end{aligned}
$$

since $B^\beta r^{1+\beta/N} < R$. This implies

$$\mathcal{H}^{n-\beta}(E) = \lim_{k\to\infty} \mathcal{H}^{n-\beta}_{nNL_{k,N}}(E) \le \liminf_{k\to\infty} \mathcal{H}^{n-\beta}_{nNL_{k,N}} \left(\bigcup_{j=k}^{\infty} E_j \right) = 0,$$

whence dim $E \le n - \beta$. Letting $\beta \to \sigma$ completes the proof. $\qquad\square$

The estimates (18.8) and (18.9) immediately follow from (18.5) once we have proved the following proposition, which we also formulate in higher dimensions:

Proposition 18.9 *Let (Ω, d) be a compact metric space, let $U \subset \mathbb{R}^n$ be open, and let*

$$\pi_\lambda : \Omega \to \mathbb{R}^m, \quad \lambda \in U.$$

Assume that the mapping $\lambda \mapsto \pi_\lambda(x)$ is in $C^\infty(U)$ for every fixed $x \in \Omega$ and

$$|\partial_\lambda^\eta \pi_\lambda(x)| \le C(\eta), \quad \lambda \in U,$$

for all multi-indices $\eta \in \mathbb{N}_0^n$. If $\mu \in \mathcal{M}(\Omega), s > 0$ and

$$\int_U \mathcal{I}_s(\mu_\lambda)\,d\lambda < \infty,$$

where $\mu_\lambda = \pi_{\lambda\sharp}\mu$, then for $0 \le t \le s$,

$$\dim\{\lambda \in U : \dim_S \mu_\lambda < t\} \le n - s + t \quad \text{if} \quad n - s + t \ge 0, \qquad (18.25)$$

and

$$\dim_S \mu_\lambda \ge s - n \quad \text{for all } \lambda \in U. \qquad (18.26)$$

Proof We may assume that $t < s$. Let $\psi \in \mathcal{S}(\mathbb{R}^m)$ be the function given by Lemma 18.6 and $\psi_{2^{-j}}(x) = 2^{jm}\psi(2^j x)$ so that $\widehat{\psi_{2^{-j}}}(x) = \widehat{\psi}(2^{-j}x)$. For $j \in \mathbb{Z}$, define the functions $h_j \in C^\infty(J)$ by

$$h_j(\lambda) = 2^{-jm} \int_{\mathbb{R}^m} \psi_{2^{-j}} * \mu_\lambda \, d\mu_\lambda = \int_\Omega \int_\Omega \psi(2^j(\pi_\lambda(x) - \pi_\lambda(y)))\,d\mu x\, d\mu y.$$

By Parseval's theorem, (3.27),

$$h_j(\lambda) = 2^{-jm} \int_{\mathbb{R}^m} \widehat{\psi}(2^j x)|\widehat{\mu_\lambda}(x)|^2\, dx \ge 0,$$

since $\widehat{\psi} \ge 0$.

Part (iii) of Lemma 18.6 tells us that for any $\sigma \in \mathbb{R}$,

$$\mathcal{I}_\sigma(\mu_\lambda) \approx \sum_{j\in\mathbb{Z}} 2^{j(\sigma-m)} \int_{\mathbb{R}^m} \psi_{2^{-j}} * \mu_\lambda\, d\mu_\lambda = \sum_{j\in\mathbb{Z}} 2^{j\sigma} h_j(\lambda). \qquad (18.27)$$

Therefore

$$\sum_{j \in \mathbb{Z}} \int_U 2^{js} h_j(\lambda)\, d\lambda \approx \int_U \mathcal{I}_s(\mu_\lambda)\, d\lambda < \infty,$$

so

$$\|h_j\|_1 \lesssim 2^{-js}.$$

For the partial derivatives of h_j we have

$$\|\partial^\eta h_j\|_\infty \lesssim_\eta 2^{j|\eta|}.$$

For any $\varepsilon > 0$ we have by (18.4) and (18.27)

$$\{\lambda \in J : \dim_S \mu_\lambda \le t\} \subset \{\lambda \in J : \mathcal{I}_{t+\varepsilon}(\mu_\lambda) = \infty\}$$

$$= \left\{\lambda \in J : \sum_{j \in \mathbb{Z}} 2^{j(t+\varepsilon)} h_j(\lambda) = \infty\right\}$$

$$= \left\{\lambda \in J : \sum_{j \in \mathbb{N}} 2^{j(t+\varepsilon)} h_j(\lambda) = \infty\right\},$$

since the functions h_j are uniformly bounded.

Suppose $n - s + t \ge 0$. Let $\varepsilon > 0$ with $t + \varepsilon < s$ and apply Lemma 18.8 with $B = 2, r = 2^{t+\varepsilon}$, $R = 2^s$ and $\sigma = s - t - \varepsilon$ to conclude that the last set has Hausdorff dimension at most $n - s + t + \varepsilon$. Letting $\varepsilon \to 0$ completes the proof of (18.25). Finally, if $t < s - n$, $\sum_{j \in \mathbb{N}} 2^{jt} h_j(\lambda) < \infty$ for all $\lambda \in U$ by Lemma 18.8, which gives (18.26). $\qquad\square$

18.2 Transversality of degree β

We shall now consider a weaker concept of transversality involving a non-negative parameter β. The case $\beta = 0$ is the one we have studied so far.

Definition 18.10 Let π_λ and Φ_λ be as in Definition 18.1 satisfying (18.1). The family π_λ, $\lambda \in J$, is said to satisfy *transversality of degree $\beta \ge 0$*, if there exists a positive constant C_β such that

$$|\Phi_\lambda(x, y)| \le C_\beta d(x, y)^\beta \implies |\partial_\lambda \Phi_\lambda(x, y)| \ge C_\beta d(x, y)^\beta \qquad (18.28)$$

for $\lambda \in J$ and $x, y \in \Omega$. The family π_λ, $\lambda \in J$, is said to satisfy *regularity of degree $\beta \ge 0$*, if to every $l \in \mathbb{N}$ there corresponds a positive constant $C_{\beta,l}$ such that

$$|\Phi_\lambda(x, y)| \leq C_\beta d(x, y)^\beta \implies |\partial_\lambda^l \Phi_\lambda(x, y)| \leq C_{\beta,l} d(x, y)^{-\beta l} \quad (18.29)$$

for $\lambda \in J$ and $x, y \in \Omega$.

The main theorem in this case is

Theorem 18.11 *There exists an absolute constant $b > 0$ such that the following holds. Let $\mu \in \mathcal{M}(\Omega)$ with $I_s(\mu) < \infty$ for some $s > 0$. Assume that the mappings π_λ, $\lambda \in J$, satisfy the transversality and regularity conditions of Definition 18.10 for some $\beta \geq 0$. Then*

$$\int_I \mathcal{I}_t(\mu_\lambda) \, d\lambda \leq C(\beta, s, t, I) I_s(\mu) \quad (18.30)$$

for compact intervals $I \subset J$ and for $0 < (1 + b\beta)t \leq s$. Therefore,

$$\dim_S \mu_\lambda \geq s/(1 + b\beta) \quad \text{for almost all } \lambda \in J. \quad (18.31)$$

Furthermore, for any $t \in (0, s - 3\beta]$ we have the estimate

$$\dim\{\lambda \in J : \dim_S \mu_\lambda < t\} \leq t. \quad (18.32)$$

For any $t \in (0, s]$ we have the estimate

$$\dim\{\lambda \in J : \dim_S \mu_\lambda < t\} \leq 1 + t - \frac{s}{1 + b\beta} \ \text{if} \ 1 + t - \frac{s}{1 + b\beta} \geq 0, \quad (18.33)$$

and

$$\dim_S \mu_\lambda \geq \frac{s}{1 + b\beta} - 1 \quad \text{for all } \lambda \in J. \quad (18.34)$$

The proof differs from that of Theorem 18.3 only in some technicalities. First Lemma 18.5 is replaced by:

Lemma 18.12 *Fix $x, y \in \Omega, x \neq y$, write $r = d(x, y)$ and suppose that (18.28) and (18.29) hold. Then*

$$\{\lambda \in J : |\Phi_\lambda(x, y)| < C_\beta r^\beta\} = \bigcup_{j=1}^N I_j,$$

where the I_j are disjoint open subintervals of J such that:

(i) *$\mathcal{L}^1(I_j) \leq 2$ for all j, $\mathcal{L}^1(I_j) \geq (2C_0/C_1)r^{2\beta}$ for all but at most two indices j, and $N \leq (C_1/2C_0)r^{-2\beta}\mathcal{L}^1(I_j)$.*

(ii) *The function $\lambda \mapsto \Phi_\lambda(x, y)$ is strictly monotone on any interval I_j.*

(iii) *There exist points $\lambda_j \in \overline{I_j}$, which satisfy: if $\lambda \in \overline{I_j}$, then $|\Phi_\lambda(x, y)| \geq |\Phi_{\lambda_j}(x, y)|$ and $|\Phi_\lambda(x, y)| \geq C_\beta r^\beta |\lambda - \lambda_j|$.*

(iv) *There exists a constant* $\delta > 0$ *depending only on* C_0 *and* β *with the following properties. Except for at most two exceptional values of* j, $\Phi_{\lambda_j}(x, y) = 0$ *and* $|\Phi_\lambda(x, y)| \leq C_\beta r^\beta/2$ *for* $|\lambda - \lambda_j| \leq \delta$. *For each of the two possible exceptional values of* j, *either* $|\Phi_\lambda(x, y)| \geq C_\beta r^\beta/4$ *for all* $\lambda \in I_j$ *or* $|\Phi_\lambda(x, y)| \leq C_\beta r^\beta/2$ *for all* $\lambda \in J$ *with* $|\lambda - \lambda_j| \leq \delta$. *For all* j, $J \cap (\lambda_j - \delta, \lambda_j + \delta) \subset I_j$.

The proof of this is about as simple as that of Lemma 18.5. The derivative of $\Phi_\lambda(x, y)$ satisfies now $r^\beta \lesssim |\Phi_\lambda(x, y)| \lesssim r^{-\beta}$ on the intervals I_j.

Secondly, the proof (18.32) is essentially the same as that of (18.7). The main complications arise in the proof of (18.30). The estimate (18.15) is replaced by

$$\left| \int_{\mathbb{R}} \varrho(\lambda) \psi(2^j(\pi_\lambda(x) - \pi_\lambda(y))) \, d\lambda \right| \lesssim (1 + 2^j d(x, y)^{1+C\beta})^{-q}. \tag{18.35}$$

Once this is established, the rest of the proof is practically the same as before. The deduction of (18.30) from (18.35) involves only inserting β (or β multiplied by a constant) in appropriate places. Lemma 18.8 is completely independent of β and in its application to get (18.33) β comes into play only in the ranges of parameters.

The steps to prove (18.15) can be used to prove also (18.35). The splitting in (18.18) and (18.19) is replaced by

$$\int_{\mathbb{R}} \varrho(\lambda) \psi(2^j[\pi_\lambda(x) - \pi_\lambda(y)]) \, d\lambda$$

$$= \int_{\mathbb{R}} \varrho(\lambda) \psi(2^j r \Phi(\lambda)) \varphi\big(C_\beta^{-1} r^{-\beta} \Phi(\lambda)\big) \, d\lambda$$

$$+ \int_{\mathbb{R}} \varrho(\lambda) \psi(2^j r \Phi(\lambda))\big(1 - \varphi\big(C_\beta^{-1} r^{-\beta} \Phi(\lambda)\big)\big) \, d\lambda,$$

and the splitting in (18.21) and (18.22) is replaced by

$$\int_{\mathbb{R}} \varrho(\lambda) \psi(2^j r \Phi(\lambda)) \varphi\big(C_\beta^{-1} r^{-\beta} \Phi(\lambda)\big) \, d\lambda$$

$$= \sum_{i=1}^{N} \int_{\mathbb{R}} \varrho(\lambda) \chi(r^{-2\beta}(\lambda - \lambda_i)) \psi(2^j r \Phi(\lambda)) \varphi\big(C_\beta^{-1} r^{-\beta} \Phi(\lambda)\big) \, d\lambda$$

$$+ \int_{\mathbb{R}} \varrho(\lambda) \left(1 - \sum_{i=1}^{N} \chi(r^{-2\beta}(\lambda - \lambda_i))\right) \psi(2^j r \Phi(\lambda)) \varphi(C_\beta^{-1} r^{-\beta} \Phi(\lambda)) \, d\lambda.$$

In the final calculations and estimations of derivatives of composite and inverse functions one has to be more careful, since the constants now depend on $d(x, y)$ and its powers. The details can be found in Peres and Schlag [2000].

18.3 Generalized projections in higher dimensions

In this section we discuss higher dimensional versions of the previous results. This means that the parameter space can be an open subset of a higher dimensional Euclidean space and the generalized projections can be vector valued. We still have (Ω, d) a compact metric space, now $Q \subset \mathbb{R}^n$ is an open connected set, and we have the mappings

$$\pi_\lambda : \Omega \to \mathbb{R}^m, \quad \lambda \in Q,$$

such that the mapping $\lambda \mapsto \pi_\lambda(x)$ is in $C^\infty(Q)$ for every fixed $x \in \Omega$, and to every compact $K \subset Q$ and any multi-index $\eta = (\eta_1, \ldots, \eta_n) \in \mathbb{N}_0^n$ there corresponds a positive constant $C_{\eta,K}$ such that

$$|\partial_\lambda^\eta \pi_\lambda(x)| \leq C_{\eta,K}, \quad \lambda \in K. \tag{18.36}$$

We shall give detailed proofs only in the case where the mappings are real-valued and satisfy the following strong transversality condition:

$$|\nabla_\lambda \Phi_\lambda(x, y)| \geq c > 0 \quad \text{for all } \lambda \in Q, x, y \in \Omega, x \neq y, \tag{18.37}$$

where again

$$\Phi_\lambda(x, y) = \frac{\pi_\lambda(x) - \pi_\lambda(y)}{d(x, y)} \quad \text{for } \lambda \in Q, \ x, y \in \Omega, x \neq y.$$

Theorem 18.13 *If $m = 1$ and (18.37) holds and $\mu \in \mathcal{M}(\Omega)$ with $I_s(\mu) < \infty$, then for any compact $K \subset Q$,*

$$\int_K |\widehat{\mu_\lambda}(u)|^2 \, d\lambda \leq C(s, K) I_s(\mu) |u|^{-s} \quad \text{for all } u \in \mathbb{R}^m. \tag{18.38}$$

In particular,

$$\int_K \mathcal{I}_t(\mu_\lambda) \, d\lambda \leq C(s, t, K) I_s(\mu) \tag{18.39}$$

for $0 < t < s$. Therefore,

$$\dim_S \mu_\lambda \geq s \quad \text{for almost all } \lambda \in Q. \tag{18.40}$$

Furthermore, for any $t \in (0, s]$ we have the estimates

$$\dim\{\lambda \in Q : \dim_S \mu_\lambda < t\} \leq n - 1 + t, \tag{18.41}$$

and

$$\dim\{\lambda \in Q : \dim_S \mu_\lambda < t\} \leq n - s + t \quad \text{if } n + t - s \geq 0. \tag{18.42}$$

Proof The proof of (18.38) is the same as that of (18.13); we just use Theorem 14.4 in place of Theorem 14.1. The estimate (18.42) follows from Proposition 18.9.

The proof (18.41) is simple. Define for a compact set $K \subset Q$, for $x, y \in \Omega$, $x \neq y$, and for $\delta > 0$,

$$S(K, x, y, \delta) = \{\lambda \in K : \Phi_\lambda(x, y) < \delta\}.$$

It follows from (18.37) that $\{\lambda \in Q : \Phi_\lambda(x, y) = 0\}$ is a smooth hypersurface such that for $x, y \in \Omega$, $x \neq y$, and $\delta > 0$, $S(K, x, y, \delta)$ can be covered with $C_K \delta^{1-n}$ balls of radius δ where C_K is independent of x, y and $\delta > 0$. Now we proceed as in the proof of the estimate (18.7) of Theorem 18.3. Suppose that (18.41) fails for some $t \in (0, s]$. Then for some compact set $K \subset Q$ and some $0 < \tau < t$,

$$\mathcal{H}^{n-1+t}(\{\lambda \in K : \dim_S \mu_\lambda < \tau\}) > 0.$$

Hence by Frostman's lemma there is a measure $\nu \in \mathcal{M}(\{\lambda \in K : \dim_S \mu_\lambda < \tau\})$ with $\nu(B(x, r)) \leq r^{n-1+t}$ for $x \in \mathbb{R}^n$ and $r > 0$. Thus

$$\nu(S(K, x, y, \delta)) \lesssim \delta^t.$$

We write now

$$\int_K I_\tau(\mu_\lambda) \, d\nu\lambda = \int_\Omega \int_\Omega \left(\int_K |\Phi_\lambda(x, y)|^{-\tau} \, d\nu\lambda \right) d(x, y)^{-\tau} \, d\mu x \, d\mu y.$$

The inner integral is bounded:

$$\int_K |\Phi_\lambda(x, y)|^{-\tau} \, d\nu\lambda = \int_0^\infty \nu(\{\lambda \in K : |\Phi_\lambda(x, y)|^{-\tau} > u\}) \, du$$

$$= \int_0^\infty \nu(S(K, x, y, u^{-1/\tau})) \, du \lesssim 1 + \int_1^\infty u^{-t/\tau} \, du \lesssim 1.$$

Therefore,

$$\int_K I_\tau(\mu_\lambda) \, d\nu\lambda \lesssim I_\tau(\mu) \lesssim I_s(\mu) < \infty.$$

This implies that $\dim_S \mu_\lambda \geq \tau$ for ν almost all $\lambda \in J$ which contradicts the choice of ν and finishes the proof. $\qquad \Box$

We have again the corollary:

Corollary 18.14 *Suppose again that $m = 1$ and (18.37) holds. Let $A \subset \Omega$ be a Borel set and $s = \dim A$.*

(a) *If $s \leq 1$ and $t \in (0, s]$, then*

$$\dim\{\lambda \in Q : \dim \pi_\lambda(A) < t\} \leq n - 1 + t.$$

(b) *If $1 < s \leq n + 1$, then*

$$\dim\{\lambda \in Q : \dim \pi_\lambda(A) < t\} \leq n - s + t \quad \text{if } n - s + t \geq 0$$

and

$$\dim\{\lambda \in Q : \mathcal{L}^1(\pi_\lambda(A)) = 0\} \leq n - s + 1.$$

(c) *If $2 < s \leq n + 2$, then*

$$\dim\{\lambda \in Q : \text{the interior of } \pi_\lambda(A) \text{ is empty}\} \leq n - s + 2.$$

Now we give the general definition of transversality in higher dimensions.

Definition 18.15 The family $\{\pi_\lambda, \lambda \in Q\}$, is said to satisfy *transversality of degree $\beta \geq 0$*, if there exists a positive constant C_β such that

$$|\Phi_\lambda(x, y)| \leq C_\beta d(x, y)^\beta \implies \det(D_\lambda\Phi_\lambda(x, y)(D_\lambda\Phi_\lambda(x, y)^t)) \geq C_\beta d(x, y)^\beta \tag{18.43}$$

for $\lambda \in Q$ and $x, y \in \Omega, x \neq y$. The family π_λ, $\lambda \in J$, is said to satisfy *regularity of degree $\beta \geq 0$*, if to every multi-index $\eta = (\eta_1, \ldots, \eta_n) \in \mathbb{N}^n$ there corresponds a positive constant $C_{\beta,\eta}$ such that

$$|\Phi_\lambda(x, y)| \leq C_\beta d(x, y)^\beta \implies |\partial_\lambda^\eta \Phi_\lambda(x, y)| \leq C_{\beta,\eta} d(x, y)^{-\beta|\eta|} \tag{18.44}$$

for $\lambda \in Q$ and $x, y \in \Omega, x \neq y$.

Here is the general theorem in higher dimensions:

Theorem 18.16 *There exists a constant $b > 0$ depending only on m and n such that the following holds. Let $\mu \in \mathcal{M}(\Omega)$ with $I_s(\mu) < \infty$ for some $s > 0$. Assume that the mappings π_λ, $\lambda \in Q$, satisfy the transversality and regularity conditions of Definition 18.15 for some $\beta \geq 0$. Then*

$$\int_K \mathcal{I}_t(\mu_\lambda)\, d\lambda \leq C(\beta, s, t, K) I_s(\mu) \tag{18.45}$$

for compact sets $K \subset Q$ and for $0 < (1 + b\beta)t \leq s$. Therefore,

$$\dim_S \mu_\lambda \geq s/(1 + b\beta) \quad \text{for almost all } \lambda \in Q. \tag{18.46}$$

Furthermore, for any $t \in (0, s - b\beta]$ we have the estimate

$$\dim\{\lambda \in Q : \dim_S \mu_\lambda < t\} \leq n + t - m. \tag{18.47}$$

For any $t \in (0, s]$ we have the estimate

$$\dim\{\lambda \in Q : \dim_S \mu_\lambda < t\} \leq \max\left\{0, n + t - \frac{s}{1 + b\beta}\right\}, \qquad (18.48)$$

and

$$\dim_S \mu_\lambda \geq \frac{s}{1 + b\beta} - n \quad for \ \lambda \in Q. \qquad (18.49)$$

The proof of Theorem 18.16 runs along similar lines as that of Theorem 18.3, or of Theorem 18.11 when $\beta > 0$. I only sketch here some parts in the case $\beta = 0$. First, in the proof of Lemma 18.5, which was needed for the change of variables in the proof of (18.5), we could use very elementary arguments based on monotonicity. In higher dimensions when $m = n$ we still have that if $\Phi_\lambda(x, y)$ is small its Jacobian determinant (with respect to λ) is non-zero and we have a quantitative estimate for it. Then the inverse function theorem (in a suitable quantitative form) gives useful local inverse mappings. When $m < n$ this is of course impossible, but then one can invert local restrictions of $\Phi_\lambda(x, y)$ to translates of some $n - m$ coordinate m-planes. Here is a higher dimensional version of Lemma 18.5:

Lemma 18.17 *Fix $x, y \in \Omega$ and write $r = d(x, y)$. Suppose that the transversality and regularity conditions of Definition 18.15 hold for $\beta = 0$. Let U be open with compact closure in Q. Then there exist positive constants C_1 and C_2 such that for some $\lambda_1, \ldots, \lambda_N \in U$ depending on x and y,*

$$\{\lambda \in U : |\Phi_\lambda(x, y)| < C_0\} \subset \bigcup_{j=1}^{N} B(\lambda_j, C_1),$$

where $N \leq C_2$ and the open balls $B_j = U(\lambda_j, 2C_1) \subset Q$ have the following properties. For each $j = 1, \ldots, N$, we can select $n - m$ coordinate directions $i_1 < \cdots < i_{n-m}$ such that for every $\kappa = (\kappa_1, \ldots, \kappa_{n-m})$ the restriction of $\lambda \mapsto \Phi_\lambda(x, y)$ to $\{\lambda \in B_j : \lambda_{i_1} = \kappa_1, \ldots, \lambda_{i_{n-m}} = \kappa_{n-m}\}$, say F_κ, is a diffeomorphism with

$$|\det(DF_\kappa)^{-1}| \leq C_2 \quad and \quad \|(DF_\kappa)^{-1}\| \leq C_2.$$

Let $V = \{\lambda \in U : |\Phi_\lambda(x, y)| < C_0\}$. To begin the proof, choose $C_1 < d(U, \partial Q)/2$ small enough so that we have by (18.43)

$$\det(D_\lambda \Phi_\lambda(x, y)(D_\lambda \Phi_\lambda(x, y)^t)) \geq C_0/2$$

when $d(\lambda, V) < 2C_1$. Then any balls $B(\lambda_j, C_1)$, $\lambda_j \in V$, covering V for which $\lambda_j \in V$ and the balls $B(\lambda_j, C_1/5)$ are disjoint will do; their existence follows by a standard covering lemma. The required coordinate directions are found by some linear algebra, namely by the Cauchy–Binet formula.

The main technical difficulty comes in the proof of the analogue of (18.15). The splitting into pieces goes still in the same way as before using Lemma 18.17, but the estimation of integrals such as in (18.21) requires more delicate higher dimensional calculus. For the details we refer to Peres and Schlag [2000].

18.4 Applications

18.4.1 Bernoulli convolutions

Recall from Chapter 9 that for $0 < \lambda < 1$ the Bernoulli convolution ν_λ is the probability distribution of

$$\sum_{j=0}^{\infty} \pm \lambda^j$$

where the signs are chosen independently with probability $1/2$. In that chapter we already presented them in a way that readily fits to the scheme of generalized projections. Namely,

$$\nu_\lambda = \Pi_{\lambda\sharp}\mu,$$

where

$$\Omega = \{-1, 1\}^{\mathbb{N}_0} = \{(\omega_j) : \omega_j = 1 \text{ or } \omega_j = -1, j = 0, 1, \dots\},$$

μ is the infinite product of the probability measure $(\delta_{-1} + \delta_1)/2$ with itself and

$$\Pi_\lambda : \Omega \to \mathbb{R}, \quad \Pi_\lambda(\omega) = \sum_{j=0}^{\infty} \omega_j \lambda^j.$$

There are many natural ways to make Ω a compact metric space. We fix an interval $J = (\lambda_0, \lambda_1), 0 < \lambda_0 < \lambda_1 < 1$, and use the metric

$$d(\omega, \tau) = \lambda_1^k \quad \text{where } k = \min\{j : \omega_j \neq \tau_j\},$$

when $\omega \neq \tau$. Then it is easy to check that

$$I_s(\mu) < \infty \iff \lambda_1^s > 1/2. \tag{18.50}$$

We discovered before that ν_λ is singular for $0 < \lambda < 1/2$ and it is absolutely continuous with L^2 density for almost all $\lambda \in (1/2, 1)$. The results of this chapter allow us to sharpen the information on the interval $(1/2, 1)$ by estimating the Hausdorff dimension of the exceptional set of λ. However, a lot will remain unanswered as it is not known if the exceptional set could be countable.

We set again

$$\Phi_\lambda(\omega, \tau) = \frac{\Pi_\lambda(\omega) - \Pi_\lambda(\tau)}{d(\omega, \tau)}, \quad \lambda \in (0, 1), \; \omega, \tau \in \Omega, \omega \neq \tau.$$

In order to apply the general results of this chapter we should verify the appropriate regularity and transversality conditions. In particular, we shall relate $\Phi_\lambda(\omega, \tau)$ to power series and β transversality to the δ transversality of power series as considered in Chapter 9.

Let $\omega, \tau \in \Omega$, $\omega \neq \tau$, and let k be the smallest integer j such that $\omega_j \neq \tau_j$. Taking into account that $d(\omega, \tau) = \lambda_1^k$ we can write as in the proof of Theorem 9.1,

$$\Phi_\lambda(\omega, \tau) = 2(\lambda^k / \lambda_1^k) g(\lambda), \tag{18.51}$$

where g is of the form (assuming that $\omega_k > \tau_k$)

$$g(\lambda) = 1 + \sum_{j=1}^\infty b_j \lambda^j \quad \text{with } b_j \in \{-1, 0, 1\}. \tag{18.52}$$

The derivatives of g are bounded in absolute value by those of $\sum_{j=0}^\infty \lambda^j = 1/(1 - \lambda)$. Hence for all $l = 0, 1, 2, \ldots$,

$$|g^{(l)}(\lambda)| \lesssim_l (1 - \lambda_1)^{-l} \quad \text{for } \lambda \in J.$$

Differentiating (18.51) we have for any $\beta > 0$ that $\partial_\lambda^{(l)} \Phi_\lambda(\omega, \tau)$ is a sum of 2^l terms of the form $2\lambda_1^{-k} k(k-1)\cdots(k-j)\lambda^{k-j} g^{(j)}(\lambda), 0 \leq j \leq l$, each of them in absolute value $\lesssim_{l,\lambda_1} k^l$. Hence

$$\left|\partial_\lambda^{(l)} \Phi_\lambda(\omega, \tau)\right| \lesssim k^l \lesssim \lambda_1^{-\beta l k} = d(\omega, \tau)^{-\beta l},$$

where the second inequality uses only the facts $0 < \lambda_1 < 1$ and $\beta > 0$. So we have the derivative bounds required by (18.29). For the transversality we shall use the following lemma:

Lemma 18.18 *Suppose that $J = [\lambda_0, \lambda_1]$, $\lambda_0 < \lambda_1$, is an interval of δ transversality in the sense of (9.6). If $\beta > 0$ and $\lambda_0 > \lambda_1^{1+\beta}$, then J is an interval of transversality of degree β.*

Proof Using the above notation, suppose $|\Phi_\lambda(\omega, \tau)| < C_\beta d(\omega, \tau)^\beta = \delta b_\beta \lambda_1^{\beta k}$ with $C_\beta = \delta b_\beta$, where the constant b_β will be determined below. Then by (18.51) for $\lambda \in J$,

$$2(\lambda_0^k / \lambda_1^k)|g(\lambda)| \leq 2(\lambda^k / \lambda_1^k)|g(\lambda)| \leq \delta b_\beta \lambda_1^{\beta k},$$

whence $|g(\lambda)| \leq \delta b_\beta (\lambda_0^{-1} \lambda_1^{1+\beta})^k / 2$. We choose b_β so that

$$b_\beta \leq \lambda_0 (\lambda_0^{-1} \lambda_1^{1+\beta})^{-k} / k \quad \text{for all } k \in \mathbb{N}.$$

Then $|g(\lambda)| \leq \delta/2 < \delta$ and so by δ tranversality $|g'(\lambda)| > \delta$. This gives

$$\begin{aligned}
|\partial_\lambda \Phi_\lambda(\omega, \tau)| &= \left| 2(\lambda_1^{-1}\lambda)^k (g'(\lambda) + k\lambda^{-1} g(\lambda)) \right| \\
&\geq 2(\lambda_1^{-1}\lambda_0)^k \left(|g'(\lambda)| - k\lambda_0^{-1} |g(\lambda)| \right) \\
&\geq 2\lambda_1^{\beta k}(\delta - \delta/2) = \delta \lambda_1^{\beta k} \geq C_\beta d(\omega, \tau)^\beta.
\end{aligned}$$

The last inequality is valid when we also choose $b_\beta \leq 1$. Thus the transversality condition (18.28) holds. $\qquad \square$

Theorem 18.19 *Suppose that $J = [\lambda_0, \lambda_0'] \subset (1/2, 1)$ is an interval of δ transversality in the sense of Chapter 9. Then*

$$\dim_s \nu_\lambda \geq \log 2 / (-\log \lambda) \quad \text{for almost all } \lambda \in J.$$

Moreover,

$$\dim\{\lambda \in J : \nu_\lambda \notin L^2(\mathbb{R})\} \leq 2 - \frac{\log 2}{-\log \lambda_0}.$$

Proof Let $\beta > 0$ be small and cover J with intervals $J_i = [\lambda_i, \lambda_{i+1}] \subset J, i = 1, \ldots, m$, such that $\lambda_i \geq (1 + \beta)\lambda_{i+1}^{1+\beta}$; this can be done when $\beta > 0$ is sufficiently small (depending on λ_0'). By Lemma 18.18 these are intervals of transversality of degree β, when we use the metric $d_i, d_i(\omega, \tau) = \lambda_{i+1}^k$ with k the smallest j such that $\omega_j \neq \tau_j$. Let $1 < \alpha_i < \frac{\log 2}{-\log \lambda_i}$, that is, $\lambda_{i+1}^{\alpha_i} > 1/2$. Then $I_{\alpha_i}(\mu) < \infty$ by (18.50) and Theorem 18.11 implies that

$$\dim_s \nu_\lambda \geq \alpha_i (1 + b\beta)^{-1} \quad \text{for almost all } \lambda \in J_i$$

and

$$\dim\{\lambda \in J_i : \nu_\lambda \notin L^2(\mathbb{R})\} \leq 2 - \alpha_i (1 + b\beta)^{-1}.$$

Letting $\alpha_i \to \frac{\log 2}{-\log \lambda_{i+1}}, \beta \to 0$ and observing that $\frac{\log 2}{-\log \lambda_0} \leq \frac{\log 2}{-\log \lambda_{i+1}}$ finishes the proof. $\qquad \square$

Recalling the discussion in Chapter 9, closed subintervals J of $[2^{-1}, 2^{-2/3})$ are intervals of δ transversality so the theorem applies to them. But, as remarked in Chapter 9, the theorem can only apply up to some $\lambda_0 < 1/\sqrt{2}$, in particular the upper bound in the second inequality is positive. One can proceed further, see Peres and Schlag [2000], and obtain estimates for the whole interval $(1/2, 1)$ and information about high order derivatives for λ close to 1:

Theorem 18.20 *For any $\lambda_0 > 1/2$ there are $\varepsilon(\lambda_0) > 0$ and $s(\lambda_0) > 0$ such that*

$$\dim\{\lambda \in (\lambda_0, 1) : \nu_\lambda \notin L^2(\mathbb{R})\} \leq 1 - \varepsilon(\lambda_0),$$

$\dim_s \nu_\lambda \geq s(\lambda_0)$ *for almost all $\lambda \in (\lambda_0, 1)$ and $s(\lambda_0) \to \infty$ as $\lambda_0 \to 1$.*

In their paper Peres and Schlag considered also asymmetric Bernoulli convolutions where the signs $+$ and $-$ in $\sum_{j=0}^{\infty} \pm\lambda^j$ are chosen with probabilities p and $1 - p$ for a given $0 < p < 1$. Recall also from Section 9.2 the paper of Shmerkin and Solomyak [2014] proving that for λ outside a set of dimension zero ν_λ belongs to L^p for some $p > 1$.

18.4.2 Pinned distance sets

Recall that the distance set of a Borel set $A \subset \mathbb{R}^n$ is by definition the following subset of the reals:

$$D(A) = \{|x - y| : x, y \in A\}.$$

The question we have discussed before is: what is the least number $c(n) > 0$ such that $\dim A > c(n)$ implies $\mathcal{L}^1(D(A)) > 0$? In Chapter 4 we gave a relatively simple proof yielding Falconer's estimate $c(n) \leq n/2 + 1/2$. In Chapter 16 we gave a very delicate proof for the best known result $c(n) \leq n/2 + 1/3$ due to Wolff and Erdoğan. We also saw in Chapter 4 that $c(n) \geq n/2$.

The distance sets are related to generalized projections via the mappings

$$d_y : \mathbb{R}^n \to \mathbb{R}, \quad d_y(x) = |x - y|, \quad y \in \mathbb{R}^n.$$

Then

$$D(A) = \bigcup_{y \in A} d_y(A).$$

The generalized projection theorems give us as a special case that $\dim A > (n + 1)/2$ implies $\mathcal{L}^1(D(A)) > 0$. But they give more, since they yield information about the *pinned distance sets*

$$D_y(A) = \{|x - y| : x \in A\}.$$

The required conditions are now easy to check. To obtain smoother maps we switch from d_y to π_λ:

$$\pi_\lambda : \mathbb{R}^n \to \mathbb{R}, \quad \pi_\lambda(x) = |x - \lambda|^2, \quad \lambda \in \mathbb{R}^n,$$

which of course does not change our problems. The regularity conditions are obvious. We now have, following our earlier notation,

$$\Phi_\lambda(x, y) = \frac{\pi_\lambda(x) - \pi_\lambda(y)}{|x - y|} = \frac{|\lambda - x|^2 - |\lambda - y|^2}{|x - y|}$$

$$= \frac{|x|^2 - |y|^2 + 2\lambda \cdot (y - x)}{|x - y|},$$

and thus

$$\nabla_\lambda \Phi_\lambda(x, y) = \frac{2(y - x)}{|x - y|},$$

so that

$$|\nabla_\lambda \Phi_\lambda(x, y)| = 2,$$

and the strong transversality as in (18.37) holds. Hence Corollary 18.14 gives:

Theorem 18.21 *For any Borel set $A \subset \mathbb{R}^n$,*

$$\dim\{y \in \mathbb{R}^n : \dim D_y(A) < t\} \le n + t - \max\{\dim A, 1\},$$
$$\dim\{y \in \mathbb{R}^n : \mathcal{L}^1(D_y(A)) = 0\} \le n + 1 - \dim A$$

and

$$\dim\{y \in \mathbb{R}^n : \text{Int}(D_y(A)) = \varnothing\} \le n + 2 - \dim A.$$

From this we immediately obtain:

Corollary 18.22 *Let $A \subset \mathbb{R}^n$ be a Borel set. If $\dim A > (n + 1)/2$, then there is $y \in A$ such that*

$$\mathcal{L}^1(D_y(A)) > 0.$$

If $\dim A > (n + 2)/2$, then there is $y \in A$ such that

$$\text{Int}(D_y(A)) \ne \varnothing.$$

We have also the following extension of Theorem 18.21:

Theorem 18.23 *For any Borel set $A \subset \mathbb{R}^n$ and any hyperplane $H \subset \mathbb{R}^n$,*

$$\dim\{y \in H : \dim D_y(A) < t\} \le n - 1 + t - \max\{\dim A, 1\},$$
$$\dim\{y \in H : \mathcal{L}^1(D_y(A)) = 0\} \le n - \dim A$$

and

$$\dim\{y \in H : \text{Int}(D_y(A)) = \varnothing\} \le n + 1 - \dim A.$$

To prove this we have to consider the mappings π_λ for $\lambda \in H$. Then we no longer have strong transversality, but transversality with $\beta = 0$ holds. For the proof see Peres and Schlag [2000], or the interested reader may want to check this as a rather easy exercise when $n = 2$.

18.5 Further comments

The results of this chapter are due to Peres and Schlag [2000]. As we mentioned earlier the crucial concept of transversality originates from Pollicott and Simon [1995].

Orponen [2014a] studied sliced measures under generalized projections in analogy to Chapter 6 extending some of the results of that chapter from plane sections to estimates of the dimensions of the level sets $\pi_\lambda^{-1}\{u\}$. In particular, he showed that for Bernoulli convolutions these level sets are typically uncountable, with a dimension estimate for the set of exceptional parameters λ. But good dimension estimates for the level sets themselves are lacking in this case.

Although using the above machinery we were able to extend Falconer's theorem, $\dim A > (n + 1)/2$ implies $\mathcal{L}^1(D(A)) > 0$, to pinned distance sets, we could not extend Theorem 4.6(a) to them. That is, it is an open problem whether $\dim A > (n + 1)/2$ implies $\text{Int } D_y(A) > 0$ for many, or even some, $y \in \mathbb{R}^n$. At the other extreme, it also is open whether $\dim A > n/2$ implies $\text{Int } D_y(A) > 0$ or $\mathcal{L}^1(D_y(A)) > 0$ for many, or even some, $y \in \mathbb{R}^n$, but these are open for the full distance sets $D(A)$, too.

D. M. Oberlin and R. Oberlin [2014] improved the first estimate in Theorem 18.21 to

$$\dim\{y \in \mathbb{R}^n : \dim D_y(A) < t\} \le n - 1 + 2t - \dim A.$$

They related the problem to mixed norm estimates for the spherical averaging operators S; $Sf(x, r) = \int_{S^{n-1}} f(x - rv)\,d\sigma^{n-1}v$.

Erdoğan, Hart and Iosevich [2013] proved that if $A \subset S^{n-1}$ is a Borel set with $\dim A > n/2$, then $\mathcal{L}^1(D_y(A)) > 0$ for many points $y \in A$. Again they derived this, as well as other consequences, from a projection theorem.

Shayya [2012] proved that if the Fourier transform of a finite Borel measure μ on $\mathbb{R}^n$ vanishes in the interior of a cone of opening less than π, then $\mathcal{L}^1(\{|x - y| : x \in G \cap \text{spt}\,\mu\}) > 0$ whenever $y \in \mathbb{R}^n$, $G \subset \mathbb{R}^n$ is open and $\mu(G) > 0$. He used the method of spherical averages from Chapter 15. In case the cone has opening greater than π, such a μ is absolutely continuous by classical results going back to Bochner.

Peres and Schlag [2000] gave a large number of applications of their theory. These include much stronger results on Bernoulli convolutions than described above, asymmetric Bernoulli convolutions; + and - taken with probabilities p and $1 - p$, the so-called $\{0, 1, 3\}$-problem; the Hausdorff dimension of $\{\sum_{j=0}^{\infty} a_j \lambda^j : a_j \in \{0, 1, 3\}\}$, dimensions of sums of Cantor sets, and dimensions of certain self-similar sets.

Applications to measures invariant under geodesic flow on manifolds were found by E. Järvenpää, M. Järvenpää and Leikas [2005] and continued by Hovila, E. Järvenpää, M. Järvenpää and Ledrappier [2012a] and [2012b]. This was based on a result of Ledrappier and Lindenstrauss [2003]: they proved that for two-dimensional surfaces the projection from the tangent bundle into the surface of such an invariant measure is absolutely continuous if the dimension of the measure is bigger than 2. Although there is only one projection the methods used for families of projections can be applied. The Järvenpääs and Leikas showed why this is so; they formulated the problem in terms of generalized projections and verified the required transversality. They also showed that on higher dimensional surfaces transversality is missing and, in fact, the analogous result is false.

Hovila, the Järvenpääs and Ledrappier [2012a] proved the analogue of the Besicovitch–Federer projection theorem for transversal families of generalized projections. This fundamental result of geometric measure theory says that if an $\mathcal{H}^m$ measurable set A with $\mathcal{H}^m(A) < \infty$ intersects every m-dimensional C^1 surface in zero $\mathcal{H}^m$ measure, then it projects to zero measure into almost all m-planes. Hovila [2014] verified that the proper submanifold of the Grassmannian $G(n, m)$ consisting of isotropic subspaces satisfies the transversality, and so combining with the afore-mentioned result, the Besicovitch–Federer projection theorem holds for these subspaces.

PART IV

Fourier restriction and Kakeya
type problems

19

Restriction problems

Here we introduce the restriction problem and conjecture, and we shall prove the basic Stein–Tomas restriction theorem.

19.1 The problem

When does $\widehat{f}|S^{n-1}$ make sense? If $f \in L^1(\mathbb{R}^n)$ it obviously does, since $\widehat{f}$ is a continuous function and as such defined uniquely at every point. If $f \in L^2(\mathbb{R}^n)$ it obviously does not, since the Fourier transform is an isometry of $L^2(\mathbb{R}^n)$ onto itself and consequently $\widehat{f}$ is only defined almost everywhere and nothing more can be said. In this chapter we shall see that for $f \in L^p(\mathbb{R}^n)$ the restriction $\widehat{f}|S^{n-1}$ does make sense also for some $1 < p < 2$. This follows immediately if we have for some $q < \infty$ an inequality

$$\|\widehat{f}\|_{L^q(S^{n-1})} \leq C(n, p, q)\|f\|_{L^p(\mathbb{R}^n)} \tag{19.1}$$

valid for all $f \in \mathcal{S}(\mathbb{R}^n)$. Then the linear operator $f \mapsto \widehat{f}$ has a unique continuous extension to $L^p(\mathbb{R}^n) \to L^q(S^{n-1})$ by the denseness of $\mathcal{S}(\mathbb{R}^n)$ in $L^p(\mathbb{R}^n)$. Hence the Fourier transform $\widehat{f}$ is defined as an L^q function in S^{n-1} satisfying (19.1). Later on when we write inequalities like (19.1) for $f \in L^p$, they should usually be understood in the above sense.

The restriction problems ask for which p and q (19.1) holds. This is open in full generality, but we shall prove a sharp result when $q = 2$.

By duality (19.1) is equivalent, with the same constant $C(n, p, q)$, to

$$\|\widehat{f}\|_{L^{p'}(\mathbb{R}^n)} \leq C(n, p, q)\|f\|_{L^{q'}(S^{n-1})}. \tag{19.2}$$

Here p' and q' are the conjugate exponents of p and q and $\widehat{f}$ means the Fourier transform of the measure $f\sigma^{n-1}$. The inequalities of this type are called *extension inequalities*.

269

The equivalence of (19.1) and (19.2) is contained in the more general Proposition 19.1. For the proof of this proposition notice that for any $\mu \in \mathcal{M}(\mathbb{R}^n)$ the Schwartz space $\mathcal{S}(\mathbb{R}^n)$ is dense in $L^p(\mu)$ when $1 \le p < \infty$. This follows easily by Lusin's theorem and the Weierstrass approximation theorem. Hence

$$\|g\|_{L^{p'}(\mu)} = \sup \left\{ \int gh \, d\mu : h \in \mathcal{S}(\mathbb{R}^n), \|h\|_{L^p(\mu)} \le 1 \right\}.$$

This formula holds also when $p = \infty$ as one can easily check separately.

Proposition 19.1 *Let $1 \le p, q \le \infty$ and let $\mu \in \mathcal{M}(\mathbb{R}^n)$. The following are equivalent for any $0 < C < \infty$:*

(1) $\|\widehat{f}\|_{L^q(\mu)} \le C\|f\|_{L^p(\mathbb{R}^n)}$ *for all $f \in \mathcal{S}(\mathbb{R}^n)$.*
(2) $\|\widehat{f\mu}\|_{L^{p'}(\mathbb{R}^n)} \le C\|f\|_{L^{q'}(\mu)}$ *for all $f \in \mathcal{S}(\mathbb{R}^n)$.*
   In the case $q = 2$, (1) and (2) are equivalent to
(3) $\|\widehat{\mu} * f\|_{L^{p'}(\mathbb{R}^n)} \le C^2\|f\|_{L^p(\mathbb{R}^n)}$ *for all $f \in \mathcal{S}(\mathbb{R}^n)$.*

Proof Suppose (1) holds and let $g \in \mathcal{S}(\mathbb{R}^n)$ with $\|g\|_{L^p(\mathbb{R}^n)} \le 1$. Then by (3.20)

$$\int \widehat{f}\mu g = \int f\widehat{g}d\mu \le \|f\|_{L^{q'}(\mu)}\|\widehat{g}\|_{L^q(\mu)}$$
$$\le C\|f\|_{L^{q'}(\mu)}\|g\|_{L^p(\mathbb{R}^n)} \le C\|f\|_{L^{q'}(\mu)}.$$

Taking supremum over $g \in \mathcal{S}(\mathbb{R}^n)$ with $\|g\|_{L^p(\mathbb{R}^n)} \le 1$ gives (2). Then (1) follows from (2) with a similar argument.

To deal with (3) we use the formula

$$\int (\widehat{\mu} * \overline{f})g = \int \overline{\widehat{f}\widehat{g}} \, d\mu,$$

valid for all $f, g \in \mathcal{S}(\mathbb{R}^n)$, recall (3.28). If (1) holds for $q = 2$, we have thus for $f, g \in \mathcal{S}(\mathbb{R}^n)$,

$$\left| \int (\widehat{\mu} * \overline{f})g \right| = \left| \int \overline{\widehat{f}\widehat{g}} \, d\mu \right| \le \|\widehat{f}\|_{L^2(\mu)}\|\widehat{g}\|_{L^2(\mu)} \le C^2\|f\|_{L^p(\mathbb{R}^n)}\|g\|_{L^p(\mathbb{R}^n)},$$

which yields (3). Finally, if (3) holds, (1) follows applying the above formula with $f = g$. $\square$

Remark 19.2 If one of the conditions (1)–(3) holds, then it holds for all f in the corresponding Lebesgue space, in the sense of extended operators as above. For (1) and (3) this follows from the denseness of $\mathcal{S}(\mathbb{R}^n)$. Since μ has compact support, we do not need the extension argument for (2), because then $L^{q'}(\mu) \subset L^1(\mu)$ and $\widehat{f\mu}$ is a pointwise defined continuous function.

19.2 Stein–Tomas restriction theorem

We shall now prove a sharp restriction theorem due to Tomas and Stein from the 1970s. We formulate it for general measures. When $\alpha = n - 1$ and $\beta = (n - 1)/2$ we get from Theorem 14.7 a large class of surface measures which satisfy these assumptions. But the theorem also applies to many lower dimensional surfaces and fractal measures.

Theorem 19.3 *Let $0 < \alpha < n, \beta > 0$ and let $\sigma \in \mathcal{M}(\mathbb{R}^n)$ be such that*

$$\sigma(B(x,r)) \le C(\sigma)r^\alpha \quad \text{for } x \in \mathbb{R}^n, \quad r > 0, \tag{19.3}$$

and

$$|\hat{\sigma}(\xi)| \le C(\sigma)(1 + |\xi|)^{-\beta} \quad \text{for } \xi \in \mathbb{R}^n. \tag{19.4}$$

Then we have for $f \in L^2(\sigma)$,

$$\|\widehat{f\sigma}\|_{L^q(\mathbb{R}^n)} \le C(n, q, \alpha, \beta, C(\sigma))\|f\|_{L^2(\sigma)}$$

for $q > 2(n + \beta - \alpha)/\beta$.

Notice that measures σ satisfying both assumptions can only exist if $\beta \le \alpha/2$ and the case $\beta = \alpha/2$ corresponds to the Salem set situation, recall Section 3.6.

Proof It is enough to consider $q < \infty$. By Proposition 19.1 the claim is equivalent, with $C = C(n, q, \alpha, \beta, C(\sigma))$, to

$$\|\hat{\sigma} * f\|_q \le C^2\|f\|_{q'} \quad \text{for } f \in \mathcal{S}(\mathbb{R}^n). \tag{19.5}$$

Let $\chi \in C^\infty(\mathbb{R}^n)$ be such that $\chi \ge 0, \chi(x) = 1$, when $|x| \ge 1$, and $\chi(x) = 0$, when $|x| \le 1/2$, and set

$$\varphi(x) = \chi(2x) - \chi(x).$$

Then

$$\operatorname{spt}\varphi \subset \{x \in \mathbb{R}^n : 1/4 \le |x| \le 1\},$$

and

$$\sum_{j=0}^\infty \varphi(2^{-j}x) = 1 \quad \text{when} \quad |x| \ge 1.$$

Write

$$\hat{\sigma} = K + \sum_{j=0}^{\infty} K_j,$$

$$K_j(x) = \varphi(2^{-j}x)\hat{\sigma}(x),$$

$$K(x) = \left(1 - \sum_{j=0}^{\infty} \varphi(2^{-j}x)\right)\hat{\sigma}(x).$$

Then K and K_j are Lipschitz functions with compact support, spt $K \subset B(0,1)$, and spt $K_j \subset \{x : 2^{j-2} \le |x| \le 2^j\}$. Young's inequality for convolution (see for example Grafakos [2008], Theorem 1.2.12) states that

$$\|g * h\|_q \le \|g\|_p \|h\|_r \quad \text{when } 1 \le p, q, r \le \infty, \frac{1}{q} + 1 = \frac{1}{p} + \frac{1}{r}.$$

Applying this with $g = K, h = f, p = q/2$ and $r = q'$ and using $\|K\|_p \lesssim \|K\|_\infty \lesssim 1$, we obtain

$$\|K * f\|_q \lesssim \|f\|_{q'}. \tag{19.6}$$

For $j = 0, 1, \ldots$, we have by (19.4),

$$\|K_j\|_\infty \lesssim 2^{-\beta j}.$$

Thus

$$\|K_j * f\|_\infty \lesssim 2^{-\beta j} \|f\|_1.$$

Define $\psi, \psi_j \in \mathcal{S}(\mathbb{R}^n)$ by

$$\psi = \breve{\varphi}, \ \psi_j(x) = 2^{nj}\psi(2^j x).$$

Then $\hat{\psi} = \varphi$ and $\widehat{\psi_j}(x) = \varphi(2^{-j}x)$ so that $K_j = \widehat{\psi_j}\hat{\sigma} = \widehat{\psi_j * \sigma}$ by (3.22). Thus by (3.9) $\widehat{K_j} = \tilde{\psi_j} * \sigma$ where $\tilde{g}(x) = g(-x)$. Hence

$$|\widehat{K_j}(\xi)| = \left|2^{nj}\int \psi(2^j(-\xi - \eta))\,d\sigma\eta\right| \lesssim 2^{nj}\int (1 + 2^j|\xi + \eta|)^{-n}\,d\sigma\eta,$$

since $\psi \in \mathcal{S}(\mathbb{R}^n)$. Thus

$$|\widehat{K_j}(\xi)| \lesssim 2^{nj} \left(\int_{B(-\xi, 2^{-j})} (1 + 2^j|\xi + \eta|)^{-n} \, d\sigma\eta \right.$$

$$\left. + \sum_{k=0}^{\infty} \int_{B(-\xi, 2^{k+1-j}) \setminus B(-\xi, 2^{k-j})} (1 + 2^j|\xi + \eta|)^{-n} \, d\sigma\eta \right)$$

$$\leq 2^{nj} \left(\sigma(B(-\xi, 2^{-j})) + \sum_{k=0}^{\infty} 2^{-nk} \sigma(B(-\xi, 2^{k+1-j})) \right)$$

$$\lesssim 2^{nj} \left(2^{-\alpha j} + \sum_{k=0}^{\infty} 2^{-nk} 2^{\alpha(k-j)} \right) = C(n, \alpha) 2^{(n-\alpha)j}.$$

This gives for $f \in \mathcal{S}(\mathbb{R}^n)$,

$$\|K_j * f\|_2 = \|\widehat{K_j * f}\|_2 = \|\widehat{K_j}\widehat{f}\|_2 \lesssim 2^{(n-\alpha)j}\|f\|_2.$$

Above we had

$$\|K_j * f\|_\infty \lesssim 2^{-\beta j}\|f\|_1.$$

Let $\theta \in (0, 1)$ be defined by $\theta/2 + (1 - \theta)/\infty = 1/q$, that is, $\theta = 2/q$. Then by the Riesz–Thorin interpolation theorem 2.12,

$$\|K_j * f\|_q \lesssim 2^{(n-\alpha)j\theta} 2^{-\beta j(1-\theta)}\|f\|_{q'} = 2^{j(2(n+\beta-\alpha)/q-\beta)}\|f\|_{q'}.$$

Since $q > 2(n + \beta - \alpha)/\beta$, we have $2(n + \beta - \alpha)/q - \beta < 0$, so

$$\sum_{j=0}^{\infty} \|K_j * f\|_q \lesssim \|f\|_{q'}.$$

By (19.6) we also have,

$$\|K * f\|_q \lesssim \|f\|_{q'}.$$

This and the representation $\widehat{\sigma} = K + \sum_{j=0}^{\infty} K_j$ give the required inequality (19.5). $\qquad\square$

We now state the result for the sphere:

Theorem 19.4 *We have for $f \in L^2(S^{n-1})$,*

$$\|\widehat{f}\|_{L^q(\mathbb{R}^n)} \leq C(n, q)\|f\|_{L^2(S^{n-1})}$$

for $q \geq 2(n + 1)/(n - 1)$. The lower bound $2(n + 1)/(n - 1)$ is the best possible.

Proof For $q > 2(n + 1)/(n - 1)$ this follows from Theorem 19.3. For the end point result, see Stein [1993], Section IX.2.1; we shall give a sketch for this in the next chapter.

We prove the sharpness using the Knapp example from Lemma 3.18. So let $e_n = (0, \ldots, 0, 1) \in \mathbb{R}^n$, $0 < \delta < 1$,

$$C_\delta = \{x \in S^{n-1} : 1 - x \cdot e_n \leq \delta^2\},$$

and $f = \chi_{C_\delta}$. Then

$$\|f\|_{L^2(S^{n-1})} = \sigma^{n-1}(C_\delta)^{1/2} \approx \delta^{(n-1)/2}. \tag{19.7}$$

By Lemma 3.18, with $c = 1/(12n)$,

$$|\widehat{f}(\xi)| \geq \sigma^{n-1}(C_\delta)/2 \quad \text{for } \xi \in R_\delta,$$

where

$$R_\delta = \{\xi \in \mathbb{R}^n : |\xi_j| \leq c/\delta \text{ for } j = 1, \ldots, n - 1, |\xi_n| \leq c/\delta^2\}.$$

Since $\mathcal{L}^n(R_\delta) = 2^n c^n \delta^{-n-1}$, we get

$$\|\widehat{f}\|_{L^q(\mathbb{R}^n)} \geq (\sigma^{n-1}(C_\delta)/2)\mathcal{L}^n(R_\delta)^{1/q} \approx \delta^{n-1-(n+1)/q}.$$

Combining with (19.7) we see that in order to have

$$\|\widehat{f}\|_{L^q(\mathbb{R}^n)} \lesssim \|f\|_{L^2(S^{n-1})} \approx \delta^{(n-1)/2},$$

we must have $\delta^{n-1-(n+1)/q} \lesssim \delta^{(n-1)/2}$ for small δ, which means $n - 1 - (n + 1)/q \geq (n - 1)/2$, that is, $q \geq 2(n + 1)/(n - 1)$ as claimed. $\qquad\square$

The dual inequality for Theorem 19.4 is

$$\|\widehat{f}\|_{L^2(S^{n-1})} \lesssim \|f\|_{L^p(\mathbb{R}^n)}, \quad 1 \leq p \leq \frac{2(n + 1)}{n + 3},$$

which of course is also sharp. We shall illustrate the sharpness of it in the plane by a slightly different example. When $n = 2$, $\frac{2(n+1)}{n+3} = \frac{6}{5}$. For $0 < \delta < 1$, consider the annulus

$$A_\delta = \{\xi \in \mathbb{R}^2 : 1 - \delta \leq |\xi| \leq 1 + \delta\}.$$

Our inequality is equivalent to

$$\int_{A_\delta} |\widehat{f}|^2 \lesssim \delta \left(\int_{\mathbb{R}^2} |f|^p \right)^{2/p}; \tag{19.8}$$

this is easily checked, or one can consult Proposition 16.2. If $c > 0$ is small enough, the rectangle

$$R_\delta = \{\xi \in \mathbb{R}^2 : |\xi_1 - 1| \leq c\delta, |\xi_2| \leq c\sqrt{\delta}\}$$

is contained in the annulus A_δ. Let $g \in \mathcal{S}(\mathbb{R})$ with $\widehat{g}(\xi) \geq 1$ when $|\xi| \leq c$ and define f by

$$f(x_1, x_2) = g(\delta x_1) e^{-2\pi i x_1} g(\sqrt{\delta} x_2) \delta^{3/2},$$

which means that

$$\widehat{f}(\xi_1, \xi_2) = \widehat{g}((\xi_1 - 1)/\delta) \widehat{g}(\xi_2/\sqrt{\delta}).$$

Thus $\widehat{f}(\xi) \geq 1$ when $\xi \in R_\delta$. Then, if (19.8) holds,

$$\int_{R_\delta} |\widehat{f}|^2 \leq \int_{A_\delta} |\widehat{f}|^2 \lesssim \delta \left(\int_{\mathbb{R}^2} |f|^p \right)^{2/p}.$$

Plugging in the formulas for f and $\widehat{f}$ and changing variables, we derive from this

$$\delta^{3/2} \lesssim \delta \left(\int_{-\infty}^{\infty} |g(\delta x_1)|^p \, dx_1 \int_{-\infty}^{\infty} |g(\sqrt{\delta} x_2)|^p \, dx_2 \delta^{3p/2} \right)^{2/p}$$
$$\approx \delta (\delta^{-1} \delta^{-1/2} \delta^{3p/2})^{2/p} = \delta^{4-3/p},$$

which yields the desired $p \leq 6/5$.

19.3 Restriction conjecture

Let us now contemplate on which pairs (p, q), $1 \leq p, q \leq \infty$, the inequalities (19.1) and (19.2) might hold. It is enough to look at only one of them, since they are equivalent. Let us choose (19.2) and write it as

$$\|\widehat{f}\|_{L^q(\mathbb{R}^n)} \leq C(n, p, q) \|f\|_{L^p(S^{n-1})}. \tag{19.9}$$

The first easy observation is that if (19.9) holds for some pair (p, q), then it holds for every pair $(\widetilde{p}, \widetilde{q})$ with $\widetilde{p} \geq p$ and $\widetilde{q} \geq q$. For p this follows from Hölder's inequality. Since $\|\widehat{f}\|_{L^\infty(\mathbb{R}^n)} \leq \|f\|_{L^1(S^{n-1})} \lesssim \|f\|_{L^p(S^{n-1})}$ we can argue for q,

$$\int |\widehat{f}/\|\widehat{f}\|_{L^\infty(\mathbb{R}^n)}|^{\widetilde{q}} \leq \int |\widehat{f}/\|\widehat{f}\|_{L^\infty(\mathbb{R}^n)}|^q \lesssim \|\widehat{f}\|_{L^\infty(\mathbb{R}^n)}^{-q} \|f\|_{L^p(S^{n-1})}^q,$$

whence

$$\int |\widehat{f}|^{\widetilde{q}} \lesssim \|\widehat{f}\|_{L^\infty(\mathbb{R}^n)}^{\widetilde{q}-q} \|f\|_{L^p(S^{n-1})}^q \lesssim \|f\|_{L^p(S^{n-1})}^{\widetilde{q}}.$$

By (3.41) and the asymptotic formula (3.37) for the Bessel functions, $\widehat{\sigma^{n-1}} \notin L^{2n/(n-1)}(\mathbb{R}^n)$. Hence in order that (19.9) could be valid when $f \equiv 1$, we must have

$$q > \frac{2n}{n-1}. \tag{19.10}$$

A second restriction comes from the example in the proof of Theorem 19.4: if we replace there 2 with p, (19.7) is replaced by $\|f\|_{L^p(S^{n-1})} \approx \delta^{(n-1)/p}$ and we arrive at $n - 1 - (n + 1)/q \geq (n - 1)/p$, that is, $q \geq (n + 1)p'/(n - 1)$. So in order that (19.9) could be valid we must have

$$q \geq \frac{n+1}{n-1}p'. \tag{19.11}$$

When $p = 2$, $\frac{n+1}{n-1}p' = \frac{2(n+1)}{n-1}$ is the exponent of the Stein–Tomas theorem 19.4, whence (19.11) is also a sufficient condition in this case. Interpolating this (using the Riesz–Thorin interpolation theorem 2.12) with the trivial inequality

$$\|\widehat{f}\|_{L^\infty(\mathbb{R}^n)} \leq \|f\|_{L^1(S^{n-1})},$$

we find that (19.9) holds if

$$q \geq \frac{2(n+1)}{n-1} \quad \text{and} \quad q = \frac{n+1}{n-1}p', \tag{19.12}$$

or equivalently if

$$q \geq \frac{2(n+1)}{n-1} \quad \text{and} \quad q \geq \frac{n+1}{n-1}p'. \tag{19.13}$$

The *restriction conjecture* asks whether this could be extended from the range $q \geq 2(n + 1)/(n - 1)$ to the optimal range $q > 2n/(n - 1)$.

Conjecture 19.5 $\|\widehat{f}\|_{L^q(\mathbb{R}^n)} \leq C(n, q)\|f\|_{L^p(S^{n-1})}$ for $q > 2n/(n - 1)$ and $q = \frac{n+1}{n-1}p'$.

The restriction conjecture is true in the plane and we shall discuss the proof in the next chapter.

A seemingly weaker conjecture is whether this could hold with $p = \infty$:

Conjecture 19.6 $\|\widehat{f}\|_{L^q(\mathbb{R}^n)} \leq C(n, q)\|f\|_{L^\infty(S^{n-1})}$ for $q > 2n/(n - 1)$.

We shall now proceed to prove a result of Bourgain saying that, in fact, these two conjectures are equivalent. Moreover, we add one more:

Conjecture 19.7 $\|\widehat{f}\|_{L^q(\mathbb{R}^n)} \leq C(n, q)\|f\|_{L^q(S^{n-1})}$ for $q > 2n/(n - 1)$.

Theorem 19.8 *The conjectures 19.5, 19.6 and 19.7 are equivalent.*

Obviously Conjecture 19.7 implies Conjecture 19.6. Once we have proved that Conjecture 19.6 implies Conjecture 19.7, the equivalence of 19.5 and 19.6

follows by interpolation; observe that if $q = 2n/(n-1)$ and $q = \frac{n+1}{n-1}p'$, then $p = q$. We turn to the dual statements and prove the following theorem, which gives immediately Theorem 19.8 by 'dualizing back'. More precisely, as a special case of Theorem 19.9 we have that for $1 \le p \le 2$,

$$\|\widehat{f}\|_{L^1(S^{n-1})} \lesssim \|f\|_{L^p(\mathbb{R}^n)} \quad \text{implies} \quad \|\widehat{f}\|_{L^q(S^{n-1})} \lesssim \|f\|_{L^q(\mathbb{R}^n)} \quad \text{for } 1 \le q < p.$$

The dual of this is: for $2 \le p \le \infty$,

$$\|\widehat{f}\|_{L^p(\mathbb{R}^n)} \lesssim \|f\|_{L^\infty(S^{n-1})} \quad \text{implies} \quad \|\widehat{f}\|_{L^q(\mathbb{R}^n)} \lesssim \|f\|_{L^q(\mathbb{R}^n)} \text{ for } q > p.$$

Theorem 19.9 *Suppose that* $1 \le p \le 2, 0 < C_0 < \infty$, *and*

$$\sigma^{n-1}(\{x \in S^{n-1} : |\widehat{f}(x)| > \lambda\}) \le C_0 \lambda^{-1} \|f\|_{L^p(\mathbb{R}^n)} \quad \text{for } \lambda > 0, \ f \in \mathcal{S}(\mathbb{R}^n). \tag{19.14}$$

Then

$$\sigma^{n-1}(\{x \in S^{n-1} : |\widehat{f}(x)| > \lambda\}) \le C(n, p, C_0)\lambda^{-p}\|f\|_{L^p(\mathbb{R}^n)}^p$$
$$\text{for } \lambda > 0, \ f \in \mathcal{S}(\mathbb{R}^n), \tag{19.15}$$

and

$$\|\widehat{f}\|_{L^q(S^{n-1})} \le C(n, q, C_0)\|f\|_{L^q(\mathbb{R}^n)} \quad \text{for } 1 \le q < p, \ f \in L^q(\mathbb{R}^n). \tag{19.16}$$

Proof Let $\sigma = \sigma^{n-1}/\sigma^{n-1}(S^{n-1})$ be the normalized surface measure on the unit sphere. The estimate (19.16) follows from (19.15) by the Marcinkiewicz interpolation theorem 2.13; interpolate (19.15) with the trivial estimate $\|\widehat{f}\|_{L^1(\sigma)} \le \|f\|_{L^1(\mathbb{R}^n)}$. Suppose now that (19.14) holds. We first prove that if $\lambda > 0$ and $f_1, f_2, \ldots \in \mathcal{S}(\mathbb{R}^n)$ with $\sum_{j=1}^\infty \|f_j\|_{L^p(\mathbb{R}^n)}^p \le 1$, then

$$\sigma\left(\left\{x \in S^{n-1} : \sup_j |\widehat{f_j}(x)| > \lambda\right\}\right) \le C_1 \lambda^{-p/(1+p)} \tag{19.17}$$

with C_1 depending only on C_0, n and p. We shall use Khintchine's inequality for this. It is enough to prove the asserted inequality for finitely many functions $f_1, \ldots, f_m \in \mathcal{S}(\mathbb{R}^n)$. Let, as in Section 2.8, $\Omega = \{-1, 1\}^{\mathbb{N}}$ and let P be the infinite product of the measures $\frac{1}{2}(\delta_{-1} + \delta_1)$. For $\omega \in \Omega$ define g_ω by

$$g_\omega(x) = \sum_{j=1}^m \omega_j f_j(x), \quad x \in \mathbb{R}^n,$$

so that $\widehat{g_\omega}(x) = \sum_{j=1}^m \omega_j \widehat{f_j}(x)$. For fixed $x \in S^{n-1}$ and $j = 1, \ldots, m$, we have $P(\{\omega : |\widehat{g_\omega}(x)| \ge |\widehat{f_j}(x)|\}) \ge 1/2$. This follows from the fact that for any complex numbers a and b, either $|a + b| \ge |b|$ or $|a - b| \ge |b|$. Hence for a fixed x,

$$P\left(\left\{\omega : |\widehat{g_\omega}(x)| \geq \sup_j |\widehat{f_j}(x)|\right\}\right) \geq 1/2. \tag{19.18}$$

Set

$$E = \{\omega \in \Omega : \|\widehat{g_\omega}\|_{L^p(\mathbb{R}^n)} > \lambda^\theta\} \quad \text{with} \quad \theta = \frac{1}{p+1}.$$

Then using (19.18) and the hypothesis (19.14) we obtain

$$\sigma(\{x \in S^{n-1} : \sup_j |\widehat{f_j}(x)| > \lambda\})$$

$$\leq 2 \int_{\{x : \sup_j |\widehat{f_j}(x)| > \lambda\}} P\left(\left\{\omega : |\widehat{g_\omega}(x)| \geq \sup_j |\widehat{f_j}(x)|\right\}\right) d\sigma x$$

$$\leq 2 \int P(\{\omega : |\widehat{g_\omega}(x)| > \lambda\}) d\sigma x$$

$$= 2 \int \sigma(\{x : |\widehat{g_\omega}(x)| > \lambda\}) dP\omega$$

$$\leq 2 \int_E dP\omega + 2C_0 \sigma^{n-1}(S^{n-1})^{-1} \lambda^{-1} \int_{\Omega \setminus E} \|g_\omega\|_{L^p(\mathbb{R}^n)} dP\omega$$

$$\leq 2\lambda^{-\theta p} \int \|g_\omega\|_{L^p(\mathbb{R}^n)}^p dP\omega + 2C_0 \sigma^{n-1}(S^{n-1})^{-1} \lambda^{\theta-1}.$$

Now we shall use Khintchine's inequalities 2.14 which give, since $p \leq 2$,

$$\int \|g_\omega\|_{L^p(\mathbb{R}^n)}^p dP\omega = \iint \left|\sum_{j=1}^m \omega_j f_j(x)\right|^p dx \, dP\omega$$

$$\approx \int \left(\sum_{j=1}^m |f_j(x)|^2\right)^{p/2} dx \leq \int \sum_{j=1}^m |f_j(x)|^p dx \leq 1.$$

Plugging this into the previous inequality and recalling that $\theta = \frac{1}{p+1}$ we get (19.17).

Next we shall show that if $0 < \eta < 1$ there exists a Borel set $B \subset S^{n-1}$ such that $\sigma(B) \leq \eta$ and

$$\sigma(\{x \in S^{n-1} \setminus B : |\widehat{f}(x)| > \lambda\}) \leq (\eta/C_1)^{-p-1} \lambda^{-p} \|f\|_{L^p(\mathbb{R}^n)}^p$$

$$\text{for } \lambda > 0, \, f \in \mathcal{S}(\mathbb{R}^n). \tag{19.19}$$

We shall apply this with $\eta = 1/2$, so we then have the required inequality valid on half of the sphere. To prove (19.19) define $M > 0$ by

$$C_1 M^{-\frac{p}{p+1}} = \eta, \quad \text{that is,} \quad M^p = (\eta/C_1)^{-p-1}.$$

Let $\mathcal{B}$ be the family of all Borel sets $B \subset S^{n-1}$ such that there exists $f \in \mathcal{S}(\mathbb{R}^n)$ for which $\|f\|_p \le 1$ and $\sigma(B)|\widehat{f}(x)|^p > M^p$ for $x \in B$, and let $\mathcal{U}$ be the collection of all disjoint subfamilies of $\mathcal{B}$. We order $\mathcal{U}$ by inclusion and use Zorn's lemma to find a maximal family $\{B_j\} \in \mathcal{U}$. Let $f_j \in \mathcal{S}(\mathbb{R}^n)$ be the corresponding functions and set $c_j = \sigma(B_j)^{1/p}$ and $B = \cup_j B_j$. If $x \in B$, then $x \in B_j$ for some j, whence $c_j|\widehat{f}_j(x)| > M$ and so

$$B \subset \left\{ x : \sup_j |c_j \widehat{f}_j(x)| > M \right\}.$$

Since also

$$\sum_j \|c_j f_j\|_p^p = \sum_j \sigma(B_j)\|f_j\|_p^p \le \sum_j \sigma(B_j) \le 1,$$

we obtain by (19.17)

$$\sigma(B) \le \sigma\left(\left\{ x : \sup_j |c_j \widehat{f}_j(x)| > M \right\}\right) \le C_1 M^{-\frac{p}{p+1}} = \eta.$$

Suppose then that (19.19) is false. Then there exist $f \in \mathcal{S}(\mathbb{R}^n)$ and $\lambda > 0$ such that

$$\sigma(\{x \in S^{n-1} \setminus B : |\widehat{f}(x)| > \lambda\}) > M^p \lambda^{-p} \|f\|_p^p.$$

Define $g = f/\|f\|_p$ and $B' = \{x \in S^{n-1} \setminus B : |\widehat{g}(x)| > \lambda/\|f\|_p\}$. Then

$$\sigma(B') = \sigma(\{x \in S^{n-1} \setminus B : |\widehat{f}(x)| > \lambda\} > M^p \lambda^{-p} \|f\|_p^p,$$

so that $\sigma(B')|\widehat{g}(x)|^p > M^p$ for $x \in B'$, which is a contradiction with the maximality of $\{B_j\} \in \mathcal{U}$, since $B' \cap B_j = \varnothing$ for all j. This proves the existence of B as in (19.19).

To finish the proof of the theorem we shall use the following lemma:

Lemma 19.10 *If E and F are Borel subsets of S^{n-1}, then there is $g_0 \in O(n)$ such that*

$$\sigma(E \cap g_0(F)) = \sigma(E)\sigma(F).$$

Proof Recall that θ_n is the Haar probability measure on $O(n)$. The function $g \mapsto \sigma(E \cap g(F))$ is easily seen to be continuous, which implies that there is

$g_0 \in O(n)$ for which

$$\int \sigma(E \cap g(F)) \, d\theta_n g = \sigma(E \cap g_0(F)).$$

For any $x \in S^{n-1}$, $A \mapsto \theta_n(\{g \in O(n) : x \in g(A)\})$ is an orthogonally invariant Borel probability measure on S^{n-1}, so it agrees with σ due to the uniqueness of such measures. Hence by Fubini's theorem,

$$\sigma(E \cap g_0(F)) = \int \sigma(E \cap g(F)) \, d\theta_n g = \iint \chi_{E \cap g(F)}(x) \, d\sigma x \, d\theta_n g$$

$$= \int_E \theta_n(\{g \in O(n) : x \in g(F)\}) \, d\sigma x = \sigma(E)\sigma(F). \qquad \square$$

Let $f \in \mathcal{S}(\mathbb{R}^n)$ with $\|f\|_p = 1$, $\lambda > 0$,

$$E = \{x \in S^{n-1} : |\widehat{f}(x)| > \lambda\},$$

and let B be as in (19.19) with $\eta = 1/2$ so that $\sigma(B) \le 1/2$ and

$$\sigma(\{x \in S^{n-1} \setminus B : |\widehat{g}(x)| > \lambda\}) \lesssim \lambda^{-p} \|g\|_{L^p(\mathbb{R}^n)}^p \quad \text{for } g \in \mathcal{S}(\mathbb{R}^n). \quad (19.20)$$

Finally let $g_0 \in O(n)$ be given by Lemma 19.10 so that

$$\sigma(E \cap g_0(B)) = \sigma(E)\sigma(B) \le \sigma(E)/2,$$

whence $\sigma(E \cap g_0(S^{n-1} \setminus B)) \ge \sigma(E)/2$. Clearly, $g_0(B)$ also satisfies (19.20) in place of B, so

$$\sigma(E) \le 2\sigma(E \cap g_0(S^{n-1} \setminus B))$$
$$= 2\sigma(\{x \in S^{n-1} \setminus g_0(B) : |\widehat{f}(x)| > \lambda\}) \lesssim \lambda^{-p} \|f\|_{L^p(\mathbb{R}^n)}^p,$$

which completes the proof of the theorem. $\qquad \square$

19.4 Applications to PDEs

One of the main motivations to restriction results is their applications to partial differential equations. Here is a quick glance at that.

Consider the Schrödinger equation as in Section 17.2:

$$2\pi i \partial_t u(x, t) + \Delta_x u(x, t) = 0, \quad u(x, 0) = f(x), \quad (x, t) \in \mathbb{R}^n \times \mathbb{R},$$

where $f \in \mathcal{S}(\mathbb{R}^n)$. Its solution is given by

$$u(x, t) = \int_{\mathbb{R}^n} e^{2\pi i(x \cdot \xi - t|\xi|^2)} \widehat{f}(\xi) \, d\xi.$$

Let

$$S = \{(x, -|x|^2) : x \in \mathbb{R}^n\}$$

and let σ be the surface measure on S. Defining g by

$$\widehat{f}(\xi) = g(\xi, -|\xi|^2)\sqrt{1 + 4|\xi|^2},$$

and observing that $4|\xi|^2 = |\nabla\varphi(\xi)|^2$ with $\varphi(\xi) = -|\xi|^2$, we have

$$u(x, t) = \widehat{g\sigma}(x, t)$$

and the restriction theorems give for certain values of p,

$$\|\widehat{g\sigma}\|_{L^p(\mathbb{R}^n \times \mathbb{R})} \lesssim \|g\|_{L^2(\sigma)}.$$

But, provided f has support in a fixed bounded set,

$$\|g\|_{L^2(\sigma)} \approx \|\widehat{f}\|_{L^2(\mathbb{R}^n)} = \|f\|_{L^2(\mathbb{R}^n)},$$

so

$$\|u\|_{L^p(\mathbb{R}^n \times \mathbb{R})} \lesssim \|f\|_{L^2(\mathbb{R}^n)}.$$

This method with variations applies to many other equations. For the wave equation

$$\partial_t^2 u(x, t) = \Delta_x u(x, t), \quad u(x, 0) = 0, \quad \frac{\partial}{\partial t}u(x, 0) = f(x), \quad (x, t) \in \mathbb{R}^n \times \mathbb{R},$$

there is a similar connection with the cone $\{(x, t) : |x| = t\}$ and one needs restriction theorems for surfaces with zero Gaussian curvature.

19.5 Further comments

The presentation of this chapter is largely based on Wolff's lecture notes [2003]. This topic is also discussed in the books Grafakos [2009], Muscalu and Schlag [2013], Sogge [1993] and Stein [1993] where much more information on the restriction problem can be found.

Stein started the research on restriction problems in the 1960s by observing that curvature makes it possible to restrict the Fourier transforms of L^p functions for some $p > 1$ to sets of measure zero. The Stein–Tomas restriction theorem 19.4 was proved by Tomas [1975] for $q > 2(n + 1)/(n - 1)$ and by Stein [1986] for the end-point. We shall still discuss this further in the next chapter including the end-point result and the sharp two-dimensional result. In the last chapter we shall discuss bilinear restriction results and their applications to the linear restriction.

The version of the Stein–Tomas restriction theorem for general measures, Theorem 19.3, is due to Mitsis [2002b] and Mockenhaupt [2000]. Bak and Seeger [2011] proved that the end-point estimate holds too. Hambrook and Łaba [2013] constructed some delicate examples which show that the range of the exponents in Theorem 19.3 is sharp in $\mathbb{R}$ when $\beta = \alpha/2$; the case $\beta < \alpha/2$ was done by Chen [2014a]. On the other hand, Chen also gave conditions under which the range of exponents can be improved and Shmerkin and Suomala [2014] verified these conditions for a large class of random measures.

In addition to the sphere, typical cases of hypersurfaces studied are the paraboloid $\{(\widetilde{x}, x_n) \in \mathbb{R}^n : x_n = |\widetilde{x}|^2\}$, for which the basic results are the same as for the sphere, and the cone $\{(\widetilde{x}, x_n) \in \mathbb{R}^n : x_n = |\widetilde{x}|\}$, for which the results differ due to the fact that one of the principal curvatures is zero. For the cone the sharp restriction theorem is known for $n = 3$, due to Barceló [1985], and for $n = 4$, due to Wolff [2001].

The literature on restriction and its applications and connections to other topics is huge involving work on various other types of surfaces such as hypersurfaces with some principal curvatures vanishing; Lee and Vargas [2010] and Müller [2012], curves and other surfaces of codimension bigger than 1; Stein [1993], Section VIII.4, Bak, D. M. Oberlin and Seeger [2009], [2013], Dendrinos and Müller [2013], and work on restriction theorems with general, perhaps fractal, measures; Mockenhaupt [2000], Hambrook and Łaba [2013], Bak and Seeger [2011], Chen [2014a], [2014b] and Ham and Lee [2014]. These are just some recent sample references whose bibliographies contain many more.

Theorem 19.9 is due to Bourgain [1991a]; he observed that it follows from some general results of Pisier and others discussed in Pisier [1986]. The proof presented above is due to Vargas [1991] from her master's thesis.

20

Stationary phase and restriction

In this chapter I describe a method based on the stationary phase (recall Chapter 14) to prove restriction theorems. Since this is well covered in many sources, I shall be rather brief and the presentation will be sketchy in parts.

20.1 Stationary phase and L^2 estimates

Recall that in Chapter 14 we investigated the decay as $\lambda \to \infty$ of the integrals

$$I(\lambda) = \int e^{i\lambda\varphi(x)}\psi(x)\,dx, \quad \lambda > 0.$$

We found in Theorem 14.5 that they decay as $\lambda^{-n/2}$ provided that the critical points of φ are non-degenerate on the support of ψ. In this chapter we allow φ and ψ to depend also on ξ, we now denote them by Φ and Ψ, and we look for $L^p - L^q$ estimates for the operators

$$T_\lambda f(\xi) = \int_{\mathbb{R}^n} e^{i\lambda\Phi(x,\xi)}\Psi(x,\xi)f(x)\,dx, \quad \xi \in \mathbb{R}^n, \quad \lambda > 0. \quad (20.1)$$

This leads to restriction theorems on surfaces via local parametrizations. We shall also see how this method can be used to prove the sharp restriction theorem in the plane.

Under the non-degeneracy of the Hessian we have a fairly simple L^2 result:

Theorem 20.1 *Suppose that* $\Phi : \mathbb{R}^{2n} \to \mathbb{R}$ *and* $\Psi : \mathbb{R}^{2n} \to \mathbb{C}$ *are* C^∞-*functions,* Ψ *with compact support. If*

$$\det\left(\frac{\partial^2\Phi(x,\xi)}{\partial x_j\partial\xi_k}\right) \neq 0 \quad for\,(x,\xi) \in \mathrm{spt}\,\Psi, \quad (20.2)$$

then the operators T_λ satisfy

$$\|T_\lambda f\|_2 \lesssim \lambda^{-n/2} \|f\|_2 \quad \textit{for all } f \in L^2(\mathbb{R}^n), \quad \lambda > 0. \tag{20.3}$$

Proof We can write

$$\|T_\lambda f\|_2^2 = \iint K_\lambda(x, y) f(x) \overline{f(y)} \, dx \, dy,$$

where

$$K_\lambda(x, y) = \int e^{i\lambda(\Phi(x,\xi) - \Phi(y,\xi))} \Psi(x, \xi) \overline{\Psi(y, \xi)} \, d\xi.$$

For $|x - y| \leq 1$ we have

$$\nabla_\xi(\Phi(x, \xi) - \Phi(y, \xi)) = \left(\frac{\partial^2 \Phi(x, \xi)}{\partial x_j \partial \xi_k} \right) (x - y) + O(|x - y|^2).$$

Assuming that spt Ψ is sufficiently small we then have for some $c > 0$,

$$|\nabla_\xi(\Phi(x, \xi) - \Phi(y, \xi))| \geq c|x - y|$$

when $(x, \xi), (y, \xi) \in$ spt Ψ. We can reduce to small support for Ψ as in Chapter 14 with finite coverings. Reducing the support of Ψ further if needed we can assume that for some $j = 1, \ldots, n$,

$$\left| \frac{\partial}{\partial \xi_j}(\Phi(x, \xi) - \Phi(y, \xi)) \right| \geq c|x - y|$$

when $(x, \xi), (y, \xi) \in$ spt Ψ. Then similar partial integrations as in Chapter 14 (more precisely, checking the dependence of the constants in Theorems 14.1 and 14.4) yield

$$|K_\lambda(x, y)| \lesssim_N (1 + \lambda|x - y|)^{-N}, \quad N = 1, 2, \ldots,$$

for $x, y \in \mathbb{R}^n$. Applying this with $N = n + 1$ we find that

$$\int |K_\lambda(x, y)| \, dy \lesssim \lambda^{-n} \quad \text{for } x \in \mathbb{R}^n,$$

$$\int |K_\lambda(x, y)| \, dx \lesssim \lambda^{-n} \quad \text{for } y \in \mathbb{R}^n.$$

Defining

$$T_{K_\lambda} f(y) = \int K_\lambda(x, y) f(x) \, dx$$

we obtain from the previous inequalities and Schur's test, which we discuss below, that

$$\|T_\lambda f\|_2^2 = \int (T_{K_\lambda} f)\overline{f} \leq \|T_{K_\lambda} f\|_2 \|f\|_2 \lesssim \lambda^{-n} \|f\|_2^2,$$

as required. $\qquad\qquad\qquad\qquad\qquad\qquad\qquad\qquad\qquad\qquad\qquad$ $\square$

Schur's test is the following general very useful boundedness criterion:

Theorem 20.2 *Let* (X, μ) *and* (Y, ν) *be measure spaces and* $K : X \times Y \to \mathbb{C}$ *a* $\mu \times \nu$ *measurable function such that* $\int |K(x, y))|^2 \, d\mu x < \infty$ *for* $y \in Y$. *Suppose that*

$$\int |K(x, y))| \, d\mu x \leq A \quad \text{for } y \in Y$$

and

$$\int |K(x, y))| \, d\nu y \leq B \quad \text{for } x \in X.$$

Define

$$T_K f(y) = \int K(x, y) f(x) \, d\mu x \quad \text{for } y \in Y, \quad f \in L^2(\mu).$$

Then

$$\|T_K f\|_{L^2(\nu)} \leq \sqrt{AB} \|f\|_{L^2(\mu)} \quad \text{for } f \in L^2(\mu). \tag{20.4}$$

Proof The finiteness of the L^2 integral of K is only assumed to guarantee that $T_K f$ is pointwise defined. The inequality (20.4) follows if we can prove that

$$\iint |K(x, y) g(x) f(y)| \, d\mu x \, d\nu y \leq \sqrt{AB}$$

whenever $\|g\|_{L^2(\mu)} = 1$ and $\|f\|_{L^2(\nu)} = 1$. To verify this we use Schwartz's inequality:

$$\iint |K(x, y) g(x) f(y)| \, d\mu x \, d\nu y$$

$$\leq \left(\iint |K(x, y)| |f(y)|^2 \, d\mu x \, d\nu y \iint |K(x, y))| |g(x)|^2 \, d\mu x \, d\nu y \right)^{1/2}$$

$$= \left(\iint |K(x, y)| \, d\mu x |f(y)|^2 \, d\nu y \iint |K(x, y)| \, d\nu y |g(x)|^2 \, d\mu x \right)^{1/2}$$

$$\leq \sqrt{AB}. \qquad\qquad\qquad\qquad\qquad\qquad\qquad\qquad\qquad\qquad\qquad\qquad\qquad$ \square$$

20.2 From stationary phase to restriction

Let us now see how the stationary phase can be applied to restriction problems. We are interested in the inequalities

$$\|\widehat{f}\|_{L^q(S)} \lesssim \|f\|_{L^p(\mathbb{R}^n)} \quad \text{for } f \in \mathcal{S}(\mathbb{R}^n). \tag{20.5}$$

Here S is a smooth surface in $\mathbb{R}^n$ with non-vanishing Gaussian curvature. Assuming that S is the graph of a smooth compactly supported function φ (20.5) reduces to inequalities like

$$\left(\int |\widehat{f}(\xi, \varphi(\xi))|^q \psi(\xi)^q \, d\xi \right)^{1/q} \lesssim \|f\|_{L^p(\mathbb{R}^n)}, \tag{20.6}$$

where φ and ψ are compactly supported C^∞ functions in $\mathbb{R}^{n-1}$ with $\psi \geq 0$, $\varphi(0) = 0$, $\nabla\varphi(0) = 0$ and $h_\varphi(0) \neq 0$ (recall (14.4)). The Fourier transform of f on S is given by

$$\widehat{f}(\xi, \varphi(\xi)) = \int_{\mathbb{R}^n} e^{-2\pi i(\xi \cdot \widetilde{x} + \varphi(\xi)x_n)} f(x) \, dx, \quad \widetilde{x} = (x_1, \dots, x_{n-1}).$$

Let η be a non-negative compactly supported C^∞ function on $\mathbb{R}^n$ with $\eta(0) = 1$ and define

$$T_\lambda f(\xi) = \int e^{i\lambda\Phi(x,\xi)} \Psi(x, \xi) f(x) \, dx, \quad \xi \in \mathbb{R}^{n-1}, \quad \lambda > 0, \tag{20.7}$$

where

$$\Phi(x, \xi) = -2\pi(\xi \cdot \widetilde{x} + \varphi(\xi)x_n),$$
$$\Psi(x, \xi) = \psi(\xi)\eta(x).$$

Suppose we could prove, with $p' = p/(p-1)$,

$$\|T_\lambda f\|_{L^q(\mathbb{R}^{n-1})} \lesssim \lambda^{-n/p'} \|f\|_{L^p(\mathbb{R}^n)}. \tag{20.8}$$

Applying this to f_λ, $f_\lambda(x) = f(\lambda x)$, we get

$$\left(\int \left| \int e^{i\lambda\Phi(x,\xi)} \eta(x) f(\lambda x) dx \right|^q \psi(\xi)^q \, d\xi \right)^{1/q}$$
$$\lesssim \lambda^{-n/p'} \|f_\lambda\|_p = \lambda^{-n/p-n/p'} \|f\|_p = \lambda^{-n} \|f\|_p.$$

Change of variable $y = \lambda x$ gives, since $\lambda\Phi(x, \xi) = \Phi(\xi, \lambda x)$,

$$\left(\int \left| \int e^{i\Phi(y,\xi)} \eta(y/\lambda) f(y) dy \right|^q \psi(\xi)^q \, d\xi \right)^{1/q} \lesssim \|f\|_p.$$

When $\lambda \to \infty$, $\eta(y/\lambda) \to 1$, and the last inequality gives (20.6).

The inequality (20.8) can be proven for much more general phase functions than Φ above, and it has applications to many problems in addition to restriction. It is true for

$$1 \le p \le \frac{2(n+1)}{n+3}, \quad q = \frac{n-1}{n+1}p'.$$

For the Fourier transform this is the Stein–Tomas restriction range (recall (19.12) in the dual form). For general Φ, defined on $\mathbb{R}^{n-1} \times \mathbb{R}^n$, it is the best possible range of exponents when $n \ge 3$; see the discussion in Section 23.4. For $n = 2$ the range can be extended to $1 \le p < 4/3$, cf. Theorem 20.3. This range is sharp also for the Fourier transform.

The main part of the proof of (20.8) is for $p = \frac{2(n+1)}{n+3}, q = 2$. The rest follows by interpolation between this and the trivial case $\|T_\lambda f\|_\infty \lesssim \|f\|_1$. We can write

$$\|T_\lambda f\|^2_{L^2(\mathbb{R}^{n-1})} = \iint K_\lambda(x, y)f(x)\overline{f(y)}\,dx\,dy,$$

where

$$K_\lambda(x, y) = \int_{\mathbb{R}^{n-1}} e^{i\lambda(\Phi(x,\xi)-\Phi(y,\xi))}\Psi(x, \xi)\overline{\Psi(y, \xi)}\,d\xi.$$

Let

$$U_\lambda g(x) = \int K_\lambda(x, y)g(y)\,dy.$$

Then

$$\|T_\lambda f\|^2_{L^2(\mathbb{R}^{n-1})} = \int (U_\lambda f)\overline{f}.$$

So we need

$$\|U_\lambda f\|_{L^{p'}(\mathbb{R}^n)} \lesssim \lambda^{-2n/p'}\|f\|_{L^p(\mathbb{R}^n)}.$$

This can be obtained by fairly complicated real and complex interpolation techniques. One benefit of going from T_λ to U_λ is that we now have an operator which acts on functions in $\mathbb{R}^n$ to functions in $\mathbb{R}^n$ (not from $\mathbb{R}^n$ to $\mathbb{R}^{n-1}$ as for T_λ). The formal way to go from T_λ to U_λ is that the adjoint T_λ^* of T_λ is

$$T_\lambda^* f(x) = \int e^{-i\lambda\Phi(x,\xi)}\overline{\Psi(x, \xi)}f(\xi)\,d\xi,$$

so

$$U_\lambda = T_\lambda^* T_\lambda.$$

A serious problem with U_λ still is that the oscillating factor in its kernel K_λ depends on the variables in $\mathbb{R}^{n-1}$ and $\mathbb{R}^n$ and cannot have non-degeneracy corresponding to the earlier conditions of non-vanishing Hessian determinant. Here one needs to study the $(n - 1) \times n$ matrix

$$\left(\frac{\partial^2 \Phi(x, \xi)}{\partial x_j \partial \xi_k} \right).$$

What helps is that in many situations it has maximal rank $n - 1$, and this is one of the assumptions for a general theorem. This can be used by freezing one coordinate x_j and applying Fubini arguments or by adding to Φ an auxiliary function $\Phi_0(x, t)$, $t \in \mathbb{R}$, which gives a non-zero Hessian determinant for $\Phi(x, \xi_1, \ldots, \xi_{n-1}) + \Phi_0(x, \xi_n)$. Then results like Theorem 20.1 can be applied. Many missing details can be found in Muscalu and Schlag [2013], Stein [1993] and Sogge [1993].

Observe that the above method also gives the end-point estimate in the Stein–Tomas Theorem 19.4.

20.3 Sharp results in the plane

In this section we shall prove a sharp $L^p - L^q$-inequality for the operators T_λ in the two-dimensional case. This will solve the restriction conjecture in the plane.

First let us observe a corollary to Theorem 20.1: under the assumption (20.2), the operators T_λ of (20.1) satisfy

$$\|T_\lambda f\|_{p'} \lesssim \lambda^{-n/p'} \|f\|_p \quad \text{for all } f \in L^p(\mathbb{R}^n), \quad \lambda > 0, \quad 1 \le p \le 2. \tag{20.9}$$

This follows readily interpolating (20.3) with the trivial case $\|T_\lambda f\|_\infty \le \|f\|_1$.

We now formulate and prove in the plane a sharp result for operators as in (20.7). The variable ξ will be a real number and $x \in \mathbb{R}^2$. We shall denote derivatives with respect to ξ by $'$ and with respect to x_j with subscript x_j. So, for example, $\Phi''_{x_j}(x, \xi) = \frac{\partial^3 \Phi(x, \xi)}{\partial^2 \xi \partial x_j}$.

Theorem 20.3 *Suppose that* $\Phi : \mathbb{R}^2 \times \mathbb{R} \to \mathbb{R}$ *and* $\Psi : \mathbb{R}^2 \times \mathbb{R} \to \mathbb{C}$ *are smooth functions such that* Ψ *has compact support and*

$$\begin{vmatrix} \Phi''_{x_1}(x, \xi) & \Phi'_{x_1}(x, \xi) \\ \Phi''_{x_2}(x, \xi) & \Phi'_{x_2}(x, \xi) \end{vmatrix} \ne 0 \quad \text{for } (x, \xi) \in \operatorname{spt} \Psi. \tag{20.10}$$

Then the operators T_λ^*,

$$T_\lambda^* f(x) = \int_{\mathbb{R}} e^{i\lambda \Phi(x, \xi)} \Psi(x, \xi) f(\xi) \, d\xi, \quad x \in \mathbb{R}^2, \quad \lambda > 0,$$

satisfy

$$\|T_\lambda^* f\|_{L^q(\mathbb{R}^2)} \lesssim \lambda^{-2/q} \|f\|_{L^p(\mathbb{R})} \quad \text{for all } f \in L^p(\mathbb{R}), \lambda > 0, q = 3p', q > 4.$$
$$(20.11)$$

Remark 20.4 Observe that we have formulated the theorem for the adjoint operators of the operators T_λ,

$$T_\lambda f(\xi) = \int_{\mathbb{R}^2} e^{-i\lambda \Phi(x,\xi)} \Psi(x,\xi) f(x) \, dx, \quad \xi \in \mathbb{R}, \quad \lambda > 0,$$

that we considered before. The theorem is equivalent to

$$\|T_\lambda f\|_{L^q(\mathbb{R})} \lesssim \lambda^{-2/p'} \|f\|_{L^p} \quad \text{for all } f \in L^q(\mathbb{R}^2), \lambda > 0, 3q = p', q > 4/3.$$

Proof What will help is that we have now $q > 4 = 2 \cdot 2$. This allows us to work with

$$T_\lambda^* f(x)^2 = \int_{\mathbb{R}^2} e^{i\lambda(\Phi(x,\xi_1) + \Phi(x,\xi_2))} \Psi(x,\xi_1) \Psi(x,\xi_2) f(\xi_1) f(\xi_2) d\xi_1 \, d\xi_2,$$

$$x \in \mathbb{R}^2, \lambda > 0.$$

We would like to apply Theorem 20.1 with the weight function

$$\Xi(x,\xi) = \Psi(x,\xi_1) \Psi(x,\xi_2), \quad (x,\xi) \in \mathbb{R}^2 \times \mathbb{R}^2,$$

and with the phase function

$$\Theta(x,\xi) = \Phi(x,\xi_1) + \Phi(x,\xi_2), \quad (x,\xi) \in \mathbb{R}^2 \times \mathbb{R}^2,$$

but the determinant $\det(\frac{\partial^2 \Theta}{\partial x_j \partial \xi_k}(x,\xi))$ vanishes for $\xi_1 = \xi_2$. Computing this determinant and applying Taylor's theorem one finds that

$$\det\left(\frac{\partial^2 \Theta}{\partial x_j \partial \xi_k}(x,\xi)\right) = \begin{vmatrix} \Phi''_{x_2}(x,\xi_1) & \Phi'_{x_2}(x,\xi_1) \\ \Phi''_{x_1}(x,\xi_1) & \Phi'_{x_1}(x,\xi_1) \end{vmatrix} (\xi_2 - \xi_1) + O(|\xi_2 - \xi_1|^2).$$

Assuming as before that Ψ has small support, we have by (20.10) for some $c > 0$,

$$\left| \det\left(\frac{\partial^2 \Theta}{\partial x_j \partial \xi_k}(x,\xi)\right) \right| \geq c|\xi_2 - \xi_1| \quad \text{when} \quad (x,\xi) \in \text{spt } \Xi. \quad (20.12)$$

Now we would like to make a change of variable in ξ to get rid of the factor $|\xi_2 - \xi_1|$. We obtain this with $\zeta = (\xi_1 + \xi_2, \xi_1 \cdot \xi_2) =: g(\xi_1, \xi_2)$. The determinant of $Dg(\xi)$ is $\xi_1 - \xi_2$. Notice that g is two to one in $\{\xi : \xi_1 \neq \xi_2\}$. Moreover, $g(\xi) = g(\xi')$ if and only if $\xi = \xi'$ or $\xi_1 = \xi'_2$ and $\xi_2 = \xi'_1$. Set

$$\widetilde{\Phi}(x,\zeta) = \Theta(x,\xi) = \Phi(x,\xi_1) + \Phi(x,\xi_2),$$

and

$$\widetilde{\Xi}(x, \zeta) = \Xi(x, \xi) = \Psi(x, \xi_1) \cdot \Psi(x, \xi_2),$$

when $\zeta = g(\xi)$. Then we have the well defined functions $\widetilde{\Phi}$ and $\widetilde{\Xi}$. They are smooth because of the symmetricity of $\Phi(x, \xi_1) + \Phi(x, \xi_2)$ and $\Psi(x, \xi_1) \cdot \Psi(x, \xi_2)$ with respect to ξ_1 and ξ_2. To relate the determinant $\det(\frac{\partial^2 \widetilde{\Phi}}{\partial x_k \partial \zeta_j}(x, \zeta))$ to the determinant in (20.12), set

$$G_x(\xi) = (\partial_{x_1} \Theta(x, \xi), \partial_{x_2} \Theta(x, \xi)) \quad \text{and} \quad \widetilde{G}_x(\zeta) = (\partial_{x_1} \widetilde{\Phi}(x, \zeta), \partial_{x_2} \widetilde{\Phi}(x, \zeta)).$$

Then $G_x = \widetilde{G}_x \circ g$ and, when $\zeta = g(\xi)$,

$$\det \left(\frac{\partial^2 \Theta}{\partial x_j \partial \xi_k}(x, \xi) \right) = \det(DG_x(\xi)) = \det(D\widetilde{G}_x(g(\xi))) \det(Dg(\xi))$$

$$= \det \left(\frac{\partial^2 \widetilde{\Phi}}{\partial x_k \partial \zeta_j}(x, \zeta) \right) (\xi_1 - \xi_2).$$

Hence it follows from (20.12) that

$$\left| \det \left(\frac{\partial^2 \widetilde{\Phi}}{\partial x_k \partial \zeta_j}(x, \zeta) \right) \right| \geq c \quad \text{for } (x, \zeta) \in \text{spt } \widetilde{\Xi}.$$

Now we have (2 in front comes from the two to one property)

$$T_\lambda^* f(x)^2 = 2 \int_{\mathbb{R}^2} e^{i\lambda \widetilde{\Phi}(x,\zeta)} \widetilde{\Xi}(x, \zeta) F(\zeta) d\zeta, \quad x \in \mathbb{R}^2, \lambda > 0,$$

where, when $\zeta = g(\xi)$,

$$F(\zeta) = \frac{f(\xi_1) \cdot f(\xi_2)}{|\xi_1 - \xi_2|}.$$

So we have won by getting a non-vanishing determinant for $\widetilde{\Phi}$, but lost by getting a singularity at the diagonal for F. Define r by $2r' = q$. Assuming, as we may, that $q < \infty$, we have then $1 < r < 2$ and we can apply (20.9) getting

$$\int |T_\lambda^* f|^q = \int |(T_\lambda^* f)^2|^{r'} \lesssim \lambda^{-2} \left(\int |F(\zeta)|^r d\zeta \right)^{r'/r}.$$

Changing from ζ to ξ we have

$$\int |F(\zeta)|^r d\zeta = \frac{1}{2} \int |f(\xi_1) f(\xi_2)|^r |\xi_1 - \xi_2|^{1-r} d\xi_1 d\xi_2.$$

To estimate the last integral we use the following Hardy–Littlewood–Sobolev inequality for functions of one variable, see, for example, Stein [1993], (31) in

Chapter VIII; here again k_γ is the Riesz kernel, $k_\gamma(y) = |y|^{-\gamma}$, $y \in \mathbb{R}$:

$$\|k_\gamma * g\|_{L^t(\mathbb{R})} \lesssim \|g\|_{L^s(\mathbb{R})} \quad \text{when} \quad 0 < \gamma < 1, \quad 1 < s < t < \infty,$$
$$\frac{1}{t} = \frac{1}{s} + \gamma - 1.$$

This and Hölder's inequality yield

$$\int |g(\xi_1)g(\xi_2)||\xi_1 - \xi_2|^{-\gamma} d\xi_1\, d\xi_2 \leq \left(\int |g|^s\right)^{2/s}$$

when $0 < \gamma < 1, 1 < s < 2, \frac{2}{s} = 2 - \gamma$. We apply this with $g = |f|^r, \gamma = r - 1$. Then

$$\int |T_\lambda^* f|^q \lesssim \lambda^{-2} \left(\int |f|^{rs}\right)^{2r'/(rs)}.$$

The choices of the parameters imply $rs = p$ and $2r'/(rs) = q/p$, and the theorem follows. $\qquad\square$

If φ is a local parametrization of a curve S and

$$\Phi(x, \xi) = -2\pi(\xi x_1 + \varphi(\xi)x_2)$$

as before in the applications to restriction, then the determinant in the assumptions of Theorem 20.3 is

$$\begin{vmatrix} \Phi''_{x_1}(x, \xi) & \Phi'_{x_1}(x, \xi) \\ \Phi''_{x_2}(x, \xi) & \Phi'_{x_2}(x, \xi) \end{vmatrix} = 4\pi^2 \varphi''(\xi).$$

So the non-vanishing determinant condition means that the curve has non-zero curvature. Recalling the argument '(20.8) implies (20.5)' and checking that the conditions on exponents match we obtain from Theorem 20.3 (recall the formulation in Remark 20.4):

Theorem 20.5 *Let S be a smooth compact curve in $\mathbb{R}^2$ with non-vanishing curvature and length measure σ. Then*

$$\left(\int_S |\widehat{f}|^q\, d\sigma\right)^{1/q} \lesssim \|f\|_{L^p(\mathbb{R}^2)} \quad \text{for } f \in L^p(\mathbb{R}^2), \quad 3q = p', \quad q > 4/3.$$

This means in particular that the restriction conjecture 19.5 is valid for the circle S^1.

20.4 Further comments

The presentation of this chapter is largely based on Stein [1993], Chapter IX. Muscalu and Schlag [2013] and Sogge [1993] have also a lot on this topic. These books contain much more information on this and related matters.

Theorem 20.5 is due to Zygmund [1974]. Theorem 20.3 is due to Carleson and Sjölin [1972].

21

Fourier multipliers

This is another topic well covered by several books. I mainly wanted to include it since historically Fefferman's solution of the multiplier problem for the ball, Theorem 21.5 below, is the starting point for Kakeya type methods in Fourier analysis. We shall also discuss Bocher–Riesz multipliers.

21.1 Definition and examples

Let $m \in L^\infty(\mathbb{R}^n)$ be a bounded function. For any function f in $L^2(\mathbb{R}^n)$ we can define the following operator T_m using the Fourier transform:

$$\widehat{T_m f} = m \widehat{f}, \quad \text{that is,} \quad T_m f = (m \widehat{f})^\vee.$$

Using Plancherel's theorem we get,

$$\|T_m f\|_2 = \|m \widehat{f}\|_2 \le \|m\|_\infty \|\widehat{f}\|_2 = \|m\|_\infty \|f\|_2 \,,$$

and therefore T_m is a bounded linear operator from L^2 to L^2 with norm bounded by $\|m\|_\infty$. In fact, this norm is exactly $\|m\|_\infty$, which is an easy exercise.

As another simple exercise one can check that the operator norm of T_m is invariant under translations and dilations. That is, T_m and $T_{m_{a,r}}$, $m_{a,r}(x) = m(rx + a)$, have the same operator norms for all $a \in \mathbb{R}^n$, $r > 0$.

The function $m \in L^\infty(\mathbb{R}^n)$ is said to be an L^p-*multiplier*, $1 < p < \infty$, if the operator T_m can be extended to $L^p(\mathbb{R}^n)$ as a bounded operator from $L^p(\mathbb{R}^n)$ to $L^p(\mathbb{R}^n)$.

For a measurable set $A \subset \mathbb{R}^n$ we denote

$$T_A = T_{\chi_A}.$$

Let us look at some examples:

293

Example 21.1 Let m be the sign function sgn in $\mathbb{R}$; $\text{sgn}(x) = -1$ for $x < 0$ and $\text{sgn}(x) = 1$ for $x \geq 0$. Then

$$\widehat{Hf} = -i \, \text{sgn} \, \widehat{f},$$

where H is the Hilbert transform. So $T_{\text{sgn}} = iH$ and sgn is an L^p-multiplier for all $1 < p < \infty$ by the well-known (but highly non-trivial) results on the Hilbert transform.

The above Fourier formula can be taken as the definition of the Hilbert transform, but it can also be defined by

$$Hf(x) = \lim_{\varepsilon \to 0} \frac{1}{\pi} \int_{|x-y|>\varepsilon} \frac{f(y)}{x - y} \, dy$$

for integrable Lipschitz functions f, for example. One can consult for instance Duoandikoetxea [2001] for the properties of the Hilbert transfrom.

Example 21.2 Let T^+ be the multiplier for the half line $(0, \infty)$,

$$\widehat{T^+f} = \chi_{(0,\infty)}\widehat{f}.$$

Then, in the L^2 sense,

$$T^+ f(x) = \lim_{R \to \infty} \int_0^R \widehat{f}(\xi)e^{2\pi i x\xi} \, d\xi \quad \text{for } f \in L^2(\mathbb{R}).$$

We can easily express T^+ in terms of the Hilbert transform:

$$\widehat{T^+f} = \chi_{(0,\infty)}\widehat{f} = \frac{1}{2}(1 + \text{sgn})\widehat{f} = \mathcal{F}(\frac{\text{id} + iH}{2}f),$$

whence

$$T^+ = \frac{\text{id} + iH}{2}.$$

Similarly we can write multipliers for bounded intervals with the Hilbert transform: For the characteristic function $\chi_{[a,b]}$ of the interval $[a, b]$ let $S_{a,b} = T_{\chi_{a,b}}$ be the corresponding multiplier operator. This easily reduces to the previous example by the formula

$$S_{a,b} = \frac{i}{2}(M_a \circ H \circ M_{-a} - M_b \circ H \circ M_{-b}),$$

where M_a is the multiplication operator: $M_a f(x) = e^{2\pi i a x} f(x)$. It follows that $\chi_{[a,b]}$ is an L^p-multiplier for all $1 < p < \infty$. Moreover, its multiplier norm is $\leq C_p$ with C_p depending only p. For $a = -R, b = R$, this gives for $f \in L^p(\mathbb{R}) \cap L^2(\mathbb{R})$ (we restrict to L^2 in order to have pointwise almost everywhere defined Fourier transform),

$$f(x) = \lim_{R \to \infty} \int_{-R}^{R} e^{2\pi i x \xi} \widehat{f}(\xi) \, d\xi \quad \text{in the } L^p \text{ sense.}$$

To prove this, check first that the formula is valid for functions in $\mathcal{S}(\mathbb{R}^n)$ and then use the denseness of $\mathcal{S}(\mathbb{R}^n)$ in L^p.

Example 21.3 As in the previous example, do we also have for $f \in L^p(\mathbb{R}^n) \cap L^2(\mathbb{R}^n)$ when $n \geq 2$,

$$f(x) = \lim_{R \to \infty} \int_{B(0,R)} e^{2\pi i x \cdot \xi} \widehat{f}(\xi) \, d\xi \quad \text{in the } L^p \text{ sense?}$$

When $p = 2$ we do have. When $p \neq 2$ we do not have. This follows from the fact that in $\mathbb{R}^n$, $n \geq 2$, the characteristic function $\chi_{B(0,1)}$ is an L^p multiplier if and only if $p = 2$; one can use the Banach–Steinhaus theorem and some scaling arguments to prove that unboundedness implies non-convergence.

The proof of the unboundedness of the ball multiplier for $p \neq 2$ will be the main content of this chapter. Recall that the operator norms of $T_{B(0,1)}$ and $T_{B(a,r)}$ for any $a \in \mathbb{R}^n$, $r > 0$, are equal because of the translation and dilation invariance.

Example 21.4 Let $P \subset \mathbb{R}^n$ be a polyhedral domain. Then χ_P is an L^p-multiplier for all $1 < p < \infty$. By definition a polyhedral domain is an intersection of finitely many half-spaces. Thus the claim reduces to showing that the characteristic function of a half-space is an L^p-multiplier. This in turn reduces to the one-dimensional examples above. The details are left as an exercise.

21.2 Fefferman's example

The following result is due to Fefferman [1971]:

Theorem 21.5 *The characteristic function of the unit ball $B(0, 1)$ in $\mathbb{R}^n$, $n \geq 2$, is an L^p multiplier if and only if $p = 2$.*

Proof We shall first consider $n = 2$ and comment on the general case later. The proof is based on Kakeya type constructions. We need a lemma, Lemma 21.6, which is a modification of a lemma used to construct Besicovitch sets.

Lemma 21.6 *Given $\varepsilon > 0$, there exist an integer $N \geq 1$ and 2^{N+1} open rectangles $R_1, \ldots, R_{2^N}, R_1^*, \ldots, R_{2^N}^*$ in the plane, each with side-lengths 1 and 2^{-N}, such that:*

(i) *for each j, the rectangles R_j and R_j^* are disjoint and have one shorter side in common,*

(ii) $\mathcal{L}^2\left(\bigcup_{j=1}^{2^N} R_j\right) < \varepsilon$,

(iii) *the rectangles R_j^* are disjoint, and so*

$$\mathcal{L}^2\left(\bigcup_{j=1}^{2^N} R_j^*\right) = 1.$$

The proof of this lemma is based on elementary geometric iterative constructions, such as the Perron tree construction in Section 11.6. Technically it is the most complicated part of the proof of Theorem 21.5. We omit the proof here; it can be found in Stein [1993], Theorem X.1.1, and Bishop and Peres [2016], Section 9.2, and also in de Guzmán [1981], Section 8.2, and Grafakos [2009], Section 10.1, in slightly different versions.

Next we establish the following general inequality in the spirit that L^p boundedness for some scalar valued operators implies L^p boundedness with the same norm for vector valued operators:

Lemma 21.7 *Let $T : L^p(\mathbb{R}^n) \to L^p(\mathbb{R}^n)$, $1 \le p < \infty$, be any bounded linear operator; $\|Tf\|_p \le C_p \|f\|_p$ for all $f \in L^p(\mathbb{R}^n)$. Then for every finite sequence of functions $\{f_j\}_{j=1}^k$ in $L^p(\mathbb{R}^n)$ we have,*

$$\left\| \left(\sum_{j=1}^k |Tf_j|^2\right)^{\frac{1}{2}} \right\|_p \le C_p \left\| \left(\sum_{j=1}^k |f_j|^2\right)^{\frac{1}{2}} \right\|_p.$$

Proof Set $f = (f_1, \ldots, f_k)$ and $Sf = (Tf_1, \ldots, Tf_k)$. For $w \in S^{k-1}$ we have by the linearity of T,

$$\int_{\mathbb{R}^n} |w \cdot Sf|^p = \int_{\mathbb{R}^n} |T(w \cdot f)|^p \le C_p^p \int_{\mathbb{R}^n} |w \cdot f|^p. \qquad (21.1)$$

For any $y \in \mathbb{R}^k$ we have,

$$\int_{S^{k-1}} |w \cdot y|^p \, d\sigma^{k-1} w = c \, |y|^p \qquad (21.2)$$

where c is independent of y. Using Fubini's theorem, (21.2) and (21.1) we get

$$c \int_{\mathbb{R}^n} |Sf|^p = \int_{S^{k-1}} \left(\int_{\mathbb{R}^n} |T(w \cdot f)|^p\right) d\sigma^{k-1} w$$

$$\le C_p^p \int_{S^{k-1}} \left(\int_{\mathbb{R}^n} |w \cdot f|^p\right) d\sigma^{k-1} w = C_p^p c \int_{\mathbb{R}^n} |f|^p.$$

This is the required inequality and the proof of the lemma is finished.   □

The next lemma associates the multiplier operator of the unit ball to those of half-spaces:

Lemma 21.8 *Assume that for some* $1 < p < \infty$ *the multiplier operator* $T = T_{B(0,1)}$ *of the characteristic function of the unit ball* $B(0,1) \subset \mathbb{R}^n$ *satisfies* $\|Tf\|_p \leq C_p \|f\|_p$ *for* $f \in L^p(\mathbb{R}^n) \cap L^2(\mathbb{R}^n)$. *Let* $\{v_j\}_{j=1}^k$ *be a finite sequence of unit vectors in* $\mathbb{R}^n$. *Let* H_j *be the half-space,*

$$H_j = \{x \in \mathbb{R}^n : v_j \cdot x \geq 0\},$$

and $T_j = T_{H_j}$. *Then for any sequence* $\{f_j\}_{j=1}^k$ *in* $L^p(\mathbb{R}^n) \cap L^2(\mathbb{R}^n)$ *we have,*

$$\left\| \left(\sum_{j=1}^k |T_j f_j|^2 \right)^{\frac{1}{2}} \right\|_p \leq C_p \left\| \left(\sum_{j=1}^k |f_j|^2 \right)^{\frac{1}{2}} \right\|_p.$$

Proof We assume that $f_j \in \mathcal{S}(\mathbb{R}^n)$; the general case follows by simple approximation. Let B_j^r be the ball of centre rv_j and radius $r > 0$. The characteristic functions $\chi_{B_j^r}$ convergence pointwise to χ_{H_j} outside ∂H_j as $r \to \infty$. Let $T_j^r f = (\chi_{B_j^r} \widehat{f})^\vee$. Then for $f \in \mathcal{S}(\mathbb{R}^n)$, $T_j^r f$ converges to $T_j f$ as $r \to \infty$ both pointwise and in $L^p(\mathbb{R}^n)$. Thus it will suffice to prove that for all $r > 0$,

$$\left\| \left(\sum_{j=1}^k |T_j^r f_j|^2 \right)^{\frac{1}{2}} \right\|_p \leq C_p \left\| \left(\sum_{j=1}^k |f_j|^2 \right)^{\frac{1}{2}} \right\|_p. \tag{21.3}$$

Observe that,

$$T_j^r f(x) = e^{2\pi i r v_j \cdot x} T_r (e^{-2\pi i r v_j \cdot \xi} f)(x),$$

where T_r is the multiplier operator of the ball $B(0, r)$. Set $g_j(\xi) = e^{-2\pi i r v_j \cdot \xi} f_j(\xi)$. Then

$$\left\| \left(\sum_{j=1}^k |T_j^r f_j|^2 \right)^{\frac{1}{2}} \right\|_p = \left\| \left(\sum_{j=1}^k |T_r g_j|^2 \right)^{\frac{1}{2}} \right\|_p.$$

As mentioned before, the operator norms of T_r and T are equal. Therefore Lemma 21.7 yields

$$\left\|\left(\sum_{j=1}^{k}|T_j^r f_j|^2\right)^{\frac{1}{2}}\right\|_p = \left\|\left(\sum_{j=1}^{k}|T_r g_j|^2\right)^{\frac{1}{2}}\right\|_p$$

$$\leq C_p \left\|\left(\sum_{j=1}^{k}|g_j|^2\right)^{\frac{1}{2}}\right\|_p = C_p \left\|\left(\sum_{j=1}^{k}|f_j|^2\right)^{\frac{1}{2}}\right\|_p,$$

so that (21.3) holds and the proof of the lemma is finished.   □

The next lemma tells us how the operators T_j of the previous lemma act on some rectangles.

Lemma 21.9 *Let R, $R^* \subset \mathbb{R}^2$ be disjoint open rectangles whose longer sides are in the direction $v \in S^1$ and such that they have one shorter side in common (as in Lemma 21.6). Let H_v be the half-plane*

$$H_v = \{x \in \mathbb{R}^2 : v \cdot x \geq 0\}.$$

Then

$$|T_{H_v}(\chi_R)| \geq \frac{1}{13}\chi_{R^*}.$$

Proof By rotating and translating we may assume that $v = (0, 1)$, $R = (-a, a) \times (-b, b)$ and $R^* = (-a, a) \times (b, 3b)$ with $a \leq b$. We have

$$T_{H_v}(\chi_R)(x_1, x_2) = (\chi_{H_v}\widehat{\chi_R})^{\vee}(x_1, x_2) = \chi_{(-a,a)}(x_1)(\chi_{(0,\infty)}\widehat{\chi_{(-b,b)}})^{\vee}(x_2)$$

because $\chi_{H_v}(x_1, x_2)\widehat{\chi_R}(\xi_1, \xi_2) = \widehat{\chi_{(-a,a)}}(\xi_1)\chi_{(0,\infty)}(x_2)\widehat{\chi_{(-b,b)}}(\xi_2)$. Recalling the multiplier T^+ of $(0, \infty)$ from Example 21.2 we obtain

$$T_{H_v}(\chi_R)(x_1, x_2) = \chi_{(-a,a)}(x_1)T^+(\chi_{(-b,b)})(x_2) = \frac{i}{2}H(\chi_{(-b,b)})(x_2)$$

when $(x_1, x_2) \in R^* = (-a, a) \times (b, 3b)$. Here

$$|H(\chi_{(-b,b)})(x_2)| = \left|\frac{1}{\pi}\int_{-b}^{b}\frac{1}{x_2 - x}\,dx\right| > \frac{1}{2\pi},$$

and the lemma follows.   □

We shall now finish the proof of Theorem 21.5 when $n = 2$. Let $B = B(0, 1) \subset \mathbb{R}^2$ be the unit disc. We have by Parseval's theorem for $f, g \in L^2(\mathbb{R}^2)$,

$$\int T_B f \overline{g} = \int (T_B f)^\wedge \overline{\hat{g}} = \int \chi_B \hat{f} \, \overline{\hat{g}} = \int \hat{f} \, \overline{\chi_B \hat{g}} = \int \hat{f} \, \overline{(T_B g)^\wedge} = \int f \overline{T_B g}.$$

It follows that T_B is an L^p multiplier if and only it is an $L^{p'}$ multiplier with $p' = \dfrac{p}{p-1}$. Hence we may assume that $p < 2$. Suppose T_B were an L^p multiplier.

Let $\varepsilon > 0$ and let R_j, $j = 1, \ldots, 2^N$, be rectangles as in Lemma 21.6, $f_j = \chi_{R_j}$, let $v_j \in S^1$ be the directions of the longer sides of R_j and let T_j be the half-plane multiplier related to v_j as in Lemma 21.8.

First notice that by Lemmas 21.9 and 21.6 we have with $c_0 = 1/13$,

$$\left\| \left(\sum_{j=1}^k |T_j f_j|^2 \right)^{\frac{1}{2}} \right\|_p \geq \left\| \left(\sum_{j=1}^k (c_0 \chi_{R_j^*})^2 \right)^{\frac{1}{2}} \right\|_p = c_0 \mathcal{L}^2 \left(\bigcup_{j=1}^k R_j^* \right)^{1/p} = c_0,$$

since the rectangles R_j^* are disjoint.

Let $E = \cup_{j=1}^{2^N} R_j$ and let q be the dual exponent of $2/p$; $1/q = 1 - p/2$. By Lemma 21.8, Hölder's inequality and Lemma 21.6,

$$\left\| \left(\sum_{j=1}^k |T_j f_j|^2 \right)^{\frac{1}{2}} \right\|_p \leq C_p \left\| \left(\sum_{j=1}^k |f_j|^2 \right)^{\frac{1}{2}} \right\|_p$$

$$\leq C_p \left(\int \sum_{j=1}^k \chi_{R_j} \right)^{\frac{1}{2}} \mathcal{L}^2(E)^{1/(pq)} < C_p \varepsilon^{1/(pq)},$$

because $\int \sum_j \chi_{R_j} = \sum_j \mathcal{L}^2(R_j) = 1$. This is a contradiction for sufficiently small ε, which completes the proof for $n = 2$.

For $n > 2$, fix some nice function f on $\mathbb{R}^{n-2}$, for example the characteristic function of the unit ball. Then proceed as above using the functions f_j,

$$f_j(x_1, \ldots, x_n) = \chi_{R_j}(x_1, x_2) f(x_3, \ldots, x_n).$$

One can also prove and use a Fubini-type result stating that if m is an L^p multiplier on $\mathbb{R}^{m+n}$, then for almost every $\xi \in R^m$, $\eta \mapsto m(\xi, \eta)$ is an L^p multiplier on $\mathbb{R}^n$ with norm bounded by that of m. For this see Grafakos [2008], Theorem 2.5.16. $\square$

21.3 Bochner–Riesz multipliers

This is a brief introduction to this important topic and its connections to restriction problems. I skip here some proofs. For them and further results and comments, see Duoandikoetxea [2001], Grafakos [2009], Stein [1993] and Sogge [1993].

We now know that we do not have the convergence asked for in Example 21.3 if $n \geq 2$ and $p \neq 2$. But what about some modified type of convergence, for example,

$$f(x) = \lim_{R \to \infty} \int_{B(0,R)} \left(1 - \frac{|\xi|}{R}\right) e^{2\pi i x \cdot \xi} \widehat{f}(\xi) \, d\xi \quad \text{in the } L^p \text{ sense?}$$

This is analogous to some classical facts for Fourier series: it is easier to get the convergence for instance in the Cesàro sense, leading to the Fejér kernel, than for the usual Fourier partial sums; see, e.g., Duoandikotxea [2001]. So we are asking about results for the multiplier $(1 - |\xi|)_+$, or equivalently for $m(\xi) = (1 - |\xi|^2)_+$, instead of the ball multiplier. Here $a_+ = \max\{a, 0\}$. Raising m to a small power $\delta > 0$ we get closer to the characteristic function of the unit ball.

Definition 21.10 The *Bochner–Riesz multiplier* m_δ with parameter $\delta > 0$ is defined by

$$m_\delta(\xi) = (1 - |\xi|^2)_+^\delta, \quad \xi \in \mathbb{R}^n.$$

The corresponding multiplier operator is S_δ;

$$S_\delta f = (m_\delta \widehat{f})^\vee.$$

For $f \in \mathcal{S}(\mathbb{R}^n)$ we have

$$S_\delta f = K_\delta * f.$$

The kernel K_δ can be computed from the formula for the Fourier transform of a radial function with the aid of some Bessel function identities. It is

$$K_\delta(x) = c(n, \delta)|x|^{-n/2-\delta} J_{n/2+\delta}(2\pi |x|).$$

From the properties of Bessel functions it follows that K_δ is bounded and its asymptotic behaviour at infinity is,

$$K_\delta(x) \approx F_\delta(x)|x|^{-n/2-\delta-1/2},$$

where F_δ is a bounded trigonometric term. Consequently, $K_\delta \in L^p(\mathbb{R}^n)$ if and only if $p > \frac{2n}{n+1+2\delta}$. This implies that m_δ is not an L^p multiplier if

$p \leq \frac{2n}{n+1+2\delta}$. By duality, neither is it when $p \geq \frac{2n}{n-1-2\delta}$. The *Bochner–Riesz conjecture* believes that these are the only restrictions:

Conjecture 21.11 m_δ is an L^p multiplier if and only if

$$\frac{2n}{n+1+2\delta} < p < \frac{2n}{n-1-2\delta}. \tag{21.4}$$

Notice that the above condition is equivalent to

$$\left| \frac{1}{p} - \frac{1}{2} \right| < \frac{2\delta + 1}{2n}.$$

Again, this conjecture is open for $n \geq 3$ and true for $n = 2$. In $\mathbb{R}^2$ it is very close to the restriction conjecture and was proved by Carleson and Sjölin [1972]; Theorem 20.3 which gave the restriction conjecture gives also this, see Stein [1993], IX.5.5. Hörmander [1973] observed that both restriction and Bochner–Riesz problems have such a common approach. Tao [1999a] proved that the Bochner–Riesz conjecture implies the restriction conjecture and there are also some partial results in the opposite direction, see Tao's paper. For the parabolic case Carbery [1992] proved that in the natural ranges of the exponents the restriction implies Bochner–Riesz. This means that the sphere is replaced by the paraboloid $\{(\tilde{x}, |\tilde{x}|^2) : \tilde{x} \in \mathbb{R}^{n-1}\}$ and the multiplier is now $\chi(x)(x_n - |\tilde{x}|^2)_+^\delta$, where $\chi \in C_0^\infty(\mathbb{R}^n)$.

We illustrate the applicability of restriction theorems by the following theorem, which is rather close to the best that is known. We first prove it for the exponent $q = \frac{2(n+1)}{n-1}$ of the Stein–Tomas restriction theorem 19.4 and then discuss briefly the general case. Notice that this q is in the necessary range (21.4) under the condition (21.6). In fact, if $\delta = \frac{n-1}{2(n+1)}$, then $\frac{2n}{n-1-2\delta} = \frac{2(n+1)}{n-1}$, so this is a kind of end-point case.

Theorem 21.12 m_δ *is an L^p multiplier if*

$$\frac{2n}{n+1+2\delta} < p < \frac{2n}{n-1-2\delta}, \tag{21.5}$$

and

$$\delta > \frac{n-1}{2(n+1)}. \tag{21.6}$$

Proof The proof is easy if $\delta > \frac{n-1}{2}$, because then $K_\delta \in L^1$. It is simpler than in the full range for

$$\left| \frac{1}{p} - \frac{1}{2} \right| < \frac{\delta}{n-1},$$

for that, see Duoandikoetxea [2001]. For the full range we can write

$$m_\delta(\xi) = \sum_{k=0}^{\infty} 2^{-k\delta} \varphi_k(|\xi|),$$

where the functions φ_k are smooth, $\operatorname{spt} \varphi_k \subset (1 - 2^{1-k}, 1 - 2^{-2-k})$ for $k \geq 1$, and $|\varphi_k^{(j)}(t)| \leq C_j 2^{kj}$ for $t \in \mathbb{R}$, $j = 0, 1, 2, \ldots$. Let T_k be the multiplier operator

$$\widehat{T_k f}(\xi) = \varphi_k(|\xi|)\widehat{f}(\xi).$$

Then

$$S_\delta = \sum_{k=0}^{\infty} 2^{-k\delta} T_k.$$

To estimate $\|T_k f\|_{L^q(\mathbb{R}^n)}$ suppose first that $f \in \mathcal{S}(\mathbb{R}^n)$ has support in $B(0, 2^k)$. For such an f by the Fourier inversion formula,

$$T_k f(x) = \int_{1-2^{1-k}}^{1-2^{-2-k}} \int_{S^{n-1}} e^{2\pi i r x \cdot \zeta} \widehat{f}(r\zeta) \varphi_k(r) \, d\sigma^{n-1} \zeta \, r^{n-1} \, dr.$$

Hence by Minkowski's integral inequality,

$$\|T_k f\|_{L^q(\mathbb{R}^n)} \lesssim \int_{1-2^{1-k}}^{1-2^{-2-k}} \left\| \int_{S^{n-1}} e^{2\pi i r x \cdot \zeta} \widehat{f}(r\zeta) \, d\sigma^{n-1} \zeta \right\|_{L^q(\mathbb{R}^n)} r^{n-1} \, dr.$$

Theorem 19.4 yields then with $q = \frac{2(n+1)}{n-1}$,

$$\|T_k f\|_{L^q(\mathbb{R}^n)} \lesssim \int_{1-2^{1-k}}^{1-2^{-2-k}} \|\widehat{f}(r \cdot)\|_{L^2(S^{n-1})} r^{n-1} \, dr.$$

From this we obtain using Schwartz's inequality, the fact that $r \approx 1$ and Plancherel's theorem,

$$\|T_k f\|_{L^q(\mathbb{R}^n)} \lesssim \left(\int_{1-2^{1-k}}^{1-2^{-2-k}} \|\widehat{f}(r \cdot)\|_{L^2(S^{n-1})}^2 r^{n-1} \, dr \right)^{1/2} 2^{-k/2} \leq \|f\|_{L^2(\mathbb{R}^n)} 2^{-k/2}.$$

Since $\operatorname{spt} f \subset B(0, 2^k)$, we get by Hölder's inequality,

$$\|T_k f\|_{L^q(\mathbb{R}^n)} \lesssim 2^{-k/2} 2^{kn\frac{q-2}{2q}} \|f\|_{L^q(\mathbb{R}^n)} = 2^{k\frac{n-1}{2(n+1)}} \|f\|_{L^q(\mathbb{R}^n)}.$$

Notice now that the kernel of T_k (the Fourier transform of $\varphi_k(|\xi|)$) decays very fast for $|x| > 2^k$. This implies that the last estimate holds without the assumption $\operatorname{spt} f \subset B(0, 2^k)$; we leave to the reader some technical details needed for the verification of this. Recalling now that $S_\delta = \sum_{k=0}^{\infty} 2^{-k\delta} T_k$ and that $\delta > \frac{n-1}{2(n+1)}$, we see that S_δ is bounded from L^q to L^q for $q = \frac{2(n+1)}{n-1}$.

To prove the full theorem, we observe first that by duality, S_δ is also bounded from L^p to L^p for the dual exponent $p = \frac{2(n+1)}{n+3}$. The rest of the theorem follows by complex interpolation. In fact, one can prove more. The multipliers m_δ can be defined for complex δ with the same formula. The above argument works if (21.6) holds with the real part $\Re\delta$ in place of δ. If $\Re\delta > \frac{n-1}{2}$ the boundedness from L^1 to L^1 is trivial because K_δ is then integrable. Then interpolation and duality imply that for any complex δ satisfying $\Re\delta > \frac{n-1}{2(n+1)}$, m_δ is an L^p multiplier if (21.5) holds. See Grafakos [2009], Section 10.4.3, for a few more details and Grafakos [2008], Theorem 1.3.7, for the required complex interpolation theorem. □

To prove Theorem 21.12 Stein [1993], Section IX.2, again uses the stationary phase. The key lemma is:

Lemma 21.13 *Let ψ be a smooth function with compact support in $\mathbb{R}^n$. Define*

$$G_\lambda f(x) = \int e^{\lambda|x-y|} \psi(x-y) f(y)\, dy, \quad x \in \mathbb{R}^n, \lambda > 0.$$

Then for $1 \leq p \leq \frac{2(n+1)}{n+3}$ and $\lambda > 0$,

$$\|G_\lambda f\|_p \lesssim \lambda^{-n/p'} \|f\|_p.$$

A difference to the earlier case is that the phase function $|x - y|$ is not smooth. To overcome this one can consider

$$\widetilde{G}_\lambda f(x) = \int e^{\lambda|x-y|} \widetilde{\psi}(x-y) f(y)\, dy,$$

where the support of $\widetilde{\psi}$ does not meet the origin. For this and other details, see Stein [1993], Section IX.2.

21.4 Almost everywhere convergence and tube null sets

For $f \in L^2(\mathbb{R}^n)$ and $x \in \mathbb{R}^n$, $R > 0$, set

$$S_R f(x) = \int_{B(0,R)} \widehat{f}(x) e^{2\pi i x \cdot \xi}\, d\xi.$$

Then $S_R f \to f$ in L^2 for $f \in L^2(\mathbb{R}^n)$. By a classical result of Carleson, when $n = 1$, $S_R \to f$ almost everywhere as $R \to \infty$, and by Hunt's generalization this also holds for $f \in L^p(\mathbb{R})$ for $1 < p < \infty$; see Grafakos [2009], Chapter 11. However, it is not known if $S_R \to f$ almost everywhere when $f \in L^2(\mathbb{R}^n)$ and $n \geq 2$. Thus it is a question of great interest to find out as much information as possible on divergence sets for L^2 functions. Here is something about that.

Let us say that $A \subset \mathbb{R}^n, n \geq 2$, is *tube null* if for every $\varepsilon > 0$ there are δ_j-neighbourhoods $L_j(\delta_j)$ of lines L_j such that $A \subset \cup_{j=1}^{\infty} L_j(\delta_j)$ and $\sum_{j=1}^{\infty} \delta_j^{n-1} < \varepsilon$. Carbery, Soria and Vargas [2007] proved that if $A \subset B(0, 1)$ is tube null, then there is $f \in L^2(\mathbb{R}^n)$ such that $S_R f(x)$ fails to converge as $R \to \infty$ for all $x \in A$.

It is clear that tube null sets have Lebesgue measure zero, sets of $\mathcal{H}^{n-1}$ measure zero are tube null and that there exist tube null sets of Hausdorff dimension n; any set whose projection on a hyperplane has zero measure is tube null. This raises the question: how small a dimension can sets which are not tube null have? Shmerkin and Suomala [2012] solved this, showing by random constructions that there exist sets in $\mathbb{R}^n$ which are not tube null and have both Hausdorff and Minkowski dimension $n - 1$; see also Shmerkin and Suomala [2014]. They also considered curved tube null sets, where lines are replaced by curves, as well as the behaviour of Hausdorff measures in tubes. The latter question was also discussed by Carbery [2009] and Orponen [2013c]. Questions of covering with δ-neighbourhoods of Lipschitz graphs is central in the deep work of Alberti, Csörnyei and Preiss [2005], [2010] on the structure of Lebesgue null sets and differentiability of Lipschitz functions. Then any Lebesgue null set is tube null with respect to Lipschitz graphs.

21.5 Further comments

The material of this chapter is discussed in the books Duoandikoetxea [2001], Grafakos [2009], de Guzmán [1981], Stein [1993], Sogge [1993], and Bishop and Peres [2016]. Many more details and related results can be found there.

Lemma 21.6 is very close to the existence of Besicovitch sets and geometric constructions for them. So it goes essentially back to Besicovitch's work [1919] and [1928]; recall the discussion in Section 11.6. Fefferman [1971] used it to solve the multiplier problem for the ball, proving Theorem 21.5. This was the beginning of the geometric methods, usually called Kakeya methods, in Fourier analysis. We shall discuss these methods rather extensively in the next two chapters.

The application of restriction theorems to Bochner–Riesz multipliers, giving Theorem 21.12, is due to Fefferman [1970]. The presentation above was based on Stein [1993], Section IX.6.9, and the lecture notes of Ana Vargas. There are several later improvements based on bilinear and multilinear methods. We shall discuss them briefly in Chapter 25.

22

Kakeya problems

Recall from Section 11 that a Borel set in $\mathbb{R}^n$ is a Besicovitch set, or a Kakeya set, if it has zero Lebesgue measure and it contains a line segment of unit length in every direction. We proved that such sets, even compact, exist in every $\mathbb{R}^n$, $n \geq 2$. In this and the next chapter we shall study them and related Kakeya maximal functions. We shall also establish a connection to restriction problems. The first instance of such interplay between Kakeya methods and Fourier analysis was Fefferman's solution of the ball multiplier problem in 1971 which we presented in the previous chapter.

The dimension n of the space will be at least 2 for the rest of the book.

22.1 Kakeya maximal function

It is natural to approach these problems via a related maximal function. For $a \in \mathbb{R}^n$, $e \in S^{n-1}$ and $\delta > 0$, define the tube $T_e^{\delta}(a)$ with centre a, direction e, length 1 and radius δ:

$$T_e^{\delta}(a) = \{x \in \mathbb{R}^n : |(x-a) \cdot e| \leq 1/2, |x - a - ((x-a) \cdot e)e| \leq \delta\}.$$

Observe that $\mathcal{L}^n(T_e^{\delta}(a)) = \alpha(n-1)\delta^{n-1}$, where $\alpha(n-1)$ is the Lebesgue measure of the unit ball in $\mathbb{R}^{n-1}$.

Definition 22.1 The *Kakeya maximal function* with width δ of $f \in L^1_{loc}(\mathbb{R}^n)$ is the function

$$\mathcal{K}_{\delta}f : S^{n-1} \to [0, \infty],$$

$$\mathcal{K}_{\delta}f(e) = \sup_{a \in \mathbb{R}^n} \frac{1}{\mathcal{L}^n(T_e^{\delta}(a))} \int_{T_e^{\delta}(a)} |f| \, d\mathcal{L}^n.$$

We have the trivial but sharp proposition:

Proposition 22.2 *For all* $0 < \delta < 1$ *and* $f \in L^1_{loc}(\mathbb{R}^n)$,

$$\|\mathcal{K}_\delta f\|_{L^\infty(S^{n-1})} \leq \|f\|_{L^\infty(\mathbb{R}^n)} \quad \text{and}$$
$$\|\mathcal{K}_\delta f\|_{L^\infty(S^{n-1})} \leq \alpha(n-1)^{1-n}\delta^{1-n}\|f\|_{L^1(\mathbb{R}^n)}.$$

If $p < \infty$, there can be no inequality

$$\|\mathcal{K}_\delta f\|_{L^q(S^{n-1})} \leq C\|f\|_{L^p(\mathbb{R}^n)} \quad \text{for all } 0 < \delta < 1, \quad f \in L^p(\mathbb{R}^n),$$

with C independent of δ. This follows from the existence of Besicovitch sets: let $B \subset \mathbb{R}^n$ be such a compact set (with $\mathcal{L}^n(B) = 0$) and let

$$f = \chi_{B_\delta}, \ B(\delta) = \{x \in \mathbb{R}^n : d(x, B) < \delta\}.$$

Then $\mathcal{K}_\delta f(e) = 1$ for all $e \in S^{n-1}$, so $\|\mathcal{K}_\delta f\|_{L^q(S^{n-1})} \approx 1$ but $\|f\|_{L^p(\mathbb{R}^n)} = \mathcal{L}^n(B_\delta)^{1/p} \to 0$ as $\delta \to 0$. Consequently we look for inequalities like

$$\|\mathcal{K}_\delta f\|_{L^p(S^{n-1})} \leq C(n, p, \varepsilon)\delta^{-\varepsilon}\|f\|_{L^p(\mathbb{R}^n)} \quad \text{for all}$$

$$\varepsilon > 0, \quad 0 < \delta < 1, \quad f \in L^p(\mathbb{R}^n). \tag{22.1}$$

Even this cannot hold if $p < n$. Let $f = \chi_{B(0,\delta)}$. Since $B(0, \delta) \subset T_e^\delta(0)$, we have for all $e \in S^{n-1}$,

$$\mathcal{K}_\delta f(e) = \frac{\mathcal{L}^n(B(0, \delta))}{\mathcal{L}^n(T_e^\delta(0))} \approx \delta.$$

But

$$\|f\|_{L^p(\mathbb{R}^n)} = \mathcal{L}^n(B(0, \delta))^{1/p} \approx \delta^{n/p},$$

and δ is much bigger than $\delta^{n/p-\varepsilon}$ for small δ if $p < n$ and $n/p - \varepsilon > 1$. The *Kakeya maximal conjecture* wishes for the next best thing:

Conjecture 22.3 (22.1) holds if $p = n$, that is,

$$\|\mathcal{K}_\delta f\|_{L^n(S^{n-1})} \leq C(n, \varepsilon)\delta^{-\varepsilon}\|f\|_{L^n(\mathbb{R}^n)} \quad \text{for all}$$
$$\varepsilon > 0, \quad 0 < \delta < 1, \quad f \in L^n(\mathbb{R}^n).$$

We shall see that this holds in $\mathbb{R}^2$ even with a logarithmic factor in place of $\delta^{-\varepsilon}$. In $\mathbb{R}^n$, $n \geq 3$, the question is open. Also in higher dimensional estimates $\delta^{-\varepsilon}$ could usually be replaced by powers of $\log(1/\delta)$, but we shall not keep track of that. In any case higher dimensional estimates given later probably are never sharp.

Instead of (L^p, L^p)-inequalities (22.1) we could also search for (L^p, L^q)-inequalities of the form

$$\|\mathcal{K}_\delta f\|_{L^q(S^{n-1})} \leq C(n, p, \varepsilon)\delta^{-(n/p-1+\varepsilon)}\|f\|_{L^p(\mathbb{R}^n)} \quad \text{for all}$$
$$\varepsilon > 0, \quad 1 \leq p \leq n, \quad q = (n-1)p'. \tag{22.2}$$

This is a natural range since interpolating (cf. Section 2.7) Conjecture 22.3 with the trivial estimate $\|\mathcal{K}_\delta f\|_{L^\infty(S^{n-1})} \leq C(\varepsilon)\delta^{1-n}\|f\|_{L^1(\mathbb{R}^n)}$ gives (22.2).

We shall soon prove that the Kakeya maximal conjecture implies the Kakeya conjecture 11.4 according to which every Besicovitch set in $\mathbb{R}^n$ should have Hausdorff dimension n. Recall that this too is true for $n = 2$ and open for $n \geq 3$.

First we shall discretize and dualize the Kakeya maximal inequalities (22.1). We say that $\{e_1, \ldots, e_m\} \subset S^{n-1}$ is a δ-separated subset of S^{n-1} if $|e_j - e_k| \geq \delta$ for $j \neq k$. It is maximal if in addition for every $e \in S^{n-1}$ there is some k for which $|e - e_k| < \delta$. We call $T_1, \ldots, T_m$ δ-separated δ-tubes if $T_k = T_{e_k}^\delta(a_k), k = 1, \ldots, m$, for some δ-separated subset $\{e_1, \ldots, e_m\}$ of S^{n-1} and some $a_1, \ldots, a_m \in \mathbb{R}^n$. Clearly, $m \lesssim \delta^{1-n}$ for all δ-separated sets $\{e_1, \ldots, e_m\} \subset S^{n-1}$ and $m \approx \delta^{1-n}$ for all maximal δ-separated sets.

Later on the next three propositions will be applied with M of the form $M = \delta^{-\beta}$.

The key fact leading to the discretization is the following simple observation: if $e, e' \in S^{n-1}$ with $|e - e'| \leq \delta$, then

$$\mathcal{K}_\delta f(e) \leq C(n)\mathcal{K}_\delta f(e').$$

This holds because any $T_e^\delta(a)$ can be covered with some tubes $T_{e'}^\delta(a_j), j = 1, \ldots, N$, with N depending only on n.

Proposition 22.4 *Let* $1 < p < \infty, q = \frac{p}{p-1}, 0 < \delta < 1$ *and* $0 < M < \infty$. *Suppose that*

$$\left\|\sum_{k=1}^m t_k \chi_{T_k}\right\|_{L^q(\mathbb{R}^n)} \leq M$$

whenever $T_1, \ldots, T_m$ *are* δ-separated δ-tubes and $t_1, \ldots, t_m$ *are positive numbers with*

$$\delta^{n-1} \sum_{k=1}^m t_k^q \leq 1.$$

Then

$$\|\mathcal{K}_\delta f\|_{L^p(S^{n-1})} \leq C(n)M\|f\|_{L^p(\mathbb{R}^n)} \quad \text{for all } f \in L^p(\mathbb{R}^n).$$

Proof Let $\{e_1, \ldots, e_m\} \subset S^{n-1}$ be a maximal δ-separated subset of S^{n-1}. If $e \in S^{n-1} \cap B(e_k, \delta)$, then by the key fact mentioned above, $\mathcal{K}_\delta f(e) \leq C\mathcal{K}_\delta f(e_k)$

with C depending only on n. Hence

$$\|\mathcal{K}_\delta f\|_{L^p(S^{n-1})}^p \leq \sum_{k=1}^m \int_{B(e_k,\delta)} (K_\delta f)^p \, d\sigma^{n-1}$$

$$\leq \sum_{k=1}^m C^p \mathcal{K}_\delta f(e_k)^p \sigma^{n-1}(B(e_k,\delta)) \lesssim \sum_{k=1}^m \mathcal{K}_\delta f(e_k)^p \delta^{n-1}.$$

By the duality of l^p and l^q, for any $a_k \geq 0, k = 1, \ldots, m$,

$$\left(\sum_{k=1}^m a_k^p\right)^{1/p} = \max\left\{\sum_{k=1}^m a_k b_k : b_k \geq 0, \sum_{k=1}^m b_k^q = 1\right\}.$$

Applying this to $a_k = \delta^{(n-1)/p}\mathcal{K}_\delta f(e_k)$ we get

$$\|\mathcal{K}_\delta f\|_{L^p(S^{n-1})} \lesssim \left(\sum_{k=1}^m (\delta^{(n-1)/p}\mathcal{K}_\delta f(e_k))^p\right)^{1/p}$$

$$= \sum_{k=1}^m \delta^{(n-1)/p}\mathcal{K}_\delta f(e_k)b_k = \delta^{n-1}\sum_{k=1}^m t_k\mathcal{K}_\delta f(e_k)$$

where $\sum_{k=1}^m b_k^q = 1, t_k = \delta^{(1-n)/q}b_k$, and so $\delta^{n-1}\sum_{k=1}^m t_k^q = 1$. Therefore for some $a_k \in \mathbb{R}^n$,

$$\|\mathcal{K}_\delta f\|_{L^p(S^{n-1})} \lesssim \delta^{n-1}\sum_{k=1}^m t_k \frac{1}{\mathcal{L}^n(T_{e_k}^\delta(a_k))} \int_{T_{e_k}^\delta(a_k)} |f| \, d\mathcal{L}^n.$$

Since $\mathcal{L}^n(T_{e_k}^\delta(a_k)) \approx \delta^{n-1}$, we obtain by Hölder's inequality

$$\|\mathcal{K}_\delta f\|_{L^p(S^{n-1})} \lesssim \sum_{k=1}^m t_k \int_{T_{e_k}^\delta(a_k)} |f| \, d\mathcal{L}^n = \int \left(\sum_{k=1}^m t_k \chi_{T_{e_k}^\delta(a_k)}\right) |f| \, d\mathcal{L}^n$$

$$\leq \left\|\sum_{k=1}^m t_k \chi_{T_{e_k}^\delta(a_k)}\right\|_{L^q(\mathbb{R}^n)} \|f\|_{L^p(\mathbb{R}^n)} \leq M\|f\|_{L^p(\mathbb{R}^n)}. \qquad \square$$

Before going on along these lines we apply the previous lemma to solve the Kakeya maximal conjecture in the plane:

Theorem 22.5 *For all $0 < \delta < 1$ and $f \in L^2(\mathbb{R}^2)$,*

$$\|\mathcal{K}_\delta f\|_{L^2(S^1)} \leq C\sqrt{\log(1/\delta)}\|f\|_{L^2(\mathbb{R}^2)}$$

with some absolute constant C.

Proof Let $T_k = T_{e_k}^\delta(a_k), k = 1, \ldots, m$, be δ-separated δ-tubes and t_k positive numbers with $\delta \sum_{k=1}^m t_k^2 \leq 1$. By Proposition 22.4 we need to show that

$$\left\| \sum_{k=1}^m t_k \chi_{T_k} \right\|_{L^2(\mathbb{R}^2)} \lesssim \sqrt{\log(1/\delta)}.$$

The following elementary inequality is the key to the proof:

$$\mathcal{L}^2 \left(T_e^\delta(a) \cap T_{e'}^\delta(a') \right) \lesssim \frac{\delta^2}{|e - e'| + \delta} \tag{22.3}$$

for $e, \ e' \in S^1$, $a, a' \in \mathbb{R}^2$ with $|e - e'| \geq \delta$. We leave the verification of this as an exercise to the reader. Using (22.3) we estimate

$$\left\| \sum_{k=1}^m t_k \chi_{T_k} \right\|_{L^2(\mathbb{R}^2)}^2 = \sum_{j,k} t_j t_k \mathcal{L}^2(T_j \cap T_k) \lesssim \sum_{j,k} t_j t_k \frac{\delta^2}{|e_j - e_k| + \delta}.$$

For any fixed k, $|e_j - e_k|$ takes essentially values $i\delta, i = 1, \ldots, N_\delta \approx 1/\delta$, when $|e_j - e_k| \leq 1$. Moreover, for a given i the number of j for which $|e_j - e_k|$ is about $i\delta$ is bounded. Thus (the contribution coming from $|e_j - e_k| > 1$ is trivially bounded)

$$\sum_j \frac{\delta}{|e_j - e_k| + \delta} \lesssim \sum_{j=1}^{N_\delta} \frac{\delta}{j\delta + \delta} = \sum_{i=1}^{N_\delta} \frac{1}{i + 1} \approx \log(1/\delta).$$

Similarly with j and k interchanged. Hence we can apply Schur's test, Theorem 20.2, to conclude

$$\left\| \sum_{k=1}^m t_k \chi_{T_k} \right\|_{L^2(\mathbb{R}^2)}^2 \lesssim \sum_{j,k} \sqrt{\delta} t_j \sqrt{\delta} t_k \frac{\delta}{|e_j - e_k| + \delta}$$
$$\lesssim \log(1/\delta) \sum_k (\sqrt{\delta} t_k)^2 \leq \log(1/\delta). \qquad \square$$

Now we return to Proposition 22.4. We get rid of the coefficients t_k and obtain a discrete characterization of the Kakeya maximal inequalities. Observe that $m\delta^{n-1}$ below is essentially the L^1-norm of $\sum_{k=1}^m \chi_{T_k}$.

Proposition 22.6 *Let $1 < p < \infty, q = \frac{p}{p-1}, 1 \leq M < \infty$ and $0 < \delta < 1$. Then*

$$\|\mathcal{K}_\delta f\|_{L^p(S^{n-1})} \lesssim_{n,p,\varepsilon} M\delta^{-\varepsilon} \|f\|_{L^p(\mathbb{R}^n)} \quad \text{for all } f \in L^p(\mathbb{R}^n), \varepsilon > 0, \tag{22.4}$$

if and only if

$$\left\| \sum_{k=1}^{m} \chi_{T_k} \right\|_{L^q(\mathbb{R}^n)} \lesssim_{n,q,\varepsilon} M\delta^{-\varepsilon}(m\delta^{n-1})^{1/q} \quad \text{for all } \varepsilon > 0, \tag{22.5}$$

and for all δ-separated δ-tubes $T_1, \ldots, T_m$.

Proof Suppose we have (22.5). Let $T_1, \ldots, T_m$ be δ-separated δ-tubes and let $t_1, \ldots, t_m$ be positive numbers with $\delta^{n-1} \sum_{k=1}^{m} t_k^q \leq 1$. By Proposition 22.4 it suffices to show

$$\left\| \sum_{k=1}^{m} t_k \chi_{T_k} \right\|_{L^q(\mathbb{R}^n)} \lesssim_{n,q,\varepsilon} M\delta^{-\varepsilon}. \tag{22.6}$$

Observing that $\| \sum_{k=1}^{m} \delta^{n-1} \chi_{T_k} \|_{L^q(\mathbb{R}^n)} \lesssim 1$ and $t_k \leq \delta^{(1-n)/q}$, we see that it suffices to sum over k such that $\delta^{n-1} \leq t_k \leq \delta^{(1-n)/q}$. Split this into $\approx \log(1/\delta)$ subsums over $k \in I_j = \{k : 2^{j-1} \leq t_k < 2^j\}$ and let m_j be the cardinality of I_j. Then applying our assumption (22.5) with $\varepsilon/2$ we get

$$\left\| \sum_{k:\delta^{n-1} \leq t_k \leq \delta^{(1-n)/q}} t_k \chi_{T_k} \right\|_{L^q(\mathbb{R}^n)}$$

$$\leq \sum_j \left\| \sum_{k \in I_j} 2^j \chi_{T_k} \right\|_{L^q(\mathbb{R}^n)} = \sum_j 2^j \left\| \sum_{k \in I_j} \chi_{T_k} \right\|_{L^q(\mathbb{R}^n)}$$

$$\lesssim_{\varepsilon} \sum_j 2^j M\delta^{-\varepsilon/2}(m_j \delta^{n-1})^{1/q} \lesssim M \log(1/\delta)\delta^{-\varepsilon/2} \lesssim M\delta^{-\varepsilon},$$

because $m_j 2^{jq} \leq \sum_{k=1}^{m}(2t_k)^q \leq 2^q \delta^{1-n}$.

To prove the converse, assume that (22.4) holds and let $T_1, \ldots, T_m$ be δ-separated δ-tubes with directions $e_1, \ldots, e_m$. Let $g \in L^p(S^{n-1})$ with $\|g\|_{L^p(S^{n-1})} \leq 1$. Then by (22.4),

$$\int \sum_{k=1}^{m} \chi_{T_k} g = \sum_{k=1}^{m} \int_{T_k} g \lesssim \sum_{k=1}^{m} \mathcal{K}_\delta g(e_k)\delta^{n-1}$$

$$\lesssim \sum_{k=1}^{m} \int_{B(e_k,\delta)} \mathcal{K}_\delta g \, d\sigma^{n-1} \lesssim \int_{\cup_k B(e_k,\delta)} \mathcal{K}_\delta g \, d\sigma^{n-1}$$

$$\leq \|\mathcal{K}_\delta g\|_{L^p(S^{n-1})} \sigma^{n-1}(\cup_{k=1}^{m} B(e_k,\delta))^{1/q} \lesssim_{\varepsilon} M\delta^{-\varepsilon}(m\delta^{n-1})^{1/q}$$

by (22.4). We used here again the fact, which appeared already in the proof of Proposition 22.4, that $\mathcal{K}_\delta g(e_k) \lesssim \mathcal{K}_\delta g(e)$ for $e \in B(e_k, \delta)$. Now (22.5) follows taking supremum over such functions g. $\qquad \square$

Notice that in (22.5) $m\delta^{n-1} \lesssim 1$, and $m\delta^{n-1} \approx 1$ means that the δ-separated set $\{e_1, \ldots, e_m\} \subset S^{n-1}$ is essentially maximal. The following proposition says that it suffices to study such essentially maximal sets. The proof of the proposition uses very little geometry; only the rotational symmetry of the sphere is involved. The reader might notice some resemblance to the proof of Theorem 19.9.

Proposition 22.7 *Let* $1 < q < \infty$, $1 \le M < \infty$, *and* $0 < \delta < 1$. *Then*

$$\left\| \sum_{k=1}^{m} \chi_{T_k} \right\|_{L^q(\mathbb{R}^n)} \lesssim_{n,q,\varepsilon} M\delta^{-\varepsilon}(m\delta^{n-1})^{1/q} \quad \text{for all } \varepsilon > 0, \tag{22.7}$$

and for all δ-*separated* δ-*tubes* $T_1, \ldots, T_m$ *provided*

$$\left\| \sum_{k=1}^{m} \chi_{T_k} \right\|_{L^q(\mathbb{R}^n)} \lesssim_{n,q,\varepsilon} M\delta^{-\varepsilon} \quad \text{for all } \varepsilon > 0, \tag{22.8}$$

and for all δ-*separated tubes* $T_1, \ldots, T_m$.

Proof Let m_0 be the maximal cardinality of δ-separated subsets of S^{n-1}, then $m_0 \approx \delta^{1-n}$. For every $m = 1, \ldots, m_0$, let $c(m)$ denote the smallest constant such that

$$\left\| \sum_{k=1}^{m} \chi_{T_k} \right\|_{L^q(\mathbb{R}^n)} \le c(m)$$

for all δ-separated δ-tubes $T_1, \ldots, T_m$. We set also $c(t) = 0$ for any $t < 1$, $c(t) = m$ for any $m \le t < m + 1$, $m = 1, \ldots, m_0 - 1$, and $c(m) = c(m_0)$ for $m > m_0$. By our assumption (22.8) we know that $c(m) \lesssim M\delta^{-\varepsilon}$, and we need to improve this to

$$c(m) \lesssim M\delta^{-\varepsilon}(m\delta^{n-1})^{1/q}. \tag{22.9}$$

Fix $m \le m_0$ for a while and choose a δ-separated set $S \subset S^{n-1}$ of cardinality m and the corresponding δ tubes T_e, $e \in S$, such that

$$\left\| \sum_{e \in S} \chi_{T_e} \right\|_{L^q(\mathbb{R}^n)} = c(m)$$

(which exist by an easy compactness argument, although $2\| \sum_{e \in S} \chi_{T_e} \|_{L^q(\mathbb{R}^n)} \ge c(m)$ would be enough for us).

Now we consider rotations of S with $g \in O(n)$ such that S and $g(S)$ are disjoint, which of course is true for θ_n almost all $g \in O(n)$. Denote by $T_{g(e)}$ the rotated tube $g(T_e)$. Then we also have

$$\left\| \sum_{e \in S} \chi_{T_{g(e)}} \right\|_{L^q(\mathbb{R}^n)} = c(m).$$

From the trivial inequality $\|f + g\|_q^q \geq \|f\|_q^q + \|g\|_q^q$ for non-negative functions, we see that

$$\left\| \sum_{e \in S \cup g(S)} \chi_{T_e} \right\|_{L^q(\mathbb{R}^n)} \geq 2^{1/q} c(m). \qquad (22.10)$$

If $S \cup g(S)$ were δ-separated, we would get $c(m) \leq 2^{-1/q} c(2m)$, and iterating this we could easily finish the proof. Of course, there is no reason why we should be able to find g such that $S \cup g(S)$ is δ-separated and instead we try to find big δ-separated subsets of $S \cup g(S)$.

Define

$$a(S, g) = \#\{(e, e') \in S \times g(S) : |e - e'| \leq \delta\}.$$

Then

$$\int a(S, g) \, d\theta_n g = \sum_{e \in S} \sum_{e' \in S} f(e, e')$$

where

$$f(e, e') = \int \chi_{B(0,\delta)}(e - g(e')) \, d\theta_n g.$$

By the rotational symmetry of the sphere $f(e, e')$ is independent of e', so

$$f(e, e') = \frac{1}{\sigma^{n-1}(S^{n-1})} \iint \chi_{B(0,\delta)}(e - g(e')) \, d\sigma^{n-1} e' \, d\theta_n g.$$

The inner integral is at most $b\delta^{n-1}\sigma^{n-1}(S^{n-1})$, whence $f(e, e') \leq b\delta^{n-1}$, where b depends only on n. This gives

$$\int a(S, g) \, d\theta_n g \leq b\delta^{n-1} m^2.$$

Hence we can find $g \in O(n)$ such that $a(S, g) \leq b\delta^{n-1} m^2$. Then we can express $S \cup g(S)$ as

$$S \cup g(S) = S_1 \cup S_2,$$

where both S_1 and S_2 are δ-separated and S_2 has cardinality $\leq b\delta^{n-1}m^2$; for S_1 we only need that trivially it has cardinality at most $2m$. Indeed, we can choose

$$S_2 = \{e \in S : \exists e' \in g(S) \text{ such that } |e - e'| \leq \delta\},$$
$$S_1 = (S \setminus S_2) \cup g(S).$$

This decomposition implies by Minkowski's inequality

$$\left\| \sum_{e \in S \cup g(S)} \chi_{T_e} \right\|_{L^q(\mathbb{R}^n)} \leq c(2m) + c(b\delta^{n-1}m^2).$$

Combining this with (22.10) we get

$$2^{1/q}c(m) \leq c(2m) + c(b\delta^{n-1}m^2).$$

Above we can choose b so that $b\delta^{n-1} = 2^{-N}$ for some integer N. It is enough to prove the claim (22.9) for m of the form $m = 2^{N-k}, k = 1, \ldots, N$. Then $m = 2^{-k}b^{-1}\delta^{1-n}$. Set

$$c_k = 2^{k/q}c(2^{-k}b^{-1}\delta^{1-n}), \quad k = 1, \ldots, N.$$

Then the last inequality becomes

$$c_k \leq c_{k-1} + 2^{-(k+1)/q}c_{2k}. \tag{22.11}$$

Our claim (22.9) means now that $c_k \lesssim_{n,q,\varepsilon} M\delta^{-\varepsilon}$ for all k. This obviously holds for $k \leq k_0$ if k_0 depends only on n and q. Moreover, $c_k = 0$ if $k \gtrsim \log(1/\delta)$. We modify the sequence (c_k) slightly by defining with a suitable positive constant c,

$$d_k = (1 + c2^{-k/q})c_k.$$

If c is chosen sufficiently large, but depending only on q, and k_0 is sufficiently large, but depending only on q and c, then by a straightforward calculation, using only (22.11) and the definition of d_k,

$$d_k < d_{k-1} + 2^{-k/q}((d_{2k} - d_k) + (d_{k-1} - d_k)) \quad \text{for } k \geq k_0.$$

This implies that the maximum value of d_k is attained with some $k = k_1 \leq k_0$, so

$$c_k < d_k \leq d_{k_1} \lesssim M\delta^{-\varepsilon}$$

for all k, which completes the proof. $\qquad\square$

Combining the last two propositions we have:

Corollary 22.8 *Let* $1 < p < \infty, q = \frac{p}{p-1}, 0 < \beta < \infty,$ *and* $0 < \delta < 1$. *Then*

$$\|\mathcal{K}_\delta f\|_{L^p(S^{n-1})} \lesssim_{n,p,\varepsilon} \delta^{-\beta-\varepsilon} \|f\|_{L^p(\mathbb{R}^n)} \quad \text{for all } f \in L^p(\mathbb{R}^n), \varepsilon > 0, \quad (22.12)$$

if and only if

$$\left\|\sum_{k=1}^m \chi_{T_k}\right\|_{L^q(\mathbb{R}^n)} \lesssim_{n,p,\varepsilon} \delta^{-\beta-\varepsilon} \quad \text{for all } \varepsilon > 0, \quad (22.13)$$

and for all δ-separated δ-tubes $T_1, \ldots, T_m$. In particular, the Kakeya maximal conjecture 22.3 holds if and only if

$$\left\|\sum_{k=1}^m \chi_{T_k}\right\|_{L^{n/(n-1)}} \lesssim_{n,p,\varepsilon} \delta^{-\varepsilon} \quad \text{for all } \varepsilon > 0,$$

and for all δ-separated δ-tubes $T_1, \ldots, T_m$.

22.2 Kakeya maximal implies Kakeya

In this section we show that L^p estimates for the Kakeya maximal function imply lower bounds for the Hausdorff dimension of Besicovitch sets.

Let us first see that proving a lower bound for the Minkowski dimension from L^p estimates is almost trivial: Suppose we have for some p and $\beta > 0$ the estimate $\|\mathcal{K}_\delta f\|_{L^p(S^{n-1})} \lesssim \delta^{-\beta} \|f\|_p$ for $0 < \delta < 1$. The δ-neighbourhood $B(\delta)$ of the Besicovitch set B contains an open δ-tube in every direction, so we have for the characteristic function f of $B(\delta)$ that $\mathcal{K}_\delta f(e) = 1$ for all $e \in S^{n-1}$. Our estimate then gives

$$1 \approx \|\mathcal{K}_\delta f\|_{L^p(S^{n-1})}^p \lesssim \delta^{-\beta p} \mathcal{L}^n(B(\delta)),$$

form which it follows that $\underline{\dim}_M B \geq n - \beta p$.

We shall now extend this to Hausdorff dimension. The problem is that we have to use coverings of B with, say, balls of very different sizes. As often, the trick is to decompose such a covering into subfamilies of balls of essentially the same size.

Theorem 22.9 *Suppose that $1 < p < \infty, \beta > 0$ and $n - \beta p > 0$. If*

$$\|\mathcal{K}_\delta f\|_{L^p(S^{n-1})} \leq C(n, p, \beta)\delta^{-\beta} \|f\|_p \quad \text{for all } 0 < \delta < 1, \quad f \in L^p(\mathbb{R}^n),$$
$$(22.14)$$

then the Hausdorff dimension of every Besicovitch set in $\mathbb{R}^n$ is at least $n - \beta p$. In particular, if (22.1) holds for some $p, 1 < p < \infty$, then the Hausdorff *dimension of every Besicovitch set in $\mathbb{R}^n$ is n. Thus Conjecture 22.3 implies the Kakeya conjecture 11.4.*

Proof Let $B \subset \mathbb{R}^n$ be a Besicovitch set. Let $0 < \alpha < n - \beta p$ and $B_j = B(x_j, r_j)$, $j = 1, 2, \ldots$, be balls such that $r_j < 1$ and $B \subset \bigcup_j B_j$. It suffices to show that $\sum_j r_j^\alpha \gtrsim 1$.

For $e \in S^{n-1}$ let $I_e \subset B$ be a unit segment parallel to e. For $k = 1, 2, \ldots$, set

$$J_k = \{ j : 2^{-k} \le r_j < 2^{1-k} \},$$

and

$$S_k = \left\{ e \in S^{n-1} : \mathcal{H}^1 \left(I_e \cap \bigcup_{j \in J_k} B_j \right) \ge \frac{1}{2k^2} \right\}.$$

Since $\sum_k \frac{1}{2k^2} < 1$ and

$$\sum_k \mathcal{H}^1 \left(I_e \cap \bigcup_{j \in J_k} B_j \right) = \mathcal{H}^1(I_e) \ge 1,$$

we have

$$\bigcup_k S_k = S^{n-1};$$

if there were some $e \in S^{n-1} \setminus \bigcup_k S_k$, we would have $\mathcal{H}^1(I_e \cap \bigcup_{j \in J_k} B_j) < \frac{1}{2k^2}$ for all k, and then

$$\sum_k \mathcal{H}^1 \left(I_e \cap \bigcup_{j \in J_k} B_j \right) < \sum_k \frac{1}{2k^2} < 1,$$

which is impossible.

Let

$$f_k = \chi_{F_k} \quad \text{with} \quad F_k = \bigcup_{j \in J_k} B(x_j, 2r_j).$$

If $e \in S_k$, then, letting a_e be the mid-point of I_e, we have by simple geometry

$$\mathcal{L}^n \big(T_e^{2^{-k}}(a_e) \cap F_k \big) \gtrsim \frac{1}{k^2} \mathcal{L}^n \big(T_e^{2^{-k}}(a_e) \big),$$

whence $\mathcal{K}_{2^{-k}} f_k(e) \gtrsim 1/k^2$ for $e \in S_k$. This and assumption (22.14) give

$$\sigma^{n-1}(S_k) \lesssim k^{2p} \int (\mathcal{K}_{2^{-k}} f_k)^p \, d\sigma^{n-1}$$

$$\leq k^{2p} C(n, p, \beta)^p 2^{k\beta p} \int f_k^p = k^{2p} C(n, p, \beta)^p 2^{k\beta p} \mathcal{L}^n(F_k).$$

$$(22.15)$$

But $\mathcal{L}^n(F_k) \leq \#J_k \alpha(n) 2^{(2-k)n}$, whence

$$\sigma^{n-1}(S_k) \lesssim k^{2p} 2^{k\beta p} 2^{-kn} \#J_k = k^{2p} 2^{-k(n-\beta p)} \#J_k \lesssim 2^{-k\alpha} \#J_k,$$

and finally

$$\sum_j r_j^\alpha \geq \sum_k \#J_k 2^{-k\alpha} \gtrsim \sum_k \sigma^{n-1}(S_k) \gtrsim 1,$$

as required. $\qquad\qquad\qquad\qquad\qquad\qquad\qquad\qquad\qquad\qquad\qquad\square$

We shall now give a different, Fourier analytic, proof for Theorem 22.5. We do it in general dimensions obtaining a fairly sharp L^2 estimate.

Theorem 22.10 *For all* $0 < \delta < 1$ *and* $f \in L^2(\mathbb{R}^2)$,

$$\|\mathcal{K}_\delta f\|_{L^2(S^1)} \leq C \sqrt{\log(1/\delta)} \|f\|_{L^2(\mathbb{R}^2)},$$

with some absolute constant C.

In $\mathbb{R}^n$, $n \geq 3$, *we have for all* $0 < \delta < 1$ *and* $f \in L^2(\mathbb{R}^n)$,

$$\|\mathcal{K}_\delta f\|_{L^2(S^{n-1})} \leq C(n) \delta^{(2-n)/2} \|f\|_{L^2(\mathbb{R}^n)},$$

where the exponent $(2-n)/2$ *is the best possible.*

Proof Let $n \geq 2$. We may assume that f is non-negative and has compact support. Changing variable and using the symmetry of $T_e^\delta(0)$ we have

$$\mathcal{K}_\delta f(e) = \sup_{a \in \mathbb{R}^n} \frac{1}{\mathcal{L}^n(T_e^\delta(a))} \int_{T_e^\delta(a)} f$$

$$= \sup_{a \in \mathbb{R}^n} \frac{1}{\alpha(n-1)\delta^{n-1}} \int_{T_e^\delta(0)} f(a-x) \, dx = \sup_{a \in \mathbb{R}^n} \varrho_\delta^e * f(a),$$

where

$$\varrho_\delta^e = \frac{1}{\alpha(n-1)\delta^{n-1}} \chi_{T_e^\delta(0)}.$$

Let $\varphi \in \mathcal{S}(\mathbb{R})$ be such that spt $\widehat{\varphi} \subset [-1, 1]$, $\varphi \geq 0$ and $\varphi(x) \geq 1$ when $|x| \leq 1$. Define

$$\psi(x) = \delta^{1-n} \varphi(x_1) \varphi(|\tilde{x}|/\delta), \quad x = (x_1, \tilde{x}) \in \mathbb{R}^n, \quad x_1 \in \mathbb{R}, \quad \tilde{x} \in \mathbb{R}^{n-1}.$$

Then $\widehat{\psi}(\xi_1, \widetilde{\xi}) = \widehat{\varphi}(\xi_1)\widehat{\varphi}(\delta|\widetilde{\xi}|)$ and so

$$\operatorname{spt} \widehat{\psi} \subset [-1, 1] \times B^{n-1}(0, 1/\delta). \tag{22.16}$$

Since $\varphi(x_1) \geq 1$ and $\varphi(|\widetilde{x}|/\delta) \geq 1$ when $|x_1| \leq 1$ and $|\widetilde{x}| \leq \delta$, we have $\varrho_\delta^{e_1} \leq \psi$, with $e_1 = (1, 0, \ldots, 0)$, and so

$$\mathcal{K}_\delta f(e_1) \leq \sup_{a \in \mathbb{R}^n} \psi * f(a). \tag{22.17}$$

For $e \in S^{n-1}$, let $g_e \in O(n)$ be a rotation for which $g_e(e_1) = e$. Then $g_e(T_{e_1}^\delta(0)) = T_e^\delta(0)$. Hence defining $\psi_e = \psi \circ g_e$, we get from (22.17)

$$\mathcal{K}_\delta f(e) \leq \sup_{a \in \mathbb{R}^n} \psi_e * f(a).$$

As f has compact support, $\psi_e * f \in \mathcal{S}(\mathbb{R}^n)$ (that ψ is not differentiable on the line $\{x : \widetilde{x} = 0\}$ does not cause problems for this). Hence by the inversion formula and Schwartz's inequality the previous inequality leads to

$$\mathcal{K}_\delta f(e) \leq \|\psi_e * f\|_{L^\infty(\mathbb{R}^n)} \leq \|\widehat{\psi_e * f}\|_{L^1(\mathbb{R}^n)} = \int |\widehat{\psi_e}||\widehat{f}|$$

$$\leq \left(\int |\widehat{\psi_e}(\xi)||\widehat{f}(\xi)|^2(1 + |\xi|) \, d\xi \right)^{1/2} \left(\int \frac{|\widehat{\psi_e}(\xi)|}{1 + |\xi|} \, d\xi \right)^{1/2}.$$

Since $\widehat{\psi_e}(\xi) = \widehat{\psi \circ g_e}(\xi) = \widehat{\psi}(g_e(\xi))$, we get from (22.16)

$$\operatorname{spt} \widehat{\psi_e} \subset C_e := g_e^{-1}([-1, 1] \times B^{n-1}(0, 1/\delta)).$$

Suppose first $n = 2$. Then

$$\int \frac{|\widehat{\psi_e}(\xi)|}{1 + |\xi|} \, d\xi \lesssim \int_{C_e} \frac{1}{1 + |\xi|} \, d\xi \approx \int_{-1/\delta}^{1/\delta} \frac{1}{1 + |t|} \, dt \approx \log\left(\frac{1}{\delta}\right).$$

Thus

$$\|\mathcal{K}_\delta f\|_{L^2(S^1)}^2 \lesssim \log\left(\frac{1}{\delta}\right) \int_{S^1} \int_{\mathbb{R}^2} |\widehat{\psi_e}(\xi)||\widehat{f}(\xi)|^2(1 + |\xi|) d\xi \, d\sigma^1 e$$

$$= \log\left(\frac{1}{\delta}\right) \int_{\mathbb{R}^2} \left(\int_{S^1} |\widehat{\psi_e}(\xi)| \, d\sigma^1 e \right) |\widehat{f}(\xi)|^2(1 + |\xi|) \, d\xi.$$

Using again that $\operatorname{spt} \widehat{\psi_e} \subset C_e$ we get for all $\xi \in \mathbb{R}^2$,

$$\sigma^1(\{e \in S^1 : \widehat{\psi_e}(\xi) \neq 0\}) \leq \sigma^1(\{e \in S^1 : \xi \in C_e\}) \lesssim \frac{1}{1 + |\xi|}.$$

The last inequality is a simple geometric fact. Consequently,

$$\int |\widehat{\psi_e}(\xi)| \, d\sigma^1 e \lesssim \frac{1}{1 + |\xi|}$$

and

$$\|\mathcal{K}_\delta f\|^2_{L^2(S^1)} \lesssim \log\left(\frac{1}{\delta}\right) \int_{\mathbb{R}^2} \frac{1}{1+|\xi|} |\widehat{f}(\xi)|^2 (1+|\xi|)\, d\xi = \log\left(\frac{1}{\delta}\right) \|f\|^2_{L^2(\mathbb{R}^2)}.$$

If $n > 2$, we have

$$\int \frac{|\widehat{\psi_e}(\xi)|}{1+|\xi|}\, d\xi \lesssim \int_{C_e} \frac{1}{1+|\xi|}\, d\xi \approx \delta^{2-n}.$$

We still have the estimate

$$\sigma^{n-1}(\{e \in S^{n-1} : \xi \in C_e\}) \lesssim \frac{1}{1+|\xi|},$$

and

$$\|\mathcal{K}_\delta f\|^2_{L^2(S^{n-1})} \lesssim \delta^{2-n} \|f\|^2_{L^2(\mathbb{R}^n)}$$

follows.

Finally, that the power in $\delta^{(2-n)/2}$ cannot be improved can be seen using $f = \chi_{B(0,\delta)}$ as at the beginning of this chapter. $\qquad\square$

Combining Theorems 22.9 and 22.10 we obtain the following, which we already proved in Theorems 11.2 and 11.3 with different methods.

Corollary 22.11 *All Besicovitch sets in $\mathbb{R}^n$, $n \geq 2$, have Hausdorff dimension at least 2.*

22.3 Restriction implies Kakeya

Next we prove that the restriction conjecture

$$\|\widehat{f}\|_{L^q(\mathbb{R}^n)} \lesssim_{n,q} \|f\|_{L^q(S^{n-1})} \quad \text{for } f \in L^q(S^{n-1}), q > 2n/(n-1) \quad (22.18)$$

implies the Kakeya maximal conjecture 22.3, and hence it also implies the Kakeya conjecture 11.4. Recall Section 19.3 for discussion on the restriction conjecture.

For the proof we shall use Khintchine's inequality; we recall it from Section 2.8.

Theorem 22.12 *Suppose that $2n/(n-1) < q < \infty$ and*

$$\|\widehat{f}\|_{L^q(\mathbb{R}^n)} \lesssim_{n,q} \|f\|_{L^q(S^{n-1})} \quad \text{for } f \in L^q(S^{n-1}). \quad (22.19)$$

Then with $p = q/(q-2)$,

$$\|\mathcal{K}_\delta f\|_{L^p(S^{n-1})} \lesssim_{n,q} \delta^{4n/q - 2(n-1)} \|f\|_p \quad \text{for all } 0 < \delta < 1, \quad f \in L^p(\mathbb{R}^n).$$
$$(22.20)$$

In particular, the restriction conjecture (22.18) implies the Kakeya maximal conjecture 22.3.

Proof We first check the second statement assuming that the first part of the theorem is valid. Observe that $2(n-1) - 4n/q \to 0$ as $q \to 2n/(n-1)$. Hence for any $\varepsilon > 0$ we can choose $q > 2n/(n-1)$ for which $2(n-1) - 4n/q < \varepsilon$. Then $p = q/(q-2) < n$ and

$$\|\mathcal{K}_\delta f\|_{L^p(S^{n-1})} \lesssim \delta^{-\varepsilon} \|f\|_{L^p(\mathbb{R}^n)} \quad \text{for all } f \in L^p(\mathbb{R}^n).$$

Interpolating this with the trivial inequality

$$\|\mathcal{K}_\delta f\|_{L^\infty(\mathbb{R}^n)} \leq \|f\|_{L^\infty(\mathbb{R}^n)}$$

we get

$$\|\mathcal{K}_\delta f\|_{L^n(S^{n-1})} \lesssim \delta^{-\varepsilon} \|f\|_{L^n(\mathbb{R}^n)} \quad \text{for all } 0 < \delta < 1, \quad f \in L^n(\mathbb{R}^n),$$

as required.

To prove the first part, let $p' = p/(p-1) = q/2$, $\{e_1, \ldots, e_m\} \subset S^{n-1}$ be a δ-separated set, let $a_1, \ldots, a_m \in \mathbb{R}^n$ and $t_1, \ldots, t_m > 0$ with

$$\delta^{n-1} \sum_{k=1}^m t_k^{p'} \leq 1,$$

and let $T_k = T_{e_k}^\delta(a_k)$. We shall show that

$$\left\| \sum_{k=1}^m t_k \chi_{T_k} \right\|_{L^{p'}(\mathbb{R}^n)} \lesssim \delta^{4n/q - 2(n-1)}. \tag{22.21}$$

By Proposition 22.4 this implies (22.20).

Let τ_k be the δ^{-2} dilation of T_k: τ_k is the tube with centre $\delta^{-2}a_k$, direction e_k, length δ^{-2} and cross-section radius δ^{-1}. Let

$$S_k = \{e \in S^{n-1} : 1 - e \cdot e_k \leq C^{-2}\delta^2\}.$$

Then S_k is a spherical cap of radius $\approx C^{-1}\delta$ and centre e_k. Here C is chosen big enough to guarantee that the S_k are disjoint. Define f_k by

$$f_k(x) = e^{2\pi i \delta^{-2} a_k \cdot x} \chi_{S_k}(x).$$

Then $\|f_k\|_\infty = 1$, spt $f_k \subset S_k$ and the Fourier transform of f_k is a translate by $\delta^{-2}a_k$ of $\widehat{\chi_{S_k}}$ by (3.4); $\widehat{f_k}(\xi) = \widehat{\chi_{S_k}}(\xi - \delta^{-2}a_k)$. Provided C is sufficiently large, but still depending only on n, the Knapp example, Lemma 3.18, gives that

$$|\widehat{f_k}(\xi)| \gtrsim \delta^{n-1} \quad \text{for } \xi \in \tau_k.$$

Fix $s_k \geq 0, k = 1, \ldots, m$, and for $\omega = (\omega_1, \ldots, \omega_m) \in \{-1, 1\}^m$ let

$$f_\omega = \sum_{k=1}^m \omega_k s_k f_k.$$

We shall consider the ω_k as independent random variables taking values 1 and -1 with equal probablility, and we shall use Khintchine's inequality.

Since the functions f_k have disjoint supports,

$$\|f_\omega\|_{L^q(S^{n-1})}^q = \sum_{k=1}^m \|s_k f_k\|_{L^q(S^{n-1})}^q \approx \sum_{k=1}^m s_k^q \delta^{n-1}.$$

By Fubini's theorem and Khintchine's inequality 2.14,

$$\mathbb{E}\left(\|\widehat{f_\omega}\|_{L^q(\mathbb{R}^n)}^q\right) = \int \mathbb{E}(|\widehat{f_\omega}(\xi)|^q)\, d\xi \approx \int \left(\sum_{k=1}^m s_k^2 |\widehat{f_k}(\xi)|^2\right)^{q/2} d\xi$$

$$\gtrsim \delta^{q(n-1)} \int \left(\sum_{k=1}^m s_k^2 \chi_{\tau_k}(\xi)\right)^{q/2} d\xi,$$

since $|\widehat{f_k}| \gtrsim \delta^{n-1} \chi_{\tau_k}$.

By our assumption the restriction property (22.19) holds and we get

$$\|\widehat{f_\omega}\|_{L^q(\mathbb{R}^n)} \lesssim \|f_\omega\|_{L^q(S^{n-1})}.$$

Combining these three inequalities, we obtain

$$\delta^{q(n-1)} \int \left(\sum_{k=1}^m s_k^2 \chi_{\tau_k}\right)^{q/2} \lesssim \delta^{(n-1)} \sum_{k=1}^m s_k^q.$$

Now we choose $s_k = \sqrt{t_k}$ and have

$$\delta^{(n-1)} \sum_{k=1}^m s_k^q = \delta^{(n-1)} \sum_{k=1}^m t_k^{p'} \leq 1.$$

Thus

$$\delta^{q(n-1)} \int \left(\sum_{k=1}^m t_k \chi_{\tau_k}\right)^{p'} \lesssim 1.$$

Changing variable $y = \delta^2 x$, τ_k goes to T_k and so

$$\delta^{q(n-1)} \delta^{-2n} \int \left(\sum_{k=1}^m t_k \chi_{T_k}\right)^{p'} \lesssim 1,$$

that is,

$$\left\| \sum_{k=1}^{m} t_k \chi_{T_k} \right\|_{L^{p'}(\mathbb{R}^n)} \lesssim \delta^{4n/q - 2(n-1)},$$

as required. □

Since the Kakeya maximal conjecture implies the Kakeya conjecture we have:

Corollary 22.13 *The restriction conjecture (22.18) implies the Kakeya conjecture 11.4.*

Combining Theorems 22.9 and 22.12 we have also

Corollary 22.14 *If $2n/(n-1) < q < \infty$ and*

$$\|\widehat{f}\|_{L^q(\mathbb{R}^n)} \le C(n,q)\|f\|_{L^q(S^{n-1})} \quad \text{for } f \in L^q(S^{n-1}),$$

then $\dim B \ge (2n - (n-2)q)/(q-2)$ *for every Besicovitch set B in $\mathbb{R}^n$.*

22.4 Nikodym maximal function

Recall from Section 11.3 that a Nikodym set is a Borel subset N of $\mathbb{R}^n$ of measure zero such that for every point $x \in \mathbb{R}^n$ there is a line L containing x such that $L \cap N$ contains a unit line segment. We found that such sets exist. The related maximal function of a locally integrable function f is the *Nikodym maximal function* defined for $0 < \delta < 1$ by

$$\mathcal{N}_\delta f(x) = \sup_{x \in T} \mathcal{L}^n(T)^{-1} \int_T |f| \, d\mathcal{L}^n, \quad x \in \mathbb{R}^n,$$

where the supremum is taken over all tubes $T = T_e^\delta(a)$ containing x. In analogy to the Kakeya maximal conjecture 22.3 we have:

Conjecture 22.15 Nikodym maximal conjecture:

$$\|\mathcal{N}_\delta f\|_{L^n(\mathbb{R}^n)} \le C(n,\varepsilon)\delta^{-\varepsilon}\|f\|_{L^n(\mathbb{R}^n)} \quad \text{for all } \varepsilon > 0, \quad 0 < \delta < 1.$$

Theorem 22.9 holds for $\mathcal{N}_\delta f$ in place of $\mathcal{K}_\delta f$ and Nikodym in place of Kakeya. This follows by a straightforward modification of the proof. In particular, the Nikodym maximal conjecture implies Nikodym conjecture 11.10 that all Nikodym sets in $\mathbb{R}^n$ have Hausdorff dimension n. Recall that we proved in Theorem 11.11 that Kakeya conjecture 11.4 implies the Nikodym conjecture.

Theorem 22.16 *Kakeya maximal conjecture 22.3 and Nikodym maximal conjecture 22.15 are equivalent.*

This is due to Tao [1999a]. We only sketch the proof here.

Although it was easy to prove that the Kakeya conjecture implies the Nikodym conjecture and the converse is open, on maximal level the details for showing that the Nikodym maximal conjecture implies the Kakeya maximal conjecture are simpler than for the converse.

Going from Nikodym maximal to Kakeya maximal is based on the following pointwise inequality:

$$\delta \mathcal{K}_\delta f(e) \lesssim \mathcal{N}_{C\delta^2} f_\delta(x), \tag{22.22}$$

where $1/3 \le |x| \le 1/2$, $e = x/|x|$, $f \in L^1(\mathbb{R}^n)$ with spt $f \subset B(0, 1)$, $f_\delta(y) = f(y/\delta)$, $0 < \delta < 1$ and C is a positive constant depending only on n. We are restricted to functions in the unit ball. This suffices for proving L^p inequalities both for Kakeya and Nikodym maximal functions. We leave checking this for the reader as an exercise.

To verify (22.22), let $T = T_e^\delta(a)$ be a δ-tube intersecting $B(0, 1)$. Then

$$\int_T |f| = \delta^{-n} \int_{\delta T} |f_\delta|.$$

By simple geometry, with some constant C depending only on n, δT is contained in a $C\delta^2$-tube U which also contains x. Thus

$$\delta^{2-n} \int_T |f| \le \delta^{2-2n} \int_U |f_\delta| \le \mathcal{N}_{C\delta^2} f_\delta(x),$$

from which (22.22) follows by taking supremum over tubes T in direction e.

Applying (22.22) and integrating in polar coordinates, we obtain

$$\delta^p \int_{S^{n-1}} (\mathcal{K}_\delta f)^p \, d\sigma^{n-1} = \frac{n}{2^{-n} - 3^{-n}} \int_{1/3}^{1/2} r^{n-1} \int_{S^{n-1}} (\delta \mathcal{K}_\delta f(e))^p \, d\sigma^{n-1} e \, dr$$

$$\lesssim \int_{1/3}^{1/2} r^{n-1} \int_{S^{n-1}} (\mathcal{N}_{C\delta^2} f_\delta(re))^p \, d\sigma^{n-1} e \, dr$$

$$\le \int_{\mathbb{R}^n} (\mathcal{N}_{C\delta^2} f_\delta)^p,$$

that is,

$$\delta \| \mathcal{K}_\delta f) \|_{L^p(S^{n-1})} \lesssim \| \mathcal{N}_{C\delta^2} f_\delta \|_{L^p(\mathbb{R}^n)}.$$

If the Nikodym maximal conjecture holds, we have

$$\|\mathcal{N}_{C\delta^2} f_\delta\|_{L^n(\mathbb{R}^n)} \lesssim \delta^{-\varepsilon} \|f_\delta\|_{L^n(\mathbb{R}^n)} = \delta^{1-\varepsilon} \|f\|_{L^n(\mathbb{R}^n)}.$$

Combining the last two inequalities we get the Kakeya maximal conjecture.

For the opposite implication we use the same projective transformation that we used to prove the corresponding set implication:

$$F(\widetilde{x}, x_n) = \frac{1}{x_n}(\widetilde{x}, 1) \quad \text{for } (\widetilde{x}, x_n) \in \mathbb{R}^n, \quad x_n \neq 0.$$

With it we have the pointwise estimate:

$$\mathcal{N}_\delta f(\widetilde{x}, 0) \lesssim \mathcal{K}_{C\delta}(f \circ F) \left(\frac{(\widetilde{x}, 1)}{|(\widetilde{x}, 1)|} \right), \tag{22.23}$$

provided f has support in $\{x \in B(0, 1) : 1/2 \leq x_n \leq 1\}$. This follows by quantifying the argument of Theorem 11.11; tubes through $(\widetilde{x}, 0)$ are transformed by F to tubes in direction $\frac{(\widetilde{x},1)}{|(\widetilde{x},1)|}$. Integrating (22.23) leads to

$$\int (\mathcal{N}_\delta f(\widetilde{x}, 0))^p \, d\widetilde{x} \lesssim \int (\mathcal{K}_{C\delta} f \circ F)^p \, d\sigma^{n-1} \lesssim \int (\mathcal{K}_{C\delta} f)^p \, d\sigma^{n-1}.$$

Without the restriction spt $f \subset \{x \in B(0, 1) : 1/2 \leq x_n \leq 1\}$ we could replace 0 by any x_n, $|x_n| \leq 2$, in the above inequality and we would be done by Fubini's theorem. The general case can be reduced to this special case by decomposing the unit ball into dyadic belts $\{x \in B(0, 1) : 2^{-k} \leq x_n \leq 2^{1-k}, k = 1, 2, \ldots\}$, and using a scaling argument. We leave the details to the reader.

22.5 Summary of conjectures

I collect here the conjectures we have discussed and some relations between them:

(1) Kakeya conjecture 11.4:
Every Besicovitch set in $\mathbb{R}^n$ has Hausdorff dimension n.

(2) Kakeya maximal conjecture 22.3:

$$\|\mathcal{K}_\delta f\|_{L^n(S^{n-1})} \lesssim_{n,\varepsilon} \delta^{-\varepsilon} \|f\|_{L^n(\mathbb{R}^n)} \quad \text{for all } \varepsilon > 0, \quad 0 < \delta < 1.$$

(3) Nikodym conjecture 11.10:
Every Nikodym set in $\mathbb{R}^n$ has Hausdorff dimension n.

(4) Nikodym maximal conjecture 22.15:

$$\|\mathcal{N}_\delta f\|_{L^n(S^{n-1})} \lesssim_{n,\varepsilon} \delta^{-\varepsilon} \|f\|_{L^n(\mathbb{R}^n)} \quad \text{for all } \varepsilon > 0, \quad 0 < \delta < 1.$$

(5) Restriction conjecture 19.5–19.7:

$$\|\widehat{f}\|_{L^q(\mathbb{R}^n)} \lesssim_{n,q,\varepsilon} \|f\|_{L^\infty(S^{n-1})} \text{ (or } \lesssim_{n,q,\varepsilon} \|f\|_{L^q(S^{n-1})}) \quad \text{for } q > 2n/(n-1).$$

(6) Bochner–Riesz conjecture 21.11:
m_δ is an L^p multiplier if and only if

$$\frac{2n}{n+1+2\delta} < p < \frac{2n}{n-1-2\delta}.$$

(7) Keleti's line segment extension conjecture 11.12:
If A is the union of a family of line segments in $\mathbb{R}^n$ and B is the union of the corresponding lines, then $\dim A = \dim B$.

We proved that (1) implies (3) in Theorem 11.11, (2) implies (1) in Theorem 22.9, similarly (4) implies (3), and we proved that (5) implies (2) in Theorem 22.12. By Theorem 22.16 (2) and (4) are equivalent and (6) implies (5) by Tao [1999a]. It is not known if (5) implies (6), but, as mentioned in Section 21.3, Carbery [1992] verified this in the parabolic case. Due to Theorem 11.15 (7) implies the packing and upper Minkowski dimension versions of (1). All these conjectures are true in $\mathbb{R}^2$. As far as I know, the other implications are unknown.

There is also Sogge's *local smoothing conjecture* from Sogge [1991] for the wave equation, which is open in all dimensions $n \geq 2$ and implies all the other conjectures (1)–(6) above.

If u is a solution of the wave equation

$$\frac{\partial^2 u}{\partial^2 t} = \Delta_x u, \quad u(x,0) = f(x), \quad \frac{\partial u}{\partial t} u(x,t)_{|t=0} = g(x), \quad (x,t) \in \mathbb{R}^n \times \mathbb{R},$$

then for $p \geq \frac{2(n+1)}{n-1}$ and for all $\sigma > n(\frac{1}{2} - \frac{1}{p}) - \frac{1}{2}$,

$$\|u\|_{L^p(\mathbb{R}^n \times [1,2])} \lesssim \|f\|_{p,\sigma} + \|g\|_{p,\sigma-1}.$$

Here $\|f\|_{p,\sigma}$ is a Sobolev norm of f in the spirit of Chapter 17. Estimates of this type are called Strichartz estimates; they originate in Strichartz [1977].

This problem is related to restriction and multiplier questions for the cone $\{(x,t) \in \mathbb{R}^n \times \mathbb{R} : |x| = t\}$. Wolff [2000] proved that the estimate is valid for $n = 2$ and $p > 74$. This was extended for arbitrary $n \geq 3$, and a certain range of p, by Łaba and Wolff [2002]. Heo, Nazarov and Seeger [2011] improved the range of p when $n \geq 5$, including also the end-point $\sigma = n(\frac{1}{2} - \frac{1}{p}) - \frac{1}{2}$ when $n \geq 4$. Their method is based on an interesting characterization of radial Fourier multipliers. For further discussion, see Sogge [1993], Chapter 7, Stein [1993], XI.4.12, Wolff [2003], Section 11.4 and Tao [1999a].

Wolff's estimates in the case $n = 2$ had the consequence mentioned in Section 11.6: if the centres of a family of circles in the plane cover a set of Hausdorff dimension bigger than one, then the union of these circles must have positive Lebesgue measure.

There is also a connection between the Kakeya maximal conjecture and Montgomery's conjecture on Dirichlet sums. Let $T \leq N^2$, $a_k \in \mathbb{C}$ with $|a_k| \leq 1$, and

$$D(t) = \sum_{k=1}^{N} a_k k^{it}, \quad t \in \mathbb{R}.$$

Montgomery's conjecture asks whether it is true that for every measurable set $E \subset [0, T]$,

$$\int_E |D(t)|^2 \, dt \lesssim N^{1+\varepsilon}(N + \mathcal{L}^1(E)).$$

Bourgain [1993] proved that this implies the Kakeya conjecture and Wolff [2003], Section 11.4, showed further that it implies the Kakeya maximal conjecture, too.

22.6 Kakeya problems in finite fields

A natural setting for studying Kakeya problems is that of finite fields. A standard example of such a field is $\mathbb{Z}_p$, integers modulo p, when p is a prime. So let $\mathbb{F}$ be a field of q elements. The line in $\mathbb{F}^n$ passing through $x \in \mathbb{F}^n$ with direction $v \in \mathbb{F}^n \setminus \{0\}$ is

$$L(x, v) = \{x + tv : t \in \mathbb{F}\}.$$

The basic question now is: how big are the subsets of $\mathbb{F}^n$ which contain a line in every direction? Let us call such sets Besicovitch sets in $\mathbb{F}^n$.

Since we are in a finite setting, we measure size by cardinality. We shall now prove (modulo some facts from algebraic combinatorics) the following theorem of Dvir [2009] which essentially says that the analogue of the Kakeya conjecture in finite fields is true:

Theorem 22.17 *There exists a constant $c_n > 0$ depending only on n such that*

$$\#B \geq c_n q^n$$

for every Besicovitch set B in $\mathbb{F}^n$ for any finite field $\mathbb{F}$ of q elements.

Proof Study of polynomials in $\mathbb{F}^n$ forms the main ingredient of the proof. We pay attention to the obvious fact that two different polynomials may have the same values, for example x and x^p in $\mathbb{Z}_p$. In particular, the zero polynomial is the polynomial with all coefficients zero. The two basic facts used are: if the degree d of the polynomials considered is fixed, then, firstly, for any small set A there exists a non-zero polynomial of degree d vanishing on A, and, secondly, if a non-zero polynomial of degree d vanishes on a big set, then it must the zero polynomial.

The first fact is the following lemma:

Lemma 22.18 *Let d be a non-negative integer and $A \subset \mathbb{F}^n$ with $\#A < \binom{n+d}{n}$. Then there exists a non-zero polynomial P on $\mathbb{F}^n$ of degree at most d which vanishes on A.*

Proof Let V be the vector space over $\mathbb{F}$ of polynomials on $\mathbb{F}^n$ with degree at most d. Then the dimension of V is $\binom{n+d}{n}$; this is an easy combinatorial result. On the other hand the dimension of the vector space of all functions $f : A \to \mathbb{F}$ is $\#A < \dim V$. So the map $P \mapsto P|A$ is not injective. Thus there exist two different polynomials $P_1, P_2 \in V$ for which $P_1|A = P_2|A$. Hence $P = P_1 - P_2$ is what we want. $\square$

The second fact is part of the standard factor theorem:

Lemma 22.19 *If P is a non-zero polynomial on $\mathbb{F}$ of degree d, then*

$$\#\{x \in \mathbb{F} : P(x) = 0\} \leq d.$$

Now we combine these lemmas to prove a third lemma:

Lemma 22.20 *Let $B \subset \mathbb{F}^n$ be a Besicovitch set. If P is a polynomial on $\mathbb{F}^n$ of degree at most $\#F - 1$ and P vanishes on B, then P is the zero polynomial.*

Proof Suppose that P has degree d and it is not the zero polynomial. Write $P = \sum_{j=1}^{d} P_j$, where P_j is a homogeneous polynomial of degree j. Then P_d is not zero. For $v \in \mathbb{F}^n \setminus \{0\}$ let $x_v \in \mathbb{F}^n$ be such that $L(x_v, v) \subset B$. Then the polynomial P_v of one variable,

$$P_v(t) = P(x_v + tv), \quad t \in \mathbb{F},$$

vanishes on $\mathbb{F}$. Since $\deg P_v < \#F$, P_v is the zero polynomial by Lemma 22.19. We have $P_v(t) = P_d(v)t^d +$ lower order terms, whence $P_d(v) = 0$ for all $v \in \mathbb{F}^n \setminus \{0\}$. Then $P_d(v) = 0$ for all $v \in \mathbb{F}^n$, since P_d is homogeneous. Fixing $x_2, \ldots, x_n \in \mathbb{F}$, $x \mapsto P_d(x, x_2, \ldots, x_n)$, $x \in \mathbb{F}$, is a polynomial in $\mathbb{F}$ of degree at most $\#F - 1$ which vanishes identically, whence it is the zero polynomial by Lemma 22.19. Repeating this with the remaining $n - 1$ variables, we infer that all coefficients of P_d are zero. This is a contradiction which proves Lemma 22.20. $\square$

Combination of Lemmas 22.18 and 22.20 immediately yields

Corollary 22.21 *If $B \subset \mathbb{F}^n$ is a Besicovitch set, then*

$$\#B \geq \binom{\#F + n - 1}{n}.$$

Theorem 22.17 follows from this since $\binom{q+n-1}{n} = \frac{1}{n!}q^n +$ lower order terms.

$\square$

22.7 Further comments

The presentation of this chapter is largely based on Wolff's [2003] lecture notes. Kakeya maximal functions are also discussed in the books Grafakos [2009], Stein [1993] and Sogge [1993] . The proof of Proposition 22.7 was taken from Tao's [1999b] UCLA lecture notes.

Kakeya maximal inequalities were pioneered by Córdoba [1977]. In particular, he proved Theorem 22.5 with the geometric argument we also used. In fact, instead of Kakeya maximal functions Córdoba studied Nikodym maximal functions $\mathcal{N}_\delta f$, which we introduced in Section 22.4, but his methods work also for Kakeya. Keich [1999] showed that the factor $\sqrt{\log(1/\delta)}$ in Theorem 22.5 is sharp. Córdoba applied his results to multiplier estimates. See also Grafakos's book [2009], Sections 10.2 and 10.3, for such applications and other estimates on Nikodym maximal functions including results on the sharpness of the constants.

Bourgain [1991a] introduced Kakeya maximal functions and gave the Fourier-analytic proof for Theorem 22.10. This and Córdoba's proof can also be found in Wolff [2003]. Fefferman's [1971] result on ball multipliers, Theorem 21.5, is already close to a 'restriction implies Kakeya' type statement. More explicitly for restriction such a result was proved by Beckner, Carbery, Semmes and Soria [1989]. Finally, essentially Theorem 22.12 was proved by Bourgain [1991a] and discussed also in Bourgain [1995].

Bourgain [1991b] investigated curved Kakeya sets and related maximal functions, curves in place of line segments, and their relations to estimates on oscillating integrals. Many further interesting results on these were proven by Wisewell [2005].

Katz [1996] and Bateman and Katz [2008] studied Kakeya maximal functions where the directions of the tubes are restricted to certain Cantor sets. They proved that even then there is no boundedness with constants independent of δ. Further recent related results are due to Bateman [2009], Kroc and Pramanik [2014a] and [2014b] and to Parcet and Rogers [2013]. In particular, the papers of Bateman [2009] in the plane and of Kroc and Pramanik [2014b] in higher dimensions give interesting equivalent conditions on unboundedness of maximal operators, existence of Kakeya type constructions and lack of lacunarity, all related to a given set of directions. Katz [1999] and Demeter [2012] proved sharp bounds for a Nikodym type maximal operator over a finite set of directions.

Kim [2009] and [2012] proved Kakeya maximal function estimates when the line segments are restricted to lie in a smooth field of planes in $\mathbb{R}^3$. Under certain conditions estimates are essentially the same as in the Euclidean plane, but in some cases essentially sharp estimates are much worse. Such situations arise from Heisenberg groups.

Theorem 22.17, and a more general form of it, was proved by Dvir [2009]. The presentation here also used Tao's [2008a] blog. The study of Kakeya problems in finite fields was proposed by Wolff [2003] with some preliminary results. Since then several people have contributed to this topic; see the references given by Dvir [2009]. One motivation for this is that understanding easier questions in a discrete setting might help to understand more difficult questions in Euclidean spaces. But it is not only that; Kakeya-type problems in finite fields have interesting relations to many other combinatorial problems, see for example the papers of Bourgain, Katz and Tao [2004], of Guth and Katz [2010], and of Dvir and Wigderson [2011]. Ellenberg, R. Oberlin and Tao [2010] applied Dvir's method to Kakeya problems in algebraic varieties over finite fields.

Tao [2014] discusses the polynomial method, an example of which is the proof of Dvir's theorem, in relation to a large number of topics. Guth [2014a] used the polynomial method to prove the best known restriction estimate in $\mathbb{R}^3$:

$$\|\widehat{f}\|_{L^q(\mathbb{R}^3)} \lesssim \|f\|_{L^\infty(S^2)} \quad \text{for } f \in L^\infty(S^2), \quad q > 3.25.$$

Very likely, many more applications of this method will be found to problems in Euclidean spaces.

23

Dimension of Besicovitch sets and Kakeya maximal inequalities

Since we cannot solve the Kakeya conjecture, we could at least try to find lower bounds for the Hausdorff dimension of Besicovitch sets. The trivial one is 1. We have also the lower bound 2 from Theorem 11.2. In this chapter we improve this in dimensions bigger than two and we prove Kakeya maximal inequalities.

23.1 Bourgain's bushes and lower bound $(n + 1)/2$

Here we shall derive the lower bound $\frac{n+1}{2}$. The results of this section will be improved in the next one, but it might be useful to look at the ideas in a simpler case first.

Theorem 23.1 *Suppose that for some $1 \leq p < \infty$ and $\beta > 0$,*

$$\sigma^{n-1}(\{e \in S^{n-1} : \mathcal{K}_\delta(\chi_E)(e) > \lambda\}) \leq C(n, p, \beta)\delta^{-\beta p}\lambda^{-p}\mathcal{L}^n(E) \quad (23.1)$$

for all Lebesgue measurable sets $E \subset \mathbb{R}^n$ and for all $0 < \delta < 1$ and $\lambda > 0$. Then the Hausdorff dimension of every Besicovitch set in $\mathbb{R}^n$ is at least $n - \beta p$.

This follows from the proof of Theorem 22.9 since (22.15) is a consequence of (23.1).

Our next plan is to verify (23.1) for $p = (n + 1)/2$ and $\beta = \frac{n-1}{n+1}$ to get the lower bound $(n + 1)/2$ for the Hausdorff dimension of Besicovitch sets. Before doing this let us contemplate a little what (23.1) means. It is a restricted weak type inequality (restricted since it only deals with characteristic functions) which would follow immediately by Chebyshev's inequality from the corresponding strong type inequality (if we knew it):

$$\|\mathcal{K}_\delta f\|_{L^p(S^{n-1})} \lesssim \delta^{-\beta}\|f\|_p.$$

The converse is not true, but if we have restricted weak type inequalities for pairs (p_1, q_1) and (p_2, q_2) we have the strong type inequality for the appropriate

329

pairs (p, q) between (p_1, q_1) and (p_2, q_2) by the interpolation results discussed in Section 2.7.

Recall the Kakeya maximal conjecture 22.3:

$$\|\mathcal{K}_\delta f\|_{L^n(S^{n-1})} \leq C(n, \varepsilon)\delta^{-\varepsilon}\|f\|_n \quad \text{for all } \varepsilon > 0,$$

and the equivalent conjecture (22.2) obtained by interpolation:

$$\|\mathcal{K}_\delta f\|_{L^q(S^{n-1})} \leq C(n, p, \varepsilon)\delta^{-(n/p-1+\varepsilon)}\|f\|_p$$
$$\text{for all} \quad \varepsilon > 0, \quad 1 \leq p \leq n, \quad q = (n-1)p'.$$

In the next theorem we shall prove the restricted weak type version of this corresponding to $p = (n + 1)/2, q = n + 1$.

Theorem 23.2 *For all Lebesgue measurable sets $E \subset \mathbb{R}^n$,*

$$\sigma^{n-1}(\{e \in S^{n-1} : \mathcal{K}_\delta(\chi_E)(e) > \lambda\}) \leq C(n)\delta^{1-n}\lambda^{-n-1}\mathcal{L}^n(E)^2 \qquad (23.2)$$

for all $0 < \delta < 1$ and $\lambda > 0$. In particular, the Hausdorff dimension of every Besicovitch set in $\mathbb{R}^n$ is at least $(n + 1)/2$.

Proof Clearly the inequality (23.2) implies

$$\sigma^{n-1}(\{e \in S^{n-1} : \mathcal{K}_\delta(\chi_E)(e) > \lambda\}) \lesssim \delta^{(1-n)/2}\lambda^{-(n+1)/2}\mathcal{L}^n(E),$$

which means that the assumption (23.1) holds with $p = \frac{n+1}{2}$ and $\beta = \frac{n-1}{n+1}$ so that the statement about Besicovitch sets follows from Theorem 23.1.

Let $S_+^{n-1} = \{e \in S^{n-1} : e_n > 1/2\}$, just to avoid antipodal points, and

$$A = \left\{e \in S_+^{n-1} : \mathcal{K}_\delta(\chi_E)(e) > \lambda\right\}.$$

To prove (23.2) it is enough to estimate the measure of A. We can choose a δ-separated set $\{e_1, \ldots, e_N\} \subset A$ such that $N \gtrsim \delta^{1-n}\sigma^{n-1}(A)$ and tubes $T_j = T_{e_j}^\delta(a_j)$ for which

$$\mathcal{L}^n(E \cap T_j) > \lambda\mathcal{L}^n(T_j) \approx \lambda\delta^{n-1}. \qquad (23.3)$$

It then suffices to show that

$$\mathcal{L}^n(E) \gtrsim \sqrt{N}\delta^{n-1}\lambda^{\frac{n+1}{2}}. \qquad (23.4)$$

Let m be the smallest integer such that every point of E belongs to at most m tubes T_j. This means that

$$\sum_j \chi_{E \cap T_j} \leq m \qquad (23.5)$$

and there is $x \in E$ which belongs to m tubes T_j. Integration of (23.5) over E gives by (23.3) that

$$\mathcal{L}^n(E) \gtrsim \sum_j m^{-1} \mathcal{L}^n(E \cap T_j) \gtrsim m^{-1} N \lambda \delta^{n-1}. \qquad (23.6)$$

To make use of x assume that it belongs to the first m tubes T_j; $x \in T_j$ for $j = 1, \ldots, m$. Let c be a positive constant depending only on n such that

$$\mathcal{L}^n\big(B(x, c\lambda) \cap T_e^\delta(a)\big) \le \frac{\lambda}{2} \mathcal{L}^n\big(T_e^\delta(a)\big)$$

for every $e \in S^{n-1}$, $a \in \mathbb{R}^n$; the existence of such a constant is an easy exercise. Then by (23.3) for $j = 1, \ldots, m$,

$$\mathcal{L}^n(E \cap T_j \setminus B(x, c\lambda)) > \frac{\lambda}{2} \mathcal{L}^n(T_j) \approx \lambda \delta^{n-1}. \qquad (23.7)$$

By simple elementary plane geometry (this is again Córdoba's inequality (22.3)) there is an absolute constant $b \ge c$ such that for any $e, e' \in S_+^{n-1}$, $a, a' \in \mathbb{R}^n$,

$$d\big(T_e^\delta(a) \cap T_{e'}^\delta(a')\big) \le \frac{b\delta}{|e - e'|}. \qquad (23.8)$$

Let $e_1', \ldots, e_{m'}'$ be a maximal $\frac{b\delta}{c\lambda}$-separated subset of $e_1, \ldots, e_m$. Here $\frac{b\delta}{c\lambda} \ge \delta$ when we assume, as we of course may, that $\lambda \le 1$. The balls $B(e_k', \frac{2b\delta}{c\lambda})$, $k = 1, \ldots, m'$, cover the disjoint balls $B(e_j, \delta/3)$, $j = 1, \ldots, m$. Thus

$$m\delta^{n-1} \approx \sigma^{n-1}\left(\bigcup_{j=1}^m B(e_j, \delta/3)\right) \le \sigma^{n-1}\left(\bigcup_{k=1}^{m'} B\left(e_k', \frac{2b\delta}{c\lambda}\right)\right) \lesssim m'(\delta/\lambda)^{n-1},$$

whence $m' \gtrsim \lambda^{n-1} m$. By (23.8) the sets $E \cap T_k' \setminus B(x, c\lambda)$, $k = 1, \ldots, m'$, (T_k' corresponds to e_k') are disjoint. Therefore by (23.7),

$$\mathcal{L}^n(E) \gtrsim \lambda \delta^{n-1} m' \gtrsim \lambda^n \delta^{n-1} m. \qquad (23.9)$$

Now both inequalities (23.6) and (23.9) hold. Consequently

$$\mathcal{L}^n(E) \gtrsim \max\{\lambda^n \delta^{n-1} m, m^{-1} N \lambda \delta^{n-1}\}$$
$$\ge \sqrt{(\lambda^n \delta^{n-1} m)(m^{-1} N \lambda \delta^{n-1})} = \sqrt{N} \delta^{n-1} \lambda^{\frac{n+1}{2}}$$

and (23.4) follows. $\qquad \square$

23.2 Wolff's hairbrushes and lower bound $(n + 2)/2$

Bourgain's bushes in the above proof are bunches of tubes containing a common point. Replacing these with Wolff's hairbrushes, many tubes spreading out from a fixed tube, will improve the bound $(n + 1)/2$ to $(n + 2)/2$. We give two proofs for this. The second is a little more complicated than the first, but it gives a better L^q estimate.

The geometric fact behind both proofs is the following lemma:

Lemma 23.3 *Let* $\alpha, \beta, \gamma, \delta \in (0, 1)$ *be positive numbers, and let* $T = T_e^\delta(a), T_j = T_{e_j}^\delta(a_j), j = 1, \ldots, N,$ *be δ-tubes in* $\mathbb{R}^n$. *Suppose that the tubes* $T_j, j = 1, \ldots, N,$ *are δ-separated and for all* $j = 1, \ldots, N, T_j \cap T \neq \emptyset$ *and* $|e_j - e| \geq \alpha\beta$. *Then for all* $j = 1, \ldots, N,$

$$\#\{i : |e_i - e_j| \leq \beta, T_i \cap T_j \neq \emptyset, d(T_j \cap T_i, T_j \cap T) \geq \gamma\}$$
$$\leq C(n, \alpha)\beta\delta^{-1}\gamma^{2-n}. \tag{23.10}$$

Proof As $\#\{i : |e_i - e_j| \leq \beta\} \lesssim \beta^{n-1}\delta^{1-n}$, we may assume that δ is very small as compared to γ. Keeping this in mind should help the reader to form the proper geometric picture of the situation.

Denote by I the index set whose size we should estimate. The tubes T and T_j contain some line segments l and l_j of unit length which intersect at an angle $\gtrsim \beta$ at some point, say at the origin. We can assume that l and l_j generate the (x_1, x_2)-plane. For $i \in I$, the tube T_i meets both tubes T and T_j in a way that the angle between T_i and T_j is at most constant, depending on α, times the angle between T and T_j. It follows by simple plane geometry from this and the fact $d(T_j \cap T_i, T_j \cap T) \geq \gamma$ (which is much bigger than δ) that T_i intersects T_j outside the $c\gamma$-neighbourhood of $T \cap T_i$ for some positive constant c depending only on n and α. This implies that the central unit segment of T_i makes an angle $\lesssim \delta/\gamma$ with the (x_1, x_2)-plane. Moreover, $e_i \in B(e_j, \beta)$. From this one concludes that

$$e_i \in B(e_j, \beta) \cap \{x \in S^{n-1} : |x_k| \leq c'\delta/\gamma, k = 3, \ldots, n\} \quad \text{for } i \in I,$$

where c' depends only on n. The surface measure of this set is $\lesssim \beta(\delta/\gamma)^{n-2}$ so it contains $\lesssim \beta\delta^{-1}\gamma^{2-n}$ δ-separated points. This implies (23.10). $\qquad\square$

Let us say that a collection $T_j, j = 1, \ldots, N$, of δ-separated δ-tubes is an (N, δ)-*hairbrush* if there is a δ-tube T such that $T_j \cap T \neq \emptyset$ for all $j = 1, \ldots, N$.

Lemma 23.4 *Suppose that T_j, $j = 1, \ldots, N$, form an (N, δ)-hairbrush. Then for all $\varepsilon > 0$ and $n/(n-1) \le p \le 2$,*

$$\int \left(\sum_{i=1}^{N} \chi_{T_i} \right)^p \le C(n, p, \varepsilon) \delta^{n-(n-1)p-\varepsilon} N \delta^{n-1}. \tag{23.11}$$

Proof We may assume that $|e_i - e_j| < 1$ for all i and j, mainly in order to avoid that far away directions would correspond to nearby tubes. Let $T_j = T_{e_j}^{\delta}(a_j)$ and let $T = T_e^{\delta}(a)$ be the base tube which all the others intersect. For non-negative integers k with $\delta \le 2^{-k} \le 2$, set

$$I(k) = \{ i \in I : 2^{-k} < |e_i - e| \le 2^{1-k} \}.$$

Since there are only about $\log(1/\delta)$ values of k to consider, it suffices to show that for each k,

$$\int \left(\sum_{i \in I(k)} \chi_{T_i} \right)^p \lesssim \delta^{n-(n-1)p-\varepsilon} N \delta^{n-1}. \tag{23.12}$$

Writing

$$\int \left(\sum_{i \in I(k)} \chi_{T_i} \right)^p = \int \left(\sum_{j \in I(k)} \chi_{T_j} \right) \left(\sum_{i \in I(k)} \chi_{T_i} \right)^{p-1} = \sum_{j \in I(k)} \int_{T_j} \left(\sum_{i \in I(k)} \chi_{T_i} \right)^{p-1},$$

(23.12) becomes

$$\sum_{j \in I(k)} \int_{T_j} \left(\sum_{i \in I(k)} \chi_{T_i} \right)^{p-1} \lesssim \delta^{n-(n-1)p-\varepsilon} N \delta^{n-1}.$$

Hence it suffices to show that for each $j \in I(k)$,

$$\delta^{1-n} \int_{T_j} \left(\sum_{i \in I(k)} \chi_{T_i} \right)^{p-1} \lesssim \delta^{n-(n-1)p-\varepsilon}.$$

Fix k and $j \in I(k)$ and for positive integers $l \ge k - 2$ and m with 2^{-l}, $2^{-m} \ge \delta/2$, define

$$I(k, j, l, m) = \{ i \in I(k) : 2^{-l} < |e_i - e_j| \le 2^{1-l}, T_j \cap T_i \ne \varnothing,$$
$$\delta 2^{m+l-1} < d(T_j \cap T_i, T_j \cap T) \le \delta 2^{m+l} \},$$

and for $m = 0$,

$$I(k, j, l, 0) = \{i \in I(k) : 2^{-l} < |e_i - e_j| \leq 2^{1-l}, T_j \cap T_i \neq \varnothing,$$
$$d(T_j \cap T_i, T_j \cap T) \leq \delta 2^l\}.$$

We only consider $l \geq k - 2$, because otherwise these sets are empty. Then by Lemma 23.3 for $m \geq 1$ and for $m = 0$ trivially,

$$\#I(k, j, l, m) \lesssim 2^{-l}\delta^{-1}(\delta 2^{m+l})^{2-n} = 2^{-l}(2^{m+l})^{2-n}\delta^{1-n}. \tag{23.13}$$

Again there are only about $\log(1/\delta)$ possible values of l and m and it suffices to show that for fixed k, j, l, m,

$$\delta^{1-n} \int_{T_j} \left(\sum_{i \in I(k,j,l,m)} \chi_{T_i} \right)^{p-1} \lesssim \delta^{n-(n-1)p}.$$

For $i \in I(k, j, l, m)$ the diameter of $T_j \cap T_i$ is at most $c2^l\delta$ for some positive constant c depending only on n. Hence, by the definition of $I(k, j, l, m)$, we only need to integrate over $T_j(l, m, \delta) := \{x \in T_j : d(x, T_j \cap T) \leq (1 + c)\delta 2^{m+l}\}$ and for this set we have $\mathcal{L}^n(T_j(l, m, \delta)) \lesssim 2^{m+l}\delta^n$. Observing also that $\mathcal{L}^n(T_j \cap T_i) \lesssim 2^l\delta^n$ when $i \in I(j, k, l, m)$ we argue using Hölder's inequality and (23.13)

$$\delta^{1-n} \int_{T_j} \left(\sum_{i \in I(k,j,l,m)} \chi_{T_i} \right)^{p-1} \leq \delta^{1-n} \left(\int_{T_j} \sum_{i \in I(k,j,l,m)} \chi_{T_i} \right)^{p-1} \mathcal{L}^n(T_j(l, m, \delta))^{2-p}$$

$$\lesssim \delta^{1-n} \left(\sum_{i \in I(k,j,l,m)} \mathcal{L}^n(T_j \cap T_i) \right)^{p-1} (2^{m+l}\delta^n)^{2-p}$$

$$\lesssim \delta^{1-n} \left(2^l\delta^n \#I(k, j, l, m) \right)^{p-1} (2^{m+l}\delta^n)^{2-p}$$

$$\lesssim \delta^{1-n} \left(2^l\delta^n 2^{-l}(2^{m+l})^{2-n}\delta^{1-n} \right)^{p-1} (2^{m+l}\delta^n)^{2-p}$$

$$= 2^{(n-(n-1)p)(m+l)}\delta^{n-(n-1)p} \leq \delta^{n-(n-1)p}. \qquad \square$$

Theorem 23.5 *Let $0 < \delta < 1$. Then for $f \in L^n(\mathbb{R}^n)$,*

$$\|\mathcal{K}_\delta f\|_{L^n(S^{n-1})} \leq C(n, \varepsilon)\delta^{\frac{2-n}{2n}-\varepsilon}\|f\|_{L^n(\mathbb{R}^n)} \tag{23.14}$$

for all $\varepsilon > 0$. In particular, the Hausdorff dimension of every Besicovitch set in $\mathbb{R}^n$ is at least $(n + 2)/2$.

Proof The statement about Besicovitch sets follows immediately from Theorem 22.9.

Let $T_j = T_{e_j}^\delta(a_j)$, $j \in I = \{1, \ldots, m\}$, where $\{e_1, \ldots, e_m\}$ is a δ-separated subset of S^{n-1} (with $|e_i - e_j| < 1$ for all i and j as before). By Proposition 22.6 it suffices to show that

$$\int \left(\sum_{j=1}^m \chi_{T_j} \right)^{n/(n-1)} \lesssim \delta^{\frac{2-n}{2(n-1)} - \varepsilon} m \delta^{n-1}. \tag{23.15}$$

As in the proof of Lemma 23.4 this is reduced to

$$\sum_{j=1}^m \int_{T_j} \left(\sum_{i=1}^m \chi_{T_i} \right)^{1/(n-1)} \lesssim \delta^{\frac{2-n}{2(n-1)} - \varepsilon} m \delta^{n-1}. \tag{23.16}$$

Let

$$I(j, k) = \{ i \in I : 2^{-k-1} < |e_i - e_j| \leq 2^{-k} \}$$

when $j \in I$ and $\delta \leq 2^{-k} \leq 1$. Using the elementary inequality $(a + b)^\alpha \leq a^\alpha + b^\alpha$ for positive numbers a, b and α with $\alpha \leq 1$ we see that

$$\int_{T_j} \left(\sum_{i=1}^m \chi_{T_i} \right)^{1/(n-1)} \leq \sum_{k=1}^{N_\delta} \int_{T_j} \left(\sum_{i \in I(j,k)} \chi_{T_i} \right)^{1/(n-1)} + \mathcal{L}^n(T_j),$$

where $N_\delta \approx \log(1/\delta)$. The last summand is harmless, since $\sum_j \mathcal{L}^n(T_j) \lesssim m\delta^{n-1}$. Fix k and cover S^{n-1} with balls $B(v_l, 2^{-k})$ so that the balls $B(v_l, 2^{1-k})$ have bounded overlap with a constant depending only on n. If $i \in I(j, k)$, then e_i and e_j belong to the same ball $B(v_l, 2^{1-k})$ for some l. Thus fixing a ball B of radius 2^{1-k}, our claim is reduced to showing

$$\sum_{j \in I(B)} \int_{T_j} \left(\sum_{i \in I(B)} \chi_{T_i} \right)^{1/(n-1)} \lesssim \delta^{\frac{2-n}{2(n-1)} - \varepsilon} \delta^{n-1} \#I(B), \tag{23.17}$$

where $I(B) = \{ i \in I : e_i \in B \}$.

Let N be a positive integer to be fixed later. Now we want to extract as many (N, δ)-hairbrushes as possible from the tubes indexed by $I(B)$. Pick one such hairbrush $\mathcal{H}_1$ (if any exists) and let $H_1 \subset I(B)$ be the corresponding index set. Next choose a hairbrush $\mathcal{H}_2$ with indices in $H_2 \subset I(B) \setminus H_1$, and so on. In this way we find the hairbrushes $\mathcal{H}_l = \{ T_i : i \in H_l \}$, $l = 1, \ldots, M$, so that settting $H = H_1 \cup \cdots \cup H_M$ and $K = I(B) \setminus H$, the collection of the tubes T_i, $i \in K$, contains no (N, δ)-hairbrushes. This means that for any δ-tube T,

$$\#\{ i \in K : T_i \cap T \neq \varnothing \} < N. \tag{23.18}$$

Since $\#I(B) \lesssim 2^{-(n-1)k}\delta^{1-n}$, we have

$$M \lesssim 2^{-(n-1)k}\delta^{1-n}/N. \tag{23.19}$$

We can estimate the sum in (23.17) with

$$\sum_{j \in I(B)} \int_{T_j} \left(\sum_{i \in I(B)} \chi_{T_i} \right)^{1/(n-1)} \leq S(H, H) + S(H, K) + S(K, H) + S(K, K), \tag{23.20}$$

where

$$S(H, H) = \sum_{j \in H} \int_{T_j} \left(\sum_{i \in H} \chi_{T_i} \right)^{1/(n-1)} = \int \left(\sum_{i \in H} \chi_{T_i} \right)^{n/(n-1)},$$

$$S(K, H) = \sum_{j \in K} \int_{T_j} \left(\sum_{i \in H} \chi_{T_i} \right)^{1/(n-1)},$$

$$S(H, K) = \sum_{j \in H} \int_{T_j} \left(\sum_{i \in K} \chi_{T_i} \right)^{1/(n-1)},$$

$$S(K, K) = \sum_{j \in K} \int_{T_j} \left(\sum_{i \in K} \chi_{T_i} \right)^{1/(n-1)}.$$

The first term is estimated by the hairbrush lemma 23.4. For every $l \in \{1, \ldots, M\}$ we have

$$\int \left(\sum_{i \in H_l} \chi_{T_i} \right)^{n/(n-1)} \lesssim \delta^{-\varepsilon} \#H_l \delta^{n-1},$$

whence by Minkowski's inequality

$$S(H, H)^{(n-1)/n} = \left\| \sum_{i \in H} \chi_{T_i} \right\|_{n/(n-1)} \leq \sum_{l=1}^{M} \left\| \sum_{i \in H_l} \chi_{T_i} \right\|_{n/(n-1)}$$

$$\lesssim \sum_{l=1}^{M} (\delta^{-\varepsilon} \#H_l \delta^{n-1})^{(n-1)/n},$$

and so by Hölder's inequality

$$S(H, H) \lesssim \left(\sum_{l=1}^{M} (\delta^{-\varepsilon} \#H_l \delta^{n-1})^{(n-1)/n} \right)^{n/(n-1)} \leq M^{1/(n-1)} \delta^{-\varepsilon} \#H \delta^{n-1}.$$

Inserting (23.19) we get

$$S(H, H) \lesssim (2^{-(n-1)k}\delta^{1-n}/N)^{1/(n-1)}\delta^{-\varepsilon} \#I(B)\delta^{n-1}. \qquad (23.21)$$

To estimate the second term we use Hölder's inequality twice, the fact that the directions of T_i and T_j make an angle roughly 2^{-k} and (23.18) to obtain

$$S(K, H) \lesssim \sum_{j \in K} \left(\int_{T_j} \sum_{i \in H} \chi_{T_i} \right)^{1/(n-1)} \cdot \delta^{n-2}$$

$$\leq (\#K)^{(n-2)/(n-1)} \left(\sum_{j \in K} \int_{T_j} \sum_{i \in H} \chi_{T_i} \right)^{1/(n-1)} \cdot \delta^{n-2}$$

$$= (\#K)^{(n-2)/(n-1)} \left(\sum_{i \in H} \sum_{j \in K} \mathcal{L}^n(T_i \cap T_j) \right)^{1/(n-1)} \cdot \delta^{n-2},$$

$$\lesssim (\#K)^{(n-2)/(n-1)} \left(\sum_{i \in H} \sum_{j \in K, T_i \cap T_j \neq \varnothing} 2^k \delta^n \right)^{1/(n-1)} \cdot \delta^{n-2}$$

$$\leq (\#K)^{(n-2)/(n-1)} \left(\sum_{i \in H} N 2^k \delta \right)^{1/(n-1)} \cdot \delta^{n-1}$$

$$= (\#K)^{(n-2)/(n-1)} \left(\#H N 2^k \delta \right)^{1/(n-1)} \cdot \delta^{n-1} \leq \#I(B) \left(N 2^k \delta \right)^{1/(n-1)} \cdot \delta^{n-1}.$$

Finally the third and fourth terms can be estimated in the same way to get

$$S(H, K) + S(K, K) \lesssim \#I(B) \left(N 2^k \delta \right)^{1/(n-1)} \delta^{n-1}.$$

Choosing $N \approx 2^{-kn/2}\delta^{-n/2}$ all the above upper bounds yield

$$S(K, H) + S(K, H) + S(H, K) + S(K, K) \leq \#I(B)\delta^{\frac{2-n}{2(n-1)}-\varepsilon}\delta^{n-1},$$

as required for (23.17). $\square$

We now give a different argument. It does not improve the dimension bound, but it gives a better maximal function estimate; (23.14) follows interpolating (23.22) with the trivial estimate $\|\mathcal{K}_\delta f\|_{L^\infty(S^{n-1})} \leq \|f\|_{L^\infty(\mathbb{R}^n)}$. It should also give further insight into the situation.

Theorem 23.6 *Let* $0 < \delta < 1$. *Then for* $f \in L^{\frac{n+2}{2}}(\mathbb{R}^n)$,

$$\|\mathcal{K}_\delta f\|_{L^{\frac{n+2}{2}}(S^{n-1})} \leq C(n, \varepsilon)\delta^{\frac{2-n}{2+n}-\varepsilon} \|f\|_{L^{\frac{n+2}{2}}(\mathbb{R}^n)} \qquad (23.22)$$

for all $\varepsilon > 0$. *In particular, the Hausdorff dimension of every Besicovitch set in* $\mathbb{R}^n$ *is at least* $(n + 2)/2$.

Proof As before the statement about Besicovitch sets follows from Theorem 22.9.

The proof of (23.22) is long, but much of it consists of simple reductions. Let $T_j = T_{e_j}^\delta(a_j)$, $j = 1, \ldots, m$, where $\{e_1, \ldots, e_m\}$ is a δ-separated subset of S^{n-1}. By Corollary 22.8 we need to show that

$$\int \left(\sum_{j=1}^{m} \chi_{T_j} \right)^q \lesssim \delta^{\frac{2-n}{n} - \varepsilon} \qquad (23.23)$$

for all $\varepsilon > 0$ with $q = (n+2)/n \le 2$.

We may assume that $|e_i - e_j| < 1/4$ for all i and j in order to avoid that far away directions would correspond to nearby tubes and for slight technical convenience later. We shall use a 'bilinear approach', that is, we write the qth power of the left hand side of (23.23) as

$$\int \left(\sum_{j=1}^{m} \chi_{T_j} \right)^q = \int \left(\left(\sum_{j=1}^{m} \chi_{T_j} \right)^2 \right)^{q/2} = \int \left(\sum_{i,j} \chi_{T_i} \chi_{T_j} \right)^{q/2}.$$

Next we split this double sum into parts according to the distance (or angle) between the directions. Let N be the smallest integer such that $2^{-N} < \delta$ and set

$$I_0 = \{1, \ldots, m\},$$
$$J_k = \{(i,j) \in I_0 \times I_0 : 2^{-k} \le |e_i - e_j| < 2^{1-k}\}, \quad k = 1, \ldots, N.$$

Now we have

$$\sum_{i,j} \chi_{T_i} \chi_{T_j} = \sum_{k=1}^{N} \sum_{(i,j) \in J_k} \chi_{T_i} \chi_{T_j} + 2 \sum_{i \in I_0} \chi_{T_i}.$$

Since $q/2 \le 1$, we have the elementary inequality $(a+b)^{q/2} \le a^{q/2} + b^{q/2}$, $a, b \ge 0$. Applying this we obtain

$$\int \left(\sum_{i=1}^{m} \chi_{T_i} \right)^q \le \sum_{k=1}^{N} \int \left(\sum_{(i,j) \in J_k} \chi_{T_i} \chi_{T_j} \right)^{q/2} + 2 \int \left(\sum_{i \in I_0} \chi_{T_i} \right)^{q/2}.$$

Since there are about $\log(1/\delta)$ values of k, the theorem will follow if we can prove for every $k = 1, \ldots, N$,

$$\int \left(\sum_{(i,j) \in J_k} \chi_{T_i} \chi_{T_j} \right)^{q/2} \lesssim \delta^{\frac{2-n}{n} - \varepsilon}, \qquad (23.24)$$

because the estimate for the sum corresponding to I_0 is trivial: as $q/2 \leq 1$,

$$\int \left(\sum_{i \in I_0} \chi_{T_i} \right)^{q/2} \leq \int \sum_{i \in I_0} \chi_{T_i} \lesssim 1.$$

So fix $k \in \{1, \ldots, N\}$. Cover S^{n-1} with balls $B(v_l, 2^{-k}), l = 1, \ldots, N_k \approx 2^{(n-1)k}$. Then for every pair $(i, j) \in J_k$, $e_i, e_j \in B_l := B(v_l, 2^{2-k})$ for some l. It follows that

$$\int \left(\sum_{(i,j) \in J_k} \chi_{T_i} \chi_{T_j} \right)^{q/2} \lesssim \sum_{l=1}^{N_k} \int \left(\sum_{(i,j) \in J_k, e_i, e_j \in B_l} \chi_{T_i} \chi_{T_j} \right)^{q/2}.$$

As $N_k \approx 2^{(n-1)k}$ we are reduced to showing for every l,

$$\int \left(\sum_{(i,j) \in J_k, e_i, e_j \in B_l} \chi_{T_i} \chi_{T_j} \right)^{q/2} \lesssim 2^{-(n-1)k} \delta^{\frac{2-n}{n} - \varepsilon}. \tag{23.25}$$

Our next step will be to reduce this essentially to the case $k = 1$, that is, $|e_i - e_j| \approx 1$. Thus we claim that it suffices to prove that

$$\int \left(\sum_{(i,j) \in K} \chi_{T_i} \chi_{T_j} \right)^{q/2} \lesssim \delta^{\frac{2-n}{n} - \varepsilon}, \tag{23.26}$$

provided

$$K = \{(i, j) : |e_i - e_j| > 2c_0\},$$

where c_0 is a positive constant depending only on n, $T_i = T_{e_i}^{\delta}(a_i), i = 1, \ldots, m$, and $\{e_1, \ldots, e_m\}$ is a δ-separated subset of S^{n-1}.

So suppose we know (23.26). Let $k \geq 3$ and l be as above; we may assume that $v_l = (0, \ldots, 0, 1)$. Consider the linear mapping $L, L(x) = (2^{k-1}x_1, \ldots, 2^{k-1}x_{n-1}, 2^{-1}x_n)$. Then $\det L = 2^{(n-1)(k-1)-1}$ and $\chi_{T_j} \circ L^{-1} = \chi_{L(T_j)}$. By change of variable,

$$\int \left(\sum_{(i,j) \in J_k, e_i, e_j \in B_l} \chi_{T_i} \chi_{T_j} \right)^{q/2} = 2^{1-(n-1)(k-1)} \int \left(\sum_{(i,j) \in J_k, e_i, e_j \in B_l} \chi_{L(T_i)} \chi_{L(T_j)} \right)^{q/2}.$$

For a sufficiently small absolute constant $c_0' > 0$ the sets $L(T_i)$ are contained in $2^{k-1}\delta$-tubes whose directions e_i' satisfy $|e_i' - e_j'| > c_0'$ for the pairs (i, j) which appear in the above sum. We can therefore apply our assumption (23.26) to get

$$
\int \left(\sum_{(i,j)\in J_k, e_i, e_j \in B_l} \chi_{T_i} \chi_{T_j} \right)^{q/2} \lesssim 2^{-(n-1)k} (2^k \delta)^{\frac{2-n}{n} - \varepsilon} \lesssim 2^{-(n-1)k} \delta^{\frac{2-n}{n} - \varepsilon}.
$$

Let us make one more reduction: partition S^{n-1} into disjoint subsets S_l, $l = 1, \ldots, N(n)$, of diameter less than $c_0/2$. Then for any $(i, j) \in K$ there are k and l, $k \neq l$, such that $e_i \in S_k$ and $e_j \in S_l$. To prove (23.26) it suffices to consider each such pair (k, l) separately. That is, it suffices to prove that

$$
\int \left(\left(\sum_{i \in I} \chi_{T_i} \right) \left(\sum_{j \in J} \chi_{T_j} \right) \right)^{q/2} = \int \left(\sum_{i \in I, j \in J} \chi_{T_i} \chi_{T_j} \right)^{q/2} \lesssim \delta^{\frac{2-n}{n} - \varepsilon}
$$

(23.27)

where $I, J \subset \{1, \ldots, m\}$ such that $|e_i - e_j| > c_0$ when $i \in I$, $j \in J$ and $m \lesssim \delta^{1-n}$.

For $\mu, \nu \in \{1, \ldots, m\}$, set

$$
E_{\mu,\nu} = \left\{ x : \mu \leq \sum_{i \in I} \chi_{T_i}(x) < 2\mu, \ \nu \leq \sum_{j \in J} \chi_{T_j}(x) < 2\nu \right\}.
$$

Then we have for the left hand side of (23.27)

$$
\int \left(\left(\sum_{i \in I} \chi_{T_i} \right) \left(\sum_{j \in J} \chi_{T_j} \right) \right)^{q/2} = \sum_{\mu,\nu} \int_{E_{\mu,\nu}} \left(\left(\sum_{i \in I} \chi_{T_i} \right) \left(\sum_{j \in J} \chi_{T_j} \right) \right)^{q/2}
$$

$$
\leq \sum_{\mu,\nu} (4\mu\nu)^{q/2} \mathcal{L}^n(E_{\mu,\nu}),
$$

where the summation is over the dyadic integers μ and ν of the form $2^l \leq m$, $l \geq 0$. There are only $\lesssim \log(1/\delta)^2$ pairs of them. Thus we can find such a pair (μ, ν) for which

$$
\int \left(\left(\sum_{j \in I} \chi_{T_i} \right) \left(\sum_{j \in J} \chi_{T_j} \right) \right)^{q/2} \lesssim \delta^{-\varepsilon} (\mu\nu)^{q/2} \mathcal{L}^n(E_{\mu,\nu}).
$$

Taking also into account that $q = (n + 2)/n$, the required inequality (23.27) is now reduced to

$$
(\mu\nu)^{(n+2)/(2n)} \mathcal{L}^n(E_{\mu,\nu}) \lesssim \delta^{(2-n)/n - \varepsilon}.
$$

(23.28)

Keeping fixed the pair (μ, ν) which we found, we define for dyadic rationals κ and λ of the form 2^{-l}, $l = 0, 1, \ldots$,

$$
I_\kappa = \{i \in I : (\kappa/2)\mathcal{L}^n(T_i) < \mathcal{L}^n(T_i \cap E_{\mu,\nu}) \leq \kappa \mathcal{L}^n(T_i)\},
$$
$$
J_\lambda = \{j \in J : (\lambda/2)\mathcal{L}^n(T_j) < \mathcal{L}^n(T_j \cap E_{\mu,\nu}) \leq \lambda \mathcal{L}^n(T_j)\}.
$$

By the definition of $E_{\mu,\nu}$,

$$\int_{E_{\mu,\nu}} \sum_{i\in I} \sum_{j\in J} \chi_{T_i} \chi_{T_j} \approx \mu\nu\mathcal{L}^n(E_{\mu,\nu}).$$

We can write this as

$$\sum_{\kappa,\lambda} \int_{E_{\mu,\nu}} \sum_{i\in I_\kappa} \sum_{j\in J_\lambda} \chi_{T_i} \chi_{T_j} \approx \mu\nu\mathcal{L}^n(E_{\mu,\nu}),$$

where the summation in κ and λ is over dyadic rationals as above. To prove (23.28) we may assume $\mu\nu\mathcal{L}^n(E_{\mu,\nu}) \geq 1$. Then we can restrict κ and λ to be at least δ^n, since, for example,

$$\sum_{\kappa\leq\delta^n} \int_{E_{\mu,\nu}} \sum_{i\in I_\kappa} \sum_{j\in J_\lambda} \chi_{T_i} \chi_{T_j} \leq \sum_{l=1}^{\infty} \sum_{2^{-l}\delta^n \leq \kappa \leq 2^{1-l}\delta^n} \#J \int_{E_{\mu,\nu}} \sum_{i\in I_\kappa} \chi_{T_i}$$

$$\lesssim \delta^{1-n} \sum_{l=1}^{\infty} \sum_{i\in I} 2^{1-l}\delta^n \mathcal{L}^n(T_i) \lesssim \delta.$$

Thus we again have only $\approx \log(1/\delta)$ values to consider and we find and fix κ and λ for which

$$\mu\nu\mathcal{L}^n(E_{\mu,\nu}) \lesssim \delta^{-\varepsilon} \int_{E_{\mu,\nu}} \sum_{i\in I_\kappa} \sum_{j\in J_\lambda} \chi_{T_i} \chi_{T_j}. \tag{23.29}$$

Then by the definition of $E_{\mu,\nu}$,

$$\mu\nu\mathcal{L}^n(E_{\mu,\nu}) \lesssim \delta^{-\varepsilon}\nu \int_{E_{\mu,\nu}} \sum_{i\in I_\kappa} \chi_{T_i} = \delta^{-\varepsilon}\nu \sum_{i\in I_\kappa} \mathcal{L}^n(E_{\mu,\nu} \cap T_i)$$

$$\leq \delta^{-\varepsilon}\nu\kappa \sum_{i\in I_\kappa} \mathcal{L}^n(T_i) \approx \delta^{-\varepsilon}\nu\kappa \#I_\kappa \delta^{n-1} \lesssim \delta^{-\varepsilon}\nu\kappa.$$

Thus

$$\mu\mathcal{L}^n(E_{\mu,\nu}) \lesssim \delta^{-\varepsilon}\kappa. \tag{23.30}$$

By (23.29) we find and fix $j \in J_\lambda$ such that

$$\delta^{n-1}\mu\nu\mathcal{L}^n(E_{\mu,\nu}) \lesssim \delta^{-\varepsilon} \int \sum_{i\in I_\kappa} \chi_{T_i} \chi_{T_j} = \delta^{-\varepsilon} \sum_{i\in I_\kappa} \mathcal{L}^n(T_i \cap T_j).$$

Since above the directions of T_i and T_j are separated by c_0, it follows that $\mathcal{L}^n(T_i \cap T_j) \lesssim \delta^n$, and we conclude

$$\delta^{n-1}\mu\nu\mathcal{L}^n(E_{\mu,\nu}) \lesssim \delta^{n-\varepsilon}\#\{i \in I_\kappa : T_i \cap T_j \neq \varnothing\}.$$

Now we have found a useful hairbrush: tubes T_i, on the number of which we have a good lower bound, intersecting a fixed tube T_j. Next we shall make use of this in a somewhat similar manner as we used Bourgain's bushes in the proof of Theorem 23.2.

So now we have fixed μ, ν, κ, λ and $j \in J_\lambda$. Let

$$\widetilde{I} = \{i \in I_\kappa : T_i \cap T_j \neq \varnothing\}$$

so that

$$\delta^{-1}\mu\nu\mathcal{L}^n(E_{\mu,\nu}) \lesssim \delta^{-\varepsilon}\#\widetilde{I}. \tag{23.31}$$

Then for $i \in \widetilde{I}$, $\mathcal{L}^n(T_i \cap E_{\mu,\nu}) > (\kappa/2)\mathcal{L}^n(T_i)$. Let L_j be the line containing the central segment of T_j. By simple geometry there is a positive constant b depending only on n such that when we set

$$U = \{x \in \mathbb{R}^n : d(x, L_j) > 2b\kappa\},$$

we have for $i \in \widetilde{I}$, $\mathcal{L}^n(T_i \setminus U) < (\kappa/4)\mathcal{L}^n(T_i)$; recall that the directions e_i and e_j of T_i and T_j satisfy $|e_i - e_j| > c_0$. Therefore

$$\mathcal{L}^n(T_i \cap E_{\mu,\nu} \cap U) \geq (\kappa/4)\mathcal{L}^n(T_i).$$

Summing over i gives

$$\int_{E_{\mu,\nu}} \sum_{i\in\widetilde{I}} \chi_{T_i\cap U} \gtrsim \kappa\delta^{n-1}\#\widetilde{I}.$$

By Schwartz's inequality,

$$\kappa\delta^{n-1}\#\widetilde{I} \lesssim \left\|\sum_{i\in\widetilde{I}} \chi_{T_i\cap U\cap E_{\mu,\nu}}\right\|_2 \mathcal{L}^n(E_{\mu,\nu})^{1/2}.$$

We shall prove that

$$\left\|\sum_{i\in\widetilde{I}} \chi_{T_i\cap U\cap E_{\mu,\nu}}\right\|_2 \lesssim (\kappa^{2-n}\delta^{n-1-\varepsilon}\#\widetilde{I})^{1/2}. \tag{23.32}$$

Let us first see how we can complete the proof of the theorem from this.

Combining (23.32) with the previous inequality, we obtain

$$\kappa^n\delta^{n-1-\varepsilon}\#\widetilde{I} \lesssim \mathcal{L}^n(E_{\mu,\nu}).$$

Bringing in (23.31) we get

$$\kappa^n\delta^{n-2}\mu\nu \lesssim \delta^{-\varepsilon}.$$

Recalling also (23.30) this gives

$$\mu^{n+1}\nu\mathcal{L}^n(E_{\mu,\nu})^n \lesssim \delta^{2-n-\varepsilon}.$$

Interchanging μ and ν,

$$\mu \nu^{n+1} \mathcal{L}^n (E_{\mu,\nu})^n \lesssim \delta^{2-n-\varepsilon}.$$

Thus

$$\mathcal{L}^n (E_{\mu,\nu})^n \lesssim \sqrt{(\mu^{-n-1} \nu^{-1} \delta^{2-n-\varepsilon})(\mu^{-1} \nu^{-n-1} \delta^{2-n-\varepsilon})} = (\mu \nu)^{-(n+2)/2} \delta^{2-n-\varepsilon},$$

which is the desired inequality (23.28).

We still have left to prove (23.32). The square of the left hand side of it is

$$\int \left(\sum_{i \in \widetilde{I}} \chi_{T_i \cap U \cap E_{\mu,\nu}} \right)^2 = \sum_{i,i' \in \widetilde{I}} \mathcal{L}^n (T_i \cap T_{i'} \cap U \cap E_{\mu,\nu}).$$

If $\kappa \lesssim \delta$,

$$\sum_{i,i' \in \widetilde{I}} \mathcal{L}^n (T_i \cap T_{i'} \cap U \cap E_{\mu,\nu}) \lesssim \kappa \delta^{n-1} (\#\widetilde{I})^2 \lesssim \kappa \#\widetilde{I} \lesssim \kappa^{2-n} \delta^{n-1} \#\widetilde{I},$$

and (23.32) follows. Hence we assume from now on that $b\kappa \geq 2\delta$, where b is as before. Then (23.32) follows provided we can show for every $i' \in \widetilde{I}$,

$$\sum_{i \in \widetilde{I}} \mathcal{L}^n (T_i \cap T_{i'} \cap U) \lesssim \kappa^{2-n} \delta^{n-1}. \tag{23.33}$$

Obviously it suffices to sum over $i \neq i'$. We split this into the sums over

$$\widetilde{I}_k = \{ i \in \widetilde{I} : 2^{-k} \leq |e_i - e_{i'}| < 2^{1-k}, T_i \cap T_{i'} \cap U \neq \varnothing \},$$

$$k = 1, \dots, N \approx \log(1/\delta):$$

$$\sum_{i \in \widetilde{I}} \mathcal{L}^n (T_i \cap T_{i'} \cap U) = \sum_{k=1}^{N} \sum_{i \in \widetilde{I}_k} \mathcal{L}^n (T_i \cap T_{i'} \cap U) \lesssim \sum_{k=1}^{N} \#\widetilde{I}_k 2^k \delta^n,$$

since, as before, by simple geometry, $\mathcal{L}^n (T_i \cap T_{i'} \cap U) \lesssim 2^k \delta^n$ for $i \in \widetilde{I}_k$. Once more we use the fact that there are no more than about $\log(1/\delta)$ terms in this sum to reduce (23.33) to

$$\#\widetilde{I}_k 2^k \delta^n \lesssim \kappa^{2-n} \delta^{n-1}. \tag{23.34}$$

To see where this geometric fact follows from let us recall the situation. We have fixed the two tubes $T_j = T^\delta_{e_j}(a_j)$ and $T_{i'} = T^\delta_{e_{i'}}(a_{i'})$ which intersect at an angle ≈ 1. For $i \in \widetilde{I}_k$, the tube $T_i = T^\delta_{e_i}(a_i)$ meets both tubes T_j and $T_{i'}$. It intersects $T_{i'}$ in U. Since $b\kappa \geq 2\delta$ it therefore intersects $T_{i'}$ outside the $b\kappa$-neighbourhood of T_j by the definition of U. Moreover, $|e_i - e_{i'}| < 2^{1-k}$. Thus Lemma 23.3 implies (23.34) and completes the proof of the theorem. $\square$

23.3 Bourgain's arithmetic method and lower
bound $cn + 1 - c$

In this section we use an arithmetic method introduced by Bourgain and developed further by Katz and Tao to improve the dimension bounds for the Besicovitch sets in high dimensions. The new bounds behave for large n like cn with some constant $c > 1/2$, while the earlier bounds behave like $n/2$.

Recall from Definition 2.1 that the lower Minkowski dimension of a bounded set $A \subset \mathbb{R}^n$ is

$$\underline{\dim}_M A = \inf\{s > 0 : \liminf_{\delta \to 0} \delta^{s-n} \mathcal{L}^n(A(\delta)) = 0\},$$

where $A(\delta) = \{x : d(x, A) < \delta\}$ is the open δ-neighbourhood of A.

We first prove a lower bound for the lower Minkowski dimension of Besicovitch sets. Then, using some deep number theoretic results which we do not prove, we extend this to Hausdorff dimension.

Theorem 23.7 *For any bounded Besicovitch set B in $\mathbb{R}^n$, $\underline{\dim}_M B \geq 6n/11 +$ 5/11.*

Note that this improves, for the Minkowski dimension, Wolff's lower bound $(n + 2)/2$ when $n > 12$.

Proof We shall prove the theorem for slightly modified Besicovitch sets. We leave it to the reader to check that the proof can be modifed for the general case. Namely, we assume that for every $v \in [0, 1]^{n-1}$ there is $x \in [0, 1]^{n-1}$ such that B contains the line segment

$$I(x, v) := \{(x, 0) + t(v, 1) : 0 \leq t \leq 1\}.$$

We make the counterassumption that $\underline{\dim}_M B < cn + 1 - c$ for some $c < 6/11$ and try to achieve a contradiction. By the definition of the Minkowski dimension

$$\mathcal{L}^n(B(2\delta)) \leq \delta^{(1-c)(n-1)}$$

for some arbitrarily small $\delta > 0$, which we now fix for a moment. For any $A \subset \mathbb{R}^n$ let

$$A(t) = A \cap (\mathbb{R}^{n-1} \times \{t\})$$

be the horizontal slice of A at the level t. By Fubini's theorem

$$\int_0^1 \mathcal{L}^{n-1}(B(2\delta)(t))dt \leq \delta^{(1-c)(n-1)},$$

so Chebyshev's inequality gives,

$$\mathcal{L}^1(\{t \in [0, 1] : \mathcal{L}^{n-1}(B(2\delta)(t)) > 100\delta^{(1-c)(n-1)}\}) < 1/100.$$

Setting

$$A = \{t \in [0, 1] : \mathcal{L}^{n-1}(B(2\delta)(t)) \leq 100\delta^{(1-c)(n-1)}\}$$

we have $\mathcal{L}^1(A) > 99/100$. From this it follows that there are $s, s + d, s + 2d \in A$ with $d = 1/10$; otherwise $[0, 1/2]$ would be covered with the complements of A, $A - d$ and $A - 2d$, which is impossible. In the rest of the argument we only use these three slices and we can assume that $s = 0$ and $d = 1/2$ so that our numbers are now $0, 1/2$ and 1.

For $t \in [0, 1]$ set

$$B[t] = \{i \in \delta\mathbb{Z}^{n-1} : (i, t) \in B(\delta)\}.$$

Then the balls $B((i, t), \delta/3), i \in B[t]$, are disjoint and contained in $B(2\delta)$. Combining this with the fact that $0, 1/2, 1 \in A$ we obtain by a simple measure comparison

$$\#B[0], \ \#B[1/2], \ \#B[1] \lesssim \delta^{c(1-n)}. \tag{23.35}$$

Define for $u, v \in \mathbb{R}^{n-1}$ the δ-tubes $T_\delta(u, v) = \{y \in \mathbb{R}^n : d(y, I(u, v)) < \delta\}$, modified to our situation, and

$$G = \{(x, y) \in B[0] \times B[1] : (x, 0), (y, 1) \in T_\delta(u, v) \subset B(\delta) \text{ for some}$$
$$u, v \in [0, 1]^{n-1}\}.$$

Then

$$\#\{x + y \in G : (x, y) \in G\} \lesssim \delta^{c(1-n)}$$

and

$$\#\{x - y \in G : (x, y) \in G\} \gtrsim \delta^{1-n}.$$

To check the first of these inequalities observe that for $(x, y) \in G$, $((x + y)/2, 1/2)$ belongs to the same tube as $(x, 0)$ and $(x, 1)$, so it belongs to $B(\delta)$. Since it also belongs to $\frac{1}{2}\delta\mathbb{Z}^{n-1}$, the cardinality of $\{x + y \in G : (x, y) \in G\}$ is dominated by the cardinality of $B[1/2]$, and the first inequality follows. The second inequality is a consequence of the Besicovitch property of B: there are roughly δ^{1-n} δ-tubes with δ-separated directions contained in $B(\delta)$, each of these contains points $(x, 0)$ and $(y, 1)$ for some $(x, y) \in G$ and for different tubes the differences $x - y$, essentially the directions of these tubes, are different.

So the sum set of G is small and its difference set is large. From this we immediately obtain a contradiction using the following proposition:

Proposition 23.8 *Let* $\varepsilon_0 = 1/6$. *Suppose that A and B are finite subsets of* $\lambda\mathbb{Z}^m$ *for some* $m \in \mathbb{N}$ *and* $\lambda > 0$, $\#A \leq N$ *and* $\#B \leq N$. *Suppose also that* $G \subset A \times B$ *and*

$$\#\{x + y \in G : (x, y) \in G\} \leq N. \tag{23.36}$$

Then

$$\#\{x - y \in G : (x, y) \in G\} \leq N^{2-\varepsilon_0}. \tag{23.37}$$

This is a purely combinatorial proposition and, as will be clear from the proof, it holds for any free Abelian group in place of $\lambda\mathbb{Z}^m$. Theorem 23.7 follows applying the proposition to what we did before with $N = \delta^{6(1-n)/11}$ if δ is sufficiently small.

Observe that the proposition is trivial for $\varepsilon_0 = 0$. The application of this gives anyway $\underline{\dim}_M B \geq (n + 1)/2$, which is not completely trivial but much less than we already know. In general, the above argument gives that if the proposition is valid with ε_0, then $\underline{\dim}_M B \geq (n + 1 - \varepsilon_0)/(2 - \varepsilon_0)$ for every Besicovitch set B in $\mathbb{R}^n$.

If this proposition holds for all $\varepsilon_0 < 1$, the Kakeya conjecture would follow. But this is not so: one cannot take ε_0 larger than $2 - \log 6/\log 3 = 0.39907\ldots$ This follows from the example where

$$A = B = \{0, 1, 3\} \subset \mathbb{Z}$$

and

$$G = \{(0, 1), (0, 3), (1, 0), (1, 3), (3, 0), (3, 1)\}.$$

Then $\#A = \#B = 3 = N$,

$$\#\{x + y \in G : (x, y) \in G\} = \{1, 3, 4\} = N,$$

and

$$\#\{x - y \in G : (x, y) \in G\} = \{-3, -2, -1, 1, 2, 3\} = 6.$$

In order to have $6 \leq 3^{2-\varepsilon_0}$ we need $\varepsilon_0 \leq 2 - \log 6/\log 3$.

In the applications we only need $\#\{x - y \in G : (x, y) \in G\} \lesssim N^{2-\varepsilon_0}$ for large sets, but the same restriction is needed even then: replace A, B and G with A^M, B^M and G^M with a large integer M.

Proof of Proposition 23.8 We begin the proof of Proposition 23.8 with the following combinatorial lemma:

Lemma 23.9 *Let X and $A_1, \ldots, A_m$ be non-empty finite sets and let $f_j : X \to A_j, j = 1, \ldots, m$, be arbitrary functions. Then*

$$\#\{(x_0, \ldots, x_m) \in X^{m+1} : f_j(x_{j-1}) = f_j(x_j) \text{ for all } j = 1, \ldots, m\}$$

$$\geq \frac{(\#X)^{m+1}}{\Pi_{j=1}^m \#A_j}. \tag{23.38}$$

Proof We prove this by induction on m. The lemma is trivial for $m = 0$, but we could also begin the induction from $m = 1$, because by Schwartz's inequality

$$(\#X)^2 = \left(\sum_{a \in A_1} \#\{x : f_1(x) = a\} \right)^2 \leq \sum_{a \in A_1} (\#\{x : f_1(x) = a\})^2 \#A_1$$

$$= \sum_{a \in A_1} \#\{(x, y) : f_1(x) = f_1(y) = a\} \#A_1 = \#\{(x, y) : f_1(x) = f_1(y)\} \#A_1.$$

Suppose then that (23.38) holds for some $m - 1$ in place of m.

We say that $a \in A_m$ is popular if

$$\#\{x \in X : f_m(x) = a\} \geq \frac{\#X}{2\#A_m}.$$

Let

$$X' = \{x \in X' : f_m(x) \text{ is popular}\}.$$

For every unpopular $a \in A_m$ there are at most $\#X/(2\#A_m)$ elements $x \in X$ with $f_m(x) = a$, whence at most $\#X/2$ are mapped to unpopular elements and the rest to popular ones. This means that

$$\#X' \geq \#X/2. \tag{23.39}$$

Applying the induction hypothesis to X' we have

$$\#\{(x_0, \ldots, x_{m-1}) \in (X')^m : f_j(x_{j-1}) = f_j(x_j) \text{ for } j = 1, \ldots, m - 1\}$$

$$\geq \frac{(\#X')^m}{\Pi_{j=1}^{m-1} \#A_j}.$$

Since here $x_{m-1} \in X'$ and so $f_m(x_{m-1})$ is popular, we get

$$\#\{(x_0, \ldots, x_m) \in (X')^m \times X : f_j(x_{j-1}) = f_j(x_j) \text{ for } j = 1, \ldots, m\}$$

$$\geq \frac{(\#X')^m}{\Pi_{j=1}^{m-1} \#A_j} \frac{\#X}{2\#A_m}.$$

Combined with (23.39) this gives

$$\#\{(x_0, \ldots, x_m) \in X^{m+1} : f_j(x_{j-1}) = f_j(x_j) \text{ for } j = 1, \ldots, m\}$$
$$\geq 2^{-m-1} \frac{(\#X)^{m+1}}{\Pi_{j=1}^m \#A_j}.$$

We need to get rid of the factor 2^{-m-1}. To do this we choose a large integer M and apply what we proved so far with X, A_j replaced by X^M, A_j^M and f_j replaced by $f_j^M : X^M \to A_j^M$ defined by

$$f_j^M(x^1, \ldots, x^M) = (f_j(x^1), \ldots, f_j(x^M)).$$

Then

$$(\#\{(x_0, \ldots, x_m) \in X^{m+1} : f_j(x_{j-1}) = f_j(x_j) \text{ for } j = 1, \ldots, m\})^M$$
$$\geq 2^{-m-1} \left(\frac{(\#X)^{m+1}}{\Pi_{j=1}^m \#A_j} \right)^M.$$

Taking the Mth root, and letting $M \to \infty$ completes the proof of the lemma. $\qquad\square$

Using this lemma we now prove Proposition 23.8. Set

$$C = \{x + y \in G : (x, y) \in G\}.$$

By removing some elements from G we may assume that the map $(a, b) \mapsto a - b$ is injective on G. Thus we have that A, B and C have cardinalities at most N and we have to show that $\#G \leq N^{11/6}$. Define

$$V = \{(a, b, b') \in A \times B \times B : (a, b), (a, b') \in G\}.$$

Applying Lemma 23.9 with $m = 1$ and $f_1 : G \to A$, $f_1(a, b) = a$, we find that

$$\#V \geq \frac{(\#G)^2}{N}. \tag{23.40}$$

Next we shall apply Lemma 23.9 with $m = 3$ and

$$f_1 : V \to C \times C, \quad f_1(a, b, b') = (a + b, a + b'),$$
$$f_2 : V \to B \times B, \quad f_2(a, b, b') = (b, b'),$$
$$f_3 : V \to C \times B, \quad f_3(a, b, b') = (a + b, b').$$

Define

$$S = \{(v_0, v_1, v_2, v_3) \in V^4 : f_1(v_0) = f_1(v_1), f_2(v_1) = f_2(v_2),$$
$$f_3(v_2) = f_3(v_3)\}.$$

Then by Lemma 23.9,

$$\#S \geq \frac{(\#V)^4}{N^6}. \tag{23.41}$$

Write $v_i = (a_i, b_i, b_i')$ for $i = 0, 1, 2, 3$ and define

$$g : S \to V \times A \times B, \quad g(v_0, v_1, v_2, v_3) = (v_0, a_2, b_3).$$

We claim that g is injective. To prove this observe first that if $(v_0, v_1, v_2, v_3) \in S$, then

$$a_0 + b_0 = a_1 + b_1, \quad a_0 + b_0' = a_1 + b_1', \tag{23.42}$$

because $f_1(v_0) = f_1(v_1)$,

$$b_1 = b_2, \quad b_1' = b_2', \tag{23.43}$$

because $f_2(v_1) = f_2(v_2)$,

$$a_2 + b_2 = a_3 + b_3, \quad b_2' = b_3', \tag{23.44}$$

because $f_3(v_0) = f_3(v_1)$. Using (23.42) and (23.43) we get

$$b_0 - b_0' = (a_1 + b_1 - a_0) - (a_1 + b_1' - a_0) = b_1 - b_1' = b_2 - b_2'. \tag{23.45}$$

By (23.44) and (23.45),

$$a_3 - b_3' = a_2 + b_2 - b_3 - b_3' = a_2 + b_2 - b_2' - b_3 = a_2 + b_0 - b_0' - b_3. \tag{23.46}$$

Suppose now that $(v_0, v_1, v_2, v_3), (\tilde{v}_0, \tilde{v}_1, \tilde{v}_2, \tilde{v}_2) \in S$ with $g(v_0, v_1, v_2, v_3) = g(\tilde{v}_0, \tilde{v}_1, \tilde{v}_2, \tilde{v}_2)$. We want to show that $v_i = \tilde{v}_i$ for $i = 0, 1, 2, 3$. By the definition of g,

$$v_0 = \tilde{v}_0, \quad a_2 = \tilde{a}_2, \quad b_3 = \tilde{b}_3. \tag{23.47}$$

Using this and the analogue of (23.46) for the $\tilde{v}_i$ we obtain

$$a_3 - b_3' = \tilde{a}_3 - \tilde{b}_3'.$$

Since $(a, b) \mapsto a - b$ is injective on G and $(a_3, b_3'), (\tilde{a}_3, \tilde{b}_3') \in G$, we get $(a_3, b_3') = (\tilde{a}_3, \tilde{b}_3')$, so $v_3 = \tilde{v}_3$. Then (23.44) and (23.47) give $v_2 = \tilde{v}_2$. Finally (23.42) and (23.43) lead to $v_1 = \tilde{v}_1$.

Now that g is injective, we have

$$\#S \leq \#(V \times A \times B) \leq N^2 \#V.$$

Combining this with (23.41), we get

$$\#V \leq N^{8/3},$$

and then (23.40) yields the desired inequality

$$\#G \leq \sqrt{N \#V} \leq N^{11/6}. \qquad \square$$

Recalling what we said earlier, this completes the proof of Theorem 23.7.

$$\square$$

We shall now extend Theorem 23.7 to Hausdorff dimension. For that we need the following deep number theoretic result which we shall not prove here.

Proposition 23.10 *There are positive numbers M_0 and c with the following property: if $M > M_0$ and $S \subset \{1, \ldots, M\}$ has cardinality at least $M/(\log M)^c$, then S contains an arithmetic progression of length 3; there are $i, i + j, i + 2j \in S$.*

This was proved by Heath-Brown [1987]. It generalized a classical result of Roth [1953] which required the lower bound $M/\log \log M$.

Theorem 23.11 *For any Besicovitch set B in $\mathbb{R}^n$, $\dim B \geq 6n/11 + 5/11$.*

Proof We assume that $B \subset [0, 1]^n$. We can make this reduction because the proof will only use that for some $S \subset S^{n-1}$ with $\sigma^{n-1}(S) > 0$ and some $d > 0$ the set B contains for every $e \in S$ a line segment of length d in direction e, and B can be written as a countable union of such sets with diameter less than 1.

Let $s > \dim B$, $0 < \varepsilon < 1$ and $0 < \eta < 1$. Set

$$\delta_k = 2^{-2^{\eta k}}, \quad k = 1, 2, \ldots,$$

so $\log \log(1/\delta_k) = \eta k \log 2 + \log \log 2$. We choose η so small that $2^\eta < 1 + \varepsilon/s$, then $\delta_k^{s+\varepsilon} < \delta_{k+1}^s$. For $k_0 \in \mathbb{N}$ we can cover B with open balls B_i such that $d(B_i) < \delta_{k_0}$ and $\sum_i d(B_i)^s < 1$. Writing for $k = k_0, k_0 + 1, \ldots$,

$$I_k = \{i : \delta_{k+1} < d(B_i) \leq \delta_k\} \quad \text{and} \quad N_k = \#I_k,$$

we have $N_k \delta_k^{s+\varepsilon} < 1$. Let

$$E_k = \bigcup_{i \in I_k} B_i.$$

Then

$$N(E_k, \delta_k) \leq N_k < \delta_k^{-s-\varepsilon}, \qquad (23.48)$$

where $N(A, \delta)$ denotes the smallest number of balls of radius δ needed to cover the set A.

For every $e \in S^{n-1}$ there is a unit line segment $I_e \subset \cup_k E_k$ with direction e. It is easy to see that we can choose these segments in such a way that

$e \mapsto \mathcal{H}^1(E_k \cap I_e)$ is a Borel function for every k. We only use the segments I_e with $e \cdot e_n > 1/2$ where $e_n = (0, \ldots, 0, 1)$. If k_0 is large enough so that $\sum_{k=k_0}^{\infty} 1/k^2$ is smaller than $\sigma^{n-1}(S^{n-1})$ and 1, it follows that there are $k \geq k_0$ and a closed set $S \subset \{e \in S^{n-1} : e \cdot e_n > 1/2\}$ such that

$$\mathcal{H}^1(E_k \cap I_e) > 1/k^2 \quad \text{for } e \in S \quad \text{and} \quad 1/k^2 < \sigma^{n-1}(S) < 1.$$

Otherwise $\sigma^{n-1}(\{e : \mathcal{H}^1(E_k \cap I_e) > 1/k^2\}) \leq 1/k^2$ for all $k \geq k_0$, whence

$$\sigma^{n-1}(\{e : \mathcal{H}^1(E_k \cap I_e) > 1/k^2 \text{ for some } k \geq k_0\}) \leq \sum_{k=k_0}^{\infty} 1/k^2 < \sigma^{n-1}(S^{n-1}),$$

and we could find $e \in S^{n-1}$ such that $\mathcal{H}^1(E_k \cap I_e) \leq 1/k^2$ for all $k \geq k_0$, which is impossible since I_e is covered with the sets E_k.

Now we fix this k, set $\delta = \delta_k$ and let N be the integer for which $N \leq \delta^{n-1} < N + 1$ and M an integer such that $MN\delta > 1$ and $M \approx \delta^{-\eta}$. Define for $i = 0, 1, \ldots, M$, and $j = 0, 1, \ldots, N - 1$,

$$A_{j,i} = \{x \in \mathbb{R}^n : j\delta + iN\delta \leq x_n < j\delta + iN\delta + \delta\},$$

$$A_j = \bigcup_{i=0}^{M} A_{j,i},$$

$$E_{k,j} = E_k \cap A_j,$$

$$I_{e,j} = I_e \cap A_j \text{ for } e \in S.$$

Then, as $e \cdot e_n > 1/2$,

$$\mathcal{H}^1(I_{e,j}) \approx M\delta \approx \delta^{1-\eta} \approx \frac{1}{N},$$

and

$$\int_S \sum_{j=0}^{N-1} \mathcal{H}^1(E_{k,j} \cap I_{e,j}) d\sigma^{n-1} e = \int_S \mathcal{H}^1(E_k \cap I_e) d\sigma^{n-1} e > 1/k^4.$$

Let

$$\mathcal{J} = \left\{ j \in \{0, 1, \ldots, N - 1\} : \int_S \mathcal{H}^1(E_{k,j} \cap I_{e,j}) d\sigma^{n-1} e \geq 1/(2Nk^4) \right\}.$$

Then

$$1/k^4 < \sum_{j=0}^{N-1} \int_S \mathcal{H}^1(E_{k,j} \cap I_{e,j}) d\sigma^{n-1} e \leq \frac{1}{N} \# \mathcal{J} + \frac{1}{2Nk^4} N,$$

whence

$$\# \mathcal{J} > N/(2k^4) \approx \delta^{\eta-1}/k^4 \gtrsim \delta^{2\eta-1}. \tag{23.49}$$

Similarly for $j \in \mathcal{J}$ there is a closed $S_j \subset S$ for which

$$\mathcal{H}^1(E_{k,j} \cap I_{e,j}) > 1/(4Nk^4) \quad \text{for} \quad e \in S_j \quad \text{and} \quad \sigma^{n-1}(S_j) > 1/(4k^4).$$

For $e \in S_j$ set

$$\mathcal{I}_{e,j} = \{i \in \{1, \dots, M\} : E_{k,j} \cap I_{e,j} \cap A_{j,i} \neq \varnothing\}.$$

Then

$$1/(4Nk^4) < \mathcal{H}^1(E_{k,j} \cap I_{e,j}) \leq \#\mathcal{I}_{e,j} 2\delta,$$

so

$$\#\mathcal{I}_{e,j} > 1/(8\delta Nk^4) \approx \delta^{-\eta}/k^4 \approx M/k^4.$$

Recall that $k \approx_\eta \log\log(1/\delta) \approx_\eta \log\log M$. Hence, if δ is sufficiently small, we can use Proposition 23.10 to find $i, i + i', i + 2i' \in \mathcal{I}_{e,j}$. This is the reason we used the double dyadic power; 2^{-ck} would not have been enough. Consequently, for every $e \in S_j$ there are

$$a_e, b_e \in E_{k,j}(n\delta) \cap I_{e,j}(n\delta) \cap \delta\mathbb{Z}^n \quad \text{such that}$$
$$(a_e + b_e)/2 \in E_{k,j}(n\delta) \cap I_{e,j}(n\delta) \cap \delta\mathbb{Z}^n$$

and a_e and b_e belong to different sets $A_{j,i}$. The sets $A_{j,i}$ for different indices i are at least distance $N\delta - \delta \approx \delta^\eta$ apart. Thus if δ is sufficiently small, $|a_e - b_e| > \delta^\eta/2$.

We now apply Proposition 23.8 with $A = \{a_e : e \in S_j\}$, $B = \{b_e : e \in S_j\}$ and $G = A \times B$. Then

$$\#A, \#B, \#\{x + y : (x, y) \in G\} \leq \#(E_{k,j}(n\delta) \cap \delta\mathbb{Z}^n) \lesssim N(E_{k,j}, \delta),$$

where the last inequality is easily checked. Thus Proposition 23.8 gives

$$\#\{x - y : (x, y) \in G\} \lesssim N(E_{k,j}, \delta)^{11/6}.$$

Since a_e and b_e are in the $n\delta$-neighbourhood of I_e and $|a_e - b_e| > \delta^\eta/2$ for $e \in S_j$, it follows that balls roughly of radius $\delta^{1-\eta}$ centred at the unit vectors $(a_e - b_e)/|a_e - b_e|$, $e \in S_j$, cover S_j. This implies, as $\sigma^{n-1}(S_j) > 1/(4k^4)$,

$$\#\{a_e - b_e : e \in S_j\} \gtrsim \delta^{(\eta-1)(n-1)}/k^4 \gtrsim \delta^{(2\eta-1)(n-1)}.$$

We have also $\#\{a_e - b_e : e \in S_j\} = \#\{x - y : (x, y) \in G\}$, whence

$$\delta^{(2\eta-1)(n-1)} \lesssim \#\{x - y : (x, y) \in G\} \lesssim N(E_{k,j}, \delta)^{11/6}.$$

The sets $E_{k,j}$, $E_{k,j'}$ for $|j - j'| > 2$ are separated by a distance bigger than 2δ. Therefore by (23.49),

$$\delta^{(6/11)(2\eta-1)(n-1)+2\eta-1} \lesssim \delta^{(6/11)(2\eta-1)(n-1)} \# \mathcal{J} \leq \sum_{j \in \mathcal{J}} N(E_{k,j}, \delta) \lesssim N(E_k, \delta).$$

Recalling (23.48) we obtain $\delta^{(6/11)(2\eta-1)(n-1)+2\eta-1} \lesssim \delta^{-s-\varepsilon}$, which gives $(6/11)(1 - 2\eta)(n - 1) - 2\eta + 1 \leq s + \varepsilon$. We can choose ε and η as small as we wish, so $s \geq (6/11)(n - 1) + 1 = (6/11)n + 5/11$. Letting $s \to \dim B$, the theorem follows. $\qquad\square$

23.4 Further comments

The lecture notes of Tao [1999b] and of Iosevich [2000] have been very helpful for the presentation of this chapter.

The lower bound $(n + 1)/2$ for the Hausdorff dimension of Besicovitch sets is due to Drury [1983], although he did not state it explicitly. The explicit $L^{(n+1)/2}$ estimate for the Nikodym maximal function was proved by Christ, Duoandikoetxea and Rubio de Francia [1986].

Drury proved the following estimate for the X-ray transform

$$Xf(L) = \int_L f, \quad L \subset \mathbb{R}^n, L \text{ a line in } \mathbb{R}^n :$$

$$\|Xf\|_{L^q(\lambda)} \lesssim \|f\|_{L^p(\mathbb{R}^n)} \quad \text{for } 1 \leq p < (n + 1)/2, n/p - (n - 1)/q = 1.$$
$$(23.50)$$

The measure λ on the space of lines can be defined for example as

$$\int F(L) d\lambda L = \int_{S^{n-1}} \int_{e^\perp} F(L_{e,v}) d\mathcal{H}^{n-1} v d\sigma^{n-1} e,$$

where $L_{e,v} = \{v + te : t \in \mathbb{R}\}$.

Such estimates are very close to Kakeya estimates. We shall discuss them a bit more in Chapter 24. Let us quickly see how (23.50) yields the lower bound $(n + 1)/2$ for the lower Minkowski dimension of Besicovitch sets in $\mathbb{R}^n$. Arguments as in the proof of Theorem 22.9 then give it for the Hausdorff dimension, too.

If $\dim B < s$ for some $s > 0$, we can find arbitrarily small $\delta > 0$ such that for the δ-neighbourhood $B(\delta)$ of B, $\mathcal{L}^n(B(\delta)) \leq \delta^{n-s}$. Let f be the characteristic function of $B(\delta)$. Then $\|f\|_{L^p(\mathbb{R}^n)} \leq \delta^{(n-s)/p}$. For each $e \in S^{n-1}$ the set of $v \in e^\perp$ for which $\mathcal{H}^1(B(\delta) \cap L_{e,v}) \geq 1$ has measure $\gtrsim \delta^{n-1}$. This gives $\|Xf\|_{L^q(\lambda)} \gtrsim \delta^{(n-1)/q}$. Combining these estimates with (23.50) with p close to $(n + 1)/2$, thus q close to $n + 1$, gives that s cannot be much smaller than $(n + 1)/2$, as desired.

The method with 'bushes', bunches of tubes containing a common point, is due to Bourgain [1991a]. Bourgain improved the lower bound $(n + 1)/2$ and the corresponding Kakeya and Nikodym maximal function estimates. This method also led to the non-existence of (n, k) Besicovitch sets for $2^{k-1} + k \geq n$ (which we shall discuss in the next chapter), to the first partial results on the restriction conjecture better than those following from the Stein–Tomas theorem by interpolation, and to some improvements on Bochner–Riesz multiplier estimates.

Bourgain's Kakeya estimates in $\mathbb{R}^3$ were proved again by Schlag [1998] with an interesting geometric method.

Wolff [1995] developed further Bourgain's method introducing the hairbrushes. He simultaneously derived the same estimate for the Nikodym maximal function by axiomatizing the situation. Thus he proved Theorem 23.6 which still to date gives the best known lower bound $(n + 2)/2$ for the Hausdorff dimension of Besicovitch sets in dimensions 3 and 4, and the best known Kakeya and Nikodym maximal function estimates in dimensions $3 \leq n \leq 8$. Employing Bourgain's [1991a] machinery he also improved the restriction estimates. These however have later been surpassed by multilinear methods which we shall discuss in the last chapter. The proof given here for Theorem 23.5 is due to Katz, it was also used by Wisewell [2005] for curved Kakeya sets. The proof of Theorem 23.6 was taken from the lecture notes of Tao [1999b]. Both of these proofs are different and simpler than Wolff's original proof.

An essential technical point in the arguments for Bourgain's theorem 23.2 and Wolff's theorem 23.6 is to avoid situations where a set under consideration would be too much concentrated in small parts of tubes. Wolff [1995] made this more explicit and this principle is often called two-ends reduction. Tao discusses it in Tao [2011], Section 4.4.

Section 23.3 is part of the recent developments in additive combinatorics and their applications to various fields. The book of Tao and Vu [2006] gives an excellent detailed overview of this topic. Bourgain [1999] introduced the arithmetic method discussed here. He proved Proposition 23.8 with $\varepsilon_0 = 1/13$. This gives by the above argument involving slicing and triples in arithmetic progression that the Hausdorff dimension of all Besicovitch sets in $\mathbb{R}^n$ is at least $\frac{13}{25}n + \frac{12}{25}$. Bourgain also used this method to get L^p estimates for the Kakeya maximal operator. Katz and Tao [1999] further improved the Hausdorff dimension estimate to $\frac{6}{11}n + \frac{5}{11}$ with the proof which we presented here; Proposition 23.8 with $\varepsilon_0 = 1/6$ is due to them. For the Minkowski dimension they got a further improvement: $\underline{\dim}_M B \geq \frac{4}{7}n + \frac{3}{7}$. For this they showed that (23.37) holds with $\varepsilon_0 = 1/4$ if one adds the assumption

$$\#\{x + 2y \in G : (x, y) \in G\} \leq N.$$

One applies this using four slices of B instead of three. The transfer to Hausdorff dimension does not work anymore as there are not sufficient estimates for arithmetic progressions of length four. The Hausdorff estimate is better than Wolff's $\frac{n+2}{2}$ if $n > 12$ and the Minkowski estimate if $n > 8$.

Later on Katz, Łaba and Tao [2000] combined arithmetic and geometric ideas to get deep structural information about Besicovitch sets. In particular, they proved that the upper Minkowski dimension of Besicovitch sets in $\mathbb{R}^3$ is greater than $5/2 + \varepsilon$ for some absolute constant $\varepsilon > 0$. For the Hausdorff dimension Wolff's bound $5/2$ is still the best that is known. Łaba and Tao [2001a] extended the $\frac{n+2}{2} + \varepsilon_n$ Minkowski bound to all n. Finally Katz and Tao [2002a] developed the arithmetic methods further with sophisticated iterations and improved the Minkowski bound when $n \geq 7$, the Hausdorff bound when $n \geq 5$, and Kakeya maximal function inequalities when $n \geq 9$, as stated below. An excellent survey on this progress can be found in Katz and Tao [2002b].

Here is a summary of the best bounds known at the moment for $n \geq 3$, ε_0 is a very small absolute constant, $\varepsilon_0 = 10^{-10}$ suffices, and α is the biggest root of the equation $\alpha^3 - 4\alpha + 2 = 0$, that is, $\alpha = 1.67513\ldots$: for a Besicovitch set $B \subset \mathbb{R}^n$,

$$\dim B \geq \frac{n+2}{2} \text{ for } n = 3, 4, \text{ Wolff [1995],}$$

$$\dim B \geq (2 - \sqrt{2})(n - 4) + 3 \text{ for } n \geq 5, \text{ Katz and Tao [2002a],}$$

$$\overline{\dim}_M B \geq \frac{n+2}{2} + \varepsilon_0 \text{ for } n = 3, 4, \text{ Katz, Łaba and Tao [2000],}$$

Łaba and Tao [2001a],

$$\overline{\dim}_M B \geq (2 - \sqrt{2})(n - 4) + 3 \text{ for } 5 \leq n \leq 23, \text{ Katz and Tao [2002a],}$$

$$\overline{\dim}_M B \geq \frac{n-1}{\alpha} + 1 \text{ for } n \geq 24, \text{ Katz and Tao [2002a],}$$

$$\|\mathcal{K}_\delta f\|_{L^{\frac{n+2}{2}}(S^{n-1})} \lesssim \delta^{\frac{2-n}{2+n} - \varepsilon} \|f\|_{L^{\frac{n+2}{2}}(\mathbb{R}^n)} \text{ for } \varepsilon > 0, 3 \leq n \leq 8, \text{ Wolff [1995],}$$

$$\|\mathcal{K}_\delta f\|_{L^{\frac{4n+3}{4}}(S^{n-1})} \lesssim \delta^{\frac{3-3n}{4n+3} - \varepsilon} \|f\|_{L^{\frac{4n+3}{7}}(\mathbb{R}^n)} \text{ for } \varepsilon > 0, n \geq 9,$$

Katz and Tao [2002a].

Carbery [2004] proved a multilinear generalization of the Cauchy–Schwarz inequality which is related to Lemma 23.9. This paper also contains an interesting extensive discussion of the context of inequalities of this type.

Minicozzi and Sogge [1997] studied Nikodym maximal functions on n-dimensional Riemannian manifolds. They proved the analogue of the $(n + 1)/2$ estimate, tubes are now taken around geodesics. They also gave examples of manifolds where $(n + 1)/2$ cannot be improved. Sogge [1999] proved the $5/2$ estimate on 3-dimensional Riemannian manifolds of constant curvature.

Let us consider the operators T_λ as in (20.1):

$$T_\lambda f(\xi) = \int_{\mathbb{R}^{n-1}} e^{i\lambda\Phi(x,\xi)} \Psi(x,\xi) f(x) \, dx, \quad \xi \in \mathbb{R}^n, \quad \lambda > 0.$$

Here again Φ and Ψ are smooth functions, Φ is real valued and Ψ has compact support. Natural conditions to assume are that the $(n - 1) \times n$ matrix $(\frac{\partial^2 \Phi(x,\xi)}{\partial x_j \partial \xi_k})$ has the maximal rank $n - 1$ and the mapping $x \mapsto (\frac{\partial \Phi(x,\xi)}{\partial x_j})$, $x \in \mathbb{R}^{n-1}$, has only non-degenerate critical points. A generalization of the Stein–Tomas restriction theorem, which we mentioned in Chapter 20 in the dual form (recall (20.8)), says that

$$\|T_\lambda f\|_q \lesssim \lambda^{-n/q} \|f\|_p \quad \text{for all } f \in L^p(\mathbb{R}^{n-1})$$

provided $q \geq 2(n + 1)/(n - 1)$ and $q = \frac{n+1}{n-1}p'$. Hörmander [1973] asked whether this could be extended to the optimal range $q > 2n/(n - 1)$. This is true for $n = 2$, as we discussed in Chapter 20. Somewhat surprisingly Bourgain [1991b] proved that when $n = 3$ the answer is negative even for very simple phase functions Φ such as

$$\Phi(x,\xi) = x_1\xi_1 + x_2\xi_2 + x_1x_2\xi_3 + x_1^2\xi_3^2/2,$$

and moreover Stein's range $q \geq 4$ is optimal for such a Φ. His proof involved curved Kakeya methods. That is, in the definitions line segments are replaced by curves and straight tubes by curved tubes. Again one can show that appropriate estimates for T_λ lead to Kakeya estimates and so counter-examples can be obtained from the failure of Kakeya estimates.

Wisewell [2005] continued this. She showed that the $(n + 1)/2$ dimension estimate for curved Besicovitch sets holds for very general curves whereas the $(n + 2)/2$ estimate may fail even for very simple quadratic curves, such as Bourgain's curve above. She also presented classes of quadratic curves for which the $(n + 2)/2$ estimate is valid and used the arithmetic method to obtain improvements in higher dimensions.

24

(n, k) Besicovitch sets

What can we say if we replace in the definition of Besicovitch sets the line segments with pieces of k-dimensional planes?

As before we denote by $G(n.k)$ the space of k-dimensional linear subspaces of $\mathbb{R}^n$ and by $\gamma_{n,k}$ its unique orthogonally invariant Borel probability measure. Recall that it is defined by

$$\gamma_{n,k}(A) = \theta_n(\{g \in O(n) : g(V_0) \in A\}), \quad A \subset G(n, k),$$

where θ_n is the Haar probability measure on the orthogonal group $O(n)$ and $V_0 \in G(n, k)$ is any fixed k-plane. For $k = 1$ and $k = n - 1$ we can reduce this measure to the surface measure on S^{n-1}; setting $L_v = \{tv : t \in \mathbb{R}\}$,

$$\gamma_{n,1}(A) = c(n)\sigma^{n-1}(\{v \in S^{n-1} : L_v \in A\}), \quad A \subset G(n, 1),$$

$$\gamma_{n,n-1}(A) = c(n)\sigma^{n-1}(\{v \in S^{n-1} : L_v^{\perp} \in A\}), \quad A \subset G(n, n - 1).$$

Definition 24.1 A Borel set $B \subset \mathbb{R}^n$ is said to be an (n, k) *Besicovitch set* if $\mathcal{L}^n(B) = 0$ and for every $V \in G(n, k)$ there is $a \in \mathbb{R}^n$ such that $B(a, 1) \cap (V + a) \subset B$.

24.1 Marstrand and the case $n = 3, k = 2$

The first question is: do they exist if $k > 1$? Probably not, at least no such pair (n, k) is known. We now prove in three different ways that they do not exist when k is sufficiently large as compared to n. We begin with Marstrand's [1979] geometric argument for $n = 3, k = 2$:

Theorem 24.2 *There are no* $(3, 2)$ *Besicovitch sets. More precisely, if $E \subset \mathbb{R}^3$ and $\mathcal{L}^3(E) = 0$, then for $\gamma_{3,2}$ almost all $V \in G(3, 2)$, $\mathcal{H}^2(E \cap (V + a)) = 0$ for all $a \in \mathbb{R}^3$.*

Proof Clearly we can assume that $E \subset B(0, 1/2)$. Set for $v \in S^2$ and $A \subset B(0, 1)$,

$$f(A, v) = \sup\{\mathcal{H}^2(A \cap (L_v^{\perp} + a)) : a \in \mathbb{R}^3\}.$$

357

We shall prove that

$$\left(\int^* f(A, v) \, d\sigma^2 v \right)^2 \lesssim \mathcal{L}^3(A),$$ (24.1)

where $\int^*$ is the upper integral. The theorem clearly follows from this. Obviously it suffices to prove (24.1) for open sets A. It is easy to check that if $B_i \subset \mathbb{R}^3$ is an increasing sequence of Borel sets with $B = \cup_i B_i$, then $f(B, v) = \lim_{i \to \infty} f(B_i, v)$. Therefore it is enough to prove (24.1) for disjoint finite unions of cubes of the same side-length with sides parallel to the coordinate axis. Thus let $B = \cup_{i=1}^m Q_j \subset B(0, 1)$ where the cubes Q_j are disjoint with side-length δ.

For every $v \in S^2$ the function $a \mapsto \mathcal{H}^2(B \cap (L_v^\perp + a))$ attains its supremum for some $a \in \mathbb{R}^3$; except for the vectors v orthogonal to coordinate planes it is a continuous function of a and for these six exceptional vectors it takes only finitely many values. Choose for every $v \in S^2$ some $a \in \mathbb{R}^3$ such that with $A(v) = L_v^\perp + a$ we have $f(B, v) = \mathcal{H}^2(B \cap A(v))$. Clearly this choice can be made so that the function $v \mapsto f(B, v)$ is a Borel function.

We can now estimate using Schwartz's inequality and Fubini's theorem,

$$\left(\int f(B, v) \, d\sigma^2 v \right)^2 = \left(\int \mathcal{H}^2(B \cap A(v)) \, d\sigma^2 v \right)^2$$

$$= \left(\sum_{j=1}^m \int \mathcal{H}^2(Q_j \cap A(v)) \, d\sigma^2 v \right)^2 \leq m \sum_{j=1}^m \left(\int \mathcal{H}^2(Q_j \cap A(v)) \, d\sigma^2 v \right)^2$$

$$= m \sum_{j=1}^m \int_{S^2 \times S^2} \mathcal{H}^2 \times \mathcal{H}^2((Q_j \times Q_j) \cap (A(v) \times A(w))) \, d(\sigma^2 \times \sigma^2)(v, w)$$

$$= m \int_{S^2 \times S^2} \mathcal{H}^2 \times \mathcal{H}^2(\cup_{j=1}^m (Q_j \times Q_j) \cap (A(v) \times A(w))) \, d(\sigma^2 \times \sigma^2)(v, w)$$

$$\leq m \int_{S^2 \times S^2} \mathcal{H}^2 \times \mathcal{H}^2(\{(x, y) \in A(v) \times A(w) : |x|, |y| \leq 1, |x - y| \leq \sqrt{3}\delta\})$$
$$\times d(\sigma^2 \times \sigma^2)(v, w)$$

$$\leq m \int_{S^2 \times S^2} \int_{B(0,1) \cap A(v)} \mathcal{H}^2(B(x, \sqrt{3}\delta) \cap A(w)) \, d\mathcal{H}^2 x \, d(\sigma^2 \times \sigma^2)(v, w)$$

$$\leq 3\pi \delta^2 m \int_{S^2 \times S^2} \mathcal{H}^2(\{x \in B(0, 1) \cap A(v) : d(x, A(w)) \leq \sqrt{3}\delta\})$$
$$\times d(\sigma^2 \times \sigma^2)(v, w).$$

We estimate the last integrand by elementary geometry. For this we may assume $v \neq \pm w$ and that the planes $A(v)$ and $A(w)$ go through the origin. Then $A(v) \neq A(w)$ and $A(v)$ and $A(w)$ intersect along a line $L \in G(3, 1)$. Denote by $\alpha(v, w)$ the angle between v and w. Then if $x \in A(v) \cap B(0, 1)$ and $d(x, A(w)) \leq \sqrt{3}\delta$, we must have $|x| \leq \frac{\sqrt{3}\delta}{\sin(\alpha(v,w))}$. This implies that our set is contained in a rectangle with side-lengths $\frac{2\sqrt{3}\delta}{\sin(\alpha(v,w))}$ and 2. This gives

$$\mathcal{H}^2(\{x \in B(0, 1) \cap A(v) : d(x, A(w)) \leq \sqrt{3}\delta\}) \leq \frac{4\sqrt{3}\delta}{\sin(\alpha(v, w))},$$

and

$$\left(\int f(B, v) \, d\sigma^2 v \right)^2 \leq 3\pi\delta^2 m \int \frac{4\sqrt{3}\delta}{\sin(\alpha(v, w))} \, d(\sigma^2 \times \sigma^2)(v, w).$$

For any fixed $w \in S^2$ we have, for example by (3.31),

$$\int \sin(\alpha(v, w))^{-1} \, d\sigma^2 v \approx 1.$$

Combining these we conclude

$$\left(\int f(B, v) \, d\sigma^2 v \right)^2 \lesssim m\delta^3 = \mathcal{L}^3(B),$$

as required. $\qquad\qquad\qquad\qquad\qquad\qquad\qquad\qquad\qquad\qquad\qquad$ $\square$

24.2 Falconer and the case $k > n/2$

It is easy to modify the above proof for $k = n - 1$ for any $n \geq 3$. Now we extend this to $k > n/2$ using the argument of Falconer [1980a]:

Theorem 24.3 *There are no (n, k) Besicovitch sets for $k > n/2$. More precisely, if $k > n/2$ and $E \subset \mathbb{R}^n$ with $\mathcal{L}^n(E) = 0$, then for $\gamma_{n,k}$ almost all $V \in G(n, k)$,*

$$\mathcal{H}^k(E \cap (V + a)) = 0 \quad \text{for all } a \in \mathbb{R}^n.$$

Proof We shall use the following formula, say for non-negative Borel functions f:

$$\int_{G(n,k)} \int_{V^\perp} |x|^k f(x) \, d\mathcal{H}^{n-k} x \, d\gamma_{n,k} V = c(n, k) \int_{\mathbb{R}^n} f \, d\mathcal{L}^n. \tag{24.2}$$

To prove this, integrate the left hand side in the spherical coordinates of $V^{\perp}$:

$$
\int_{G(n,k)} \int_{V^{\perp}} |x|^k f(x)\, d\mathcal{H}^{n-k} x\, d\gamma_{n,k} V
$$

$$
= \int_{G(n,k)} \int_0^{\infty} \int_{V^{\perp} \cap S^{n-1}} r^k f(rv)\, d\sigma^{n-k-1} v r^{n-k-1}\, dr\, d\gamma_{n,k} V
$$

$$
= \int_0^{\infty} r^{n-1} \int_{G(n,k)} \int_{V^{\perp} \cap S^{n-1}} f(rv)\, d\sigma^{n-k-1} v\, d\gamma_{n,k} V\, dr.
$$

For non-negative Borel functions on S^{n-1},

$$
\int_{G(n,k)} \int_{V^{\perp} \cap S^{n-1}} g(v)\, d\sigma^{n-k-1} v\, d\gamma_{n,k} V = c(n,k) \int_{S^{n-1}} g\, d\sigma^{n-1},
$$

because the left hand side defines an orthogonally invariant measure on S^{n-1} and such a measure is unique up to multiplication by a constant. Thus

$$
\int_{G(n,k)} \int_{V^{\perp}} |x|^k f(x)\, d\mathcal{H}^{n-k} x\, d\gamma_{n,k} V
$$

$$
= c(n,k) \int_0^{\infty} r^{n-1} \int_{S^{n-1}} f(rv)\, d\sigma^{n-1} v\, dr = c(n,k) \int_{\mathbb{R}^n} f\, d\mathcal{L}^n.
$$

Suppose now $f \in L^1(\mathbb{R}^n) \cap L^2(\mathbb{R}^n)$. Let $V \in G(n,k)$. If $\xi \in V^{\perp}$, then, writing for a moment $x = x_V + x_V'$, $x_V \in V$, $x_V' \in V^{\perp}$, we have by Fubini's theorem,

$$
\widehat{f}(\xi) = \int_{V^{\perp}} e^{-2\pi i \xi \cdot x_V'} \int_{V + x_V'} f\, d\mathcal{H}^k\, dx_V' = \widehat{F_V}(\xi),
$$

where

$$
F_V(x_V') = \int_{V + x_V'} f\, d\mathcal{H}^k \quad \text{for } x_V' \in V^{\perp}.
$$

Thus by (24.2) and Schwartz's inequality,

$$
\int_{G(n,k)} \int_{\{\xi \in V^{\perp} : |\xi| \geq 1\}} |\widehat{F_V}(\xi)|\, d\mathcal{H}^{n-k} \xi\, d\gamma_{n,k} V
$$

$$
= \int_{G(n,k)} \int_{\{\xi \in V^{\perp} : |\xi| \geq 1\}} |\widehat{f}(\xi)|\, d\mathcal{H}^{n-k} \xi\, d\gamma_{n,k} V
$$

$$
= c(n,k) \int_{\{\xi \in \mathbb{R}^n : |\xi| \geq 1\}} |\widehat{f}(\xi)| |\xi|^{-k}\, d\xi
$$

$$
\leq c(n,k) \left(\int |\widehat{f}(\xi)|^2\, d\xi \right)^{1/2} \left(\int_{\{\xi \in \mathbb{R}^n : |\xi| \geq 1\}} |\xi|^{-2k}\, d\xi \right)^{1/2}
$$

$$
= c'(n,k) \|f\|_2,
$$

where $c'(n, k) < \infty$ since $2k > n$. As $\|\widehat{f}\|_\infty \leq \|f\|_1$ and $k < n$, we also have

$$\int_{G(n,k)} \int_{\{\xi \in V^\perp : |\xi| \leq 1\}} |\widehat{F_V}(\xi)| \, d\mathcal{H}^{n-k}\xi \, d\gamma_{n,k}V$$

$$= c(n, k) \int_{\{\xi \in \mathbb{R}^n : |\xi| \leq 1\}} |\widehat{f}(\xi)| |\xi|^{-k} \, d\xi \lesssim \|f\|_1.$$

So we see that for almost all $V \in G(n, k)$, $\widehat{F_V} \in \mathcal{L}^1(V^\perp)$. By Fubini's theorem also $F_V \in \mathcal{L}^1(V^\perp)$ for all $V \in G(n, k)$. Thus the Fourier inversion formula implies $\|F_V\|_{L^\infty(V^\perp)} \leq \|\widehat{F_V}\|_{L^1(V^\perp)}$ for almost all $V \in G(n, k)$. Consequently,

$$\int_{G(n,k)} \|F_V\|_{L^\infty(V^\perp)} \, d\gamma_{n,k}V \leq \int_{G(n,k)} \int_{V^\perp} |\widehat{F_V}(\xi)| \, d\mathcal{H}^{n-k}\xi \, d\gamma_{n,k}V \lesssim \|f\|_1 + \|f\|_2.$$

Suppose now that f is a continuous function with compact support. Then F_V is also continuous and the above inequality turns into

$$\int_{G(n,k)} \mathcal{M}^k f(V) \, d\gamma_{n,k}V \lesssim \|f\|_1 + \|f\|_2 \tag{24.3}$$

where $\mathcal{M}^k f$ is the maximal k-plane transform,

$$\mathcal{M}^k f(V) = \sup_{x'_V \in V^\perp} \left| \int_{V + x'_V} f \, d\mathcal{H}^k \right| \quad \text{for } V \in G(n, k). \tag{24.4}$$

By easy approximation (24.3) extends from continuous functions to all $f \in L^1 \cap L^2$. To see this observe first that if (24.3) holds for all f_j in an increasing sequence of non-negative functions in $L^1 \cap L^2$, then it holds for $\lim_{j \to \infty} f_j$. Thus it is enough to verify (24.3) for simple functions $\sum_{j=1}^m a_j \chi_{A_j}$. Approximating each A_j with open sets, we are reduced to the case where the sets A_j are open. Such functions $\sum_{j=1}^m a_j \chi_{A_j}$ are increasing limits of continuous functions with compact support and (24.3) follows for all $f \in L^1 \cap L^2$. Applying (24.3) to the characteristic functions of bounded measurable sets completes the proof of the theorem. $\qquad\square$

24.3 Bourgain and the case $k > (n+1)/3$

Next we get a further extension by proving a result of Bourgain [1991a]. The proof makes use of Kakeya maximal function inequalities.

Theorem 24.4 *If $k > (n+1)/3$, there are no (n, k) Besicovitch sets. More precisely, if $E \subset \mathbb{R}^n$ with $\mathcal{L}^n(E) = 0$, then for almost all $V \in G(n, k)$,*

$$\mathcal{H}^k(E \cap (V + a)) = 0 \quad \text{for all } a \in \mathbb{R}^n.$$

Proof We may assume that $E \subset B(0, 1)$. We begin with the following corollary of the formula (24.2): if $g \in \mathcal{S}(\mathbb{R}^n)$, $R > 0$ and $\mathrm{spt}\,\widehat{g} \subset B(0, 4R) \setminus B(0, R)$, then for $l = 1, \ldots, n - 1$,

$$\frac{c(n, l)}{(4R)^l} \int_{\mathbb{R}^n} |g|^2 \, d\mathcal{L}^n \leq \int_{G(n,l)} \int_{V^\perp} |g_V|^2 \, d\mathcal{H}^{n-l} \, d\gamma_{n,l} V \leq \frac{c(n, l)}{R^l} \int_{\mathbb{R}^n} |g|^2 \, d\mathcal{L}^n,$$
(24.5)

where g_V is defined by

$$g_V(x) = \int_V g(x + v) \, dv \quad \text{for } x \in V^\perp.$$

To prove this let $V \in G(n, l)$ and observe that

$$\widehat{g_V}(x) = \widehat{g}(x) \quad \text{for } x \in V^\perp,$$

where $\widehat{g_V}$ is the Fourier transform of g_V in the $(n - l)$-dimensional Euclidean space $V^\perp$. Applying Plancherel's theorem both in $V^\perp$ and in $\mathbb{R}^n$ and using also (24.2) we obtain

$$\int_{G(n,l)} \int_{V^\perp} |g_V|^2 \, d\mathcal{H}^{n-l} \, d\gamma_{n,l} V = \int_{G(n,l)} \int_{V^\perp} |\widehat{g_V}|^2 \, d\mathcal{H}^{n-l} \, d\gamma_{n,l} V$$

$$= \int_{G(n,l)} \int_{V^\perp} |\widehat{g}|^2 \, d\mathcal{H}^{n-l} d\gamma_{n,l} V \leq R^{-l} \int_{G(n,l)} \int_{V^\perp} |x|^l |\widehat{g}(x)|^2 \, d\mathcal{H}^{n-l} d\gamma_{n,l} V$$

$$= c(n, l) R^{-l} \int_{\mathbb{R}^n} |\widehat{g}|^2 \, d\mathcal{L}^n = c(n, l) R^{-l} \int_{\mathbb{R}^n} |g|^2 \, d\mathcal{L}^n.$$

The other inequality follows in the same way.

Let f be a non-negative continuous function with support in $B(0, 1)$. Choose $\varphi \in \mathcal{S}(\mathbb{R}^n)$ such that

$$\widehat{\varphi}(x) = 1 \quad \text{for } x \in B(0, 1) \quad \text{and} \quad \widehat{\varphi}(x) = 0 \quad \text{for } x \in \mathbb{R}^n \setminus B(0, 2),$$

and define for $j = 1, 2, \ldots,$

$$\varphi_j(x) = 2^{jn} \varphi(2^j x) \quad \text{for } x \in \mathbb{R}^n,$$

for which $\widehat{\varphi}_j(x) = \widehat{\varphi}(2^{-j} x)$. Define also

$$f_j = f * \varphi_j - f * \varphi_{j-1},$$

where $\varphi_0 = 0$. Then

$$f(x) = \sum_j f_j(x) \quad \text{for } x \in \mathbb{R}^n,$$

and

$$\widehat{f}_j = \widehat{f}(\widehat{\varphi}_j - \widehat{\varphi}_{j-1}) \quad \text{with} \quad \mathrm{spt}\,\widehat{f}_j \subset \{x : 2^{j-1} \leq |x| \leq 2^{j+1}\}.$$

Fix j for a while and let $g = f_j$ and $\delta = 2^{-j}$. Define also as above

$$g_V(x) = \int_V g(x+v)\,dv \quad \text{for } x \in \mathbb{R}^n, V \in G(n,k),$$

and let $\mathcal{M}^k g$ be the maximal k-plane transform of g as in (24.4). Using the fact that spt $f \subset B(0,1)$ it follows by Fubini's theorem that

$$|g_V(x)| \leq 2\alpha(k)\|\varphi\|_1\|f\|_\infty, \tag{24.6}$$

where $\alpha(k)$ is the volume of the k-dimensional unit ball. We claim that with $V_e = \{v + te : v \in V, t \in \mathbb{R}\}$,

$$\mathcal{M}^k g(V_e) \lesssim \mathcal{K}_\delta(g_V)(e) \quad \text{for } e \in V^\perp, V \in G(n,k). \tag{24.7}$$

To prove this, notice first that $g_V = \psi_j * f_V$ with $\psi_j = \varphi_j - \varphi_{j-1}$ and

$$\int_{V_e+x} g = \int_\mathbb{R}\int_V g(x+v+te)\,dv\,dt = \int_\mathbb{R} g_V(x+te)\,dt.$$

Fixing V and e define

$$F(x) = \int_\mathbb{R} f_V(x+te)\,dt \quad \text{and} \quad G(x) = \int_\mathbb{R} g_V(x+te)\,dt \quad \text{for } x \in \mathbb{R}^n. \tag{24.8}$$

Then $G = \psi_j * F$ and to get (24.7) from this we first prove:

Lemma 24.5 *Let $\psi \in \mathcal{S}(\mathbb{R}^n)$ with spt $\widehat{\psi}$ compact and let $F \in L^1(\mathbb{R}^n)$. Defining for $\delta > 0$, $\psi_\delta = \delta^{-n}\psi(x/\delta)$, we have for all $\delta > 0, x \in \mathbb{R}^n$,*

$$|\psi_\delta * F(x)| \leq C(\psi) \sup_{z\in\mathbb{R}^n} \delta^{-n} \int_{B(z,\delta)} |\psi_\delta * F(y)|\,dy.$$

Proof By change of variable we may assume $\delta = 1$. Set

$$s = \sup_{z\in\mathbb{R}^n} \int_{B(z,1)} |\psi * F(y)|\,dy.$$

Choose $\varphi \in \mathcal{S}(\mathbb{R}^n)$ with $\widehat{\varphi} = 1$ on spt $\widehat{\psi}$. Then $\widehat{\psi} = \widehat{\varphi}\widehat{\psi} = \widehat{\varphi * \psi}$, so $\psi = \varphi * \psi$. Thus $\psi * F(x) = \varphi * (\psi * F)(x)$ so that

$$|\psi * F(x)| \lesssim \int_{B(x,1)} |\psi * F(y)|\,dy + \int_{\mathbb{R}^n\setminus B(x,1)} |\varphi(x-y)\psi * F(y)|\,dy$$

$$\leq s + \int_{\mathbb{R}^n\setminus B(x,1)} |\varphi(x-y)\psi * F(y)|\,dy.$$

We estimate the remaining integral by dividing $\mathbb{R}^n \setminus B(x,1)$ into dyadic annuli $B(x,2^j) \setminus B(x,2^{j-1})$, $j = 1, 2, \ldots$, and covering each such annulus with roughly 2^{jn} balls $B_{j,i}$ of radius 1. Then using the fast decay of φ,

$$\int_{\mathbb{R}^n \setminus B(x,1)} |\varphi(x-y)\psi * F(y)|\,dy \leq \sum_{j=1}^{\infty}\sum_{i} \int_{B_{j,i}} |\varphi(x-y)\psi * F(y)|\,dy$$

$$\lesssim \sum_{j=1}^{\infty}\sum_{i} 2^{-j(n+1)} \int_{B_{j,i}} |\psi * F(y)|\,dy \leq \sum_{j=1}^{\infty}\sum_{i} 2^{-j(n+1)}s \lesssim \sum_{j=1}^{\infty} 2^{-j}s = s.$$

$\square$

Returning to $G = \psi_j * F$ as in (24.8) and setting $\widetilde{T}_e^{\delta}(a) = \{y + te : |y - a| \leq \delta/2, t \in \mathbb{R}\}$ (recall that $\delta = 2^{-j}$), Lemma 24.5 gives that

$$\left| \int_{V_e + x} g \right| = |G(x)| \lesssim \sup_{z \in \mathbb{R}^n} \delta^{-n} \int_{B(z,\delta)} |G(y)|\,dy \lesssim \sup_{a \in \mathbb{R}^n} \delta^{1-n} \int_{\widetilde{T}_e^{\delta}(a)} |g_V(y)|\,dy,$$

where the last inequality is easy to check.

Finally to get to $\mathcal{K}_{\delta}(g_V)(e)$ we need to show that the averages over infinite tubes in the last term are dominated by averages over tubes $T_e^{\delta}(a)$ of length one. For this we can use the same trick as in the proof of Lemma 24.5. Choose $\varphi \in \mathcal{S}(\mathbb{R}^n)$ with $\widehat{\varphi} = 1$ on $\mathrm{spt}\,\widehat{\psi_j}$ and write again $g_V = \psi_j * f_V = \varphi * \psi_j * f_V = \varphi * g_V$. Then

$$\int_{\widetilde{T}_e^{\delta}(a)} |g_V(y)|\,dy \lesssim \sup_{b \in T_e^{\delta}(a)} \int_{T_e^{\delta}(b)} |g_V(y)|\,dy$$

follows by the fast decay of φ, as in the proof of Lemma 24.5. This completes the proof of (24.7).

The measure $\gamma_{n,k}$ can be written as

$$\int h\,d\gamma_{n,k} = c \int_{G(n,k-1)} \int_{S_V} h(V_e)\,d\sigma_V e\,d\gamma_{n,k-1} V$$

for non-negative Borel functions h on $G(n, k)$, where $S_V = V^{\perp} \cap S^{n-1}$ and σ_V is the surface measure on S_V. This follows from the fact that with a proper normalization constant c the right hand side defines an orthogonally invariant Borel probability measure on $G(n, k)$.

Let $\varepsilon > 0$ be such that $(3k - n - 1)/2 - \varepsilon > 0$. Applying (24.7) and the Kakeya maximal inequality (23.22) on $V^{\perp}$, $V \in G(n, k-1)$, we get with $p = (n + 3 - k)/2$,

$$\int_{S_V} |\mathcal{M}^k g(V_e)|^p\,d\sigma_V e \lesssim \int_{S_V} \mathcal{K}_{\delta}(g_V)(e)^p\,d\sigma_V e$$

$$\lesssim \delta^{(k-n+1)/2-\varepsilon} \int_{V^{\perp}} |g_V|^p\,d\mathcal{H}^{n-k+1}.$$

Integrating over $G(n, k - 1)$ and using (24.6) and the above formula for $\gamma_{n,k}$ we get

$$\int (\mathcal{M}^k g)^p \, d\gamma_{n,k} \lesssim \delta^{(k-n+1)/2-\varepsilon} \|f\|_\infty^{p-2} \int_{G(n,k-1)} \int_{V^\perp} |g_V|^2 \, d\mathcal{H}^{n-k+1} \, d\gamma_{n,k-1} V.$$

Recalling (24.5) and the fact that $\operatorname{spt} \widehat{g} \subset B(0, 2/\delta) \setminus B(0, 1/(2\delta))$ we obtain

$$\int (\mathcal{M}^k f)^p \, d\gamma_{n,k} \lesssim \delta^{(3k-n-1)/2-\varepsilon} \|f\|_\infty^{p-2} \int |g|^2 \, d\mathcal{L}^n.$$

Returning to $f = \sum_j f_j$, and replacing back $g = f_j$ and $\delta = 2^{-j}$, we deduce

$$\|\mathcal{M}^k f\|_{L^p(\gamma_{n,k})} \leq \sum_j \|\mathcal{M}^k f_j\|_{L^p(\gamma_{n,k})}$$

$$\lesssim \|f\|_\infty^{1-2/p} \|f\|_2^{2/p} \sum_j 2^{j((n+1-3k)/2+\varepsilon)/p)} \lesssim \|f\|_\infty^{1-2/p} \|f\|_2^{2/p},$$

because $\|f_j\|_2 \lesssim \|f\|_2$ and $(n + 1 - 3k)/2 + \varepsilon < 0$.

We have now proved the inequality

$$\|\mathcal{M}^k f\|_{L^p(\gamma_{n,k})} \lesssim \|f\|_\infty^{1-2/p} \|f\|_2^{2/p} \tag{24.9}$$

for continuous functions with support in $B(0, 1)$. The same approximation that we used at the end of the proof of Theorem 24.3 yields it for all $f \in L^\infty \cap L^p$ with $\operatorname{spt} f \subset B(0, 1)$, and the theorem follows again applying this to characteristic functions. $\qquad\square$

24.4 Further comments

Theorem 24.2 was proved by Marstrand [1979] and Theorem 24.3 by Falconer [1980a]. Falconer [1985a], Theorem 7.12, gave a duality proof for Theorem 24.2, similar in spirit to the one we gave for Theorem 11.2. The above proof of Theorem 24.3 shows that for almost all $V \in G(n, k)$ the functions F_V agree almost everywhere with continuous functions. It can be developed to give more information about the differentiability properties of these functions for $f \in L^p$; see Falconer [1980a].

Falconer [1980b] related the problem of the existence of (n, k) Besicovitch sets for $k \geq 2$ to certain projection theorems for lower dimensional families of the Grassmannians. Unfortunately there is a gap in the proof and it remains open whether this type of approach could be used. Recall Section 5.4 for a discussion on some such restricted projection theorems, but these are far from being applicable to (n, k) Besicovitch sets.

Theorem 24.4 is due to Bourgain [1991a]. In fact, Bourgain proved with a rather complicated induction argument the stronger result that there exist no (n, k) Besicovitch sets if $2^{k-1} + k \geq n$. R. Oberlin [2010] extended this for $(1 + \sqrt{2})^{k-1} + k > n$. It is an open question whether there exist (n, k) Besicovitch sets for any $k > 1$.

We can also get a maximal inequality for the k-plane transform: if $p > (n + 3 - k)/2$ and $k > (n + 1)/3$, then

$$\|\mathcal{M}^k f\|_{L^p(\gamma_{n,k})} \lesssim \|f\|_p \tag{24.10}$$

for $L^p(\mathbb{R}^n)$ with spt $f \subset B(0, 1)$. This follows by interpolation combining (24.9) and results of Stein, see Theorem 1 and its corollaries in Stein [1961]. Falconer [1980a] obtained such inequalities for certain values of p when $k > n/2$ and Bourgain [1991a] when $2^{k-1} + k \geq n$. R. Oberlin [2007] and [2010] extended them further using the Kakeya maximal function estimates of Katz and Tao [2002a].

This kind of estimate immediately tells us that we can foliate the space by parallel planes none of which intersects a given set in a large measure. For example, already (24.3) gives the following: if $A \subset B(0, 1)$ (in $\mathbb{R}^n$) is Lebesgue measurable and $k > n/2$, then there is $V \in G(n, k)$ such $\mathcal{H}^k(A \cap (V + x)) \lesssim \sqrt{\mathcal{L}^n(A)}$ for all $x \in \mathbb{R}^n$. Of course, because of Besicovitch sets such inequalities are false for $k = 1$, and again they are open for small $k > 1$. However Guth [2007] was able to find a good foliation with curved surfaces for all $k \geq 1$. Gromov and Guth [2012] applied inequalities of this type to embeddings of simplicial complexes into Euclidean spaces.

Estimates for k-plane transforms have been studied and applied extensively. The cases $k = 1$ (X-ray transform) and $k = n - 1$ (Radon transform) are particularly important. Fix a k, $1 \leq k \leq n - 1$, and let

$$Tf(W) = \int_W f, \quad W \text{ a } k\text{-plane in } \mathbb{R}^n.$$

Parametrizing the k-planes as

$$W_{V,w} = \{v + w : v \in V\}, \quad V \in G(n, k), w \in V^\perp,$$

one is led to search for mixed norm estimates

$$\left(\int \left(\int_{V^\perp} |Tf(W_{V,w})|^q \, d\mathcal{H}^{n-k} w \right)^{r/q} d\gamma_{n,k} V \right)^{1/r} \lesssim \|f\|_{L^p(\mathbb{R}^n)} \tag{24.11}$$

for various values of p, q and r, with obvious modifications if q or r is ∞. The norm on the right hand side could also be replaced by a Sobolev norm.

I do not go here into details on the known and conjectured (the full solution is still missing for all k) ranges of exponents. I only make a few comments and the reader can complete the picture from the references given. The case $q = \infty$ corresponds to maximal transforms we just discussed. When $q = \infty$ and $k = 1$, (24.11) is false because of the existence of Besicovitch sets. For other values of q the case $k = 1$ is close to properties of Besicovitch sets and Kakeya maximal function estimates. We already mentioned Drury's X-ray estimate (with $k = 1, r = q$) in Section 23.4 and its application to the $(n + 1)/2$ bound. Its range of exponents was improved by Christ [1984], who also proved estimates for general k. Wolff [1998] developed further his geometric methods from Wolff [1995] in $\mathbb{R}^3$ to improve known mixed X-ray estimates. Łaba and Tao [2001b] generalized this to $\mathbb{R}^n$. In these two papers relations between Kakeya methods and mixed estimates are pursued in a deep way.

Mixed norm estimates as in (24.11) are closely related to estimates on maximal k-plane Kakeya functions

$$\mathcal{K}_{k,\delta} f(V) = \sup_{a \in \mathbb{R}^n} \frac{1}{\mathcal{L}^n(T_V^\delta(a))} \int_{T_V^\delta(a)} |f| \, d\mathcal{L}^n, \quad V \in G(n, k),$$

where $T_V^\delta(a)$ is the δ-neighbourhood of $(V + a) \cap B(a, 1)$. Such estimates were proven by Mitsis [2005] and R. Oberlin [2007], [2010]. They give again lower bounds for the Hausdorff dimension of (n, k) Besicovitch sets. Mitsis [2004a] proved that the Hausdorff dimension of $(n, 2)$ Besicovitch sets (if they exist) is at least $2n/3 + 1$. R. Oberlin [2010] proved that

$$\dim B \geq n - \frac{n - k}{(1 + \sqrt{2})^k}$$

for all (n, k) Besicovitch sets.

It is not always necessary to consider all planes in the Grassmannian: if $G \subset G(n, n - 1)$ is a Borel set and if a Borel set $A \subset \mathbb{R}^n$ intersects a translate of every plane $V \in G$ in a set of positive $n - 1$ measure, then $\mathcal{L}^n(A) > 0$, if $\dim G > 1$, and $\dim A \geq n - 1 + \dim G$, if $0 \leq \dim G \leq 1$. This was shown by D. M. Oberlin [2007] who first proved a restricted weak type inequality for the maximal Radon transform involving measures with finite energy on the space of hyperplanes. It also follows from Falconer and Mattila [2015] with a duality method. See Mitsis [2003b] and Oberlin [2006a] for results preceding these. Oberlin [2007] proved similar estimates for families of spheres, too. In Oberlin [2014a] he obtained analogous, but weaker, results for k planes when $1 \leq k < n - 1$. Rogers [2006] proved estimates for the Hausdorff dimension of other restricted (n, k) sets; the planes considered form a smooth submanifold of the Grassmannian. For instance, he showed that if a subset of R^3 contains a translate of every plane in a sufficiently curved one-dimensional submanifold of $G(3, 2)$, then it must have Hausdorff dimension 3.

By Falconer's result in Falconer [1986], for any $n \geq 2$ there exist $(n, n - 1)$ Nikodym sets, that is, Borel sets $N \subset \mathbb{R}^n$ of Lebesgue measure zero such that for every $x \in \mathbb{R}^n \setminus N$ there is a hyperplane V through x for which $V \setminus \{x\} \subset N$. Mitsis [2004b] showed that they have Hausdorff dimension n.

25

Bilinear restriction

In this chapter we prove a sharp bilinear restriction theorem and we show how it can be used to improve the Stein–Tomas restriction theorem.

25.1 Bilinear vs. linear restriction

Earlier we studied the restriction and extension inequalities such as

$$\|\widehat{f}\|_{L^q(\mathbb{R}^n)} \lesssim \|f\|_{L^p(S^{n-1})} \quad \text{for } f \in L^p(S^{n-1}).$$

Recall that by $\widehat{f}$ we mean here the Fourier transform of the measure $f\sigma^{n-1}$. By Schwartz's inequality we can write this in an equivalent form

$$\|\widehat{f_1}\widehat{f_2}\|_{L^{q/2}(\mathbb{R}^n)} \lesssim \|f_1\|_{L^p(S^{n-1})}\|f_2\|_{L^p(S^{n-1})} \quad \text{for } f_1, f_2 \in L^p(S^{n-1}). \quad (25.1)$$

As such there is not much gain but if f_1 and f_2 are supported in different parts of the sphere, we can get something better. Let us look at the case $p = 2, q = 4$. By Plancherel's theorem the inequality

$$\|\widehat{f_1}\widehat{f_2}\|_{L^2(\mathbb{R}^n)} \lesssim \|f_1\|_{L^2(S^{n-1})}\|f_2\|_{L^2(S^{n-1})} \quad (25.2)$$

reduces to the non-Fourier statement

$$\|(f_1\sigma^{n-1}) * (f_2\sigma^{n-1})\|_{L^2(\mathbb{R}^n)} \lesssim \|f_1\|_{L^2(S^{n-1})}\|f_2\|_{L^2(S^{n-1})}. \quad (25.3)$$

If $n = 2$, this inequality fails when $f_1 = f_2 = 1$, because $\sigma^1 * \sigma^1$ behaves like $|2 - |x||^{-1/2}$ when $|x| \approx 2$. But if the distance between the supports of f_1 and f_2 is greater than some constant $c > 0$, its validity is rather easy to verify. We have the following more general theorem:

Theorem 25.1 *Let S_1 and S_2 be compact C^1 hypersurfaces in $\mathbb{R}^n$ such that their unit normals $n_j(x_j)$ at $x_j \in S_j$ satisfy $d(n_1(x_1), n_2(x_2)) \geq c$ for all*

$x_j \in S_j$, $j = 1, 2$, *and for some positive constant c. Then*

$$\|\widehat{f_1}\widehat{f_2}\|_{L^2(\mathbb{R}^n)} \leq C(S_1, S_2)\|f_1\|_{L^2(S_1)}\|f_2\|_{L^2(S_2)}$$

for all $f_j \in L^2(S_j)$, $j = 1, 2$.

Proof Let $0 < \delta < 1$ and let $S_j(\delta)$ be the δ-neighbourhood of S_j. Due to the transversality assumption we have

$$\mathcal{L}^n((-S_1(\delta) + x) \cap S_2(\delta)) \lesssim \delta^2.$$

Let $g_j \in L^2(\mathbb{R}^n)$ with spt $g_j \subset S_j(\delta)$. Then for all $x \in \mathbb{R}^n$, $g_1(x - y)g_2(y) \neq 0$ implies $y \in (-S_1(\delta) + x) \cap S_2(\delta)$. Hence by Schwartz's inequality

$$\int |g_1 * g_2(x)|^2 \, dx \leq \iint |g_1(x - y)|^2 |g_2(y)|^2 \, dy \mathcal{L}^n((-S_1(\delta) + x) \cap S_2(\delta)) dx$$

$$\lesssim \delta^2 \int |g_1|^2 \int |g_2|^2.$$

The theorem now follows approximating functions $f_j \in L^2(S_j)$ by functions $g_j = g_j(\delta) \in L^2(S_j(\delta))$ and letting $\delta \to 0$. $\qquad\square$

In the plane this theorem is sharp, but not in higher dimensions. The main purpose of this chapter is to prove a sharp theorem in every $\mathbb{R}^n$, $n \geq 3$. But then we shall also need curvature assumptions in addition to transversality.

The bilinear restriction problem on the sphere asks for what exponents p and q the inequality (25.1) holds for $f_j \in L^p(S^{n-1})$, $j = 1, 2$, or for $f_j \in \mathcal{S}(\mathbb{R}^n)$, if spt $f_j \subset S_j$ and S_1 and S_2 are transversal (normals pointing to separated directions) subsurfaces of S^{n-1}. More generally, S_1 and S_2 can be some other type of surfaces (pieces of paraboloids, cones, etc.). The essential conditions required are usually a certain amount of curvature and that the surfaces are transversal.

The point in bilinear estimates is not only, nor mainly, in getting new types of inequalities, but it is in their applications. In particular, they can be used to improve the linear estimates, and that is what we are going to discuss soon. One way (and equivalent to others we have met) to state the restriction conjecture is (recall Conjecture 19.5):

Conjecture 25.2

$$\|\widehat{f}\|_{L^q(\mathbb{R}^n)} \leq C(n, q)\|f\|_{L^p(S^{n-1})} \quad \text{for} \quad f \in L^p(S^{n-1}),$$

$$p' \leq \frac{n-1}{n+1}q, q > 2n/(n-1).$$

By the Stein–Tomas theorem this is valid for $p = 2$, $q = 2(n + 1)/(n - 1)$, and as observed in Section 19.3 also for $q \geq 2(n + 1)/(n - 1)$. The Kakeya methods developed by Bourgain and Wolff give some improvements for this, but still better results can be obtained via bilinear restriction: we reach $q > 2(n + 2)/n$. This is based on two facts: a general result of Tao, Vargas and Vega [1998] of the type 'bilinear restriction estimates imply linear ones' and a bilinear restriction theorem of Tao [2003]. The latter is the following:

Theorem 25.3 *Let $c > 0$ and let $S_j \subset \{x \in S^{n-1} : x_n > c\}, j = 1, 2$, with $d(S_1, S_2) \geq c > 0$. Then*

$$\|\widehat{f_1}\widehat{f_2}\|_{L^q(\mathbb{R}^n)} \leq C(n, q, c)\|f_1\|_{L^2(S_1)}\|f_2\|_{L^2(S_2)} \quad \text{for } q > (n + 2)/n$$

and for all $f_j \in L^2(S_j)$ with spt $f_j \subset S_j, j = 1, 2$.

The lower bound $(n + 2)/n$ is the best possible. This can been seen using the second part of Lemma 3.18 in the same way as we used the first part to prove the sharpness of Stein–Tomas theorem 19.4. More precisely, let $0 < \delta < 1, e_{n-1} = (0, \ldots, 0, 1, 0), e_n = (0, \ldots, 0, 1) \in \mathbb{R}^n, c = 1/(12n)$ and

$$D_1 = \{x \in S^{n-1} : |x_{n-1}| \leq \delta^2, 1 - x \cdot e_n \leq \delta^2\},$$
$$D_2 = \{x \in S^{n-1} : |x_n| \leq \delta^2, 1 - x \cdot e_{n-1} \leq \delta^2\}$$

with $\sigma^{n-1}(D_j) \approx \delta^n$. Then we have as in Lemma 3.18 for $g_j = \chi_{D_j}$,

$$|\widehat{g_j}(\xi)| \gtrsim \delta^n \quad \text{for } \xi \in S_\delta,$$

where

$$S_\delta = \{\xi \in \mathbb{R}^n : |\xi_j| \leq c/\delta \text{ for } j = 1, \ldots, n - 2, |\xi_{n-1}| \leq c/\delta^2, |\xi_n| \leq c/\delta^2\}$$

and $\mathcal{L}^n(S_\delta) \approx \delta^{-n-2}$. If the estimate of Theorem 25.3 is valid for some q we get

$$\delta^{2n-(n+2)/q} \lesssim \|\widehat{g_1}\widehat{g_2}\|_{L^q(\mathbb{R}^n)} \lesssim \|g_1\|_{L^2(S_1)}\|g_2\|_{L^2(S_2)} \approx \delta^n.$$

Letting $\delta \to 0$, we obtain $q \geq (n + 2)/n$.

Theorem 25.3 is not quite enough to get improvements for the linear restriction inequalities; we need such estimates for more general surfaces. We need these also for the application to distance sets, and for that we need a version for more general measures on the left hand side. So let us now present the setting where we shall prove a bilinear restriction theorem.

25.2 Setting for the bilinear restriction theorem

We have positive constants C_0, c_0, ε_0 and R_0 and we have for $j = 1, 2$, bounded open sets $V_j \subset \mathbb{R}^{n-1} \subset B(0, R_0)$, $\widetilde{V}_j$ is the ε_0-neighbourhood of V_j, V_j^* is the

$4\varepsilon_0$-neighbourhood of $\widetilde{V}_j$, C^2-functions $\varphi_j : V_j^* \to \mathbb{R}$ satisfying: the maps $\nabla\varphi_j$ are diffeomorphisms such that for all $v_j \in \widetilde{V}_j$, $\det(D(\nabla\varphi_j)(v_j)) \neq 0$ and

$$|\nabla\varphi_j)(v_j)| \leq C_0, \tag{25.4}$$

$$|D(\nabla\varphi_j)(v_j)(x)| \geq c_0|x| \quad \text{for } x \in \mathbb{R}^{n-1}, \tag{25.5}$$

$$|D(\nabla\varphi_1)(v_1)^{-1}(\nabla\varphi_2(v_2) - \nabla\varphi_1(v_1)) \cdot (\nabla\varphi_2(v_2) - \nabla\varphi_1(v_1))| \geq c_0, \tag{25.6}$$

$$|D(\nabla\varphi_2)(v_2)^{-1}(\nabla\varphi_1(v_1) - \nabla\varphi_2(v_2)) \cdot (\nabla\varphi_1(v_1) - \nabla\varphi_2(v_2))| \geq c_0, \tag{25.7}$$

$S_j = \{(x, \varphi_j(x)) : x \in V_j\}$, $j = 1, 2$, are the corresponding surfaces, s and q are positive numbers with $s \leq n$ and

$$q > q_0 = \max\left\{1, \min\left\{\frac{4s}{n+2s-2}, \frac{n+2}{n}\right\}\right\}, \tag{25.8}$$

$\omega \in L^\infty(\mathbb{R}^n)$ such that

$$\omega \geq 0, \|\omega\|_\infty \leq 1 \quad \text{and} \quad \omega(B(x,r)) \leq r^s \quad \text{for all } x \in \mathbb{R}^n, \quad r > 0. \tag{25.9}$$

Here, as before, we identify ω with a measure and $\omega(A)$ means $\int_A \omega$.

μ is a Borel measure on $\mathbb{R}^n$ such that $\mu(B(x,r)) \leq r^s$ for all $x \in \mathbb{R}^n, r > 0$. (25.10)

Notice that these inequalities yield that there is a positive constant c_1, depending only on C_0 and c_0, such that

$$|\nabla\varphi_1(v_1) - \nabla\varphi_2(v_2)| \geq c_1 \tag{25.11}$$

for all $v_1 \in \widetilde{V}_1$, $v_2 \in \widetilde{V}_2$. Since $(\nabla\varphi_1(v_1), 1)$ and $(\nabla\varphi_2(v_2), 1)$ give the normal directions of the surfaces S_1 and S_2, these surfaces are transversal.

If the eigenvalues of the Hessians $D(\nabla\varphi_j)$, whose matrix elements are the second order partial derivatives $\partial_k\partial_l\varphi_j$, are all positive, which means that the principal curvatures of the surfaces S_j are positive, then (25.11) is equivalent to the conditions (25.6) and (25.7), at least if we restrict to sufficiently small subdomains of V_1 and V_2. This is easy to check. In general, (25.11) does not imply (25.6) and (25.7). We have formulated here the conditions (25.6) and (25.7) not only because of greater generality but because they appear quite naturally at the end of the proof.

25.3 Bilinear restriction theorems

We shall prove the bilinear restriction theorem for $q > q_0$. Of course, when $\mu = \omega$ is Lebesgue measure, $s = n$. Then $q_0 = \frac{n+2}{n}$ and we have the same range as in Theorem 25.3. We have $q_0 = \frac{4s}{n+2s-2}$ if and only if $\frac{n-2}{2} \leq s \leq \frac{n+2}{2}$; this range is all, and even more, than we need for applications to distance sets.

Compact spherical subcaps of open half-spheres can be parametrized as above. Later we shall need surfaces which are obtained by scaling small spherical caps to unit size. If we scale all directions by the same factor, we would get flatter and flatter surfaces from smaller and smaller caps. In order to have uniformly curving surfaces we need to scale caps of size η by $1/\eta$ in the tangential directions and by $1/\eta^2$ in the normal direction. The following example presents this more precisely.

Example 25.4 Let η be a small positive number and C_1 and C_2 spherical caps in S^{n-1} with $d(C_j) \approx d(C_1, C_2) \approx \eta$. We could as well assume that $C_j = S^{n-1} \cap B(v_j, \eta)$ such that $v_1 = \left(2\eta, 0, \ldots, 0, \sqrt{1 - (2\eta)^2}\right)$ and $v_2 = \left(v_{2,1}, 0, \ldots, 0, \sqrt{1 - v_{2,1}^2}\right)$ with $(4 + c)\eta \le v_{2,1} \le C\eta$ with some positive constants c and C. Let φ, $\varphi(x) = \sqrt{1 - |x|^2}$, $x \in B^{n-1}(0, 1/2)$, parametrize these caps.

We use the linear map $T : \mathbb{R}^n \to \mathbb{R}^n$,

$$Tx = \eta^{-1}(x_1, \ldots, x_{n-1}, \eta^{-1}x_n), \quad x \in \mathbb{R}^n,$$

to scale the caps C_1 and C_2. Define

$$S_j = \{(x, \varphi_j(x)) : x \in V_j\}, \quad j = 1, 2,$$

where

$$V_j = B(\eta^{-1}u_j, 1) \subset \mathbb{R}^{n-1} \quad \text{with} \quad v_j = (u_j, v_{j,n})$$

and

$$\varphi_j(x) = \eta^{-2}(1 - \eta^2|x|^2)^{1/2} \quad \text{for } x \in V_j.$$

Then we have $T(C_j) \subset S_j$, $V_j \subset B(0, C + 1) \setminus B(0, 1)$ and $d(V_1, V_2) \ge c$. Moreover for $x \in V_j$,

$$\nabla\varphi_j(x) = \frac{-x}{(1 - \eta^2|x|^2)^{1/2}},$$

the matrix of $D(\nabla\varphi_j)(x)$ is $\left(\dfrac{-\delta_{k,l}}{(1 - \eta^2|x|^2)^{1/2}} - \dfrac{\eta^2 x_k x_l}{(1 - \eta^2|x|^2)^{3/2}} \right).$

If η_n is sufficiently small and $0 < \eta < \eta_n$, the inequalities (25.4)–(25.7) are easily checked with constants depending only on n.

The bilinear restriction theorem we shall prove is the following:

Theorem 25.5 *Suppose that the assumptions of Section 25.2 are satisfied. Then*

$$\|\widehat{f_1}\widehat{f_2}\|_{L^q(\mu)} \le C\|f_1\|_{L^2(S_1)}\|f_2\|_{L^2(S_2)} \quad \text{for } f_j \in L^2(S_j), \quad j = 1, 2,$$

where the constant C only depends on the numbers n, s, q, C_0, c_0, R_0 and ε_0 in Section 25.2.

Remark 25.6 As in Section 19.3 once we have this theorem for some q, it follows for any larger q. In particular, it suffices to consider $q < \frac{n+1}{n-1}$, just for some technical reasons which will appear later.

Here as before the Fourier transform $\widehat{f_j}$ means that of the measure $f_j \sigma_j$ where σ_j is the surface measure on S_j. It is equivalent to consider the actual surface measure, that is, the Hausdorff $(n-1)$-dimensional measure restricted to S_j, or Lebesgue measure lifted from $\mathbb{R}^{n-1}$ onto the graph. It will be more convenient to use the latter and we shall denote it by σ_j. Then the integrals over S_j (and similarly for other graphs that will appear) mean

$$\int_{S_j} g \, d\sigma_j = \int_{V_j} g(x, \varphi_j(x)) \, dx. \tag{25.12}$$

In particular, $\widehat{f_j}$ means

$$\widehat{f_j}(x, t) = \widehat{f_j \sigma_j}(x, t) = \int_{V_j} e^{-2\pi i(x \cdot v + t\varphi_j(v))} f_j(v, \varphi_j(v)) \, dv,$$

$$(x, t) \in \mathbb{R}^{n-1} \times \mathbb{R}. \tag{25.13}$$

At some instances, in particular in Section 25.5, it will be more convenient to work with the bounded weight ω instead of the measure μ, so we shall now reduce Theorem 25.5 to the following:

Theorem 25.7 *Suppose that the assumptions of Section 25.2 are satisfied. Then*

$$\|\widehat{f_1}\widehat{f_2}\|_{L^q(\omega)} \le C \|f_1\|_{L^2(S_1)} \|f_2\|_{L^2(S_2)} \quad \text{for } f_j \in L^2(S_j), \quad j = 1, 2,$$

where the constant C only depends on the numbers n, s, q, C_0, c_0, R_0 and ε_0 in Section 25.2.

Proof that Theorem 25.7 implies Theorem 25.5 Choose $\varphi \in \mathcal{S}(\mathbb{R}^n)$ such that $\varphi(-x) = \varphi(x)$, $\widecheck{\varphi} = 1$ on $\text{spt}(f_1 * f_2)$. Then $\widehat{f_1}\widehat{f_2} = (\widehat{f_1 f_2}) * \varphi$ and by Hölder's inequality with $q' = q/(q-1)$ we have for $x \in \mathbb{R}^n$,

$$|(\widehat{f_1}\widehat{f_2})(x)| \le \int |(\widehat{f_1}\widehat{f_2})(y)| |\varphi(x - y)| \, dy$$

$$\le \left(\int |\widehat{f_1}\widehat{f_2})(y)|^q |\varphi(y - x)| \, dy \right)^{1/q} \left(\int |\varphi(y - x)| \, dy \right)^{1/q'}$$

$$\lesssim \left(\int |\widehat{f_1}\widehat{f_2})(y)|^q |\varphi(y - x)| \, dy \right)^{1/q}.$$

It follows that

$$\int |\widehat{f_1}\widehat{f_2})|^q \, d\mu \lesssim \iint |\widehat{f_1}\widehat{f_2})(y)|^q |\varphi(y - x)| \, dy \, d\mu x = \int |\widehat{f_1}\widehat{f_2})|^q \omega$$

with $\omega = |\varphi| * \mu$. Using the fast decay of φ and the growth condition (25.10) for μ it is easy to check that $\|\omega\|_\infty \le C(s, \varphi)$ and $\omega(B(x, r)) \le C(s, \varphi)r^s$ for all $x \in \mathbb{R}^n$, $r > 0$. Consequently, Theorem 25.7 implies Theorem 25.5. $\square$

25.4 Bilinear restriction implies restriction

Before starting to prove Theorem 25.7, let us go to the theorem of Tao, Vargas and Vega mentioned above.

Theorem 25.8 *Let $M > 0, 1 < p, q < \infty, q > 2n/(n - 1)$ and $p' \le \frac{n-1}{n+1}q$. If the estimate*

$$\|\widehat{f_1}\widehat{f_2}\|_{L^{q/2}(\mathbb{R}^n)} \le M\|f_1\|_{L^p(S_1)}\|f_2\|_{L^p(S_2)} \tag{25.14}$$

holds for all surfaces S_1 and S_2 as in 25.2, then also the estimate

$$\|\widehat{f}\|_{L^q(\mathbb{R}^n)} \le C(n, q)M\|f\|_{L^p(S^{n-1})}$$

holds.

Combining Theorems 25.3 and 25.8 , we obtain

Theorem 25.9 *The restriction conjecture holds for $q > 2(n + 2)/n$:*

$$\|\widehat{f}\|_{L^q(\mathbb{R}^n)} \le C(n, q)\|f\|_{L^p(S^{n-1})} \quad for \quad f \in L^p(S^{n-1}),$$

$$p' \le \frac{n - 1}{n + 1}q, q > 2(n + 2)/n.$$

To check this, we only need to consider the case where q is smaller than the Stein–Tomas exponent $2(n + 1)/(n - 1)$. Then $p > 2$ and we can apply Theorem 25.3 with p in place of 2.

Observe that these results again give the restriction conjecture in the plane.

Proof of Theorem 25.8 We give the proof only for $p' < \frac{n-1}{n+1}q$, allowing the constant to depend also on p. For the end-point result, see Tao, Vargas and Vega [1998].

We only consider the case where $q \le 4$. This is actually enough by the Stein–Tomas theorem and the fact that the restriction conjecture is valid in the plane.

It is enough to consider $f \in \mathcal{S}(\mathbb{R}^n)$. Moreover we may and shall assume that $S^{n-1} \cap$ spt f lies in a part of S^{n-1} which has a parametrization $(v, \varphi(v))$, $v \in Q$, where Q is a cube in $\mathbb{R}^{n-1}$ and $\varphi > 0$. Then we can write the Fourier transform of f as

$$\widehat{f}(x, t) = \int_Q e^{-2\pi i(x \cdot v + t\varphi(v))} f(v, \varphi(v)) \, dv, \quad (x, t) \in \mathbb{R}^{n-1} \times \mathbb{R}.$$

Next we write

$$\|\widehat{f}\|_{L^q(\mathbb{R}^n)}^2 = \|(\widehat{f})^2\|_{L^{q/2}(\mathbb{R}^n)},$$

and

$$(\widehat{f})^2(x, t) = \int_Q \int_Q e^{-2\pi i(x \cdot v + t\varphi(v))} f(v, \varphi(v)) e^{-2\pi i(x \cdot w + t\varphi(w))} f(w, \varphi(w)) \, dv \, dw.$$

As in Chapter 16 we introduce a Whitney decomposition of $Q \times Q \setminus \Delta$, $\Delta = \{(v, w) : v = w\}$, into disjoint cubes $I \times J \in \mathcal{Q}_k$, $k = k_0, k_0 + 1, \ldots$, where I and J are dyadic subcubes of Q such that $d(I) = d(J) = 2^{-k} d(Q) \approx d(I, J)$ when $I \times J \in \mathcal{Q}_k$. Let $f_I(v, \varphi(v)) = f(v, \varphi(v)) \chi_I(v)$. Then we have

$$\|\widehat{f}\|_{L^q(\mathbb{R}^n)}^2 = \left\| \sum_k \sum_{I \times J \in \mathcal{Q}_k} \widehat{f_I} \widehat{f_J} \right\|_{L^{q/2}(\mathbb{R}^n)} \leq \sum_k \left\| \sum_{I \times J \in \mathcal{Q}_k} \widehat{f_I} \widehat{f_J} \right\|_{L^{q/2}(\mathbb{R}^n)}.$$

$$(25.15)$$

The Fourier transform of the measure $f_I * f_J$ is $\widehat{f_I * f_J} = \widehat{f_I} \widehat{f_J}$. We have

$$\text{spt } f_I * f_J \subset S(I, J) := \{(v, t) : v \in \overline{I} + \overline{J}, 0 \leq t \leq 2\}.$$

Denoting by $2I$ the cube with the same centre as I and with the double side-length, we have that for $I \times J \in \mathcal{Q}_k$, $2I + 2J$ lies in a $C2^{-k}$-neighbourhood of $2I + 2I$ for some constant C depending only on n, so the sets $S(2I, 2J)$, $I \times J \in \mathcal{Q}_k$, have for each fixed k bounded overlap with a constant independent of k. Choose smooth compactly supported functions $0 \leq \psi(I, J) \leq 1$ such that $\psi(I, J) = 1$ on $S(I, J)$, spt $\psi(I, J) \subset S(2I, 2J)$ and $\|\widehat{\psi(I, J)}\|_1 \approx 1$, and define the operators $T_{I,J}$ by

$$T_{I,J} g = \widehat{\psi(I, J)} * g.$$

Using the bounded overlap of the supports of the functions $\widehat{\psi(I, J)}$, $\widehat{\psi(I, J)}(x) = \psi(I, J)(-x)$, Plancherel's theorem gives the L^2 estimate for arbitrary L^2 functions $g_{I,J}$,

$$\left\| \sum_{I \times J \in Q_k} T_{I,J} g_{I,J} \right\|^2_{L^2(\mathbb{R}^n)} = \left\| \sum_{I \times J \in Q_k} \widehat{\psi(I,J)} \widehat{g_{I,J}} \right\|^2_{L^2(\mathbb{R}^n)}$$

$$\lesssim \sum_{I \times J \in Q_k} \|\widehat{g_{I,J}}\|^2_{L^2(\mathbb{R}^n)} = \sum_{I \times J \in Q_k} \|g_{I,J}\|^2_{L^2(\mathbb{R}^n)}.$$

The L^1-estimate

$$\left\| \sum_{I \times J \in Q_k} T_{I,J} g_{I,J} \right\|_{L^1(\mathbb{R}^n)} \lesssim \sum_{I \times J \in Q_k} \|g_{I,J}\|_{L^1(\mathbb{R}^n)}$$

for arbitrary L^1-functions $g_{I,J}$ follows by $\|T_{I,J} g_{I,J}\|_1 \lesssim \|g_{I,J}\|_1$ and triangle inequality. These two inequalities tell us that the operator T_k, $T_k(g_{I,J}) = \sum_{I \times J \in Q_k} T_{I,J} g_{I,J}$, acting on vector valued functions is bounded from $L^r(\mathbb{R}^n, l^r)$ to $L^r(\mathbb{R}^n)$ for $r = 1$ and $r = 2$. By the Riesz–Thorin interpolation theorem for such operators, see, e.g., Grafakos [2008], Section 4.5, T_k is also bounded from $L^{q/2}(\mathbb{R}^n, l^{q/2})$ to $L^{q/2}(\mathbb{R}^n)$, since $1 \le q/2 \le 2$ (all of course with norms independent of k). Thus

$$\left\| \sum_{I \times J \in Q_k} T_{I,J} g_{I,J} \right\|^{q/2}_{L^{q/2}(\mathbb{R}^n)} \lesssim \sum_{I \times J \in Q_k} \|g_{I,J}\|^{q/2}_{L^{q/2}(\mathbb{R}^n)}.$$

Recall that $\psi(I, J) = 1$ on the support of $f_I * f_J$, so

$$T_{I,J}(\widehat{f_I} \widehat{f_J}) = \widehat{\psi(I,J)} * (\widehat{f_I} \widehat{f_J}) = \mathcal{F}(\psi(I,J)(f_I * f_J)) = \widehat{f_I} \widehat{f_J},$$

whence

$$\left\| \sum_{I \times J \in Q_k} \widehat{f_I} \widehat{f_J} \right\|^{q/2}_{L^{q/2}(\mathbb{R}^n)} \lesssim \sum_{I \times J \in Q_k} \|\widehat{f_I} \widehat{f_J}\|^{q/2}_{L^{q/2}(\mathbb{R}^n)}. \qquad (25.16)$$

In order to apply our bilinear assumption we have to scale f_I and f_J back to the unit scale. Let $I \times J \in Q_k$. After a translation and rotation we may assume that $I \cup J \subset B(0, C2^{-k}) \subset \mathbb{R}^{n-1}$. Then the appropriate scaling is $(v, \varphi(v)) \mapsto (2^k v, 2^{2k} \varphi(v)) := (w, \psi(w))$. Define, with this notation,

$$g_I(w, \psi(w)) = f_I(v, \varphi(v)), \quad g_J(w, \psi(w)) = f_J(v, \varphi(v)).$$

The change of variable formulas give ($\widehat{g_I}$ and $\widehat{g_J}$ are now of course with respect to the graph G_ψ of ψ, recall our convention (25.12)),

$$\widehat{f_I}(x,t) = 2^{-k(n-1)}\widehat{g_I}(2^{-k}x, 2^{-2k}t), \; \widehat{f_J}(x,t) = 2^{-k(n-1)}\widehat{g_J}(2^{-k}x, 2^{-2k}t),$$

$$\int |\widehat{f_I}\widehat{f_J}|^{q/2} = 2^{-k(q(n-1)-(n+1))} \int |\widehat{g_I}\widehat{g_J}|^{q/2},$$

$$\int |g_I|^p = 2^{k(n-1)} \int |f_I|^p, \quad \int |g_I|^p = 2^{k(n-1)} \int |f_I|^p.$$

By Example 25.4 we can apply (25.14) to get,

$$\|\widehat{g_I}\widehat{g_J}\|_{L^{q/2}(\mathbb{R}^n)}^{q/2} \le M^{q/2} \|g_I\|_{L^p(G_\psi)}^{q/2} \|g_J\|_{L^p(G_\psi)}^{q/2}.$$

Combining these statements we find

$$\|\widehat{f_I}\widehat{f_J}\|_{L^{q/2}(\mathbb{R}^n)}^{q/2} \le M^{q/2} 2^{-k\left(\frac{(n-1)}{p'} - \frac{(n+1)}{q}\right)q} \|f_I\|_{L^p(S^{n-1})}^{q/2} \|f_J\|_{L^p(S^{n-1})}^{q/2}.$$

Recalling (25.15), inserting the last estimate into (25.16), and using the fact that for each I there are only boundedly many J such that $I \times J \in \mathcal{Q}_k$, we obtain

$$\|\widehat{f}\|_{L^q(\mathbb{R}^n)}^2 \lesssim \sum_k \left(\sum_{I \times J \in \mathcal{Q}_k} M^{q/2} 2^{-k\left(\frac{(n-1)}{p'} - \frac{(n+1)}{q}\right)q} \|f_I\|_{L^p(S^{n-1})}^{q/2} \|f_J\|_{L^p(S^{n-1})}^{q/2} \right)^{2/q}$$

$$\lesssim \sum_k M 2^{-2k\left(\frac{(n-1)}{p'} - \frac{(n+1)}{q}\right)} \left(\sum_{I \in \mathcal{D}_k} \|f_I\|_{L^p(S^{n-1})}^q \right)^{2/q}.$$

Here $\mathcal{D}_k$ is the collection of all dyadic subcubes of Q of diameter $2^{-k}d(Q)$. The factor $\frac{(n-1)}{p'} - \frac{(n+1)}{q}$ is positive by our assumptions. So the theorem follows if we have

$$\sum_{I \in \mathcal{D}_k} \|f_I\|_{L^p(S^{n-1})}^q \le \|f\|_{L^p(S^{n-1})}^q.$$

This is true if $q/p \ge 1$. Choosing p' sufficiently close to $\frac{n-1}{n+1}q$ we do have $q/p \ge 1$ due to the assumption $q > 2n/(n-1)$; if $p' = \frac{n-1}{n+1}q$, then $q/p = q - \frac{n+1}{n-1} > 1$. Moreover, getting the result for some p gives it also for larger p (and smaller p'). $\qquad\square$

25.5 Localization

We now proceed towards the proof of Theorem 25.7 proving first a localization theorem of Tao and Vargas. In the following theorem the relations between p and q probably are not sharp, but all that is really needed is that if the assumption holds for all $\alpha > 0$, then the assertion holds for all $p > q$. In this section we shall assume that S_1 and S_2 are compact $(n-1)$-dimensional graphs of C^2 functions φ_1 and φ_2 with non-vanishing Gaussian curvature. More precisely,

$$S_j = \{(x, \varphi_j(x)) : x \in K_j\},$$

where $\varphi_j : V_j \to \mathbb{R}$ is a C^2 function, V_j is open, $K_j \subset V_j$ is compact and the assumptions of Theorem 14.7 are satisfied. We do not need here the transversality assumptions (25.6) and (25.7).

Theorem 25.10 *Let $f_j \in L^2(S_j)$, $j = 1, 2$. Suppose that $\omega \in L^\infty(\mathbb{R}^n)$ with $\omega \geq 0$, $\|\omega\|_\infty \leq 1$ and $1 < p < \frac{n+1}{n-1}$. If $\alpha > 0$, $\frac{1}{p}(1 + \frac{4\alpha}{n-1}) < \frac{1}{q} + \frac{2\alpha}{n+1}$, $M_\alpha \geq 1$ and*

$$\|\widehat{f_1}\widehat{f_2}\|_{L^q(\omega, B(x,R))} \leq M_\alpha R^\alpha \|f_1\|_{L^2(S_1)} \|f_2\|_{L^2(S_2)} \tag{25.17}$$

for $x \in \mathbb{R}^n$, $R > 1$, $f_j \in L^2(S_j)$, $j = 1, 2$, then

$$\|\widehat{f_1}\widehat{f_2}\|_{L^p(\omega)} \leq C M_\alpha \|f_1\|_{L^2(S_1)} \|f_2\|_{L^2(S_2)} \quad \text{for } f_j \in L^2(S_j), \quad j = 1, 2, \tag{25.18}$$

where C depends only on the structure constants of Section 25.2.

By Theorem 14.7 the Fourier transform of the surface measure σ_j on S_j satisfies

$$|\widehat{\sigma_j}(x)| \leq C_1(1 + |x|)^{-(n-1)/2}, \quad x \in \mathbb{R}^n. \tag{25.19}$$

We shall use this information in order to be able to apply the Stein–Tomas restriction theorem 19.4:

$$\|\widehat{g_j}\|_{\frac{2(n+1)}{n-1}} \leq C_2 \|g_j\|_{L^2(S_j)}. \tag{25.20}$$

We need that the constant C_2, and so also C_1, only depends on the structure constants given in Section 25.2. Theorem 14.7 as given and proved is not quite enough for that, but one can for example use the argument of Stein [1993], Section VIII.3.1, to get the sufficient estimate for the constant. On the other hand, just for proving the bilinear restriction theorem 25.3 for the sphere and for the application to the distance sets, Corollary 25.25, only the surfaces in Example 25.4 are needed and for these the required dependence of the constants is immediate. We do not really need the rather delicate end-point result of Theorem 19.4, since we have open conditions for the exponents in Theorem 25.10. We could in place of (25.20) use

$$\|\widehat{g_j}\|_r \leq C_2 \|g_j\|_{L^2(S_j)} \tag{25.21}$$

with any $r > \frac{2(n+1)}{n-1}$ sufficiently close to $\frac{2(n+1)}{n-1}$.

Our assumptions on p and q in terms of $\beta = (n-1)/2$ read as

$$1 < p < \frac{\beta+1}{\beta}, \quad \frac{1}{p}\left(1 + \frac{2\alpha}{\beta}\right) < \frac{1}{q} + \frac{\alpha}{1+\beta}.$$

Instead of applying Theorem 14.7 and assuming non-vanishing Gaussian curvature, it would be enough to assume the decay condition

$$|\widehat{\sigma_j}(x)| \lesssim (1+|x|)^{-\beta}, \quad x \in \mathbb{R}^n,$$

with some $\beta > 0$. Then by the general form of Stein–Tomas restriction theorem, Theorem 19.3, (25.21) holds for $r > \frac{2(\beta+1)}{\beta}$. The proof below works under these conditions. Thus the method also gives a version of the theorem for example for conical hypersurfaces.

The proof of Theorem 25.10 will be based on three lemmas. The first of these says that the hypothesis (25.17) yields a similar statement if the functions live in neighbourhoods of the surfaces. Recall that $A(r)$ stands for the r-neighbourhood $\{x : d(x, A) < r\}$ of a set A and C_0 is the upper bound for $\|\nabla\varphi_j\|_\infty$ given in (25.4). By $L^p(A)$ we shall now mean the space of functions in $L^p(\mathbb{R}^n)$ which vanish outside A.

Lemma 25.11 *Let $1 \le q < \infty$, let M be a positive number and let μ be a Borel measure on $\mathbb{R}^n$.*
(a) If

$$\|\widehat{f}\|_{L^q(\mu)} \le M\|f\|_{L^2(S_j)} \quad \text{for } f \in L^2(S_j), \quad j = 1, 2, \tag{25.22}$$

then for all $r > 0$,

$$\|\widehat{f}\|_{L^q(\mu)} \le C(C_0)M\sqrt{r}\|f\|_2 \quad \text{for } f \in L^2(S_j(r)). \tag{25.23}$$

(b) If

$$\|\widehat{f_1}\widehat{f_2}\|_{L^q(\mu)} \le M\|f_1\|_{L^2(S_1)}\|f_2\|_{L^2(S_2)} \quad \text{for } f_j \in L^2(S_j), \quad j = 1, 2, \tag{25.24}$$

then for all $r > 0$,

$$\|\widehat{f_1}\widehat{f_2}\|_{L^q(\mu)} \le C(C_0)Mr\|f_1\|_2\|f_2\|_2 \quad \text{for } f_j \in L^2(S_j(r)), \quad j = 1, 2. \tag{25.25}$$

Proof We prove only (b). The proof of (a) is similar. Let $S_{j,t} = \{(x, \varphi_j(x) + t) : x \in V_j\}$ and $S_j^r = \cup_{|t| < r} S_{j,t}$. It is enough to prove the lemma for S_j^r in place of $S_j(r)$.

Let $f_j \in L^2(S_j^r)$ and $f_{j,t}(z) = f_j(x, \varphi_j(x) + t)$ for $z = (x, \varphi_j(x)) \in S_j$, $|t| < r$. Then by change of variable $u + \varphi_j(y) = s$ and Fubini's theorem,

$$
\begin{aligned}
|\widehat{f_j}(x, t)| &= \left| \iint e^{-2\pi i (x \cdot y + ts)} f_j(y, s)\, ds\, dy \right| \\
&= \left| \iint_{-r}^{r} e^{-2\pi i (x \cdot y + t(u + \varphi_j(y)))} f_j(y, u + \varphi_j(y))\, du\, dy \right| \\
&= \left| \int_{-r}^{r} \int e^{-2\pi i (x \cdot y + t(u + \varphi_j(y)))} f_{j,u}(y, \varphi_j(y))\, dy\, du \right| \\
&= \left| \int_{-r}^{r} e^{-2\pi i t u} \widehat{f_{j,u}}(x, t)\, du \right| \leq \int_{-r}^{r} |\widehat{f_{j,u}}(x, t)|\, du.
\end{aligned}
$$

Thus using Minkowski's integral inequality, (25.24), Schwartz's inequality and Fubini's theorem,

$$
\begin{aligned}
\|\widehat{f_1}\widehat{f_2}\|_{L^q(\mu)} &\leq \left(\int \left(\int_{-r}^{r} |\widehat{f_{1,u}}(z)|\, du \int_{-r}^{r} |\widehat{f_{2,v}}(z)|\, dv \right)^q d\mu z \right)^{1/q} \\
&= \left(\int \left(\int_{-r}^{r} \int_{-r}^{r} |\widehat{f_{1,u}}(z)\widehat{f_{2,v}}(z)|\, du\, dv \right)^q d\mu z \right)^{1/q} \\
&\leq \int_{-r}^{r} \int_{-r}^{r} \|\widehat{f_{1,u}}\widehat{f_{2,v}}\|_{L^q(\mu)}\, du\, dv \\
&\leq M \int_{-r}^{r} \int_{-r}^{r} \|f_{1,u}\|_{L^2(S_1)} \|f_{2,v}\|_{L^2(S_2)}\, du\, dv \\
&\leq M \sqrt{2r} \left(\int_{-r}^{r} \int_{S_1} |f_{1,u}|^2\, du \right)^{1/2} \sqrt{2r} \left(\int_{-r}^{r} \int_{S_2} |f_{2,v}|^2\, dv \right)^{1/2} \\
&\approx Mr \|f_1\|_2 \|f_2\|_2.
\end{aligned}
$$

$\square$

The second lemma shows that for functions living in neighbourhoods of the surfaces S_j, a local hypothesis, namely (25.26), gives a global estimate.

Lemma 25.12 *Let* $1 \leq q < \infty$, *let* M *and* R *be positive numbers and let* μ *be a Borel measure on* $\mathbb{R}^n$ *such that*

$$
\|\widehat{f_1}\widehat{f_2}\|_{L^q(\mu, B(x,R))} \leq M \|f_1\|_2 \|f\|_2 \quad \text{for } x \in \mathbb{R}^n,
$$
$$
f_j \in L^2(S_j(2/R)), \quad j = 1, 2. \tag{25.26}
$$

Then

$$
\|\widehat{f_1}\widehat{f_2}\|_{L^q(\mu)} \leq C(n, q) M \|f_1\|_2 \|f_2\|_2 \quad \text{for } f_j \in L^2(S_j(1/R)), \quad j = 1, 2. \tag{25.27}
$$

Proof Let $f_j \in L^2(S_j(1/R))$, $j = 1, 2$. Let $\psi \in \mathcal{S}(\mathbb{R}^n)$ be such that $0 \le \psi \le 1$, $\psi \approx 1$ on $B(0, 1)$ and spt $\check{\psi} \subset B(0, 1)$. Cover $\mathbb{R}^n$ with balls $B(x_k, R/2)$, $k = 1, 2, \ldots$, such that $\sum_k \chi_{B(x_k, R)} \approx 1$ and define $\psi_k(x) = \psi((x - x_k)/R)$. Then $\sum_k \psi_k \approx \sum_k \psi_k^{2q} \approx \sum_k \chi_{B(x_k, R)} \psi_k^{2q} \approx 1$. Moreover spt $\check{\psi}_k \subset B(0, 1/R)$, whence spt $\check{\psi}_k * f_j \subset S_j(2/R)$. Applying (25.26) and Plancherel's theorem, we obtain

$$\|\psi_k^2 \widehat{f_1}\widehat{f_2}\|_{L^q(\mu, B(x_k, R))} = \|\widetilde{\check{\psi}_k * f_1}\widetilde{\check{\psi}_k * f_2}\|_{L^q(\mu, B(x_k, R))}$$
$$\le M\|\check{\psi}_k * f_1\|_2 \|\check{\psi}_k * f_2\|_2 = M\|\psi_k \widehat{f_1}\|_2 \|\psi_k \widehat{f_2}\|_2.$$

Summing over k we get by Schwartz's inequality,

$$\|\widehat{f_1}\widehat{f_2}\|_{L^q(\mu)} \lesssim \sum_k \|\chi_{B(x_k, R)} \psi_k^2 \widehat{f_1}\widehat{f_2}\|_{L^q(\mu)}$$
$$= \sum_k \|\psi_k^2 \widehat{f_1}\widehat{f_2}\|_{L^q(\mu, B(x_k, R))} \le M \sum_k \|\psi_k \widehat{f_1}\|_2 \|\psi_k \widehat{f_2}\|_2$$
$$\le M \left(\sum_k \|\psi_k \widehat{f_1}\|_2^2\right)^{1/2} \left(\sum_k \|\psi_k \widehat{f_2}\|_2^2\right)^{1/2}$$
$$\approx M\|\widehat{f_1}\|_2 \|\widehat{f_2}\|_2 = M\|f_1\|_2 \|f_2\|_2. \qquad \square$$

Corollary 25.13 *Assuming (25.17) we have for all $R > 1$,*

$$\|\widehat{f_1}\widehat{f_2}\|_{L^q(\omega)} \le C(n, q, C_0) M_\alpha R^{\alpha-1} \|f_1\|_2 \|f_2\|_2$$
$$\text{for } f_j \in L^2(S_j(1/R)), \quad j = 1, 2. \tag{25.28}$$

Proof Applying the assumption (25.17) and Lemma 25.11 with $\mu = \omega \mathcal{L}^n \llcorner B(x, R)$ we have

$$\|\widehat{f_1}\widehat{f_2}\|_{L^q(\omega, B(x, R))} \lesssim M_\alpha R^{\alpha-1} \|f_1\|_2 \|f\|_2 \quad \text{for } x \in \mathbb{R}^n, \quad f_j \in L^2(S_j(2/R)),$$
$$j = 1, 2.$$

Hence the corollary follows by Lemma 25.12. $\qquad \square$

The third lemma tells us how estimates in the neighbourhoods of the surfaces S_j lead to estimates for functions defined on the surfaces themselves. Recall that C_2 is the Stein–Tomas constant in (25.20).

Lemma 25.14 *For any $F \in L^\infty(\mathbb{R}^n) \cap L^1(\mathbb{R}^n)$ with $\|F\|_\infty \le 1$ and any $N, R \ge 1$,*

$$\left|\int F \widehat{g_1}\widehat{g_2}\right|^2 \le C(n, C_2, N) R^{-(n-1)/2} \|F\|_{\frac{2(n+1)}{n+3}}^2 + \sum_{k=0}^\infty R2^{-kN} \|\widehat{F}\widehat{g_2}\|_{L^2(S_1(2^k/R))}^2, \tag{25.29}$$

for all $g_j \in L^2(S_j)$, $j = 1, 2$, with $\|g_j\|_{L^2(S_j)} \leq 1$, and

$$\left| \int F \widehat{g_1} \widehat{h_2} \right|^2 \leq C(n, C_0, C_2, N) \lambda R^{-(n-1)/2-1} \|F\|^2_{\frac{2(n+1)}{n+3}}$$

$$+ \sum_{k=0}^{\infty} R2^{-kN} \|\widehat{Fh_2}\|^2_{L^2(S_1(2^k/R))}, \qquad (25.30)$$

for all $g_1 \in L^2(S_1)$, $h_2 \in L^2(S_2(\lambda/R))$, $\lambda > 0$ with $\|g_1\|_{L^2(S_1)} \leq 1$, $\|h_2\|_2 \leq 1$.

Proof By the product formula (3.20), Schwartz's inequality and (3.28) (we leave it to the reader to check here and below that these formulas hold in the needed generality),

$$\left| \int F \widehat{g_1} \widehat{g_2} \right|^2 = \left| \int \widehat{F\widehat{g_2}} g_1 \, d\sigma_1 \right|^2 \leq \|\widehat{F\widehat{g_2}}\|^2_{L^2(S_1)} \|g_1\|^2_{L^2(S_1)}$$

$$= \int (\overline{\widehat{F\widehat{g_2}}} * \widehat{\sigma_1}) F\widehat{g_2} \|g_1\|^2_{L^2(S_1)} \leq \int (\overline{\widehat{F\widehat{g_2}}} * \widehat{\sigma_1}) F\widehat{g_2}.$$

By Hölder's inequality and (25.20),

$$\|F\widehat{g_2}\|_1 \leq \|F\|_{\frac{2(n+1)}{n+3}} \|\widehat{g_2}\|_{\frac{2(n+1)}{n-1}} \leq C_2 \|F\|_{\frac{2(n+1)}{n+3}}. \qquad (25.31)$$

Choose $\varphi \in \mathcal{S}(\mathbb{R}^n)$ such that $\varphi = 1$ on $B(0, 1)$, φ vanishes outside $B(0, 2)$ and write σ_1 as

$$\sigma_1 = \tau_1 + \tau_2 \quad \text{with} \quad \widehat{\tau_1}(x) = \varphi(x/R)\widehat{\sigma_1}(x).$$

Then

$$\|\widehat{\tau_2}\|_\infty \lesssim R^{-(n-1)/2},$$

and so by (25.31)

$$\left| \iint (\overline{\widehat{F\widehat{g_2}}} * \widehat{\tau_2}) F\widehat{g_2} \right| \leq \|\widehat{\tau_2}\|_\infty \|F\widehat{g_2}\|^2_1 \lesssim R^{-(n-1)/2} \|F\|^2_{\frac{2(n+1)}{n+3}}. \qquad (25.32)$$

Next we estimate $|\iint (\overline{\widehat{F\widehat{g_2}}} * \widehat{\tau_1}) F\widehat{g_2}|$. By (3.28) (note that $\tau_1 \in L^2$)

$$\left| \iint (\overline{\widehat{F\widehat{g_2}}} * \widehat{\tau_1}) F\widehat{g_2} \right| \lesssim \int |\widehat{F\widehat{g_2}}|^2 |\tau_1|.$$

Using the rapid decay of $\widecheck{\varphi}$ one checks easily that

$$|\tau_1(x)| = \left| R^n \int \widecheck{\varphi}(R(y - x)) \, d\sigma_1 y \right| \lesssim_N R(1 + d(x, S_1))^{-N}.$$

Hence

$$\left| \iint (\widehat{F g_2} * \widehat{\tau_1}) F \widehat{g_2} \right| \lesssim_N \sum_{k=0}^{\infty} R 2^{-kN} \|\widehat{F g_2}\|_{L^2(S_1(2^k/R))}^2.$$

This proves (25.29). To prove (25.30) we argue in the same way but use the Stein–Tomas theorem in combination with Lemma 25.11(a) to have

$$\|\widehat{h_2}\|_{\frac{2(n+1)}{n-1}} \leq C(C_0, C_2) \lambda^{1/2} R^{-1/2}. \qquad \square$$

In order to complete the proof of Theorem 25.10 we shall prove that for any measurable set $A \subset \mathbb{R}^n$ with $1 \leq \omega(A) < \infty$,

$$\|\chi_A \widehat{g_1} \widehat{g_2}\|_{L^1(\omega)} \lesssim M_\alpha \omega(A)^{1/p'} \|g_1\|_{L^2(S_1)} \|g_2\|_{L^2(S_2)} \quad \text{for } g_j \in L^2(S_j), \ j = 1, 2. \tag{25.33}$$

Let us first see how this implies the theorem. Let p, q and α be as in the assumptions of Theorem 25.10 and choose $p_1 < p$ so that also p_1 satisfies these conditions. Fix $f_j \in L^2(S_j)$, $j = 1, 2$, with $\|f_j\|_{L^2(S_j)} = 1$. Apply (25.33) with p_1 in place of p and with

$$A = \{x : |\widehat{f_1} \widehat{f_2}(x)| > \lambda\}, \quad \lambda > 0.$$

Note that $\omega(A) < \infty$ because $\widehat{f_j} \in L^{p_0}(\mathbb{R}^n)$ for some $p_0 < \infty$ by the Stein–Tomas restriction theorem and ω is bounded. Then by (25.33), if $\omega(A) \geq 1$,

$$\lambda \omega(A) \leq \|\chi_A \widehat{f_1} \widehat{f_2}\|_{L^1(\omega)} \lesssim M_\alpha \omega(A)^{1/p_1'},$$

which gives

$$\omega(\{x : |\widehat{f_1} \widehat{f_2}(x)| > \lambda\}) \lesssim M_\alpha^p \max\{\lambda^{-p_1}, 1\}.$$

Combining this weak type inequality with the trivial inequality

$$\|\widehat{f_1} \widehat{f_2}\|_\infty \leq C_3,$$

with C_3 depending only on the structure constants, (25.18) follows from

$$\int |\widehat{f_1} \widehat{f_2}|^p \omega = \int_0^{C_3^p} \omega(\{x : |\widehat{f_1} \widehat{f_2}(x)|^p > \lambda\}) \, d\lambda$$

$$\lesssim M_\alpha^p \int_0^{C_3^p} \lambda^{-p_1/p} \, d\lambda < \infty.$$

It remains to verify (25.33). For this it suffices to show that if ζ is a measurable function with $|\zeta| = 1$ and if we set $\zeta_A = \zeta \chi_A$, then

$$\left| \int \zeta_A \widehat{g_1} \widehat{g_2} \omega \right| \lesssim M_\alpha \omega(A)^{1/p'} \|g_1\|_{L^2(S_1)} \|g_2\|_{L^2(S_2)}. \tag{25.34}$$

Applying Lemma 25.14 with $N = 3$ and choosing (notice that the exponent below is positive as $1 < p < \frac{n+1}{n-1}$)

$$R = \omega(A)^{\frac{2}{n-1}\left(\frac{n+3}{n+1} - \frac{2}{p'}\right)}, \qquad (25.35)$$

we obtain

$$\left|\int \zeta_A \widehat{g_1} \widehat{g_2} \omega\right|^2 \lesssim R^{-(n-1)/2} \omega(A)^{\frac{n+3}{n+1}} + \sum_{k=0}^{\infty} R2^{-3k} \|\widehat{\zeta_A \omega \widehat{g_2}}\|^2_{L^2(S_1(2^k/R))}$$

$$= \omega(A)^{2/p'} + \sum_{k=0}^{\infty} R2^{-3k} \|\widehat{\zeta_A \omega \widehat{g_2}}\|^2_{L^2(S_1(2^k/R))}.$$

Here

$$\|\widehat{\zeta_A \omega \widehat{g_2}}\|_{L^2(S_1(2^k/R))} = \sup_{\|h_{1,k}\|_{L^2(S_1(2^k/R))} \leq 1} \left|\int \widehat{\zeta_A \omega \widehat{g_2}} h_{1,k}\right|$$

$$= \sup_{\|h_{1,k}\|_{L^2(S_1(2^k/R))} \leq 1} \left|\int \zeta_A \widehat{g_2} \widehat{h_{1,k}} \omega\right|.$$

We can repeat the above argument with g_2 playing the role of g_1 and $h_{1,k}$ playing the role of g_2. Now $h_{1,k}$ is in $L^2(S_1(2^k/R))$ with norm at most 1 and we have by (25.30)

$$\left|\int \zeta_A \widehat{g_2} \widehat{h_{1,k}} \omega\right|^2 \lesssim 2^k R^{-1} \omega(A)^{2/p'} + \sum_{l=0}^{\infty} R2^{-3l} \|\widehat{\zeta_A \omega \widehat{h_{1,k}}}\|^2_{L^2(S_1(2^l/R))}.$$

Again

$$\|\widehat{\zeta_A \omega \widehat{h_{1,k}}}\|_{L^2(S_1(2^l/R))} = \sup_{\|h_{2,l}\|_{L^2(S_2(2^l/R))} \leq 1} \left|\int \widehat{\zeta_A \omega \widehat{h_{1,k}}} h_{2,l}\right|$$

$$\leq \sup_{\|h_{2,l}\|_{L^2(S_2(2^l/R))} \leq 1} \|\zeta_A \widehat{h_{1,k}} \widehat{h_{2,l}}\|_{L^1(\omega)}.$$

By Hölder's inequality

$$\|\zeta_A \widehat{h_{1,k}} \widehat{h_{2,l}}\|_{L^1(\omega)} \leq \omega(A)^{1/q'} \|\widehat{h_{1,k}} \widehat{h_{2,l}}\|_{L^q(\omega)}.$$

By Corollary 25.13 we have for $k \leq l$,

$$\|\widehat{h_{1,k}} \widehat{h_{2,l}}\|_{L^q(\omega)} \lesssim M_\alpha (R/2^l)^{\alpha-1} \leq M_\alpha R^{\alpha-1} 2^l.$$

Combining these inequalities,

$$\left| \int \zeta_A \widehat{g_1} \widehat{g_2} \omega \right|^2$$

$$\lesssim \omega(A)^{2/p'} + M_\alpha^2 \sum_{k=0}^{\infty} R2^{-3k} \sum_{l=0}^{\infty} R2^{-3l} \omega(A)^{2/q'} R^{2\alpha-2} 2^{2l}$$

$$\approx \omega(A)^{2/p'} + M_\alpha^2 \omega(A)^{2/q'} R^{2\alpha}.$$

Recalling how we chose R in (25.35) we see that

$$\omega(A)^{2/q'} R^{2\alpha} = \omega(A)^{2/q' + \frac{4\alpha}{n-1}\left(\frac{n+3}{n+1} - \frac{2}{p'}\right)}.$$

Since by the assumption of the theorem,

$$2/q' + \frac{4\alpha}{n-1}\left(\frac{n+3}{n+1} - \frac{2}{p'}\right) < 2/p',$$

the desired inequality (25.34) follows and the proof of the theorem is complete.

25.6 Induction on scales

The second crucial idea is an induction on scales argument due to Wolff:

Proposition 25.15 *Suppose that the assumptions of Section 25.2 are satisfied. Then there is a constant $c > 0$ such the following holds. Assume that (25.17) holds for some $\alpha > 0$:*

$$\|\widehat{f_1}\widehat{f_2}\|_{L^q(\omega, B(x,R))} \le M_\alpha R^\alpha \|f_1\|_{L^2(S_1)} \|f_2\|_{L^2(S_2)} \tag{25.36}$$

for $x \in \mathbb{R}^n$, $R > 1$ and $f_j \in L^2(S_j)$, $j = 1, 2$. Then for all $0 < \delta, \varepsilon < 1$,

$$\|\widehat{f_1}\widehat{f_2}\|_{L^q(\omega, B(x,R))} \le C R^{\max\{\alpha(1-\delta), c\delta\} + \varepsilon} \|f_1\|_{L^2(S_1)} \|f_2\|_{L^2(S_2)} \tag{25.37}$$

for $x \in \mathbb{R}^n$, $R > 1$ and $f_j \in L^2(S_j)$, $j = 1, 2$, where the constant C depends only on the structure constants of Section 25.2 and on M_α, δ and ε.

The point here is that once we have this proposition we can argue inductively to get down to arbitrarily small α. That is, Proposition 25.15 implies (25.17) for all $\alpha > 0$. To see this note that $\|\widehat{f_j}\|_\infty \lesssim \|f_j\|_{L^2(S_j)}$ by Schwartz's inequality, whence (25.17) holds for $\alpha = \alpha_0 = s/q$. Fix $\varepsilon > 0$ and define

$$\alpha_{j+1} = c\alpha_j/(\alpha_j + c) + \varepsilon, \quad j = 0, 1, 2, \ldots.$$

Suppose (25.17) holds for $\alpha = \alpha_j$ for some j. Apply Proposition 25.15 with $\delta = \delta_j = \alpha_j/(\alpha_j + c)$. Then

$$\max\{\alpha_j(1 - \delta), c\delta\} = c\alpha_j/(\alpha_j + c),$$

and it follows that (25.17) holds for $\alpha = \alpha_{j+1}$. It is easy to check that if ε is chosen small enough, the sequence (α_j) is decreasing and

$$\alpha_j \to \left(\varepsilon + \sqrt{\varepsilon^2 + 4c\varepsilon}\right)/2.$$

Since we can choose ε arbitrarily small, (25.17) holds for all $\alpha > 0$.

So Proposition 25.15 together with Theorem 25.10 yield Theorem 25.7. Before giving the details for the proof of Proposition 25.15 we give a sketch of the main ideas in the case $\omega = 1$.

25.7 Sketch of the proof of Theorem 25.7

The proof of Proposition 25.15, which is the core of the whole argument, uses the third basic tool: the wavepacket decomposition. Fix $R > 1$ and let $f_j \in L^2(S_j)$ with $\|f_j\|_{L^2(S_j)} \leq 1$. The wavepacket decomposition allows us to write $\widehat{f_j}$ as a sum of functions p_{y,v_j} which together with their Fourier transforms are well localized:

$$\widehat{f_j}(x, t) = \Sigma_{w_j} p_{w_j}(x, t), \quad j = 1, 2. \tag{25.38}$$

The indices w_j (where w_1 always is related to f_1 and w_2 to f_2) are of the form (y_j, v_j) where the v_j run through a $1/\sqrt{R}$-separated set in V_j and the y_j run through a $\sqrt{R}$-separated set in $\mathbb{R}^{n-1}$. The functions p_{w_j} are essentially supported in the tubes (that is, they decay very fast off them)

$$T_{w_j} = \{(x, t) : |t| \leq R, |x - (y_j - t\nabla\varphi_j(v_j))| \leq R^{1/2}\}$$

and their Fourier transforms have supports in $S_j \cap B((v_j, \varphi_j(v_j)), 2/\sqrt{R})$. The transversality assumptions on S_1 and S_2 guarantee that any two tubes T_{w_1} and T_{w_2} are transversal.

The proof of the wavepacket decomposition involves several technicalities, but in principle it is not very difficult. Here are the main ideas. First find C^∞-functions η and ψ on $\mathbb{R}^{n-1}$ such that

$$\operatorname{spt} \widehat{\eta} \subset B(0, 1), \operatorname{spt} \psi \subset B(0, n),$$

$$\sum_{k\in\mathbb{Z}^{n-1}} \eta(x - k) = \sum_{k\in\mathbb{Z}^{n-1}} \psi(x - k) = 1 \text{ for } x \in \mathbb{R}^{n-1}.$$

Define for $y_j \in R^{1/2}\mathbb{Z}^{n-1}$ and $v_j \in R^{-1/2}\mathbb{Z}^{n-1} \cap V_j$,

$$\eta_{y_j}(x) = \eta\left(\frac{x + y_j}{\sqrt{R}}\right), \quad \psi_{v_j}(v) = \psi(\sqrt{R}(v - v_j)), \quad x, v \in \mathbb{R}^{n-1}.$$

Then

$$\widehat{\eta_{y_j}}(v) = R^{(n-1)/2}e^{2\pi i y_j \cdot v}\widehat{\eta}(\sqrt{R}v), \quad \mathrm{spt}\,\widehat{\eta_{y_j}} \subset B(0, 1/\sqrt{R}),$$

$$\mathrm{spt}\,\psi_{v_j} \subset B(v_j, n/\sqrt{R}).$$

Defining g_j on V_j by $g_j(v) = f_j(v, \varphi(v))$, we have

$$\sum_{y_j} \eta_{y_j} = 1 \quad \text{and} \quad g_j = \sum_{v_j} \psi_{v_j} g_j.$$

Thus

$$g_j = \sum_{v_j, y_j} \mathcal{F}^{-1}(\widehat{\psi_{v_j} g_j}\eta_{y_j}).$$

Now the functions p_{y, v_j},

$$p_{y_j, v_j}(x, t) = \int_{V_j} e^{-2\pi i(x \cdot v + t\varphi_j(v))}\mathcal{F}^{-1}(\widehat{\psi_{v_j} g_j}\eta_{y_j})(v)\,dv, \quad (x, t) \in \mathbb{R}^{n-1} \times \mathbb{R},$$

have the required properties. The decomposition $\widehat{f_j}(x, t) = \Sigma_{w_j} p_{w_j}(x, t)$ and the fact $\mathrm{spt}\,\widehat{p_{w_j}} \subset S_j \cap B((v_j, \varphi_j(v_j)), 2/\sqrt{R})$ are easily checked. The fast decay of p_{w_j} outside T_{w_j} follows by stationary phase estimates, more precisely, by Theorem 14.4.

In order to prove Proposition 25.15 (when $\omega = 1$), we need, by (25.38), the estimate

$$\|\Sigma_{w_j} p_{w_1} p_{w_2}\|_{L^q(Q(R))} \lesssim R^\varepsilon(R^{(1-\delta)\alpha} + R^{c\delta}).$$

Here $Q(R)$ is the cube of side-length R centred at the origin. Some pigeonholing arguments and normalizations of the functions p_{w_j} reduce this to

$$\|\Sigma_{w_j \in W_j} p_{w_1} p_{w_2}\|_{L^q(Q(R))} \lesssim R^\varepsilon(R^{(1-\delta)\alpha} + R^{c\delta})\sqrt{\#W_1 \#W_2}$$

for arbitrary subsets W_j of the index sets under the conditions

$$\|p_{w_j}\|_\infty \lesssim R^{(1-n)/4}.$$

Next the cube $Q(R)$ is decomposed into cubes $Q \in \mathcal{Q}$ of side-length $R^{1-\delta}$. Then

$$\|\Sigma_{w_j \in W_j} p_{w_1} p_{w_2}\|_{L^q(Q(R))} \le \Sigma_{Q \in \mathcal{Q}}\|\Sigma_{w_j \in W_j} p_{w_1} p_{w_2}\|_{L^q(Q)},$$

and the problem easily reduces to the estimation of each Q summand on the right hand side. For a fixed $Q \in \mathcal{Q}$ the sum over w_1 and w_2 is split to the local part, denoted $w_1 \sim Q$ and $w_2 \sim Q$, and the far-away part, $w_1 \not\sim Q$ or $w_2 \not\sim Q$. Local here means that for a given w_j the cubes Q with $w_j \sim Q$ are contained in some cube with side-length $\approx R^{1-\delta}$ which allows us to use the induction hypothesis (25.36) to get the upper bound $R^\varepsilon R^{(1-\delta)\alpha} \sqrt{\#W_1 \#W_2}$ for this part of the sum.

The far-away part will be estimated by $R^\varepsilon R^{c\delta} \sqrt{\#W_1 \#W_2}$. First there is the L^1-estimate

$$\left\| \sum_{w_1 \in W_1, w_2 \in W_2, w_1 \not\sim Q \text{ or } w_2 \not\sim Q} p_{w_1} p_{w_2} \right\|_{L^1(Q)} \lesssim R(\#W_1)^{1/2}(\#W_2)^{1/2},$$

which follows by some L^2 estimates for the functions p_{w_j}. Hence by interpolation the required estimate is reduced to showing that for every $Q \in \mathcal{Q}$,

$$\left\| \sum_{w_1 \in W_1, w_2 \in W_2, w_1 \not\sim Q \text{ or } w_2 \not\sim Q} p_{w_1} p_{w_2} \right\|_{L^2(Q)} \lesssim R^{c\delta - (n-2)/4}(\#W_1)^{1/2}(\#W_2)^{1/2}.$$

Next $Q(R)$ is split into cubes $P \in \mathcal{P}$ of side-length $\sqrt{R}$. We are led to show that for any $Q \in \mathcal{Q}$,

$$\sum_{P \in \mathcal{P}, P \subset 2Q} \left\| \sum_{w_j \in W_j, R^\delta P \cap T_{w_j} \neq \varnothing, w_1 \not\sim Q \text{ or } w_2 \not\sim Q} p_{w_1} p_{w_2} \right\|_2^2 \lesssim R^{c\delta - (n-2)/2}(\#W_1)(\#W_2).$$

The reduction to $R^\delta P \cap T_{w_j} \neq \varnothing$ follows from the fast decay of p_{w_j} outside T_{w_j}. Writing

$$\left\| \sum_{w_1 \in U_1, w_2 \in U_2} p_{w_1} p_{w_2} \right\|_2^2 = \sum_{w_1, w_1' \in U_1, w_2, w_2' \in U_2} \int p_{w_1} p_{w_2} \overline{p_{w_1'} p_{w_2'}},$$

and

$$\int p_{w_1} p_{w_2} \overline{p_{w_1'} p_{w_2'}} = \int \widehat{p_{w_1} p_{w_2}} \, \overline{\widehat{p_{w_1'} p_{w_2'}}} = \int (\widehat{p_{w_1}} * \widehat{p_{w_2}}) \overline{\widehat{p_{w_1'}} * \widehat{p_{w_2'}}},$$

the support properties of the Fourier transforms $\widehat{p_{w_j}}$ are used to estimate

$$\left| \int p_{w_1} p_{w_2} \overline{p_{w_1'} p_{w_2'}} \right| \lesssim R^{-(n-2)/2}.$$

Furthermore, the support properties yield that if we fix w_1 and w_2' and if w_1' is such that $\int p_{w_1} p_{w_2} \overline{p_{w_1'} p_{w_2'}} \neq 0$ for some w_2, then v_1' lies in an $R^{-1/2}$-neighbourhood of a smooth hypersurface depending on w_1 and w_2'. The geometry of this surface is well understood because of the initial transversality and curvature assumptions for the surfaces S_j. These and the transversality of the tubes T_{w_1} and T_{w_2} lead to good estimates on the number of indices for which $\int p_{w_1} p_{w_2} \overline{p_{w_1'} p_{w_2'}} \neq 0$, which together with some combinatorial arguments will complete the proof.

25.8 Extension operators

We shall now go to the remaining details of the proof of Theorem 25.7. Recall that $\widehat{f_j}$ means

$$\widehat{f_j}(x, t) = \widehat{f_j \sigma_j}(x, t)$$
$$= \int_{V_j} e^{-2\pi i (x \cdot v + t \varphi_j(v))} f_j(v, \varphi_j(v)) \, dv, \quad (x, t) \in \mathbb{R}^{n-1} \times \mathbb{R}.$$

Just as a change of notation, instead of functions on S_j we can, and shall, as well consider functions on V_j, which we always extend as 0 outside V_j, and we set (the change of sign in the exponential is irrelevant and only for slight later convenience),

$$E_j f_j(x, t) = \int_{V_j} e^{2\pi i (x \cdot v + t \varphi_j(v))} f_j(v) \, dv, \quad (x, t) \in \mathbb{R}^{n-1} \times \mathbb{R}.$$

The operators E_j are called Fourier extension operators. Theorem 25.7 now reads as

Theorem 25.16 *Suppose the assumptions of 25.2 are satisfied. Then*

$$\|E_1 f_1 E_2 f_2\|_{L^q(\omega)} \leq C \|f_1\|_2 \|f_2\|_2 \quad \text{for } f_j \in L^2(V_j), \, j = 1, 2,$$

where the constant C depends only on the structure constants of Section 25.2.

By Theorem 25.10 and Remark 25.6 it will be enough to prove the following localized version: for all $\alpha > 0$,

$$\|E_1 f_1 E_2 f_2\|_{L^q(\omega, Q(x, R))} \lesssim_\alpha R^\alpha \|f_1\|_2 \|f\|_2 \quad \text{for } x \in \mathbb{R}^n, \, R > 1,$$
$$f_j \in L^2(V_j), \, j = 1, 2. \quad (25.39)$$

Here $Q(x, R)$ is the cube with centre x and side-length $2R$. Notice that (25.39) holds for $R > 1$ if and only if it holds for $R > R_\alpha$ with some $R_\alpha > 0$. We shall obtain (25.39) by the induction on scales argument, recall Proposition 25.15. That is, we shall prove

Proposition 25.17 *There is a constant $c > 0$ such the following holds. Suppose that (25.39) holds for some $\alpha > 0$:*

$$\| E_1 f_1 E_2 f_2 \|_{L^q(\omega, Q(x,R))} \leq M_\alpha R^\alpha \| f_1 \|_2 \| f_2 \|_2 \qquad (25.40)$$

for $x \in \mathbb{R}^n$, $R > 1$ and $f_j \in L^2(V_j)$, $j = 1, 2$. Then for all $0 < \delta, \varepsilon < 1$,

$$\| E_1 f_1 E_2 f_2 \|_{L^q(\omega, Q(a,R))} \leq C R^{\max\{\alpha(1-\delta), c\delta\} + \varepsilon} \| f_1 \|_2 \| f_2 \|_2 \qquad (25.41)$$

for $a \in \mathbb{R}^n$, $R > 1$ and $f_j \in L^2(V_j)$, $j = 1, 2$, where the constant C depends only on the structure constants of Section 25.2 and on M_α, δ and ε.

Now we begin the long proof of Proposition 25.17. As stated before, this will complete the proof of Theorem 25.7. Suppose $\alpha > 0$ is such that (25.40) holds. Fix $R > 1$, which we can choose later as big as we want. To prove (25.41) we may assume $a = 0$ and $nR^{-1/2} < \varepsilon_0$. Recall that $\widetilde{V}_j$ is the ε_0-neighbourhood of V_j.

Notation: Until the end of the proof of Proposition 25.17 the notation $\lesssim$ will mean that the implicit constant depends only on the structure constants of Section 25.2 and on M_α, δ and ε. Other dependencies will be denoted with subindices, for example $\lesssim_N$.

25.9 Wavepacket decomposition

Set

$$\mathcal{Y} = R^{1/2}\mathbb{Z}^{n-1},$$
$$\mathcal{V}_j = R^{-1/2}\mathbb{Z}^{n-1} \cap \widetilde{V}_j,$$
$$\mathcal{W}_j = \mathcal{Y} \times \mathcal{V}_j.$$

For each $w_j = (y_j, v_j) \in \mathcal{W}_j$ define

$$T_{w_j} = \{(x,t) : |t| \leq R, |x - (y_j - t\nabla\varphi_j(v_j))| \leq R^{1/2}\}. \qquad (25.42)$$

Then T_{w_j} is a tube with centre $(y_j, 0)$ and direction $(\nabla\varphi_j(v_j), 1)$. Notice that $\#\mathcal{V}_j \lesssim R^{(1-n)/2}$ and for a fixed v_j the tubes T_{y,v_j}, $y \in \mathcal{Y}$, have bounded overlap.

The main tool for the proof of Proposition 25.17 is the following wavepacket decomposition of $E_j f_j$, $j = 1, 2$, in terms of functions which are essentially localized in the tubes T_{w_j} and whose Fourier transforms in x-variable are localized in the balls $B(v_j, CR^{-1/2})$, $v_j \in \mathcal{V}_j$.

Lemma 25.18 *Let C_0 be as (25.4). Let $f_j \in L^2(V_j)$. Then there are functions $p_{w_j} \in L^\infty(\mathbb{R}^n)$ and non-negative constants C_{w_j}, $w_j \in \mathcal{W}_j$, $j = 1, 2$, with the following properties for $(x, t) \in \mathbb{R}^{n-1} \times \mathbb{R}$:*

(i) $E_j f_j(x, t) = \Sigma_{w_j \in \mathcal{W}_j} C_{w_j} p_{w_j}(x, t).$

(ii) $p_{w_j} = E_j(\widehat{p_{w_j}}(\cdot, 0)).$

(iii) $\|p_{w_j}\|_\infty \lesssim R^{(1-n)/4}.$

(iv) $\mathrm{spt}\, \widehat{p_{w_j}}(\cdot, t) \subset B(v_j, 2nR^{-1/2}).$

(v) $\widehat{p_{w_j}}$ *is a measure in* $\mathcal{M}(\mathbb{R}^n)$ *with*

$$\mathrm{spt}\, \widehat{p_{w_j}} \subset S_j \cap \{(x, t) : |x - v_j| \le 2nR^{-1/2}\}$$
$$\subset B((v_j, \varphi(v_j)), 2n(1 + C_0)R^{-1/2}).$$

(vi) $\sum_{w_j \in \mathcal{W}_j} |C_{w_j}|^2 \lesssim \|f_j\|_2^2.$

(vii) *If L is a sufficiently large constant and $|t| \le R$ or $|x - (y_j - t\nabla\varphi_j(v_j))| > LR^{-1/2}|t|$, then*

$$|p_{w_j}(x, t)| \lesssim_N R^{(1-n)/4} \left(1 + \frac{|x - (y_j - t\nabla\varphi_j(v_j))|}{\sqrt{R}}\right)^{-N} \quad \textit{for all } N \in \mathbb{N}.$$

In particular, if $|t| \le R$ and $\lambda \ge 1$,

$$|p_{w_j}(x, t)| \lesssim_\delta R^{-10n} \quad \textit{if} \quad d((x, t), T_{w_j}) \ge R^{\delta + 1/2},$$

$$|p_{w_j}(x, t)| \lesssim (\lambda R)^{-10n} \quad \textit{if} \quad d((x, t), T_{w_j}) \ge \lambda R.$$

(viii) *If $|t| \le R$, then for any $W \subset \mathcal{W}_j$,*

$$\left\|\sum_{w_j \in W} p_{w_j}(\cdot, t)\right\|_2^2 \lesssim \#W.$$

(ix) *The product $p_{w_1} p_{w_2} \in L^2(\mathbb{R}^n)$.*

The notation is not quite correct: p_{w_1} could be different from p_{w_2} although $w_1 = w_2$ and similarly for C_{w_j} but this should not cause any confusion; we prefer not to complicate notation writing p_{1,w_1}, for example.

The main estimates are those for $|t| \le R$. We only need the estimate for $|t| > R$ in (vii) to get (ix), and only to be able to use Plancherel's theorem for $p_{w_1} p_{w_2}$.

Proof We can choose C^∞-functions η and ψ on $\mathbb{R}^{n-1}$ such that $\check{\eta} = \widehat{\eta}$,

$$\mathrm{spt}\, \widehat{\eta} \subset B(0, 1), \mathrm{spt}\, \psi \subset B(0, n),$$

$$\sum_{k \in \mathbb{Z}^{n-1}} \eta(x - k) = \sum_{k \in \mathbb{Z}^{n-1}} \psi(x - k) = 1 \quad \textit{for} \quad x \in \mathbb{R}^{n-1}.$$

By the Poisson summation formula, Corollary 3.20, we can take for η any radial C^∞-function such that $\int \eta = 1$ and spt $\widehat{\eta} \subset B(0, 1)$. For ψ we can take any non-negative C^∞-function of the form $\psi(x) = g(x)/(\sum_{k \in \mathbb{Z}^{n-1}} g(x - k))$, where spt $g \subset B(0, n)$ and $g(x) > 0$ for $x \in [-1, 1]^{n-1}$.

For $y \in \mathcal{Y}$ and $v_j \in \mathcal{V}_j$, define

$$\eta_y(x) = \eta\left(\frac{x + y}{\sqrt{R}}\right), \quad \psi_{v_j}(v) = \psi(\sqrt{R}(v - v_j)), \quad x, v \in \mathbb{R}^{n-1}.$$

Then

$$\widehat{\eta_y}(v) = R^{(n-1)/2} e^{2\pi i y \cdot v} \widehat{\eta}(\sqrt{R}v), \quad \text{spt } \widehat{\eta_y} \subset B(0, 1/\sqrt{R}),$$

$$\text{spt } \psi_{v_j} \subset B(v_j, n/\sqrt{R}). \tag{25.43}$$

We have

$$\sum_{y \in \mathcal{Y}} \eta_y = 1 \quad \text{and} \quad f_j = \sum_{v_j \in \mathbb{R}^{-1/2} \mathbb{Z}^{n-1}} \psi_{v_j} f_j = \sum_{v_j \in \mathcal{V}_j} \psi_{v_j} f_j,$$

since f_j vanishes outside V_j and $nR^{-1/2} < \varepsilon_0$, and so $\psi_{v_j} f_j = 0$ when $v_j \in R^{-1/2}\mathbb{Z}^{n-1} \setminus \mathcal{V}_j$. Thus

$$f_j = \sum_{v_j \in \mathcal{V}_j, y \in \mathcal{Y}} \mathcal{F}^{-1}(\widehat{\psi_{v_j} f_j} \eta_y),$$

whence

$$E_j f_j = \sum_{v_j \in \mathcal{V}_j, y \in \mathcal{Y}} q_{y,v_j}, \tag{25.44}$$

where

$$q_{y,v_j} = E_j(\mathcal{F}^{-1}(\widehat{\psi_{v_j} f_j} \eta_y)),$$

that is,

$$q_{y,v_j}(x, t) = \int_{V_j} e^{2\pi i(x \cdot v + t\varphi_j(v))} \mathcal{F}^{-1}(\widehat{\psi_{v_j} f_j} \eta_y)(v)\, dv, \quad (x, t) \in \mathbb{R}^{n-1} \times \mathbb{R}.$$

Then $q_{y,v_j}(\cdot, 0) = \mathcal{F}^{-1}(\mathcal{F}^{-1}(\widehat{\psi_{v_j} f_j} \eta_y))$, so $\widehat{q_{y,v_j}(\cdot, 0)} = \mathcal{F}^{-1}(\widehat{\psi_v f_j} \eta_y)$ and thus

$$q_{y,v_j} = E_j(\widehat{q_{y,v_j}(\cdot, 0)}). \tag{25.45}$$

We define the Hardy–Littlewood maximal function Mg in $\mathbb{R}^{n-1}$ by

$$Mg(x) = \sup_{r>0} r^{1-n} \int_{B(x,r)} |g|\, d\mathcal{L}^{n-1}.$$

We also set $L(\varphi_j) = 3n\mathrm{Lip}(\nabla\varphi_j)$:

$$|\nabla\varphi_j(x) - \nabla\varphi_j(y)| \leq L(\varphi_j)|x - y|/(3n).$$

Now we show

Lemma 25.19 *Let* $y \in \mathcal{Y}$, $v_j \in \mathcal{V}_j$, $j = 1, 2$. *For all* $(x, t) \in \mathbb{R}^{n-1} \times \mathbb{R}$,

$$|q_{y,v_j}(x, t)| \lesssim M(\widehat{\psi_{v_j} f_j})(-y). \tag{25.46}$$

If $|t| \leq R$ *or* $|x - y + t\nabla\varphi_j(v_j)| \geq 4L(\varphi_j)R^{-1/2}|t|$, *then for any* $N \in \mathbb{N}$,

$$|q_{y,v_j}(x, t)| \lesssim_N M(\widehat{\psi_{v_j} f_j})(-y)\left(1 + \frac{|x - (y - t\nabla\varphi(v_j))|}{\sqrt{R}}\right)^{-N}. \tag{25.47}$$

Proof Since $\mathrm{spt}\,\mathcal{F}^{-1}(\widehat{\psi_{v_j} f_j}\eta_y) \subset \mathrm{spt}\,\psi_{v_j} + \mathrm{spt}\,\mathcal{F}^{-1}(\eta_y) \subset B(v_j, 2n/\sqrt{R})$, we find a C^∞-function $\widetilde{\psi}$ such that $\mathrm{spt}\,\widetilde{\psi} \subset B(0, 3n)$ and $\widetilde{\psi}_{v_j} = 1$ on $\mathrm{spt}\,\mathcal{F}^{-1}(\widehat{\psi_{v_j} f_j}\eta_y)$ where $\widetilde{\psi}_{v_j}(v) = \widetilde{\psi}(\sqrt{R}(v - v_j))$. Set $F_{v_j} = \widehat{\psi_{v_j} f_j}$. Then, changing $\sqrt{R}(v - v_j)$ to v,

$$q_{y,v_j}(x, t) = \int e^{2\pi i(x\cdot v + t\varphi_j(v))}\mathcal{F}_{v_j}^{-1}(\widehat{\psi_{v_j} f_j}\eta_y)(v)\widetilde{\psi}_{v_j}(v)\,dv$$

$$= \iint e^{2\pi i(x\cdot v + t\varphi_j(v) + z\cdot v)}F_{v_j}(z)\eta_y(z)\widetilde{\psi}_{v_j}(v)\,dz\,dv$$

$$= R^{(1-n)/2}\int K(x + z, t)F_{v_j}(z)\eta(R^{-1/2}(z + y))\,dz,$$

where

$$K(x, t) = \int e^{2\pi i(R^{-1/2}x\cdot v + x\cdot v_j + t\varphi_j(R^{-1/2}v + v_j))}\widetilde{\psi}(v)\,dv.$$

Using the fast decay of η we conclude

$$|q_{y,v_j}(x, t)| \lesssim R^{(1-n)/2}\int |F_{v_j}(z)\eta(R^{-1/2}(z + y))|\,dz$$

$$\lesssim R^{(1-n)/2}\int_{B(-y,\sqrt{R})}|F_{v_j}| + \sum_{j=1}^{\infty}R^{(1-n)/2}$$

$$\times \int_{B(-y,2^j\sqrt{R})\setminus B(-y,2^{j-1}\sqrt{R})}2^{-jn}|F_{v_j}|$$

$$\leq M(F_{v_j})(-y) + \sum_{j=1}^{\infty}2^{-j}M(F_{v_j})(-y) = 2M(F_{v_j})(-y),$$

and (25.46) follows.

Suppose that $|t| \leq R$. If $v \in \mathrm{spt}\, \widetilde{\psi}$, then

$$R^{-1/2}|t\nabla\varphi_j(R^{-1/2}v + v_j) - t\nabla\varphi_j(v_j)| \leq \mathrm{Lip}(\nabla\varphi_j)|v| \leq L(\varphi_j).$$

Hence if $|t| \leq R$ and $|x + t\nabla\varphi_j(v_j)| \geq 2L(\varphi_j)R^{1/2}$,

$$|\nabla_v(R^{-1/2}x \cdot v + x \cdot v_j + t\varphi_j(R^{-1/2}v + v_j))| = R^{-1/2}|x + t\nabla\varphi_j(R^{-1/2}v + v_j)|$$
$$\geq R^{-1/2}|x + t\nabla\varphi_j(v_j)| - R^{-1/2}|t\nabla\varphi_j(v_j) - t\nabla\varphi_j(R^{-1/2}v + v_j)|$$
$$\geq R^{-1/2}|x + t\nabla\varphi_j(v_j)|/2.$$

Thus by Theorem 14.4,

$$|K(x,t)| \lesssim_N \left(1 + \frac{|x + t\nabla\varphi(v_j)|}{\sqrt{R}}\right)^{-N}.$$

This holds trivially if $|x + t\nabla\varphi_j(v_j)| < 2L(\varphi_j)R^{1/2}$, so

$$|q_{y,v_j}(x,t)| \lesssim R^{(1-n)/2} \int |K(x + z - y, t)\eta(R^{-1/2}z)F_{v_j}(z - y)|\, dz$$
$$\lesssim R^{(1-n)/2} \int \left(1 + \frac{|a + z|}{\sqrt{R}}\right)^{-N} |\eta(R^{-1/2}z)F_{v_j}(z - y)|\, dz$$

with $a = x - (y - t\nabla\varphi_j(v_j))$.

Thus to prove (25.47) it suffices to show that for $\lambda > 1, a \in \mathbb{R}^{n-1}, F \in L^1_{loc}(\mathbb{R}^{n-1})$,

$$I := \lambda^{1-n} \int \left(1 + \frac{|a + z|}{\lambda}\right)^{-N} |\eta(z/\lambda)F(z)|\, dz \lesssim_N \left(1 + \frac{|a|}{\lambda}\right)^{-N} MF(0).$$
$$(25.48)$$

If $|a| \leq 2\lambda$, we get as above,

$$I \leq \lambda^{1-n} \int |\eta(z/\lambda)F(z)|\, dz \lesssim MF(0) \leq 3^N \left(1 + \frac{|a|}{\lambda}\right)^{-N} MF(0).$$

Thus we may assume that $|a| > 2\lambda$ so that $(1 + \frac{|a|}{\lambda})^{-N} \approx (\frac{|a|}{\lambda})^{-N}$. We have

$$I = \sum_{k=0}^{\infty} I_k,$$

where

$$I_0 = \lambda^{1-n} \int_{B(\lambda)} \left(1 + \frac{|a + z|}{\lambda}\right)^{-N} |\eta(z/\lambda)F(z)|\, dz,$$

$$I_k = \lambda^{1-n} \int_{B(2^k\lambda)\setminus B(2^{k-1}\lambda)} \left(1 + \frac{|a + z|}{\lambda}\right)^{-N} |\eta(z/\lambda)F(z)|\, dz, \quad k = 1, 2, \ldots.$$

If $|a| \leq 2^{k+1}\lambda$, we use the rapid decay of η, $|\eta(z/\lambda)| \lesssim_N 2^{-k(N+n)}$ for $z \in B(2^k\lambda) \setminus B(2^{k-1}\lambda)$, to get for $k \geq 1$,

$$I_k \leq \lambda^{1-n} \int_{B(2^k\lambda) \setminus B(2^{k-1}\lambda)} |\eta(z/\lambda)F(z)| \, dz$$

$$\lesssim_N 2^{-k-kN}(2^k\lambda)^{1-n} \int_{B(2^k\lambda)} |F(z)| \, dz \leq 2^{-k}2^N(|a|/\lambda)^{-N} MF(0).$$

If $|a| > 2^{k+1}\lambda$ and $z \in B(2^k\lambda) \setminus B(2^{k-1}\lambda)$, then $|a + z| \geq |a|/2$ and $|\eta(z/\lambda)| \lesssim 2^{-kn}$, whence

$$I_k \leq \lambda^{1-n}2^N(|a|/\lambda)^{-N} \int_{B(2^k\lambda) \setminus B(2^{k-1}\lambda)} |\eta(z/\lambda)F(z)| \, dz$$

$$\lesssim_N 2^{-k}2^N(|a|/\lambda)^{-N}(2^k\lambda)^{1-n} \int_{B(2^k\lambda)} |F(z)| \, dz \leq 2^{-k}2^N(|a|/\lambda)^{-N} MF(0).$$

Also

$$I_0 \lesssim 2^N(|a|/\lambda)^{-N}\lambda^{1-n} \int_{B(\lambda)} |F(z)| \, dz \leq 2^N(|a|/\lambda)^{-N} MF(0).$$

Summing over k gives (25.48) and proves (25.47) for $|t| \leq R$.

To prove the remaining part of (25.47) assume that $|t| > R$ and $|x - y + t\nabla\varphi_j(v_j)| \geq 4L(\varphi_j)R^{-1/2}|t|$. Then $|x - y + t\nabla\varphi_j(v_j)| \geq 4L(\varphi_j)R^{1/2}$. We have again as above,

$$q_{y,v_j}(x, t) = R^{(1-n)/2} \int K(x - y + z, t)\eta(R^{-1/2}z)F_{v_j}(z - y) \, dz,$$

where

$$K(x - y + z, t) = \int e^{2\pi i(R^{-1/2}(x-y+z)\cdot v + (x-y+z)\cdot v_j + t\varphi_j(R^{-1/2}v + v_j))} \widetilde{\psi}(v) \, dv.$$

Now we have for $v \in \text{spt } \widetilde{\psi}$,

$$R^{-1/2}|t\nabla\varphi_j(R^{-1/2}v + v_j) - t\nabla\varphi_j(v_j)| \leq L(\varphi_j)R^{-1}|t|.$$

If $|x - y + z + t\nabla\varphi_j(v_j)| \geq |x - y + t\nabla\varphi_j(v_j)|/2$,

$$|\nabla_v(R^{-1/2}(x - y + z) \cdot v + (x - y + z) \cdot v_j + t\varphi_j(R^{-1/2}v + v_j))|$$
$$= R^{-1/2}|x - y + z + t\nabla\varphi_j(R^{-1/2}v + v_j)|$$
$$\geq R^{-1/2}|x - y + z + t\nabla\varphi_j(v_j)| - R^{-1/2}|t\nabla\varphi_j(R^{-1/2}v + v_j) - t\nabla\varphi_j(v_j)|$$
$$\geq R^{-1/2}|x - y + z + t\nabla\varphi_j(v_j)| - L(\varphi_j)R^{-1}|t|$$
$$\geq R^{-1/2}|x - y + t\nabla\varphi_j(v_j)|/4,$$

whence

$$|K(x - y + z, t)| \lesssim_N \left(1 + \frac{|x - y + t\nabla\varphi(v_j)|}{\sqrt{R}}\right)^{-N}.$$

This gives

$$|q_{y,v_j}(x, t)|$$

$$\lesssim R^{(1-n)/2} \int |K(x + z - y, t)\eta(R^{-1/2}z)F_{v_j}(z - y)| \, dz$$

$$\lesssim R^{(1-n)/2} \int \left(1 + \frac{|x - y + t\nabla\varphi_j(v_j)|}{\sqrt{R}}\right)^{-N} |\eta(R^{-1/2}z)F_{v_j}(z - y)| \, dz$$

$$+ R^{(1-n)/2} \int_{B(y-x-t\nabla\varphi_j(v_j), |y-x-t\nabla\varphi_j(v_j)|/2)} |\eta(R^{-1/2}z)F_{v_j}(z - y)| \, dz.$$

The first term is dominated by $M(\widehat{\psi_{v_j} f_j})(-y)(1 + \frac{|x-(y-t\nabla\varphi_j(v_j))|}{\sqrt{R}})^{-N}$ as in the case $|t| \leq R$. Setting $u = y - x - t\nabla\varphi_j(v_j)$ we have $|u| \gtrsim \sqrt{R}$ and so by the fast decay of η we get for the second term,

$$R^{(1-n)/2} \int_{B(u,|u|/2)} |\eta(R^{-1/2}z)F_{v_j}(z - y)| \, dz$$

$$\lesssim R^{(1-n)/2}(|u|/\sqrt{R})^{-N+1-n} \int_{B(0,2|u|)} |F_{v_j}(z - y)| \, dz$$

$$\leq (|u|/\sqrt{R})^{-N} M F_{v_j}(-y).$$

$\square$

We shall now show that

$$C_{y,v_j} = R^{(n-1)/4} M(\widehat{\psi_v f_j})(-y), \quad p_{y,v_j} = q_{y,v_j}/C_{y,v_j},$$

satisfy the claims of Lemma 25.18. Here $p_{y,v_j} = 0$ if $C_{y,v_j} = 0$, that is, if $\psi_v f_j = 0$. It follows for example from Lemma 25.20 below that $M(\widehat{\psi_v f_j})(-y)$ is finite.

First, (i) is clear by (25.44) and (ii) follows from (25.45). (iii) follows from (25.46). To see (iv) note that

$$\widehat{q_{y,v_j}(\cdot, t)}(v) = e^{2\pi i t\varphi_j(v)}\mathcal{F}^{-1}(\widehat{\psi_{v_j} f_j}\eta_y)(v) = e^{2\pi i t\varphi_j(v)}(\psi_{v_j} f_j) * \widecheck{\eta}_y(v),$$

so by (25.43)

$$\text{spt } \widehat{q_{y,v_j}(\cdot, t)} \subset \text{spt}(\psi_{v_j} f) + \text{spt } \widecheck{\eta}_y \subset B(v_j, n/\sqrt{R}) + B(0, 1/\sqrt{R})$$

from which (iv) follows. By its definition q_{y,v_j} is the inverse transform of the measure v_j for which $\int g \, dv_j = \int g(v, \varphi_j(v))\mathcal{F}^{-1}(\widehat{\psi_{v_j} f_j}\eta_y)(v) \, dv$, from which (v) follows by (25.43) and using (25.4) for the second inclusion.

For (vi) we use:

Lemma 25.20 *If $f \in L^1(\mathbb{R}^n)$ and spt $f \subset B(a, r)$, then $M(\widehat{f})(y) \approx M(\widehat{f})(y')$ when $|y - y'| < 1/r$.*

Proof Let Ψ be a C^∞-function on $\mathbb{R}^n$ such that $\Psi = 1$ on $B(0, 1)$ and spt $\Psi \subset B(0, 2)$. Define $\Psi_{a,r} = \Psi((x - a)/r)$. Then $\Psi_{a,r} = 1$ on spt f, so $\widehat{f} = \widehat{\Psi_{a,r}} * \widehat{f}$, and $\widehat{\Psi_{a,r}}(x) = r^n e^{-2\pi i a \cdot x} \widehat{\Psi}(rx)$. Let $\varrho > 0$. If $2\varrho > 1/r$, then

$$\varrho^{-n} \int_{B(y,\varrho)} |\widehat{f}| \le 3^n (3\varrho)^{-n} \int_{B(y',3\varrho)} |\widehat{f}| \le 3^n M(\widehat{f})(y'). \tag{25.49}$$

Suppose $2\varrho \le 1/r$. We have

$$\varrho^{-n} \int_{B(y,\varrho)} |\widehat{f}| = \varrho^{-n} \int_{B(y,\varrho)} \left| \int \widehat{f}(z) \widehat{\Psi_{a,r}}(x - z) \, dz \right| dx$$

$$\le \varrho^{-n} r^n \int |\widehat{f}(z)| \int_{B(y,\varrho)} |\widehat{\Psi}(r(x - z))| \, dx \, dz$$

$$\lesssim r^n \int_{B(y',2/r)} |\widehat{f}(z)| \, dz + \varrho^{-n} r^n \sum_{k=1}^{\infty} \int_{B(y',2^{k+1}/r) \setminus B(y',2^k/r)} |\widehat{f}(z)|$$

$$\times \int_{B(y,\varrho)} |\widehat{\Psi}(r(x - z))| \, dx \, dz.$$

The first summand is $\le 2^n M(\widehat{f})(y')$. In the second if $x \in B(y, \varrho)$ and $z \in B(y', 2^{k+1}/r) \setminus B(y', 2^k/r)$, then

$$r|x - z| \ge r|y' - z| - r|y' - y| - r|y - x| \ge 2^k - 1 - r\varrho \ge 2^k - 3/2 \ge 2^{k-2},$$

whence $|\Psi(r(x - z))| \lesssim 2^{-(n+1)k}$. Therefore

$$\varrho^{-n} r^n \int_{B(y',2^{k+1}/r) \setminus B(y',2^k/r)} |\widehat{f}(z)| \int_{B(y,\varrho)} |\widehat{\Psi}(r(x - z))| \, dx \, dz$$

$$\lesssim 2^{-k} (2^{k+1}/r)^{-n} \int_{B(y',2^{k+1}/r)} |\widehat{f}(z)| \, dz \le 2^{-k} M(\widehat{f})(y').$$

Summing over k we get

$$\varrho^{-n} \int_{B(y,\varrho)} |\widehat{f}| \lesssim M(\widehat{f})(y'). \tag{25.50}$$

The lemma follows from (25.49) and (25.50). □

Using Lemma 25.20, the L^2 boundedness of M and Plancherel's theorem, we have

$$\sum_{w\in\mathcal{W}_j} |C_w|^2 \approx \sum_{y\in\mathcal{Y}, v\in\mathcal{V}_j} \int_{B(y,\sqrt{R})} M(\widehat{\psi_v f_j})^2 \lesssim \sum_{v\in\mathcal{V}_j} \int M(\widehat{\psi_v f_j})^2$$

$$\lesssim \sum_{v\in\mathcal{V}_j} \int |\widehat{\psi_v f_j}|^2 = \sum_{v\in\mathcal{V}_j} \int |\psi_v f_j|^2 \lesssim \int |f_j|^2,$$

so that (vi) holds. The first statement of (vii) follows from Lemma 25.19. The second follows from the first.

(viii) follows by Plancherel's theorem, (iv) and (vii): for every $W \subset \mathcal{W}_j$,

$$\Big\| \sum_{w\in W} p_w(\cdot,t) \Big\|_2^2 = \int \Big| \sum_{v\in\mathcal{V}_j} \sum_{y\in\mathcal{Y}:(y,v)\in W} \widehat{p_{y,v}(\cdot,t)} \Big|^2$$

$$\lesssim \sum_{v\in\mathcal{V}_j} \int \Big| \sum_{y\in\mathcal{Y}:(y,v)\in W} \widehat{p_{y,v}(\cdot,t)} \Big|^2 = \sum_{v\in\mathcal{V}_j} \int \Big| \sum_{y\in\mathcal{Y}:(y,v)\in W} p_{y,v}(\cdot,t) \Big|^2$$

$$\lesssim \sum_{v\in\mathcal{V}_j} \sum_{y\in\mathcal{Y}:(y,v)\in W} \int |p_{y,v}(\cdot,t)|^2 \lesssim \#W,$$

where the last two inequalities follow from (vii); the first of them by the bounded overlap of $T_{y,v}$, $y \in \mathcal{Y}$, the second of them since (vii) implies that $\int |p_{y,v}(\cdot,t)|^2 \lesssim 1$.

The function $p_{w_1} p_{w_2}$ is bounded by (vii) and it decays very fast outside the intersection of the sets $\{(x,t) : |x - y_j + t\nabla\varphi_j(v_j)| \geq C_0|t|/\sqrt{R}\}$, $j = 1,2$. It follows from (25.11) that for sufficiently large R this intersection is a bounded set which implies that $p_{w_1} p_{w_2} \in L^2$. $\qquad\square$

25.10 Some pigeonholing

We now assume (25.40), and fix $f_j \in L^2(V_j)$ with $\|f_j\|_2 = 1$ for $j = 1,2$. Then we have the wavepacket representations as in Lemma 25.18. Recall also from (25.8) that

$$q > q_0 = \max\left\{1, \min\left\{\frac{4s}{n+2s-2}, \frac{n+2}{n}\right\}\right\}.$$

To prove Proposition 25.17, it suffices to prove that

$$\Big\| \sum_{w_1\in\mathcal{W}_1} \sum_{w_2\in\mathcal{W}_2} C_{w_1} C_{w_2} p_{w_1} p_{w_2} \Big\|_{L^q(\omega, Q(R))} \lesssim R^\varepsilon (R^{\alpha(1-\delta)} + R^{c\delta}), \qquad (25.51)$$

for some positive constant c, where $Q(R)$ is the cube in $\mathbb{R}^n$ with centre 0 and side-length R.

Below c will always depend on the setting described in 25.2, but we will often increase its value while going on.

We now make some reductions in this sum. First, it is enough to consider w_j for which

$$T_{w_j} \cap 5Q(R) \neq \varnothing \quad \text{for } j = 1, 2. \tag{25.52}$$

To see this split the rest of the sum into three parts where $T_{w_1} \cap 5Q(R) = \varnothing$ and $T_{w_2} \cap 5Q(R) \neq \varnothing$, $T_{w_2} \cap 5Q(R) = \varnothing$ and $T_{w_1} \cap 5Q(R) \neq \varnothing$, and $T_{w_1} \cap 5Q(R) = \varnothing$ and $T_{w_2} \cap 5Q(R) = \varnothing$. They can all be dealt with in the same way, so we only consider the first one. By (iii) and (vi) of Lemma 25.18, $|C_{w_1} C_{w_2} p_{w_2}| \lesssim 1$. The cardinality of $w_2 \in \mathcal{W}_2$ such that $T_{w_2} \cap 5Q(R) \neq \varnothing$ is roughly R^{n-1}. The number of $w_1 \in \mathcal{W}_1$ such that $5^k R < d(T_{w_1}, Q(R)) \leq 5^{k+1} R$ is dominated by $(\sqrt{R})^{n-1}(5^k \sqrt{R})^{n-1} = (5^k R)^{n-1}$. Thus using Lemma 25.18(vii) and (25.9), we get

$$\left\| \sum_{(w_1, w_2) \in \mathcal{W}_1 \times \mathcal{W}_2, T_{w_1} \cap 5Q(R) = \varnothing, T_{w_2} \cap 5Q(R) \neq \varnothing} C_{w_1} C_{w_2} p_{w_1} p_{w_2} \right\|_{L^q(\omega, Q(R))}$$

$$\lesssim R^{n-1} \sum_{k=0}^{\infty} \sum_{w_1 \in \mathcal{W}_1, 5^k R < d(T_{w_1}, Q(R)) \leq 5^{k+1} R} \| p_{w_1} \|_{L^q(\omega, Q(R))}$$

$$\lesssim \sum_{k=0}^{\infty} R^n \#\{w_1 \in \mathcal{W}_1 : 5^k R < d(T_{w_1}, Q(R)) \leq 5^{k+1} R\}(5^k R)^{-10n}(5^k R)^{s/q}$$

$$\lesssim \sum_{k=0}^{\infty} 5^{-k} R^{-7n} < 2R^{-7n}.$$

Secondly, the number of pairs (w_1, w_2) satisfying (25.52) is $\lesssim R^{2(n-1)}$, so the sum over such pairs of the terms $\| C_{w_1} C_{w_2} p_{w_1} p_{w_2} \|_{L^q(\omega, Q(R))}$ such that $|C_{w_1}| \leq R^{-10n}$ or $|C_{w_2}| \leq R^{-10n}$ is $\lesssim R^{-8n}$. Therefore we can assume that for some constant C,

$$R^{-10n} \leq C_{w_j} \leq C. \tag{25.53}$$

From now on we replace the sets $\mathcal{W}_j$ by their subsets which correspond to those w_j for which (25.52) and (25.53) hold. Evidently $\#\mathcal{W}_j \lesssim R^n$ and just to fix an upper bound we assume that $\#\mathcal{W}_j \leq R^{2n}$.

Thirdly, we get rid of the C_{w_j}. The number of dyadic rationals 2^j, $j \in \mathbb{Z}$, in $[R^{-10n}, C]$ is about $\log R$, so the $L^q(\omega, Q(R))$-norm of $\sum C_{w_1} C_{w_2} p_{w_1} p_{w_2}$

over (w_1, w_2) which satisfy (25.52) and (25.53) is $\lesssim$

$$(\log R)^2 \left\| \sum_{w_1 \in \mathcal{W}_1 : \kappa_1 \leq C_{w_1} \leq 2\kappa_1} \sum_{w_2 \in \mathcal{W}_2 : \kappa_2 \leq C_{w_2} \leq 2\kappa_2} C_{w_1} C_{w_2} p_{w_1} p_{w_2} \right\|_{L^q(\omega, Q(R))}$$

for some dyadic rationals κ_1 and κ_2. Writing $\widetilde{p}_{w_j} = (C_{w_j}/\kappa_j)p_{w_j}$ and W_j for the set of $w_j \in \mathcal{W}_j$ for which $\kappa_j \leq C_{w_j} \leq 2\kappa_j$ we have

$$\left\| \sum_{w_1 \in \mathcal{W}_1 : \kappa_1 \leq C_{w_1} \leq 2\kappa_1} \sum_{w_2 \in \mathcal{W}_2 : \kappa_2 \leq C_{w_2} \leq 2\kappa_2} C_{w_1} C_{w_2} p_{w_1} p_{w_2} \right\|_{L^q(\omega, Q(R))}$$

$$= \left\| \sum_{w_1 \in W_1} \sum_{w_2 \in W_2} \widetilde{p}_{w_1} \widetilde{p}_{w_2} \right\|_{L^q(\omega, Q(R))} \kappa_1 \kappa_2 .$$

Since by Lemma 25.18(vi) $\sqrt{\#W_j} \lesssim 1/\kappa_j$ and since the functions $\widetilde{p}_{w_j}$ satisfy all the conditions (ii)–(v) and (vii)–(ix) (we shall not anymore make use of (i) and (vi)), it suffices to show that

$$\left\| \sum_{w_1 \in W_1} \sum_{w_2 \in W_2} p_{w_1} p_{w_2} \right\|_{L^q(\omega, Q(R))} \lesssim R^\varepsilon (R^{\alpha(1-\delta)} + R^{c\delta}) \sqrt{\#W_1 \#W_2}, \quad (25.54)$$

for $W_1 \subset \mathcal{W}_1$, $W_2 \subset \mathcal{W}_2$.

Now we fix $W_1 \subset \mathcal{W}_1$, $W_2 \subset \mathcal{W}_2$ for the rest of the proof. Decompose $Q(R)$ into $R^{\delta n}$ cubes $Q \in \mathcal{Q}$ of side-length $R^{1-\delta}$ (without loss of generality we may assume that R^δ is an integer). Then

$$\left\| \sum_{w_1 \in W_1} \sum_{w_2 \in W_2} p_{w_1} p_{w_2} \right\|_{L^q(\omega, Q(R))} \lesssim \sum_{Q \in \mathcal{Q}} \left\| \sum_{w_1 \in W_1} \sum_{w_2 \in W_2} p_{w_1} p_{w_2} \right\|_{L^q(\omega, Q)} .$$
$$(25.55)$$

We define a relation $\sim$ between $w_j \in W_j$ and $Q \in \mathcal{Q}$ and split the sum into the parts where $w_1 \sim Q$ and $w_2 \sim Q$ and where $w_1 \not\sim Q$ or $w_2 \not\sim Q$. To define this relation we also decompose $Q(R)$ into $R^{n/2}$ cubes $P \in \mathcal{P}$ of side-length $\sqrt{R}$ (assuming that $\sqrt{R}$ too is an integer) . For $P \in \mathcal{P}$, set

$$W_j(P) = \{w_j \in W_j : T_{w_j} \cap R^\delta P \neq \varnothing\}. \quad (25.56)$$

For dyadic integers $1 \leq \kappa_1, \kappa_2 \leq R^{2n}$ set

$$\mathcal{Q}(\kappa_1, \kappa_2) = \{P \in \mathcal{P} : \kappa_1 \leq \#W_1(P) \leq 2\kappa_1, \kappa_2 \leq \#W_2(P) \leq 2\kappa_2\}, \quad (25.57)$$

and for $w_j \in W_j$,

$$\lambda(w_j, \kappa_1, \kappa_2) = \#\{P \in \mathcal{Q}(\kappa_1, \kappa_2) : T_{w_j} \cap R^\delta P \neq \varnothing\}, \quad (25.58)$$

and for dyadic integers $1 \leq \lambda \leq R^{2n}$,

$$W_j(\lambda, \kappa_1, \kappa_2) = \{w_j \in W_j : \lambda \leq \lambda(w_j, \kappa_1, \kappa_2) \leq 2\lambda\}. \tag{25.59}$$

For dyadic integers $1 \leq \lambda, \kappa_1, \kappa_2 \leq R^{2n}$ and $w_j \in W_j(\lambda, \kappa_1, \kappa_2)$ we choose a cube $Q(w_j, \lambda, \kappa_1, \kappa_2) \in \mathcal{Q}$ which maximizes the quantity

$$\#\{P \in \mathcal{Q}(\kappa_1, \kappa_2) : T_{w_j} \cap R^\delta P \neq \varnothing, P \cap Q \neq \varnothing\}$$

among the cubes $Q \in \mathcal{Q}$. Since $\#\mathcal{Q} = R^{n\delta}$, it follows that

$$\#\{P \in \mathcal{Q}(\kappa_1, \kappa_2) : T_{w_j} \cap R^\delta P \neq \varnothing, P \cap Q(w_j, \lambda, \kappa_1, \kappa_2) \neq \varnothing\} \geq \lambda R^{-n\delta}. \tag{25.60}$$

We define for $w_j \in W_j$ and $Q \in \mathcal{Q}$,

$$w_j \sim Q \text{ if for some dyadic integers } \lambda, \kappa_1, \kappa_2 \in [1, R^{2n}],$$
$$Q \cap 10Q(w_j, \lambda, \kappa_1, \kappa_2) \neq \varnothing.$$

There are roughly $(\log R)^3$ dyadic triples $(\lambda, \kappa_1, \kappa_2) \in [1, R^{2n}]^3$, so for all $w_j \in W_j$,

$$\#\{Q \in \mathcal{Q} : w_j \sim Q\} \lesssim R^\varepsilon.$$

Thus by (ii) and (viii) of Lemma 25.18, the induction hypotheses (25.40) (recall that Q is an $R^{1-\delta}$-cube), Plancherel's theorem and Schwartz's inequality,

$$\sum_{Q \in \mathcal{Q}} \left\| \sum_{w_1 \in W_1, w_1 \sim Q} \sum_{w_2 \in W_2, w_2 \sim Q} p_{w_1} p_{w_2} \right\|_{L^q(\omega, Q)}$$

$$\leq \sum_{Q \in \mathcal{Q}} \left\| E_1 \left(\sum_{w_1 \in W_1, w_1 \sim Q} \widehat{p_{w_1}(\cdot, 0)} \right) E_2 \left(\sum_{w_2 \in W_2, w_2 \sim Q} \widehat{p_{w_2}(\cdot, 0)} \right) \right\|_{L^q(\omega, Q)}$$

$$\lesssim R^{\alpha(1-\delta)} \sum_{Q \in \mathcal{Q}} \left\| \sum_{w_1 \in W_1, w_1 \sim Q} p_{w_1}(\cdot, 0) \right\|_2 \left\| \sum_{w_2 \in W_2, w_2 \sim Q} p_{w_2}(\cdot, 0) \right\|_2$$

$$\lesssim R^{\alpha(1-\delta)} \sum_{Q \in \mathcal{Q}} (\#\{w_1 \in W_1 : w_1 \sim Q\})^{1/2} (\#\{w_2 \in W_2 : w_2 \sim Q\})^{1/2}$$

$$\leq R^{\alpha(1-\delta)} \left(\sum_{w_1 \in W_1} \#\{Q \in \mathcal{Q} : w_1 \sim Q\} \right)^{1/2} \left(\sum_{w_2 \in W_2} \#\{Q \in \mathcal{Q} : w_2 \sim Q\} \right)^{1/2}$$

$$\lesssim R^\varepsilon R^{\alpha(1-\delta)} (\#W_1)^{1/2} (\#W_2)^{1/2}.$$

Hence, recalling (25.54) and (25.55), it is enough to prove

$$\sum_{Q \in \mathcal{Q}} \left\| \sum_{w_1 \in W_1, w_2 \in W_2, w_1 \not\sim Q \text{ or } w_2 \not\sim Q} p_{w_1} p_{w_2} \right\|_{L^q(\omega, Q)} \lesssim R^{c\delta}(\#W_1)^{1/2}(\#W_2)^{1/2}.$$

Since $\#\mathcal{Q} = R^{n\delta}$, it suffices to show for all $Q \in \mathcal{Q}$,

$$\left\| \sum_{w_1 \in W_1, w_2 \in W_2, w_1 \not\sim Q \text{ or } w_2 \not\sim Q} p_{w_1} p_{w_2} \right\|_{L^q(\omega, Q)} \lesssim R^{c\delta}(\#W_1)^{1/2}(\#W_2)^{1/2}. \quad (25.61)$$

25.11 Reduction to $L^2(\mathbb{R}^n)$

We shall prove

$$\left\| \sum_{w_1 \in W_1, w_2 \in W_2, w_1 \not\sim Q \text{ or } w_2 \not\sim Q} p_{w_1} p_{w_2} \right\|_{L^2(Q)} \lesssim R^{c\delta - (n-2)/4}(\#W_1)^{1/2}(\#W_2)^{1/2}.$$

$$(25.62)$$

Let us check that this implies (25.61). Suppose first that $q \geq 4s/(n - 2 + 2s)$; this corresponds to the case $s \leq (n + 2)/2$ in (25.8). Recall also from Remark 25.6 that $q < 2$. Assuming (25.62) we have by Hölder's inequality and (25.9),

$$\left\| \sum_{w_1 \in W_1, w_2 \in W_2, w_1 \not\sim Q \text{ or } w_2 \not\sim Q} p_{w_1} p_{w_2} \right\|_{L^q(\omega, Q)}$$

$$\leq \left\| \sum_{w_1 \in W_1, w_2 \in W_2, w_1 \not\sim Q \text{ or } w_2 \not\sim Q} p_{w_1} p_{w_2} \right\|_{L^2(Q)} \left(\int_Q \omega^{2/(2-q)} \right)^{(2-q)/(2q)}$$

$$\lesssim R^{c\delta - (n-2)/4}(\#W_1)^{1/2}(\#W_2)^{1/2} \left(\int_Q \omega^{2/(2-q)} \right)^{(2-q)/(2q)}$$

$$\leq R^{c\delta - (n-2)/4}(\#W_1)^{1/2}(\#W_2)^{1/2} \left(\int_Q \omega \right)^{(2-q)/(2q)}$$

$$\lesssim R^{c\delta - (n-2)/4 + (s(2-q)/(2q))(1-\delta)}(\#W_1)^{1/2}(\#W_2)^{1/2}$$

$$\leq R^{c\delta - (n-2)/4 + s(2-q)/(2q)}(\#W_1)^{1/2}(\#W_2)^{1/2} \leq R^{c\delta}(\#W_1)^{1/2}(\#W_2)^{1/2},$$

since $s(2 - q)/(2q) \leq (n - 2)/4$ when $q \geq 4s/(n - 2 + 2s)$.

Next we show that (25.62) implies (25.61) for $(n + 2)/n \leq q \leq 2$. This settles the remaining case. Here we only use that $\|\omega\|_\infty \leq 1$, that is, we switch to Lebesgue measure without making use of the fact that $\int_{B(x,r)} \omega$ is much smaller than $\mathcal{L}^n(B(x,r))$ when r is large and $s < n$. We decompose the sum into the parts $w_1 \not\sim Q$ and $w_2 \not\sim Q$, $w_1 \sim Q$ and $w_2 \not\sim Q$, and $w_1 \not\sim Q$ and $w_2 \sim Q$. They can all be treated in the same way and we consider only the first one. By Schwartz's inequality and part (viii) of Lemma 25.18,

$$\left\| \sum_{w_1 \in W_1, w_2 \in W_2, w_1 \not\sim Q \text{ and } w_2 \not\sim Q} p_{w_1} p_{w_2} \right\|_{L^1(Q)}$$

$$\lesssim \left\| \sum_{w_1 \in W_1, w_1 \not\sim Q} p_{w_1} \right\|_{L^2(Q)} \left\| \sum_{w_2 \in W_2, w_2 \not\sim Q} p_{w_2} \right\|_{L^2(Q)}$$

$$\leq \left(\int_{-R}^{R} \int_{\mathbb{R}^{n-1}} \left| \sum_{w_1 \in W_1, w_1 \not\sim Q} p_{w_1}(x,t) \right|^2 dx\, dt \right)^{1/2}$$

$$\times \left(\int_{-R}^{R} \int_{\mathbb{R}^{n-1}} \left| \sum_{w_2 \in W_2, w_2 \not\sim Q} p_{w_2}(x,t) \right|^2 dx\, dt \right)^{1/2}$$

$$\lesssim R(\#W_1)^{1/2}(\#W_2)^{1/2}.$$

Thus we have the L^1 estimate

$$\left\| \sum_{w_1 \in W_1, w_2 \in W_2, w_1 \not\sim Q \text{ or } w_2 \not\sim Q} p_{w_1} p_{w_2} \right\|_{L^1(Q)} \lesssim R(\#W_1)^{1/2}(\#W_2)^{1/2}.$$

This with (25.62) and the inequality $\|g\|_q \leq \|g\|_2^{2(q-1)/q} \|g\|_1^{(2-q)/q}$, which is an immediate consequence of Hölder's inequality, yields that the left hand side of (25.61) is bounded by $\sqrt{\#W_1 \#W_2}$ times

$$R^{(c\delta - (n-2)/4)2(q-1)/q} R^{(2-q)/q} = R^{c\delta 2(q-1)/q} R^{-((n-2)/4)2(q-1)/q + (2-q)/q}$$

$$\leq R^{c\delta 2(q-1)/q},$$

where the last inequality holds since $-((n-2)/2)(q-1)/q + (2-q)/q \leq 0$ due to the assumption $q \geq (n+2)/n$. Since we are allowed to change the value of c, (25.61) follows.

For (25.62) it is enough to show that

$$\sum_{P \in \mathcal{P}, P \subset 2Q} \int_P \left| \sum_{w_j \in W_j, w_1 \not\sim Q \text{ or } w_2 \not\sim Q} p_{w_1} p_{w_2} \right|^2 \lesssim R^{c\delta - (n-2)/2} \#W_1 \#W_2.$$

Recall the definitions of $W_j(P)$, $Q(\kappa_1, \kappa_2)$ and $W_j(\lambda, \kappa_1, \kappa_2)$ from (25.56), (25.57) and (25.59). If $w_j \notin W_j(P)$ for $j = 1$ or $j = 2$, then $|p_{w_1} p_{w_2}| \lesssim R^{-10n}$ on P by parts (iii) and (vii) of Lemma 25.18. Since $\#(W_1 \times W_2) \lesssim R^{2n}$, we have on P,

$$\sum_{\substack{w_j \in W_j, w_1 \notin W_1(P) \text{ or } w_2 \notin W_2(P)}} |p_{w_1} p_{w_2}| \lesssim R^{-8n}.$$

Writing

$$\sum_{\substack{w_j \in W_j, w_1 \not\sim Q \text{ or } w_2 \not\sim Q}} p_{w_1} p_{w_2} = g + h,$$

where g consists of the terms for which $w_j \in W_j(P)$ for $j = 1$ and $j = 2$ and h consists of the rest, we have $|g| \lesssim R^{2n}$, $|h| \lesssim R^{-8n}$ on P and

$$\int_P |g + h|^2 \leq \int_P |g|^2 + 2 \int_P |gh| + \int_P |h|^2 \lesssim \int_P |g|^2 + R^{-5n}.$$

There are $\lesssim R^n$ cubes $P \in \mathcal{P}$ with $P \subset 2Q$, so it suffices to show that

$$\sum_{P \in \mathcal{P}, P \subset 2Q} \left\| \sum_{\substack{w_j \in W_j(P), w_1 \not\sim Q \text{ or } w_2 \not\sim Q}} p_{w_1} p_{w_2} \right\|_{L^2(P)}^2 \lesssim R^{c\delta - (n-2)/2} \# W_1 \# W_2.$$

Pigeonholing as before we can reduce to sum over $P \in Q(\kappa_1, \kappa_2)$, $P \subset 2Q$, for some dyadic integers $\kappa_1, \kappa_2 \in [1, R^{2n}]$. By further pigeonholing we can replace $W_j(P)$ by $W_j(P) \cap W_j(\lambda_j, \kappa_1, \kappa_2)$ for some dyadic integers $\lambda_1, \lambda_2 \in [1, R^{2n}]$. Let us put

$$W_j^{\not\sim Q}(P, \lambda, \kappa_1, \kappa_2) = \{w_j \in W_j(P) \cap W_j(\lambda, \kappa_1, \kappa_2) : w_j \not\sim Q\},$$

and for $U_j \subset W_j$,

$$U_j(P) = \{w_j \in U_j : T_{w_j} \cap R^\delta P \neq \varnothing\}.$$

Breaking the sum over $w_1 \not\sim Q$ or $w_2 \not\sim Q$ into the three sums over $w_1 \not\sim Q$ and $w_2 \not\sim Q$, $w_1 \not\sim Q$ and $w_2 \sim Q$, and $w_1 \sim Q$ and $w_2 \not\sim Q$, it is enough to show that for any $U_2 \subset W_2$ and any dyadic integers $1 \leq \lambda, \kappa_1, \kappa_2 \leq R^{2n}$,

$$\sum_{P \in Q(\kappa_1, \kappa_2), P \subset 2Q} \left\| \sum_{w_1 \in W_1^{\not\sim Q}(P, \lambda, \kappa_1, \kappa_2)} \sum_{w_2 \in U_2(P)} p_{w_1} p_{w_2} \right\|_{L^2(P)}^2 \lesssim R^{c\delta - (n-2)/2} \# W_1 \# W_2.$$

$$(25.63)$$

25.12 Geometric arguments

Now we approach the stage where geometric properties of the surfaces are used. For this we reduce the open sets V_1 and V_2 to more convenient ones. Recall the setting from Section 25.2. Covering V_j with finitely many small cubes, say of diameter at most ε_0, we may assume that V_j is such a cube. Then we gain that

$$v_1' + v_2' - v_1 \in V_2^* \quad \text{whenever} \quad v_1, v_1' \in \tilde{V}_1 \quad \text{and} \quad v_2' \in \tilde{V}_2,$$

and vice versa with respect to V_1 and V_2. Moreover, when these cubes are sufficiently small

$$|\nabla\varphi_j(v_j') - \nabla\varphi_j(v_j)| < \varepsilon_1 \quad \text{whenever} \quad v_j, v_j' \in \tilde{V}_j, \qquad (25.64)$$

where ε_1 is a small constant that will be specified later. Define for $v_1 \in \tilde{V}_1$, $v_2' \in \tilde{V}_2$,

$$\Phi_{v_1,v_2'} : V_1 \to \mathbb{R}, \Phi_{v_1,v_2'}(v_1') = \varphi_1(v_1) + \varphi_2(v_1' + v_2' - v_1) - \varphi_1(v_1') - \varphi_2(v_2'),$$

and

$$\Pi_{v_1,v_2'} = \{v_1' \in \tilde{V}_1 : \Phi_{v_1,v_2'}(v_1') = 0\}.$$

Then by (25.11),

$$|\nabla\Phi_{v_1,v_2'}(v_1')| = |\nabla\varphi_2(v_1' + v_2' - v_1) - \nabla\varphi_1(v_1')| \geq c_1 > 0. \qquad (25.65)$$

Set for $U_1 \subset W_1$,

$$\mathcal{N}(U_1) = \sup_{v_1 \in \tilde{V}_1, v_2' \in \tilde{V}_2} \#\{w_1' \in U_1 : v_1' \in \Pi_{v_1,v_2'}(C_1 R^{-1/2})\}, \qquad (25.66)$$

where again $A(r)$ denotes the r-neighbourhood of the set A. The constant C_1 will be determined below.

Lemma 25.21 *For $P \in \mathcal{P}$ and $U_j \subset W_j(P)$, $j = 1, 2$,*

$$\left\| \sum_{w_1 \in U_1, w_2 \in U_2} p_{w_1} p_{w_2} \right\|_2^2 \lesssim R^{c\delta - (n-2)/2} \mathcal{N}(U_1) \#U_1 \#U_2. \qquad (25.67)$$

Proof Recall first that $p_{w_1} p_{w_2} \in L^2(\mathbb{R}^n)$ by Lemma 25.18(ix). We write

$$\left\| \sum_{w_1 \in U_1, w_2 \in U_2} p_{w_1} p_{w_2} \right\|_2^2 = \sum_{w_1, w_1' \in U_1, w_2' \in U_2} I_{w_1, w_1', w_2'},$$

where

$$I_{w_1, w_1', w_2'} = \sum_{w_2 \in U_2} \int p_{w_1} p_{w_2} \overline{p_{w_1'} p_{w_2'}}.$$

Now

$$\int p_{w_1} p_{w_2} \overline{p_{w_1'} p_{w_2'}} = \int \widehat{p_{w_1} p_{w_2}} \, \overline{\widehat{p_{w_1'} p_{w_2'}}} = \int (\widehat{p_{w_1}} * \widehat{p_{w_2}}) \overline{\widehat{p_{w_1'}} * \widehat{p_{w_2'}}}.$$

By Lemma 25.18(v) the $\widehat{p_{w_j}}$ are measures for which, with $C_2 = 2n(1 + C_0)$,

$$\operatorname{spt} \widehat{p_{w_1}} * \widehat{p_{w_2}} \subset B((v_1 + v_2, \varphi_1(v_1) + \varphi_2(v_2)), 2C_2 R^{-1/2}),$$
$$\operatorname{spt} \widehat{p_{w_1'}} * \widehat{p_{w_2'}} \subset B((v_1' + v_2', \varphi_1(v_1') + \varphi_2(v_2')), 2C_2 R^{-1/2}).$$

Hence $\int p_{w_1} p_{w_2} p_{w_1'} p_{w_2'} = 0$ unless

$$|v_1 + v_2 - (v_1' + v_2')| \le 4C_2 R^{-1/2} \tag{25.68}$$

and

$$|\varphi_1(v_1) + \varphi_2(v_2) - (\varphi_1(v_1') + \varphi_2(v_2'))| \le 4C_2 R^{-1/2}. \tag{25.69}$$

If $I_{w_1, w_1', w_2'} \ne 0$, then there is v_2 such that (25.68) and (25.69) hold. Thus

$$\begin{aligned}
|\Phi_{v_1, v_2'}(v_1')| &= |\varphi_1(v_1) + \varphi_2(v_1' + v_2' - v_1) - (\varphi_1(v_1') + \varphi_2(v_2'))| \\
&\le |\varphi_1(v_1) + \varphi_2(v_2) - (\varphi_1(v_1') + \varphi_2(v_2'))| \\
&\quad + |\varphi_2(v_2) - \varphi_2(v_1' + v_2' - v_1)| \\
&\le (4C_2 + 4C_2\|\nabla\varphi_2\|_\infty) R^{-1/2} \le 8C_2^2 R^{-1/2}.
\end{aligned}$$

Therefore by simple elementary analysis using (25.65) v_1' is contained in $\Pi_{v_1, v_2'}(C_1 R^{-1/2})$ where C_1 depends only on φ_1 and φ_2; this is the constant we use to define $\mathcal{N}(U_1)$ in (25.66). Hence the left hand side of (25.67) is

$$\sum_{w_1 \in U_1} \sum_{w_2' \in U_2} \sum_{w_1' \in U_1, v_1' \in \Pi_{v_1, v_2'}(C_1 R^{-1/2})} \sum_{w_2 \in U_2, v_2 \in B(v_1' + v_2' - v_1, 4C_2 R^{-1/2})} \int p_{w_1} p_{w_2} \overline{p_{w_1'} p_{w_2'}}.$$

Given w_1, w_2', w_1', there are boundedly many points v_2 in the above sum. Since all the tubes T_{w_2} meet $R^\delta P$, there are at most $O(R^{c\delta})$ points w_2 if v_2 is fixed because $y_2 \in \mathcal{Y}$ are $\sqrt{R}$-separated. By the transversality between the tubes T_{w_1} and T_{w_2} (recall (25.11)), the measure of their intersection is $\lesssim R^{n/2}$. By parts (vii) and (iii) of Lemma 25.18 the product $p_{w_1} p_{w_2} \overline{p_{w_1'} p_{w_2'}}$ decays very fast off this intersection and it is uniformly $\lesssim R^{1-n}$. These give easily

$$\left| \int p_{w_1} p_{w_2} \overline{p_{w_1'} p_{w_2'}} \right| \lesssim R^{-(n-2)/2}.$$

Therefore for fixed w_1, w_2', w_1',

$$\sum_{w_2 \in U_2, v_2 \in B(v_1' + v_2' - v_1, 4C_2 R^{-1/2})} \left| \int p_{w_1} p_{w_2} \overline{p_{w_1'} p_{w_2'}} \right| \lesssim R^{c\delta - (n-2)/2}.$$

The lemma follows from this. □

The proof of the theorem will be finished by the following lemma.

Lemma 25.22 *For any dyadic integers* $1 \leq \kappa_1, \kappa_2, \lambda \leq R^{2n}$, $Q \in \mathcal{Q}$, $P \in \mathcal{Q}(\kappa_1, \kappa_2)$, $P \subset 2Q$,

$$\mathcal{N}(W_1^{\not\sim Q}(P, \lambda, \kappa_1, \kappa_2)) \lesssim R^{c\delta} \# W_2 / (\lambda \kappa_2).$$

Let us see that this implies (25.63). For any $P \in \mathcal{Q}(\kappa_1, \kappa_2)$, $\#U_2(P) \leq \#W_2(P) \leq 2\kappa_2$. Using this, Lemma 25.21 and the definitions (25.57)–(25.59), we get

$$\sum_{P \in \mathcal{Q}(\kappa_1,\kappa_2), P \subset 2Q} \left\| \sum_{w_1 \in W_1^{\not\sim Q}(P,\lambda,\kappa_1,\kappa_2)} \sum_{w_2 \in U_2(P)} p_{w_1} p_{w_2} \right\|_{L^2(P)}^2$$

$$\lesssim R^{c\delta - (n-2)/2} \sum_{P \in \mathcal{Q}(\kappa_1,\kappa_2), P \subset 2Q} \mathcal{N}(W_1^{\not\sim Q}(P, \lambda, \kappa_1, \kappa_2)) \# W_1^{\not\sim Q}(P, \lambda, \kappa_1, \kappa_2) \# U_2(P)$$

$$\lesssim R^{2c\delta - (n-2)/2} \frac{\# W_2}{\lambda \kappa_2} \sum_{P \in \mathcal{Q}(\kappa_1,\kappa_2), P \subset 2Q} \# W_1^{\not\sim Q}(P, \lambda, \kappa_1, \kappa_2) \# U_2(P)$$

$$\leq R^{2c\delta - (n-2)/2} \frac{2 \# W_2}{\lambda} \sum_{P \in \mathcal{Q}(\kappa_1,\kappa_2), P \subset 2Q} \# W_1^{\not\sim Q}(P, \lambda, \kappa_1, \kappa_2)$$

$$\leq R^{2c\delta - (n-2)/2} \frac{2 \# W_2}{\lambda} \sum_{w_1 \in W_1(\lambda, \kappa_1, \kappa_2)} \#\{P \in \mathcal{Q}(\kappa_1, \kappa_2) : T_{w_1} \cap R^\delta P \neq \varnothing\}$$

$$\leq 4 R^{2c\delta - (n-2)/2} \# W_1 \# W_2.$$

So we have (25.63) which implies the theorem.

To prove Lemma 25.22 we need to show that for any $v_1 \in \widetilde{V}_1$, $v_2' \in \widetilde{V}_2$ and $P_0 \in \mathcal{Q}(\kappa_1, \kappa_2)$, $P_0 \subset 2Q$,

$$\#\{w_1' \in W_1^{\not\sim Q}(P_0, \kappa_1, \kappa_2, \lambda) : v_1' \in \Pi_{v_1, v_2'}(C_1 R^{-1/2})\} \lesssim R^{c\delta} \frac{\# W_2}{\lambda \kappa_2}. \tag{25.70}$$

Set

$$W_1^{\not\sim Q}(\Pi_{v_1, v_2'}) = \{w_1' \in W_1^{\not\sim Q}(P_0, \kappa_1, \kappa_2, \lambda) : v_1' \in \Pi_{v_1, v_2'}(C_1 R^{-1/2})\}.$$

Let $w_1' \in W_1^{\not\sim Q}(\Pi_{v_1, v_2'})$. Then $T_{w_1'} \cap R^\delta P_0 \neq \varnothing$ and $Q \cap 10Q(w_1', \lambda, \kappa_1, \kappa_2) = \varnothing$. Since $P_0 \subset 2Q$,

$$d(P_0, 2Q(w_1', \lambda, \kappa_1, \kappa_2)) \geq R^{1-\delta},$$

so by (25.60),

$$\#\{P \in \mathcal{Q}(\kappa_1, \kappa_2) : T_{w_1'} \cap R^\delta P \neq \varnothing, d(P, P_0) \geq R^{1-\delta}\} \geq \lambda R^{-n\delta}.$$

Since $\kappa_2 \leq \#W_2(P) \leq 2\kappa_2$ for $P \in \mathcal{Q}(\kappa_1, \kappa_2)$, we get

$$\#\{(P, w_1', w_2) \in \mathcal{Q}(\kappa_1, \kappa_2) \times W_1^{\neq Q}(\Pi_{v_1, v_2'}) \times W_2 :$$
$$T_{w_1'} \cap R^\delta P \neq \varnothing, T_{w_2} \cap R^\delta P \neq \varnothing, d(P, P_0) \geq R^{1-\delta}\} \qquad (25.71)$$
$$\geq \lambda R^{-n\delta} \# W_1^{\neq Q}(\Pi_{v_1, v_2'}) \kappa_2.$$

We shall prove Lemma 25.22 by finding an upper bound for the left hand side of this inequality. This is accomplished by

Lemma 25.23 *Let* $w_2 \in W_2$ *and set*

$$S = \{(P, w_1') \in \mathcal{Q}(\kappa_1, \kappa_2) \times W_1^{\neq Q}(\Pi_{v_1, v_2'}) : T_{w_1'} \cap R^\delta P \neq \varnothing,$$
$$T_{w_2} \cap R^\delta P \neq \varnothing, d(P, P_0) \geq R^{1-\delta}\}.$$

Then $\#S \lesssim R^{c\delta}$.

Combining this with (25.71) yields immediately (25.70) and Lemma 25.22.

Proof of Lemma 25.23 Define

$$C_{v_1, v_2'} = \{(su, s) \in \mathbb{R}^{n-1} \times \mathbb{R} : u \in \nabla\varphi_1(\Pi_{v_1, v_2'}), |s| \leq 2R\}.$$

For $w_1' \in W_1^{\neq Q}(\Pi_{v_1, v_2'})$, we have $v_1' \in \Pi_{v_1, v_2'}(C_1 R^{-1/2})$ and $T_{w_1'} \cap R^\delta P_0 \neq \varnothing$, whence

$$\bigcup_{w_1' \in W_1^{\neq Q}(\Pi_{v_1, v_2'})} T_{w_1'} \subset C_{v_1, v_2'}(C_3 R^{1/2+\delta}) + P_0,$$

for a constant $C_3 \geq 1$. If $(P, w_1') \in S$, then $T_{w_1'} \cap R^\delta P \neq \varnothing$, so

$$P \subset C_{v_1, v_2'}(C_4 R^{1/2+\delta}) + P_0.$$

Since $d(P, P_0) \geq R^{1-\delta}$ and both P and P_0 meet $T_{w_1'}$, we have

$$P \subset C_{v_1, v_2'}(R^{1/2+\delta}, R^{1-\delta}, R, P_0) \quad \text{if} \quad (P, w_1') \in S \quad \text{for some} \quad w_1',$$

where for a suitable constant $c_2 > 0$,

$$C_{v_1, v_2'}(R^{1/2+\delta}, R^{1-\delta}, R, P_0)$$
$$= C_{v_1, v_2'}(C_4 R^{1/2+\delta}) \cap \{(x, t) : c_2 R^{1-\delta} \leq |t| \leq R\} + P_0.$$

Furthermore, $T_{w_2} \cap R^\delta P \neq \varnothing$ if $(P, w_1') \in S$, so

$$\bigcup_{(P, w_1') \in S \text{ for some } w_1'} P \subset R^\delta T_{w_2} \cap C_{v_1, v_2'}(R^{1/2+\delta}, R^{1-\delta}, R, P_0),$$

where

$$R^\delta T_{w_2} = \{(x, t) : |t| \leq R, |x - (y_2 - t\nabla\varphi_2(v_2))| \leq (2 + C_0)R^{1/2+\delta}\},$$

with C_0 as in (25.4). We claim that

$$R^\delta T_{w_2} \cap C_{v_1, v_2'}(R^{1/2+\delta}, R^{1-\delta}, R, P_0) \subset B(y_0, R^{1/2+c\delta}) \qquad (25.72)$$

for some $y_0 \in \mathbb{R}^n$ and some positive constant c. This is a consequence of the fact that the tube T_{w_2} intersects transversally the surface $C_{v_1, v_2'}$ due to our basic assumptions on the functions φ_j. We shall formulate this geometric fact in Lemma 25.24 below. From (25.72) it follows that for each w_1' there are $O(R^{c\delta})$ cubes P with $(P, w_1') \in S$. Since $d(P, P_0) \geq R^{1-\delta}$ the number of possible w_1' for which $T_{w_1'}$ meets both $R^\delta P$ and $R^\delta P_0$ is also $O(R^{c\delta})$. Lemma 25.23 follows from this. $\qquad \square$

We still need to check the transversality stated in (25.72). Part of this will be done by the following lemma. For a smooth hypersurface S in $\mathbb{R}^n$ we denote by $\text{Tan}(S, p)$ the tangent space of S at p considered as an $(n-1)$-dimensional linear subspace of $\mathbb{R}^n$. Then the geometric tangent plane is $\text{Tan}(S, p) + p$.

Lemma 25.24 *Let $c > 0$ and let Π be a smooth hypersurface in $\mathbb{R}^{n-1}$ with $\Pi \subset B(0, 1)$ such that $\Pi = \{v \in \mathbb{R}^{n-1} : \Phi(v) = 0\}$ where Φ is of class C^2 and $|\nabla \Phi(v)| \geq c$ for all $v \in B(0, 1)$. Set*

$$C(\Pi) = \{s(x, 1) \in \mathbb{R}^{n-1} \times \mathbb{R} : 0 \leq s \leq 1, x \in \Pi\}.$$

For $y, v \in \mathbb{R}^{n-1}$, $v \neq 0$, let $l_{y,v}$ be the line in direction $(v, 1)$ through $(y, 0)$, that is,

$$l_{y,v} = \{(x, t) \in \mathbb{R}^{n-1} \times \mathbb{R} : x = y + vt\}.$$

Suppose for some $v \in B(0, 1)$,

$$d(v, \text{Tan}(\Pi, x) + x) \geq c \quad \text{for all } x \in \Pi.$$

Then for any $y \in \mathbb{R}^n$ and $0 < \delta < 1$,

$$l_{y,v}(\delta) \cap C(\Pi)(\delta) \subset B(y_0, C\delta) \qquad (25.73)$$

for some $y_0 \in \mathbb{R}^n$, where C depends only on c and n.

Proof We claim that for all $p \in C(\Pi)$,

$$d((v, 1), \text{Tan}(C(\Pi), p)) \geq c/2. \qquad (25.74)$$

This means that $l_{y,v}$ meets transversally $C(\Pi)$ if it meets $C(\Pi)$ at all. This gives easily (25.73) and proves the lemma. To prove (25.74), let $p = s(x, 1) \in C(\Pi), x \in \Pi$. Note that

$$\text{Tan}(C(\Pi), p) = \text{Tan}(\Pi, x) \times \{0\} + \{t(x, 1) : t \in \mathbb{R}\}.$$

Suppose $d((v, 1), \text{Tan}(C(\Pi, p))) < c/2$. Then there are $u \in \text{Tan}(\Pi, x)$ and $t \in \mathbb{R}$ such that $|(v, 1) - (u + tx, t)| < c/2$. This gives $|v - u - tx| < c/2$ and $|1 - t| < c/2$. Thus $|v - (u + x)| < c$ and so $d(v, \text{Tan}(\Pi, x) + x) < c$ giving a contradiction. This completes the proof of the lemma. □

It remains to see that Lemma 25.24 implies (25.72). Recall that the maps $\nabla \varphi_j$, $j = 1, 2$, are diffeomorphisms. Define

$$\Psi(v) = \varphi_1(v_1) + \varphi_2((\nabla \varphi_1)^{-1}(v) + v_2' - v_1) - \varphi_1((\nabla \varphi_1)^{-1}(v)) - \varphi_2(v_2')$$

when $v \in \nabla \varphi_1(V_1)$. Then

$$\nabla \varphi_1(\Pi_{v_1, v_2'}) \subset \{v \in \mathbb{R}^{n-1} : \Psi(v) = 0\}.$$

By a straightforward computation,

$$\nabla \Psi(\nabla \varphi_1(v_1')) = D(\nabla \varphi_1)(v_1')^{-1}(\nabla \varphi_2(v_1' + v_2' - v_1) - \nabla \varphi_1(v_1')).$$

The normal vector to the surface $\nabla \varphi_1(\Pi_{v_1, v_2'})$ at $\nabla \varphi_1(v_1')$ is parallel to this gradient, so the tangent space is

$$\text{Tan}(\nabla \varphi_1(\Pi_{v_1, v_2'}), \nabla \varphi_1(v_1')) = \{x : x \cdot \nabla \Psi(\nabla \varphi_1(v_1')) = 0\}.$$

Let $w_2 = (y_2, v_2) \in W_2$. Using (25.6) and choosing ε_1 in (25.64) small enough we have

$$|\nabla \Psi(\nabla \varphi_1(v_1')) \cdot (\nabla \varphi_2(v_2) - \nabla \varphi_1(v_1'))|$$
$$= |D(\nabla \varphi_1)(v_1')^{-1}(\nabla \varphi_2(v_1' + v_2' - v_1) - \nabla \varphi_1(v_1')) \cdot (\nabla \varphi_2(v_2) - \nabla \varphi_1(v_1'))|$$
$$\geq |D(\nabla \varphi_1)(v_1')^{-1}(\nabla \varphi_2(v_2) - \nabla \varphi_1(v_1')) \cdot (\nabla \varphi_2(v_2) - \nabla \varphi_1(v_1'))|$$
$$- |D(\nabla \varphi_1)(v_1')^{-1}(\nabla \varphi_2(v_1' + v_2' - v_1) - \nabla \varphi_2(v_2)) \cdot (\nabla \varphi_2(v_2) - \nabla \varphi_1(v_1'))|$$
$$\geq c_0/2.$$

Then

$$d(\nabla \varphi_2(v_2), \text{Tan}(\nabla \varphi_1(\Pi_{v_1, v_2'}), \nabla \varphi_1(v_1')) + \nabla \varphi_1(v_1'))$$
$$\approx |\nabla \Psi(\nabla \varphi_1(v_1')) \cdot (\nabla \varphi_2(v_2) - \nabla \varphi_1(v_1'))|$$
$$\geq c_0/2.$$

We now apply Lemma 25.24 to the surface $\Pi = \nabla \varphi_1(\Pi_{v_1, v_2'})$ with $v = \nabla \varphi_2(v_2)$ and δ replaced by $R^{\delta - 1/2}$. Scaling by R (25.72) follows and the theorem is proven.

For the distance sets we need the following corollary, which was already stated as Theorem 16.5. Recall that

$$A_r = \{x : r - 1 < |x| < r + 1\}.$$

Corollary 25.25 *Let* $(n-2)/2 < s < n, q > \frac{4s}{n+2s-2}, c > 0$ *and* $\mu \in \mathcal{M}(\mathbb{R}^n)$
such that

$$\mu(B(x, \varrho)) \leq \varrho^s \quad \text{for all } x \in \mathbb{R}^n, \quad \varrho > 0. \tag{25.75}$$

There is a constant $\eta_n \in (0, 1)$ *depending only on* n *such that
if* $0 < \eta < \eta_n, r > 1/\eta, f_j \in L^2(\mathbb{R}^n)$, spt $f_j \subset A_r \cap B(v_j, \eta r), |v_j| = r, j = 1, 2, c\eta r \leq d(A_r \cap B(v_1, \eta r)), A_r \cap B(v_2, \eta r)) \leq \eta r$, *then*

$$\|\widehat{f_1}\widehat{f_2}\|_{L^q(\mu)} \leq C(n, s, q, c)\eta^{-1/q}(\eta r)^{n-1-s/q}\|f_1\|_2\|f_2\|_2.$$

Proof Notice that that $\frac{4s}{n+2s-2} > 1$ and so q satisfies (25.8). We may assume that $v_1 = (2\eta r, 0, \ldots, 0, \sqrt{r^2 - (2\eta r)^2})$ and $v_2 = (v_{2,1}, 0, \ldots, 0, v_{2,n})$ with $v_{2,1} \geq (4 + c)\eta r$. Let $T : \mathbb{R}^n \to \mathbb{R}^n$ be the linear map defined by

$$Tx = (\eta r)^{-1}(x_1, \ldots, x_{n-1}, \eta^{-1}x_n).$$

Set

$$S_j = T(S^{n-1}(r) \cap B(v_j, \eta r)), \quad j = 1, 2.$$

If $0 < \eta < \eta_n$ and η_n is sufficiently small, the surfaces S_j satisfy the conditions in the setting of Theorem 25.5, as was already stated in Example 25.4.

Define

$$g_j(x) = f_j(T^{-1}(x)), \quad j = 1, 2.$$

Then

$$\widehat{f_j}(v) = \frac{1}{\det T}\widehat{g_j}(T^{-1}(v)) = \eta(\eta r)^n\widehat{g_j}(T^{-1}(v)), \quad j = 1, 2.$$

Therefore

$$\|\widehat{f_1}\widehat{f_2}\|_{L^q(\mu)} = \eta^2(\eta r)^{2n}\left(\int |\widehat{g_1}(T^{-1}(x))\widehat{g_2}(T^{-1}(x))|^q \, d\mu x\right)^{1/q}$$

$$= \eta^2(\eta r)^{2n}\left(\int |\widehat{g_1}(y)\widehat{g_2}(y)|^q \, dT_\sharp^{-1}\mu y\right)^{1/q}$$

$$= \eta^2(\eta r)^{2n}\eta^{-1/q}(\eta r)^{-s/q}\|\widehat{g_1}\widehat{g_2}\|_{L^q(\nu)}$$

$$= \eta^{2-1/q}(\eta r)^{2n-s/q}\|\widehat{g_1}\widehat{g_2}\|_{L^q(\nu)},$$

where

$$\nu = \eta(\eta r)^s T_\sharp^{-1}\mu.$$

To check the growth condition (25.10) for v, let $z \in \mathbb{R}^n$ and $\varrho > 0$. Then

$$v(B(z, \varrho)) = \eta(\eta r)^s \mu(T(B(z, \varrho)) \lesssim \eta(\eta r)^s \eta^{-1}(\eta r)^{-s} \varrho^s = \varrho^s,$$

where the last estimate follows by covering $T(B(z, \varrho))$ with roughly η^{-1} balls of radius $(\eta r)^{-1}\varrho$ and applying (25.75). Since spt g_j is contained in a $C_1/(\eta^2 r)$-neighbourhood of S_j for some positive constant C_1, we have by Theorem 25.5 and by Lemma 25.11(b),

$$\|\widehat{g_1}\widehat{g_2}\|_{L^q(v)} \lesssim (\eta^2 r)^{-1} \|g_1\|_2 \|g_2\|_2.$$

We have also

$$\|g_j\|_2 = \eta^{-1/2}(\eta r)^{-n/2} \|f_j\|_2, \quad j = 1, 2.$$

Putting all these together we obtain

$$\begin{aligned}
\|\widehat{f_1}\widehat{f_2}\|_{L^q(\mu)} &= \eta^{2-1/q}(\eta r)^{2n-s/q} \|\widehat{g_1}\widehat{g_2}\|_{L^q(v)} \\
&\lesssim \eta^{2-1/q}(\eta r)^{2n-s/q}(\eta^2 r)^{-1} \|g_1\|_2 \|g_2\|_2 \\
&= \eta^{2-1/q}(\eta r)^{2n-s/q}(\eta^2 r)^{-1}\eta^{-1}(\eta r)^{-n} \|f_1\|_2 \|f_2\|_2 \\
&= \eta^{-1/q}(\eta r)^{n-1-s/q} \|f_1\|_2 \|f_2\|_2. \quad\quad \square
\end{aligned}$$

25.13 Multilinear restriction and applications

Recall the bilinear restriction theorem 25.1 in the plane:

$$\|\widehat{f_1}\widehat{f_2}\|_{L^2(\mathbb{R}^2)} \lesssim \|f_1\|_{L^2(S_1)} \|f_2\|_{L^2(S_2)}, \tag{25.76}$$

where S_1 and S_2 are compact smooth transversal curves. We did not need any curvature assumptions for these curves whereas such assumptions are needed for the sharp bilinear restriction theorem in higher dimensions. In this spirit the following local n-linear theorem of Bennett, Carbery and Tao [2006] is a natural analogue of (25.76): for every $\varepsilon > 0$,

$$\|\widehat{f_1} \cdots \widehat{f_n}\|_{L^{2/(n-1)}(B(0,R))} \lesssim R^\varepsilon \|f_1\|_{L^2(S_1)} \cdots \|f_n\|_{L^2(S_n)}, \tag{25.77}$$

for all $R > 1$, where $S_1, \ldots, S_n$ are compact smooth hypersurfaces in $\mathbb{R}^n$ which are transversal in the sense that for all $x_j \in S_j$ their normals $n_j(x_j)$ at x_j span the whole space. To prove this result the authors use Kakeya methods. We have seen that restriction estimates imply Kakeya estimates via Khintchine's inequality. Although Bourgain and Wolff could partially reverse this, any kind of equivalence is lacking in the linear case. But in the multilinear case such an equivalence was established by Bennett, Carbery and Tao which allowed them to prove (25.77). More precisely, they first proved the following Kakeya estimate: for $q > n/(n-1)$ and for every $\varepsilon > 0$,

$$\left\| \sum_{T_1 \in \mathcal{T}_1} \chi_{T_1} \cdots \sum_{T_n \in \mathcal{T}_n} \chi_{T_n} \right\|_{L^{q/n}(\mathbb{R}^n)} \lesssim (\delta^{n/q} \# \mathcal{T}_1) \cdots (\delta^{n/q} \# \mathcal{T}_n), \qquad (25.78)$$

for any transversal families $\mathcal{T}_j$ of δ-tubes. Tranversality here means that the directions of all tubes in $\mathcal{T}_j$ are in a fixed neighbourhood of the basis vector e_j. Different tubes in any $\mathcal{T}_j$ need not be separated, they can even be parallel. Then (25.77) is derived using this Kakeya estimate.

Guth [2010] gave a different proof for these Kakeya estimates and he also established the end-point estimate for $q = n/(n-1)$. The proof uses rather heavy algebraic topology and the polynomial method of Dvir; recall Section 22.6. Carbery and Valdimarsson [2013] gave a proof avoiding algebraic topology and using the Borsuk–Ulam theorem on continuous maps on the sphere instead. See also Guth [2014b] for a short proof for a weaker version of the inequality (25.78).

Bourgain and Guth [2011] used the above results, together with other methods, to improve the restriction estimates in all dimensions greater than 2. For example, in $\mathbb{R}^3$ they showed

$$\|\widehat{f}\|_{L^q(\mathbb{R}^3)} \lesssim \|f\|_{L^\infty(S^2)} \quad \text{for } f \in L^\infty(S^2), \quad q > 33/10.$$

Recall that Tao's bilinear estimate and Theorem 25.9 gave this for $q > 10/3$. They also proved Bochner–Riesz estimates in the same range. Temur [2014] gave further improvements on the restriction exponent in $\mathbb{R}^6$. His method also works in dimensions $n = 3k, k \in \mathbb{N}$, and it is based on the ideas which Bourgain and Guth used in $\mathbb{R}^3$.

More recently, Guth [2014a] improved the restriction estimate in $\mathbb{R}^3$ to $q > 3.25$ using the polynomial method; recall Sections 22.6 and 22.7.

Bennett [2014] has an excellent survey on recent multilinear developments with many other references.

25.14 Further comments

Theorem 25.3 is due to Tao [2003]. In fact, Tao proved his results for paraboloids, but as he says in the paper, the method works for more general surfaces including spheres. Before that Wolff [2001] proved the sharp bilinear restriction theorem for the cone. Many of the ideas in Tao's proof, and presented here, originate in that paper of Wolff, in particular the induction on scales argument. The class of surfaces was further extended by Lee [2006a]. We have mostly followed Lee's presentation. The weighted version of Theorem 25.7 and its application to distance sets is due to Erdoğan [2005].

Tao, Vargas and Vega [1998] proved Theorem 25.8; getting restriction from bilinear restriction. Based on earlier work of Bourgain, Tao and Vargas [2000] proved the localization theorem 25.10.

Lee [2004] used bilinear restriction theorems to obtain improvement for Bochner–Riesz estimates; he proved them for the same range $p > 2(n + 2)/2$ as appeared in Theorem 25.9. This was also surpassed by Bourgain and Guth.

Many other results on bilinear restriction can be found in the above mentioned references and in Tao's [2004] lecture notes.

References

D. R. Adams and L. I. Hedberg [1996] *Function Spaces and Potential Theory*, Springer-Verlag.

G. Alberti, M. Csörnyei and D. Preiss [2005] *Structure of null sets in the plane and applications*, in European Congress of Mathematics, Eur. Math. Soc., Zürich, 3–22.

[2010] *Differentiability of Lipschitz functions, structure of null sets, and other problems*, in Proceedings of the International Congress of Mathematicians. Volume III, Hindustan Book Agency, New Delhi, 1379–1394.

R. Alexander [1975] *Random compact sets related to the Kakeya problem*, Proc. Amer. Math. Soc. **53**, 415–419.

G. Arutyunyants and A. Iosevich [2004] *Falconer conjecture, spherical averages and discrete analogs*, in Towards a Theory of Geometric Graphs, Contemp. Math., 342, Amer. Math. Soc., 15–24.

Y. Babichenko, Y. Peres, R. Peretz, P. Sousi and P. Winkler [2014] *Hunter, Cauchy rabbit, and optimal Kakeya sets*, Trans. Amer. Math. Soc. **366**, 5567–5586.

J-G. Bak, D. M. Oberlin and A. Seeger [2009] *Restriction of Fourier transforms to curves and related oscillatory integrals*, Amer. J. Math. **131**, 277–311.

[2013] *Restriction of Fourier transforms to curves: an endpoint estimate with affine arclength measure*, J. Reine Angew. Math. **682**, 167–205.

J-G. Bak and A. Seeger [2011] *Extensions of the Stein–Tomas theorem*, Math. Res. Lett. **18**, 767–781.

Z. M. Balogh, E. Durand Cartagena, K. Fässler, P. Mattila and J. T. Tyson [2013] *The effect of projections on dimension in the Heisenberg group*, Rev. Mat. Iberoam. **29**, 381–432.

Z. M. Balogh, K. Fässler, P. Mattila and J. T. Tyson [2012] *Projection and slicing theorems in Heisenberg groups*, Adv. Math. **231**, 569–604.

C. Bandt and S. Graf [1992] *Self-similar sets. VII. A characterization of self-similar fractals with positive Hausdorff measure*, Proc. Amer. Math. Soc. **114**, 995–1001.

R. Banuelos and P. J. Méndez-Hernández [2010] *Symmetrization of Lévy processes and applications*, J. Funct. Anal. **258**, 4026–4051.

B. Barany, A. Ferguson and K. Simon [2012] *Slicing the Sierpinski gasket*, Nonlinearity **25**, 1753–1770.

B. Barany and M. Rams [2014] *Dimension of slices of Sierpinski-like carpets*, J. Fractal Geom. **1**, 273–294.

B. Barceló [1985] *On the restriction of the Fourier transform to a conical surface*, Trans. Amer. Math. Soc. **292**, 321–333.

J. A. Barceló, J. Bennett, A. Carbery and K. M. Rogers [2011] *On the dimension of divergence sets of dispersive equations*, Math. Ann. **349**, 599–622.

J. A. Barceló, J. Bennett, A. Carbery, A. Ruiz and M. C. Vilela [2007] *Some special solutions of the Schrödinger equation*, Indiana Univ. Math. J. **56**, 1581–1593.

M. Bateman [2009] *Kakeya sets and directional maximal operators in the plane*, Duke Math. J. **147**, 55–77.

M. Bateman and N. H. Katz [2008] *Kakeya sets in Cantor directions*, Math. Res. Lett. **15**, 73–81.

M. Bateman and A. Volberg [2010] *An estimate from below for the Buffon needle probability of the four-corner Cantor set*, Math. Res. Lett. **17**, 959–967.

W. Beckner, A. Carbery, S. Semmes and F. Soria [1989] *A note on restriction of the Fourier transform to spheres*, Bull. London Math. Soc. **21**, 394–398.

I. Benjamini and Y. Peres [1991] *On the Hausdorff dimension of fibres*, Israel J. Math. **74**, 267–279.

J. Bennett [2014] *Aspects of multilinear harmonic analysis related to transversality*, Harmonic analysis and partial differential equations, Contemporary Math. **612**, 1–28.

J. Bennett, T. Carbery and T. Tao [2006] *On the multilinear restriction and Kakeya conjectures*, Acta Math. **196**, 261–302.

J. Bennett and K. M. Rogers [2012] *On the size of divergence sets for the Schrödinger equation with radial data*, Indiana Univ. Math. J. **61**, 1–13.

J. Bennett and A. Vargas [2003] *Randomised circular means of Fourier transforms of measures*, Proc. Amer. Math. Soc. **131**, 117–127.

M. Bennett, A. Iosevich and K. Taylor [2014] *Finite chains inside thin subsets of* $\mathbb{R}^d$, arXiv:1409.2581.

A. S. Besicovitch [1919] *Sur deux questions d'intégrabilité des fonctions*, J. Soc. Phys-Math. (Perm) **2**, 105–123.

[1928] *On Kakeya's problem and a similar one*, Math. Zeitschrift **27**, 312–320.

[1964] *On fundamental geometric properties of plane line–sets*, J. London Math. Soc. **39**, 441–448.

D. Betsakos [2004] *Symmetrization, symmetric stable processes, and Riesz capacities*, Trans. Amer. Math. Soc. **356**, 735–755.

C. J. Bishop and Y. Peres [2016] *Fractal Sets in Probability and Analysis*, Cambridge University Press.

C. Bluhm [1996] *Random recursive construction of Salem sets*, Ark. Mat. **34**, 51–63.

[1998] *On a theorem of Kaufman: Cantor-type construction of linear fractal Salem sets*, Ark. Mat. **36**, 307–316.

M. Bond, I. Łaba and A. Volberg [2014] *Buffon's needle estimates for rational product Cantor sets*, Amer. J. Math. **136**, 357–391.

M. Bond, I. Łaba and J. Zhai [2013] *Quantitative visibility estimates for unrectifiable sets in the plane*, to appear in Trans. Amer. Math. Soc., arXiv:1306.5469.

M. Bond and A. Volberg [2010] *Buffon needle lands in ε-neighborhood of a 1-dimensional Sierpinski gasket with probability at most* $|\log \varepsilon|^{-c}$, C. R. Math. Acad. Sci. Paris **348**, 653–656.

[2011] *Circular Favard length of the four-corner Cantor set*, J. Geom. Anal. **21**, 40–55.

[2012] *Buffon's needle landing near Besicovitch irregular self-similar sets*, Indiana Univ. Math. J. **61**, 2085–2019.

J. Bourgain [1986] *Averages in the plane over convex curves and maximal operators*, J. Anal. Math. **47**, 69–85.

[1991a] *Besicovitch type maximal operators and applications to Fourier analysis*, Geom. Funct. Anal. **1**, 147–187.

[1991b] L^p-*estimates for oscillatory integrals in several variables*, Geom. Funct. Anal. **1**, 321–374.

[1993] *On the distribution of Dirichlet sums*, J. Anal. Math. **60**, 21–32.

[1994] *Hausdorff dimension and distance sets*, Israel J. Math. **87**, 193–201.

[1995] *Some new estimates on oscillatory integrals*, in Essays on Fourier Analysis in Honor of Elias M. Stein, Princeton University Press, 83–112.

[1999] *On the dimension of Kakeya sets and related maximal inequalities*, Geom. Funct. Anal. **9**, 256–282.

[2003] *On the Erdős–Volkmann and Katz–Tao ring conjectures*, Geom. Funct. Anal. **13**, 334–365.

[2010] *The discretized sum-product and projection theorems*, J. Anal. Math. **112**, 193–236.

[2013] *On the Schrödinger maximal function in higher dimension*, Proc. Steklov Inst. Math. **280**, 46–60.

J. Bourgain and L. Guth [2011] *Bounds on oscillatory integral operators based on multilinear estimates*, Geom. Funct. Anal. **21**, 1239–1235.

J. Bourgain, N. H. Katz and T. Tao [2004] *A sum-product estimate in finite fields, and applications*, Geom. Funct. Anal. **14**, 27–57.

G. Brown and W. Moran [1974] *On orthogonality for Riesz products*, Proc. Cambridge Philos. Soc. **76**, 173–181.

A. M. Bruckner, J. B. Bruckner and B. S. Thomson [1997] *Real Analysis*, Prentice Hall.

A. Carbery [1992] *Restriction implies Bochner–Riesz for paraboloids*, Math. Proc. Cambridge Philos. Soc. **111**, 525–529.

[2004] *A multilinear generalisation of the Cauchy-Schwarz inequality*, Proc. Amer. Math. Soc. **132**, 3141–3152.

[2009] *Large sets with limited tube occupancy*, J. Lond. Math. Soc. (2) **79**, 529–543.

A. Carbery, F. Soria and A. Vargas [2007] *Localisation and weighted inequalities for spherical Fourier means*, J. Anal. Math. **103**, 133–156.

A. Carbery and S. I. Valdimarsson [2013] *The endpoint multilinear Kakeya theorem via the Borsuk–Ulam theorem*, J. Funct. Anal. **264**, 1643–1663.

L. Carleson [1967] *Selected Problems on Exceptional Sets*, Van Nostrand.

[1980] *Some analytic problems related to statistical mechanics*, in Euclidean Harmonic Analysis, Lecture Notes in Math. 779, Springer-Verlag, 5–45.

L. Carleson and P. Sjölin [1972] *Oscillatory integrals and a multiplier problem for the disc*, Studia Math. **44**, 287–299.

V. Chan, I. Łaba and M. Pramanik [2013] *Finite configurations in sparse sets*, to appear in J. Anal. Math., arXiv:1307.1174.

J. Chapman, M. B. Erdoğan, D. Hart, A. Iosevich and D. Koh [2012] *Pinned distance sets, k-simplices, Wolff's exponent in finite fields and sum-product estimates*, Math. Z. **271**, 63–93.

X. Chen [2014a] *Sets of Salem type and sharpness of the L^2-Fourier restriction theorem*, to appear in Trans. Amer. Math. Soc., arXiv:1305.5584.

[2014b] *A Fourier restriction theorem based on convolution powers*, Proc. Amer. Math. Soc. **142**, 3897–3901.

M. Christ [1984] *Estimates for the k-plane tranforms*, Indiana Univ. Math. J. **33**, 891–910.

M. Christ, J. Duoandikoetxea and J. L. Rubio de Francia [1986] *Maximal operators related to the Radon transform and the Calderón-Zygmund method of rotations*, Duke Math. J. **53**, 189–209.

A. Córdoba [1977] *The Kakeya maximal function and the spherical summation of multipliers*, Amer. J. Math. **99**, 1–22.

B. E. J. Dahlberg and C. E. Kenig [1982] *A note on the almost everywhere behavior of solutions to the Schrödinger equation*, in Harmonic Analysis, Lecture Notes in Math. 908, Springer-Verlag, 205–209.

G. David and S. Semmes [1993] *Analysis of and on Uniformly Rectifiable Sets*, Surveys and Monographs 38, Amer. Math. Soc.

R. O. Davies [1952a] *On accessibility of plane sets and differentation of functions of two real variables*, Proc. Cambridge Philos. Soc. **48**, 215–232.

[1952b] *Subsets of finite measure in analytic sets*, Indag. Math. **14**, 488–489.

[1971] *Some remarks on the Kakeya problem*, Proc. Cambridge Philos. Soc. **69**, 417–421.

C. Demeter [2012] *L^2 bounds for a Kakeya type maximal operator in $\mathbb{R}^3$*, Bull. London Math. Soc. **44**, 716–728.

S. Dendrinos and D. Müller [2013] *Uniform estimates for the local restriction of the Fourier transform to curves*, Trans. Amer. Math. Soc. **365**, 3477–3492.

C. Donoven and K.J. Falconer [2014] *Codimension formulae for the intersection of fractal subsets of Cantor spaces*, arXiv:1409.8070.

S. W. Drury [1983] *L^p estimates for the X-ray transform*, Illinois J. Math. **27**, 125–129.

J. Duoandikoetxea [2001] *Fourier Analysis*, Graduate Studies in Mathematics Volume 29, American Mathematical Society.

Z. Dvir [2009] *On the size of Kakeya sets in finite fields*, J. Amer. Math. Soc. **22**, 1093–1097.

Z. Dvir and A. Wigderson [2011] *Kakeya sets, new mergers, and old extractors*, SIAM J. Comput. **40**, 778–792.

G. A. Edgar and C. Miller [2003] *Borel subrings of the reals*, Proc. Amer. Math. Soc. **131**, 1121–1129.

F. Ekström, T. Persson, J. Schmeling [2015] *On the Fourier dimension and a modification*, to appear in J. Fractal Geom., arXiv:1406.1480.

F. Ekström [2014] *The Fourier dimension is not finitely stable*, arXiv:1410.3420.

M. Elekes, T. Keleti and A. Máthé [2010] *Self-similar and self-affine sets: measure of the intersection of two copies*, Ergodic Theory Dynam. Systems **30**, 399–440.

J. S. Ellenberg, R. Oberlin and T. Tao [2010] *The Kakeya set and maximal conjectures for algebraic varieties over finite fields*, Mathematika **56**, 1–25.

M. B. Erdoğan [2004] *A note on the Fourier transform of fractal measures*, Math. Res. Lett. **11**, 299–313.

[2005] *A bilinear Fourier extension problem and applications to the distance set problem*, Int. Math. Res. Not. **23**, 1411–1425.

[2006] *On Falconer's distance set conjecture*, Rev. Mat. Iberoam. **22**, 649–662.

M. B. Erdoğan, D. Hart and A. Iosevich [2013] *Multi-parameter projection theorems with applications to sums-products and finite point configurations in the Euclidean setting*, in Recent Advances in Harmonic Analysis and Applications, Springer Proc. Math. Stat., 25, Springer, 93–103.

M. B. Erdoğan and D. M. Oberlin [2013] *Restricting Fourier transforms of measures to curves in* $\mathbb{R}^2$, Canad. Math. Bull. **56**, 326–336.

P. Erdős [1939] *On a family of symmetric Bernoulli convolutions*, Amer. J. Math. **61**, 974–976.

[1940] *On the smoothness properties of a family of Bernoulli convolutions*, Amer. J. Math. **62**, 180–186.

[1946] *On sets of distances of n points*, Amer. Math. Monthly **53**, 248–250.

P. Erdős and B. Volkmann [1966] *Additive Gruppen mit vorgegebener Hausdorffscher Dimension*, J. Reine Angew. Math. **221**, 203–208.

S. Eswarathasan, A. Iosevich and K. Taylor [2011] *Fourier integral operators, fractal sets, and the regular value theorem*, Adv. Math. **228**, 2385–2402.

[2013] *Intersections of sets and Fourier analysis*, to appear in J. Anal. Math.

L. C. Evans [1998] *Partial Differential Equations*, Graduate Studies in Mathematics 19, Amer. Math. Soc.

L. C. Evans and R. F. Gariepy [1992] *Measure Theory and Fine Properties of Functions*, CRC Press.

K. J. Falconer [1980a] *Continuity properties of k-plane integrals and Besicovitch sets*, Math. Proc. Cambridge Philos. Soc. **87**, 221–226.

[1980b] *Sections of sets of zero Lebesgue measure*, Mathematika **27**, 90–96.

[1982] *Hausdorff dimension and the exceptional set of projections*, Mathematika **29**, 109–115.

[1984] *Rings of fractional dimension*, Mathematika **31**, 25–27.

[1985a] *Geometry of Fractal Sets*, Cambridge University Press.

[1985b] *On the Hausdorff dimension of distance sets*, Mathematika **32**, 206–212.

[1986] *Sets with prescribed projections and Nikodym sets*, Proc. London Math. Soc. (3) **53**, 48–64.

[1990] *Fractal Geometry: Mathematical Foundations and Applications*, John Wiley and Sons.

[1997] *Techniques in Fractal Geometry*, John Wiley and Sons.

[2005] *Dimensions of intersections and distance sets for polyhedral norms*, Real Anal. Exchange **30**, 719–726.

K. J. Falconer, J. Fraser and X. Jin [2014] *Sixty years of fractal projections*, arXiv:1411.3156.

K. J. Falconer and J. D. Howroyd [1996] *Projection theorems for box and packing dimensions*, Math. Proc. Cambridge Philos. Soc. **119**, 287–295.

[1997] *Packing dimensions of projections and dimension profiles*, Math. Proc. Cambridge Philos. Soc. **121**, 269–286.

K. J. Falconer and X. Jin [2014a] *Exact dimensionality and projections of random self-similar measures and sets*, J. London Math. Soc. **90**, 388–412.

[2014b] *Dimension conservation for self-similar sets and fractal percolation*, arXiv:1409.1882.

K. J. Falconer and P. Mattila [1996] *The packing dimension of projections and sections of measures*, Math. Proc. Cambridge Philos. Soc. **119**, 695–713.

[2015] *Strong Marstrand theorems and dimensions of sets formed by subsets of hyperplanes*, arXiv:1503.01284.

K. J. Falconer and T. O'Neil [1999] *Convolutions and the geometry of multifractal measures*, Math. Nachr. **204**, 61–82.

A.-H. Fan and X. Zhang [2009] *Some properties of Riesz products on the ring of p-adic integers*, J. Fourier Anal. Appl. **15**, 521–552.

A. Farkas [2014] *Projections and other images of self-similar sets with no separation condition*, arXiv:1307.2841.

K. Fässler and R. Hovila [2014] *Improved Hausdorff dimension estimate for vertical projections in the Heisenberg group*, Ann. Scuola Norm. Sup. Pisa.

K. Fässler and T. Orponen [2013] *Constancy results for special families of projections*, Math. Proc. Cambridge Philos. Soc. **154**, 549–568.

[2014] *On restricted families of projections in* $\mathbb{R}^3$, Proc. London Math. Soc. (3) **109**, 353–381.

H. Federer [1969] *Geometric Measure Theory*, Springer-Verlag.

C. Fefferman [1970] *Inequalities for strongly singular convolution operators*, Acta Math. **124**, 9–36.

[1971] *The multiplier problem for the ball*, Ann. of Math. (2) **94**, 330–336.

A. Ferguson, J. Fraser, and T. Sahlsten [2015] *Scaling scenery of* $(\times m, \times n)$ *invariant measures*, Adv. Math. **268**, 564–602.

A. Ferguson, T. Jordan and P. Shmerkin [2010] *The Hausdorff dimension of the projections of self-affine carpets*, Fund. Math. **209**, 193–213.

J. Fraser, E. J. Olson and J. C. Robinson [2014] *Some results in support of the Kakeya Conjecture*, arXiv:1407.6689.

J. Fraser, T. Orponen and T. Sahlsten [2014] *On Fourier analytic properties of graphs*, Int. Math. Res. Not. **10**, 2730–2745.

D. Freedman and J. Pitman [1990] *A measure which is singular and uniformly locally uniform*, Proc. Amer. Math. Soc. **108**, 371–381.

H. Furstenberg [1970] *Intersections of Cantor sets and transversality of semigroups*, Problems in Analysis, Sympos. Salomon Bochner, Princeton University, Princeton, N.J. 1969, Princeton University Press, 41–59.

[2008] *Ergodic fractal measures and dimension conservation*, Ergodic Theory Dynam. Systems **28**, 405–422.

J. Garibaldi, A. Iosevich and S. Senger [2011] *The Erdős distance problem*, Student Mathematical Library, 56, American Mathematical Society.

L. Grafakos [2008] *Classical Fourier Analysis*, Springer-Verlag.

[2009] *Modern Fourier Analysis*, Springer-Verlag.

L. Grafakos, A. Greenleaf, A. Iosevich and E. Palsson [2012] *Multilinear generalized Radon transforms and point configurations*, to apper in Forum Math., arXiv:1204.4429.

C. C. Graham and O. C. McGehee [1970] *Essays in Commutative Harmonic Analysis*, Springer-Verlag.

A. Greenleaf and A. Iosevich [2012] *On triangles determined by subsets of the Euclidean plane, the associated bilinear operators and applications to discrete geometry*, Anal. PDE **5**, 397–409.

A. Greenleaf, A. Iosevich, B. Liu and E. Palsson [2013] *A group-theoretic viewpoint on Erdos-Falconer problems and the Mattila integral*, to appear in Rev. Mat. Iberoam., arXiv:1306.3598.

A. Greenleaf, A. Iosevich and M. Mourgoglou [2011] *On volumes determined by subsets of Euclidean space*, arXiv:1110.6790.

A. Greenleaf, A. Iosevich and M. Pramanik [2014] *On necklaces inside thin subsets of* $\mathbb{R}^d$, arXiv:1409.2588.

M. Gromov and L. Guth [2012] *Generalizations of the Kolmogorov–Barzdin embedding estimates*, Duke Math. J. **161**, 2549–2603.

L. Guth [2007] *The width-volume inequality*, Geom. Funct. Anal. **17**, 1139–1179.

[2010] *The endpoint case of the Bennett–Carbery–Tao multilinear Kakeya conjecture*, Acta Math. **205**, 263–286.

[2014] *A restriction estimate using polynomial partitioning*, arXiv:1407.1916.

[2015] *A short proof of the multilinear Kakeya inequality*, Math. Proc. Cambridge Philos. Soc. **158**, 147–153.

L. Guth and N. Katz [2010] *Algebraic methods in discrete analogs of the Kakeya problem*, Adv. Math. **225**, 2828–2839.

[2015] *On the Erdős distinct distance problem in the plane*, Ann. of Math. (2) **181**, 155–190.

M. de Guzmán [1975] *Differentiation of Integrals in* $\mathbb{R}^n$, Lecture Notes in Mathematics 481, Springer-Verlag.

[1981] *Real Variable Methods in Fourier Analysis*, North-Holland.

S. Ham and S. Lee [2014] *Restriction estimates for space curves with respect to general measures*, Adv. Math. **254**, 251–279.

K. Hambrook and I. Łaba [2013] *On the sharpness of Mockenhaupt's restriction theorem*, Geom. Funct. Anal. **23**, 1262–1277.

V. Harangi, T. Keleti, G. Kiss, P. Maga, A. Máthé, P. Mattila and B. Strenner [2013] *How large dimension guarantees a given angle*, Monatshefte für Mathematik **171**, 169–187.

K. E. Hare, M. Parasar and M. Roginskaya [2007] *A general energy formula*, Math. Scand. **101**, 29–47.

K. E. Hare and M. Roginskaya [2002] *A Fourier series formula for energy of measures with applications to Riesz products*, Proc. Amer. Math. Soc. **131**, 165–174.

[2003] *Energy of measures on compact Riemannian manifolds*, Studia Math. **159**, 291–314.

[2004] *The energy of signed measures*, Proc. Amer. Math. Soc. **133**, 397–406.

D. Hart, A. Iosevich, D. Koh and M. Rudnev [2011] *Averages over hyperplanes, sum-product theory in vector spaces over finite fields and the Erdős-Falconer distance conjecture*, Trans. Amer. Math. Soc. **363**, 3255–3275.

V. Havin and B. Jöricke [1995] *The Uncertainty Principle in Harmonic Analysis*, Springer-Verlag.

J. Hawkes [1975] *Some algebraic properties of small sets*, Quart. J. Math. **26**, 195–201.

D. R. Heath-Brown [1987] *Integer sets containing no arithmetic progressions*, J. London Math. Soc. (2) **35**, 385–394.

Y. Heo, F. Nazarov and A. Seeger [2011] *Radial Fourier multipliers in high dimensions*, Acta Math. **206**, 55–92.

M. Hochman [2014] *On self-similar sets with overlaps and inverse theorems for entropy*, Ann. of Math. (2) **180**, 773–822.

M. Hochman and P. Shmerkin [2012] *Local entropy averages and projections of fractal measures*, Ann. of Math. (2) **175**, 1001–1059.

S. Hofmann and A. Iosevich [2005] *Circular averages and Falconer/Erdős distance conjecture in the plane for random metrics*, Proc. Amer. Math. Soc **133**, 133–143.

S. Hofmann, J. M. Martell and I. Uriarte-Tuero [2014] *Uniform rectifiability and harmonic measure II: Poisson kernels in L^p imply uniform rectifiability*, Duke Math. J. **163**, 1601–1654.

S. Hofmann and M. Mitrea and M. Taylor [2010] *Singular integrals and elliptic boundary problems on regular Semmes–Kenig–Toro domains*, Int. Math. Res. Not. **14**, 2567–2865.

L. Hörmander [1973] *Oscillatory integrals and multipliers on FL^p*, Ark. Math. **11**, 1–11.

R. Hovila [2014] *Transversality of isotropic projections, unrectifiability, and Heisenberg groups*, Rev. Math. Iberoam. **30**, 463–476.

R. Hovila, E. Järvenpää, M. Järvenpää and F. Ledrappier [2012a] *Besicovitch-Federer projection theorem and geodesic flows on Riemann surfaces*, Geom. Dedicata **161**, 51–61.

[2012b] *Singularity of projections of 2-dimensional measures invariant under the geodesic flow*, Comm. Math. Phys. **312**, 127–136.

J. E. Hutchinson [1981] *Fractals and self-similarity*, Indiana Univ. Math. J. **30**, 713–747.

A. Iosevich [2000] *Kakeya lectures*, www.math.rochester.edu/people/faculty/iosevich/expositorypapers.html.

[2001] *Curvature, Combinatorics, and the Fourier Transform*, Notices Amer. Math. Soc. **48(6)**, 577–583.

A. Iosevich and I. Łaba [2004] *Distance sets of well-distributed planar point sets*, Discrete Comput. Geom. **31**, 243–250.

[2005] *K-distance sets, Falconer conjecture and discrete analogs*, Integers **5**, 11 pp.

A. Iosevich, M. Mourgoglou and E. Palsson [2011] *On angles determined by fractal subsets of the Euclidean space via Sobolev bounds for bi-linear operators*, to appear in Math. Res. Lett., arXiv:1110.6792.

A. Iosevich, M. Mourgoglou and S. Senger [2012] *On sets of directions determined by subsets of R^d*, J. Anal. Math. **116**, 355–369.

A. Iosevich, M. Mourgoglou and K. Taylor [2012] *On the Mattila-Sjölin theorem for distance sets*, Ann. Acad. Sci Fenn. Ser. A I Math. **37**, 557–562.

A. Iosevich and M. Rudnev [2005] *Non-isotropic distance measures for lattice generated sets*, Publ. Mat. **49**, 225–247.

[2007a] *Distance measures for well-distributed sets*, Discrete Comput. Geom. **38**, 61–80.

[2007b] *The Mattila integral associated with sign indefinite measures*, J. Fourier Anal. Appl. **13**, 167–173.

[2007c] *Erdős distance problem in vector spaces over finite fields*, Trans. Amer. Math. Soc. **359**, 6127–6142.

[2009] *Freiman theorem, Fourier transform, and additive structure of measures*, J. Aust. Math. Soc. **86**, 97–109.

424   References

A. Iosevich, M. Rudnev and I. Uriarte-Tuero [2014] *Theory of dimension for large discrete sets and applications*, Math. Model. Nat. Phenom. **9**, 148–169.

A. Iosevich, E. Sawyer, K. Taylor and I. Uriarte-Tuero [2014] *Measures of polynomial growth and classical convolution inequalities*, arXiv:1410.1436.

A. Iosevich and S. Senger [2010] *On the sharpness of Falconer's distance set estimate and connections with geometric incidence theory*, arXiv:1006.1397.

V. Jarnik [1928] *Zur metrischen theorie der diophantischen approximationen*, Prace Mat. Fiz., **36**(1), 91–106.

[1931] *Über die simultanen diophantischen Approximationen*, Mat. Z. **33**, 505–543.

E. Järvenpää, M. Järvenpää and T. Keleti [2014] *Hausdorff dimension and non-degenerate families of projections*, J. Geom. Anal. **24**, 2020–2034.

E. Järvenpää, M. Järvenpää, T. Keleti and A. Máthé [2011] *Continuously parametrized Besicovitch sets in $\mathbb{R}^n$*, Ann. Acad. Sci. Fenn. Ser. A I Math. **36**, 411–421.

E. Järvenpää, M. Järvenpää, F. Ledrappier and M. Leikas [2008] *One-dimensional families of projections*, Nonlinearity **21**, 453–463.

E. Järvenpää, M. Järvenpää and M. Leikas [2005] *(Non)regularity of projections of measures invariant under geodesic flow*, Comm. Math. Phys. **254**, 695–717.

E. Järvenpää, M. Järvenpää and M. Llorente [2004] *Local dimensions of sliced measures and stability of packing dimensions of sections of sets*, Adv. Math. **183**, 127–154.

M. Järvenpää [1994] *On the upper Minkowski dimension, the packing dimension, and orthogonal projections*, Ann. Acad. Sci. Fenn. Ser. A I Math. Dissertationes **99**, 1–34.

[1997a] *Concerning the packing dimension of intersection measures*, Math. Proc. Cambridge Philos. Soc. **121**, 287–296.

[1997b] *Packing dimension, intersection measures, and isometries*, Math. Proc. Cambridge Philos. Soc. **122**, 483–490.

M. Järvenpää and P. Mattila [1998] *Hausdorff and packing dimensions and sections of measures*, Mathematika **45**, 55–77.

T. Jordan and T. Sahlsten [2013] *Fourier transforms of Gibbs measures for the Gauss map*, to appear in Math. Ann., arXiv:1312.3619.

A. Käenmäki and P. Shmerkin [2009] *Overlapping self-affine sets of Kakeya type*, Ergodic Theory Dynam. Systems **29**, 941–965.

J.-P. Kahane [1969] *Trois notes sur les ensembles parfaits linéaires*, Enseign. Math. **15**, 185–192.

[1970] *Sur certains ensembles de Salem*, Acta Math. Acad. Sci. Hungar. **21**, 87–89.

[1971] *Sur la distribution de certaines séries aléatoires*, in Colloque de Théories des Nombres (Univ. Bordeaux, Bordeaux 1969), Bull. Soc. Math. France Mém. **25**, 119–122.

[1985] *Some Random Series of Functions*, Cambridge University Press, second edition, first published 1968.

[1986] *Sur la dimension des intersections*, in Aspects of Mathematics and Applications, North-Holland Math. Library, 34, 419–430.

[2010] *Jacques Peyriére et les produits de Riesz*, arXiv:1003.4600.

[2013] *Sur un ensemble de Besicovitch*, Enseign. Math. **59**, 307–324.

J.-P. Kahane and R. Salem [1963] *Ensembles parfaits et séries trigonométriques*, Hermann.

S. Kakeya [1917] *Some problems on maxima and minima regarding ovals*, Tohoku Science Reports **6**, 71–88.

N. H. Katz [1996] *A counterexample for maximal operators over a Cantor set of directions*, Math. Res. Lett. **3**, 527-536.

[1999] *Remarks on maximal operators over arbitrary sets of directions*, Bull. London Math. Soc. **31**, 700–710.

N. H. Katz, I. Łaba and T. Tao [2000] *An improved bound on the Minkowski dimension of Besicovitch sets in* $\mathbb{R}^3$, Ann. of Math. (2) **152**, 383–446.

N. H. Katz and T. Tao [1999] *Bounds on arithmetic projections, and applications to the Kakeya conjecture*, Math. Res. Lett. **6**, 625–630.

[2001] *Some connections between Falconer's distance set conjecture and sets of Furstenburg type*, New York J. Math. **7**, 149–187.

[2002a] *New bounds for Kakeya problems*, J. Anal. Math. **87**, 231–263.

[2002b] *Recent progress on the Kakeya conjecture*, Proceedings of the 6th International Conference on Harmonic Analysis and Partial Differential Equations, El Escorial, 2000, Publ. Mat., 161–179.

Y. Katznelson [1968] *An Introduction to Harmonic Analysis*, Dover Publications.

R. Kaufman [1968] *On Hausdorff dimension of projections*, Mathematika **15**, 153–155.

[1969] *An exceptional set for Hausdorff dimension*, Mathematika **16**, 57–58.

[1973] *Planar Fourier transforms and Diophantine approximation*, Proc. Amer. Math. Soc. **40**, 199–204.

[1975] *Fourier analysis and paths of Brownian motion*, Bull. Soc. Math. France **103**, 427–432.

[1981] *On the theorem of Jarnik and Besicovitch*, Acta Arith. **39**, 265–267.

R. Kaufman and P. Mattila [1975] *Hausdorff dimension and exceptional sets of linear transformations*, Ann. Acad. Sci. Fenn. Ser. A I Math. **1**, 387–392.

A. S. Kechris and A. Louveau [1987] *Descriptive Set Theory and the Structure of Sets of Uniqueness*, London Math. Soc. Lecture Notes **128**, Cambridge University Press.

U. Keich [1999] *On* L^p *bounds for Kakeya maximal functions and the Minkowski dimension in* R^2, Bull. London Math. Soc. **31**, 213–221.

T. Keleti [1998] *A 1-dimensional subset of the reals that intersects each of its translates in at most a single point*, Real Anal. Exchange **24**, 843–844.

[2008] *Construction of one-dimensional subsets of the reals not containing similar copies of given patterns*, Anal. PDE, **1**, 29–33.

[2014] *Are lines bigger than line segments?*, arXiv:1409.5992.

T. Kempton [2013] *Sets of beta-expansions and the Hausdorff measure of slices through fractals*, to appear in J. Eur. Math. Soc., arXiv:1307.2091.

C. E. Kenig and T. Toro [2003] *Poisson kernel characterization of Reifenberg flat chord arc domains*, Ann. Sci. École Norm. Sup. **36**, 323–401.

R. Kenyon [1997] *Projecting the one-dimensional Sierpinski gasket*, Israel J. Math. **97**, 221–238.

R. Kenyon and Y. Peres [1991] *Intersecting random translates of invariant Cantor sets*, Invent. Math. **104**, 601–629.

J. Kim [2009] *Two versions of the Nikodym maximal function on the Heisenberg group*, J. Funct. Anal. **257**, 1493-1518.

[2012] *Nikodym maximal functions associated with variable planes in* $\mathbb{R}^3$, Integral Equations Operator Theory **73**, 455–480.

L. Kolasa and T. W. Wolff [1999] *On some variants of the Kakeya problem*, Pacific J. Math. **190**, 111–154.

S. Konyagin and I. Łaba [2006] *Distance sets of well-distributed planar sets for polygonal norms*, Israel J. Math. **152**, 157–179.

T. Körner [2003] *Besicovitch via Baire*, Studia Math. **158**, 65–78.

[2009] *Fourier transforms of measures and algebraic relations on their supports*, Ann. Inst. Fourier (Grenoble) **59**, 1291–1319.

[2011] *Hausdorff and Fourier dimension*, Studia Math. **206**, 37–50.

[2014] *Fourier transforms of distributions and Hausdorff measures*, J. Fourier Anal. Appl. **20**, 547–556.

G. Kozma and A. Olevskii [2013] *Singular distributions, dimension of support, and symmetry of Fourier transform*, Ann. Inst. Fourier (Grenoble) **63**, 1205–1226.

E. Kroc and M. Pramanik [2014a] *Kakeya-type sets over Cantor sets of directions in* $\mathbb{R}^{d+1}$, arXiv:1404.6235.

[2014b] *Lacunarity, Kakeya-type sets and directional maximal operators*, arXiv:1404.6241.

I. Łaba [2008] *From harmonic analysis to arithmetic combinatorics*, Bull. Amer. Math. Soc. (N.S.) **45**, 77–115.

[2012] *Recent progress on Favard length estimates for planar Cantor sets*, arXiv:1212.0247.

[2014] *Harmonic analysis and the geometry of fractals*, Proceedings of the 2014 International Congress of Mathematicians.

I. Łaba and M. Pramanik [2009] *Arithmetic progressions in sets of fractional dimension*, Geom. Funct. Anal. **19**, 429–456.

I. Łaba and T. Tao [2001a] *An improved bound for the Minkowski dimension of Besicovitch sets in medium dimension*, Geom. Funct. Anal. **11**, 773–806.

[2001b] *An X-ray transform estimate in* $\mathbb{R}^n$, Rev. Mat. Iberoam. **17**, 375–407.

I. Łaba and T. W. Wolff [2002] *A local smoothing estimate in higher dimensions*, J. Anal. Math. **88**, 149–171.

I. Łaba and K. Zhai [2010] *The Favard length of product Cantor sets*, Bull. London Math. Soc. **42**, 997–1009.

J. Lagarias and Y. Wang [1996] *Tiling the line with translates of one tile*, Invent. Math. **124**, 341–365.

N. S. Landkof [1972] *Foundations of Modern Potential Theory*, Springer-Verlag.

F. Ledrappier and E. Lindenstrauss [2003] *On the projections of measures invariant under the geodesic flow*, Int. Math. Res. Not. **9**, 511–526.

S. Lee [2004] *Improved bounds for Bochner–Riesz and maximal Bochner–Riesz operators*, Duke Math. J. **122**, 205–232.

[2006a] *Bilinear restriction estimates for surfaces with curvatures of different signs*, Trans. Amer. Math. Soc. **358**, 3511–3533.

[2006b] *On pointwise convergence of the solutions to Schrödinger equations in* $\mathbb{R}^2$, Int. Math. Res. Not., Art. ID 32597, 21 pp.

S. Lee, K. M. Rogers and A. Seeger [2013] *On space-time estimates for the Schrödinger operator*, J. Math. Pures Appl.(9) **99**, 62–85.

S. Lee and A. Vargas [2010] *Restriction estimates for some surfaces with vanishing curvatures*, J. Funct. Anal. **258**, 2884–2909.

Y. Lima and C. G. Moreira [2011] *Yet another proof of Marstrand's theorem*, Bull. Braz. Math. Soc. (N.S.) **42**, 331–345.

E. Lindenstrauss and N. de Saxcé [2014] *Hausdorff dimension and subgroups of SU(2)*, to appear in Israel J. Math.

B. Liu [2014] *On radii of spheres determined by subsets of Euclidean space*, J. Fourier Anal. Appl. **20**, 668–678.

Q. H. Liu, L. Xi and Y. F. Zhao [2007] *Dimensions of intersections of the Sierpinski carpet with lines of rational slopes*, Proc. Edinb. Math. Soc. (2) **50**, 411–427.

R. Lucà and K. M. Rogers [2015] *Average decay of the Fourier transform of measures with applications*, arXiv:1503.00105.

R. Lyons [1995] *Seventy years of Rajchman measures*, J. Fourier Anal. Appl., 363–377, Kahane Special Issue.

P. Maga [2010] *Full dimensional sets without given patterns*, Real Anal. Exchange **36** (2010/11), 79–90.

A. Manning and K. Simon [2013] *Dimension of slices through the Sierpinski carpet*, Trans. Amer. Math. Soc. **365**, 213–250.

J. M. Marstrand [1954] *Some fundamental geometrical properties of plane sets of fractional dimensions*, Proc. London Math. Soc. (3) **4**, 257–302.

[1979] *Packing planes in $\mathbb{R}^3$*, Mathematika **26**, 180–183.

[1987] *Packing circles in the plane*, Proc. London Math. Soc. **55**, 37–58.

P. Mattila [1975] *Hausdorff dimension, orthogonal projections and intersections with planes*, Ann. Acad. Sci. Fenn. Ser. A I Math. **1**, 227–244.

[1981] *Integralgeometric properties of capacities*, Trans. Amer. Math. Soc. **266**, 539–554.

[1984] *Hausdorff dimension and capacities of intersections of sets in n-space*, Acta Math. **152**, 77–105.

[1985] *On the Hausdorff dimension and capacities of intersections*, Mathematika **32**, 213–217.

[1987] *Spherical averages of Fourier transforms of measures with finite energy; dimension of intersections and distance sets*, Mathematika **34**, 207–228.

[1990] *Orthogonal projections, Riesz capacities, and Minkowski content*, Indiana Univ. Math. J. **39**, 185–198.

[1995] *Geometry of Sets and Measures in Euclidean Spaces*, Cambridge University Press.

[2004] *Hausdorff dimension, projections, and the Fourier transform*, Publ. Mat. **48**, 3–48.

[2014] *Recent progress on dimensions of projections*, in Geometry and Analysis of Fractals, D.-J. Feng and K.-S. Lau (eds.), Springer Proceedings in Mathematics and Statistics 88, Springer-Verlag, 283–301.

P. Mattila and P. Sjölin [1999] *Regularity of distance measures and sets*, Math. Nachr. **204**, 157–162.

W. Minicozzi and C. Sogge [1997] *Negative results for Nikodym maximal functions and related oscillatory integrals in curved space*, Math. Res. Lett. **4**, 221–237.

T. Mitsis [1999] *On a problem related to sphere and circle packing*, J. London Math. Soc. (2) **60**, 501–516.

[2002a] *A note on the distance set problem in the plane*, Proc. Amer. Math. Soc. **130**, 1669–1672.

[2002b] *A Stein-Tomas restriction theorem for general measures*, Publ. Math. Debrecen **60**, 89–99.

[2003a] *Topics in Harmonic Analysis*, University of Jyväskylä, Department of Mathematics and Statistics, Report 88.

[2003b] *An optimal extension of Marstrand's plane-packing theorem*, Arch. Math. **81**, 229–232.

[2004a] *(n, 2)-sets have full Hausdorff dimension*, Rev. Mat. Iberoam. **20**, 381–393, Corrigenda: Rev. Mat. Iberoam. **21** (2005), 689–692.

[2004b] *On Nikodym-type sets in high dimensions*, Studia Math. **163**, 189–192.

[2005] *Norm estimates for a Kakeya-type maximal operator*, Math. Nachr. **278**, 1054–1060.

G. Mockenhaupt [2000] *Salem sets and restriction properties of Fourier transforms*, Geom. Funct. Anal. **10**, 1579–1587.

U. Molter, and E. Rela [2010] *Improving dimension estimates for Furstenberg-type sets*, Adv. Math. **223**, 672–688.

[2012] *Furstenberg sets for a fractal set of directions*, Proc. Amer. Math. Soc. **140**, 2753–2765.

[2013] *Small Furstenberg sets*, J. Math. Anal. Appl. **400**, 475–486.

C. G. Moreira [1998] *Sums of regular Cantor sets, dynamics and applications to number theory*, Period. Math. Hungar. **37**, 55–63.

C. G. Moreira and J.–C. Yoccoz [2001] *Stable intersections of regular Cantor sets with large Hausdorff dimensions*, Ann. of Math. (2) **154**, 45–96.

P. Mörters and Y. Peres [2010] *Brownian Motion*, Cambridge University Press.

D. Müller [2012] *Problems of Harmonic Analysis related to finite type hypersurfaces in $\mathbb{R}^3$, and Newton polyhedra*, arXiv:1208.6411.

C. Muscalu and W. Schlag [2013] *Classical and Multilinear Harmonic Analysis I*, Cambridge University Press.

F. Nazarov, Y. Peres and P. Shmerkin [2012] *Convolutions of Cantor measures without resonance*, Israel J. Math. **187**, 93–116.

F. Nazarov, Y. Peres and A. Volberg [2010] *The power law for the Buffon needle probability of the four-corner Cantor set*, Algebra i Analiz **22**, 82–97; translation in St. Petersburg Math. J. **22** (2011), 61–72.

F. Nazarov, X. Tolsa and A. Volberg [2014] *On the uniform rectifiability of AD regular measures with bounded Riesz transform operator: the case of codimension 1*, Acta Math. **213**, 237–321.

E. M. Nikishin [1972] *A resonance theorem and series in eigenfunctions of the Laplace operator*, Izv. Akad. Nauk SSSR Ser. Mat. **36**, 795–813 (Russian).

O. Nikodym [1927] *Sur la measure des ensembles plans dont tous les points sont rectilinearément accessibles*, Fund. Math. **10**, 116–168.

D. M. Oberlin [2006a] *Restricted Radon transforms and unions of hyperplanes*, Rev. Mat. Iberoam. **22**, 977–992.

[2006b] *Packing spheres and fractal Strichartz estimates in $\mathbb{R}^d$ for $d \geq 3$*, Proc. Amer. Math. Soc. **134**, 3201–3209.

[2007] *Unions of hyperplanes, unions of spheres, and some related estimates*, Illinois J. Math. **51**, 1265–1274.

[2012] *Restricted Radon transforms and projections of planar sets*, Canad. Math. Bull. **55**, 815–820.

[2014a] *Exceptional sets of projections, unions of k-planes, and associated transforms*, Israel J. Math. **202**, 331–342.

[2014b] *Some toy Furstenberg sets and projections of the four-corner Cantor set*, Proc. Amer. Math. Soc. **142**, 1209–1215.

D. M. Oberlin and R. Oberlin [2013a] *Unit distance problems*, arXiv:1307.5039.

[2013b] *Application of a Fourier restriction theorem to certain families of projections in R^3*, arXiv:1307.5039.

[2014] *Spherical means and pinned distance sets*, arXiv:1411.0915.

R. Oberlin [2007] *Bounds for Kakeya-type maximal operators associated with k-planes*, Math. Res. Lett. **14**, 87–97.

[2010] *Two bounds for the X-ray transform*, Math. Z. **266**, 623–644.

T. Orponen [2012a] *On the distance sets of self-similar sets*, Nonlinearity **25**, 1919–1929.

[2012b] *On the packing dimension and category of exceptional sets of orthogonal projections*, to appear in Ann. Mat. Pur. Appl., arXiv:1204.2121.

[2014a] *Slicing sets and measures, and the dimension of exceptional parameters*, J. Geom. Anal. **24**, 47–80.

[2014b] *Non-existence of multi-line Besicovitch sets*, Publ. Mat. **58**, 213–220.

[2013a] *Hausdorff dimension estimates for some restricted families of projections in $\mathbb{R}^3$*, arXiv:1304.4955.

[2013b] *On the packing measure of slices of self-similar sets*, arXiv:1309.3896.

[2013c] *On the tube-occupancy of sets in $\mathbb{R}^d$*, to appear in Int. Math. Res. Not., arXiv:1311.7340.

[2014c] *A discretised projection theorem in the plane*, arXiv:1407.6543.

T. Orponen and T. Sahlsten [2012] *Tangent measures of non-doubling measures*, Math. Proc. Cambridge Philos. Soc. **152**, 555–569.

W. Ott, B. Hunt and V. Kaloshin [2006] *The effect of projections on fractal sets and measures in Banach spaces*, Ergodic Theory Dynam. Systems **26**, 869–891.

J. Parcet and K. M. Rogers [2013] *Differentiation of integrals in higher dimensions*, Proc. Natl. Acad. Sci. USA **110**, 4941–4944.

A. Peltomäki [1987] Licentiate thesis (in Finnish), University of Helsinki.

Y. Peres and M. Rams [2014] *Projections of the natural measure for percolation fractals*, arXiv:1406.3736.

Y. Peres and W. Schlag [2000] *Smoothness of projections, Bernoulli convolutions, and the dimension of exceptions*, Duke Math. J. **102**, 193–251.

Y. Peres, W. Schlag and B. Solomyak [2000] *Sixty years of Bernoulli convolutions*, in Fractal Geometry and Stochastics II, Birkhäuser, 39–65.

Y. Peres and P. Shmerkin [2009] *Resonance between Cantor sets*, Ergodic Theory Dynam. Systems **29**, 201–221.

Y. Peres, K. Simon and B. Solomyak [2000] *Self-similar sets of zero Hausdorff measure and positive packing measure*, Israel J. Math. **117**, 353–379.

[2003] *Fractals with positive length and zero buffon needle probability*, Amer. Math. Monthly **110**, 314–325.

Y. Peres and B. Solomyak [1996] *Absolute continuity of Bernoulli convolutions, a simple proof*, Math. Res. Lett. **3**, 231–239.

[1998] *Self-similar measures and intersections of Cantor sets*, Trans. Amer. Math. Soc. **350**, 4065–4087.

[2002] *How likely is Buffon's needle to fall near a planar Cantor set?*, Pacific J. Math. **204**, 473–496.

J. Peyriére [1975] *Étude de quelques propriétés des produits de Riesz*, Ann. Inst. Fourier Grenoble **25**, 127–169.

G. Pisier [1986] *Factorization of operators through $L_{p\infty}$ or L_{p^1} and noncommutative generalizations*, Math. Ann. **276**, 105–136.

M. Pollicott and K. Simon [1995] *The Hausdorff dimension of λ-expansions with deleted digits*, Trans. Amer. Math. Soc. **347**, 967–983.

A. Poltoratski [2012] *Spectral gaps for sets and measures*, Acta Math. **208**, 151–209.

D. Preiss [1987] *Geometry of sets and measures in* $\mathbb{R}^n$; *distribution, rectifiability, and densities*, Ann. of Math. (2) **125**, 537–643.

M. Rams and K. Simon [2014] *The dimension of projections of fractal percolations*, J. Stat. Phys. **154**, 633–655.

[2015] *Projections of fractal percolations*, Ergodic Theory Dyn. Systems **35**, 530–545.

F. Riesz [1918] *Über die Fourierkoeffizienten einer stetigen Funktion von Beschränkter Schwankung*, Math. Z. **2**, 312–315.

K. M. Rogers [2006] *On a planar variant of the Kakeya problem*, Math. Res. Lett. **13**, 199–213.

V. A. Rokhlin [1962] *On the fundamental ideas of measure theory*, Trans. Amer. Math. Soc., Series 1, **10**, 1–52.

K. Roth [1953] *On certain sets of integers*, J. London. Math. Soc. **28**, 104–109.

R. Salem [1944] *A remarkable class of algebraic integers. Proof of a conjecture by Vijayaraghavan*, Duke Math. J. **11**, 103–108.

[1951] *On singular monotonic functions whose spectrum has a given Hausdorff dimension*, Ark. Mat. **1**, 353–365.

[1963] *Algebraic Numbers and Fourier Analysis*, Heath Mathematical Monographs.

E. Sawyer [1987] *Families of plane curves having translates in a set of measure zero*, Mathematika **34**, 69–76.

N. de Saxcé [2013] *Subgroups of fractional dimension in nilpotent or solvable Lie groups*, Mathematika **59**, 497–511.

[2014] *Borelian subgroups of simple Lie groups*, arXiv:1408.1579.

W. Schlag [1998] *A geometric inequality with applications to the Kakeya problem in three dimensions*, Duke Math. J. **93**, 505–533.

B. Shayya [2011] *Measures with Fourier transforms in* L^2 *of a half-space*, Canad. Math. Bull. **54**, 172–179.

[2012] *When the cone in Bochner's theorem has an opening less than* π, Bull. London Math. Soc. **44**, 207–221.

N.-R. Shieh and Y. Xiao [2006] *Images of Gaussian random fields: Salem sets and interior points*, Studia Math. **176**, 37–60.

N.-R. Shieh and X. Zhang [2009] *Random p-adic Riesz products: Continuity, singularity, and dimension*, Proc. Amer. Math. Soc. **137**, 3477–3486.

P. Shmerkin [2014] *On the exceptional set for absolute continuity of Bernoulli convolutions*, Geom. Funct. Anal. **24**, 946–958.

P. Shmerkin and B. Solomyak [2014] *Absolute continuity of self-similar measures, their projections and convolutions*, arXiv:1406.0204.

P. Shmerkin and V. Suomala [2012] *Sets which are not tube null and intersection properties of random measures*, to appear in J. London Math. Soc., arXiv:1204.5883.

[2014] *Spatially independent martingales, intersections, and applications*, arXiv:1409.6707.

K. Simon and L. Vágó [2014] *Projections of Mandelbrot percolation in higher dimensions*, arXiv:1407.2225.

P. Sjölin [1993] *Estimates of spherical averages of Fourier transforms and dimensions of sets*, Mathematika **40**, 322–330.

[1997] *Estimates of averages of Fourier transforms of measures with finite energy*, Ann. Acad. Sci. Fenn. Ser. A I Math. **22**, 227–236.

[2002] *Spherical harmonics and spherical averages of Fourier transforms*, Rend. Sem. Mat. Univ. Padova **108**, 41–51.

[2007] *Maximal estimates for solutions to the nonelliptic Schrödinger equation*, Bull. London Math. Soc. **39**, 404–412.

[2013] *Nonlocalization of operators of Schrödinger type*, Ann. Acad. Sci. Fenn. Ser. A I Math. **38**, 141–147.

P. Sjölin and F. Soria [2003] *Estimates of averages of Fourier transforms with respect to general measures*, Proc. Royal Soc. Edinburgh A **133**, 943–950.

[2014] *Estimates for multiparameter maximal operators of Schrödinger type*, J. Math. Anal. Appl. **411**, 129–143.

C. D. Sogge [1991] *Propagation of singularities and maximal functions in the plane*, Invent. Math. **104**, 349–376.

[1993] *Fourier Integrals in Classical Analysis*, Cambridge University Press.

[1999] *Concerning Nikodym-type sets in 3-dimensional curved spaces*, J. Amer. Math. Soc. **12**, 1–31.

B. Solomyak [1995] *On the random series $\sum \pm\lambda^n$ (an Erdős problem)*, Ann. of Math. (2) **142**, 611–625.

E. M. Stein [1961] *On limits of sequences of operators*, Ann. of Math. (2) **74**, 140–170.

[1986] *Oscillatory integrals in Fourier analysis*, Beijing Lectures in Harmonic Analysis, pp. 307–355, Annals of Math. Studies 112, Princeton University Press.

[1993] *Harmonic Analysis: Real Variable Methods, Orthogonality, and Oscillatory integrals*, Princeton University Press.

E. M. Stein and G. Weiss [1971] *Introduction to Fourier Analysis on Euclidean Spaces*, Princeton University Press.

R. Strichartz [1977] *Restrictions of Fourier transforms to quadratic surfaces and decay of solutions of wave equations*, Duke Math. J. **44**, 705–714.

[1989] *Besicovitch meets Wiener–Fourier expansions and fractal measures*, Bull. Amer. Math. Soc. (N.S.) **20**, 55–59.

[1990a] *Fourier asymptotics of fractal measures*, J. Funct. Anal. **89**, 154–187.

[1990b] *Self-similar measures and their Fourier transforms I*, Indiana Univ. Math. J. **39**, 797–817.

[1993a] *Self-similar measures and their Fourier transforms II*, Trans. Amer. Math. Soc. **336**, 335–361.

[1993b] *Self-similar measures and their Fourier transforms III*, Indiana Univ. Math. J. **42**, 367–411.

[1994] *A Guide to Distribution Theory and Fourier Transforms*, Studies in Advanced Mathematics, CRC Press.

T. Tao [1999a] *Bochner–Riesz conjecture implies the restriction conjecture*, Duke Math. J. **96**, 363–375.

[1999b] *Lecture notes for the course Math 254B, Spring 1999 at UCLA*, www.math.ucla.edu/tao/254b.1.99s/

[2000] *Finite field analogues of the Erdos, Falconer, and Furstenburg problems*, www.math.ucla.edu/tao/preprints/kakeya.html.

[2001] *From rotating needles to stability of waves: emerging connections between combinatorics, analysis, and PDE*, Notices Amer. Math. Soc. **48**, 294–303.

[2003] *A sharp bilinear restriction estimate for paraboloids*, Geom. Funct. Anal. **13**, 1359–1384.

[2004] *Some recent progress on the restriction conjecture* in Fourier analysis and convexity, 217–243, Appl. Numer. Harmon. Anal., Birkhäuser Boston, Boston, MA.

[2008a] *Dvir's proof of the finite field Kakeya conjecture*, http://terrytao. wordpress.com/2008/03/24/dvirs-proof-of-the-finite-field-kakeya-conjecture, 24 March 2008.

[2008b] *A remark on the Kakeya needle problem*, http://terrytao.wordpress.com/2008/12/31/a-remark-on-the-kakeya-needle-problem.

[2009] *A quantitative version of the Besicovitch projection theorem via multiscale analysis*, Proc. London Math. Soc. **98**, 559–584.

[2011] *An epsilon of room, II: pages from year three of a mathematical blog*, Amer. Math. Soc.

[2014] *Algebraic combinatorial geometry: the polynomial method in arithmetic combinatorics, incidence combinatorics, and number theory*, EMS Surv. Math. Sci. **1**, 1–46.

T. Tao and A. Vargas [2000] *A bilinear approach to cone multipliers I. Restriction estimates*, Geom. Funct. Anal. **10**, 185–215.

T. Tao, A. Vargas and L. Vega [1998] *A bilinear approach to the restriction and Kakeya conjectures*, J. Amer. Math. Soc. **11**, 967–1000.

T. Tao and V. Vu [2006] *Additive Combinatorics*, Cambridge University Press.

F. Temur [2014] *A Fourier restriction estimate for surfaces of positive curvature in* $\mathbb{R}^6$, Rev. Mat. Iberoam. **30**, 1015–1036.

X. Tolsa [2014] *Analytic capacity, the Cauchy Transform, and Non-homogeneous Calderón-Zygmund Theory*, Birkhäuser.

P. A. Tomas [1975] *A restriction theorem for the Fourier transform*, Bull. Amer. Math Soc. **81**, 477–478.

G. Travaglini [2014] *Number Theory, Fourier Analysis and Geometric Discrepancy*, London Mathematical Society Student Texts **81**, Cambridge University Press.

A. Vargas [1991] *Operadores maximales, multiplicadores de Bochner–Riesz y teoremas de restricción*, Master's thesis, Universidad Autónoma de Madrid.

A. Volberg and V. Eiderman [2013] *Nonhomogeneous harmonic analysis: 16 years of development*, Uspekhi Mat. Nauk. **68**, 3–58; translation in Russian Math. Surveys **68**, 973–1026.

N. G. Watson [1944] *A Treatise on Bessel Functions*, Cambridge University Press.

Z.-Y. Wen, W. Wu and L. Xi [2013] *Dimension of slices through a self-similar set with initial cubical pattern*, Ann. Acad. Sci. Fenn. Ser. A I Math. **38**, 473–487.

Z.-Y. Wen and L. Xi [2010] *On the dimension of sections for the graph-directed sets*, Ann. Acad. Sci. Fenn. Ser. A I Math. **35**, 515–535.

L. Wisewell [2004] *Families of surfaces lying in a null set*, Mathematika **51**, 155–162.

[2005] *Kakeya sets of curves*, Geom. Funct. Anal. **15**, 1319–1362.

T. W. Wolff [1995] *An improved bound for Kakeya type maximal functions*, Rev. Mat. Iberoam. **11**, 651–674.

[1997] *A Kakeya-type problem for circles*, Amer. J. Math. **119**, 985–1026.

[1998] *A mixed norm estimate for the X-ray transform*, Rev. Mat. Iberoam. **14**, 561–600.

[1999] *Decay of circular means of Fourier transforms of measures*, Int. Math. Res. Not. **10**, 547–567.

[2000] *Local smoothing type estimates on L^p for large p*, Geom. Funct. Anal. **10**, 1237–1288.

[2001] *A sharp bilinear cone restriction estimate*, Ann. of Math. (2) **153**, 661–698.

[2003] *Lectures on Harmonic Analysis*, Amer. Math. Soc., University Lecture Series 29.

Y. Xiao [2013] *Recent developments on fractal properties of Gaussian random fields*, in Further Developments in Fractals and Related Fields, J. Barral and S. Seuret (eds.), Trends in Mathematics, Birkhäuser, 255–288.

Y. Xiong and J. Zhou [2005] *The Hausdorff measure of a class of Sierpinski carpets*, J. Math. Anal. Appl. **305**, 121–129.

W. P. Ziemer [1989] *Weakly Differentiable Functions*, Springer-Verlag.

A. Zygmund [1959] *Trigonometric Series, volumes I and II*, Cambridge University Press (the first edition 1935 in Warsaw).

[1974] *On Fourier coefficients and transforms of functions of two variables*, Studia Math. **50**, 189–201.

Index of basic notation

434

Author index

Subject index

Printed in the United States
By Bookmasters